Marine and Aquatic Sciences

Marine and Aquatic Sciences

Editor: Jeremy Harper

CALLISTO REFERENCE

www.callistoreference.com

Callisto Reference,
118-35 Queens Blvd., Suite 400,
Forest Hills, NY 11375, USA

Visit us on the World Wide Web at:
www.callistoreference.com

ISBN: 978-1-63239-848-2 (Hardback)

Cataloging-in-publication Data

Marine and aquatic sciences / edited by Jeremy Harper.
     p. cm.
Includes bibliographical references and index.
ISBN 978-1-63239-848-2
1. Marine biology. 2. Aquatic biology. 3. Marine sciences. 4. Aquatic sciences. 5. Oceanography. I. Harper, Jeremy.
QH91 .M37 2017
578.77--dc23

Table of Contents

Preface

The disciplines of marine and aquatic science deal with the regulation and promotion of aquatic life and the ecological regions that encompass the aquatic ecosystem, the limnological and hydrological mechanisms at work in these ecosystems, and the impact of human activity in these areas. These fields branch out into various sub-fields such as marine biology, marine chemistry, oceanography and hydrology. This book on marine and aquatic sciences discusses topics related to climate change and water content on the earth's surface and the effect of wastewater and other pollutants to marine life. It covers important aspects such as principles related to aquatic sciences, conservation and hydrographical practices. Different approaches, evaluations, methodologies and advancements in this field have also been included. This book aims to equipped students and experts with the advanced topics and upcoming concepts in this area.

This book has been a concerted effort by a group of academicians, researchers and scientists, who have contributed their research works for the realization of the book. This book has materialized in the wake of emerging advancements and innovations in this field. Therefore, the need of the hour was to compile all the required researches and disseminate the knowledge to a broad spectrum of people comprising of students, researchers and specialists of the field.

At the end of the preface, I would like to thank the authors for their brilliant chapters and the publisher for guiding us all-through the making of the book till its final stage. Also, I would like to thank my family for providing the support and encouragement throughout my academic career and research projects.

Editor

De Novo Transcriptomes of Olfactory Epithelium Reveal the Genes and Pathways for Spawning Migration in Japanese Grenadier Anchovy (*Coilia nasus*)

Guoli Zhu[1], Liangjiang Wang[2], Wenqiao Tang[1]*, Dong Liu[1], Jinquan Yang[1]

1 College of Fisheries and Life Science, Shanghai Ocean University, Shanghai, China, **2** Department of Genetics and Biochemistry, Clemson University, Clemson, South Carolina, United States of America

Abstract

Background: *Coilia nasus* (Japanese grenadier anchovy) undergoes spawning migration from the ocean to fresh water inland. Previous studies have suggested that anadromous fish use olfactory cues to perform successful migration to spawn. However, limited genomic information is available for *C. nasus*. To understand the molecular mechanisms of spawning migration, it is essential to identify the genes and pathways involved in the migratory behavior of *C. nasus*.

Results: Using *de novo* transcriptome sequencing and assembly, we constructed two transcriptomes of the olfactory epithelium from wild anadromous and non-anadromous *C. nasus*. Over 178 million high-quality clean reads were generated using Illumina sequencing technology and assembled into 176,510 unigenes (mean length: 843 bp). About 51% (89,456) of the unigenes were functionally annotated using protein databases. Gene ontology analysis of the transcriptomes indicated gene enrichment not only in signal detection and transduction, but also in regulation and enzymatic activity. The potential genes and pathways involved in the migratory behavior were identified. In addition, simple sequence repeats and single nucleotide polymorphisms were analyzed to identify potential molecular markers.

Conclusion: We, for the first time, obtained high-quality *de novo* transcriptomes of *C. nasus* using a high-throughput sequencing approach. Our study lays the foundation for further investigation of *C. nasus* spawning migration and genome evolution.

Editor: Hiroaki Matsunami, Duke University, United States of America

Funding: This study was supported by the Special Fund for Agro-Scientific Research in the Public Interest (No. 201203065), the National Natural Science Foundation of China (No. 31172407), the Ministry of Education's Doctoral Discipline Foundation (No. 20123104110006), and Shanghai Universities First-class Disciplines Project of Fisheries. The funders had no role in study design, data collection and analysis, decision to publish, or preparation of the manuscript.

Competing Interests: The authors have declared that no competing interests exist.

* Email: wqtang@shou.edu.cn

Introduction

The Japanese grenadier anchovy (*Coilia nasus*) is a small commercial fish in China, which belongs to the family of Engraulidae, order of Clupeiformes [1]. It is renowned for its delicate and tender meat. Moreover, *C. nasus* is well known for the long-distance ocean–river spawning migration of its anadromous population.

C. nasus lives in coastal ocean water for most of its lifetime, and normally reaches sexual maturity at the age of 1–2 years. *C. nasus* spawns between February and September [2]. Every year, when the spawning period arrives, thousands of mature *C. nasus* individuals undergo a long-distance migration from coastal ocean up to exorheic rivers, such as the Yangtze River, and then spawn in the lower and middle reaches of these rivers and adjacent lakes. Interestingly, the sedentary population of *C. nasus* in lakes has abandoned the long-distance migration for unknown reasons and become permanent residents there.

The ability to recognize the spawning ground is a key skill for successful reproduction. Recently, there has been a sharp decline in the population of anadromous *C. nasus* because of environmental pollution, overfishing and the destruction of spawning grounds. Therefore, the understanding of *C. nasus* spawning migration is essential for its conservation and stock management. However, little is known about the molecular basis of *C. nasus* spawning migration.

Previous studies on fish migration have mostly focused on salmonids. It has been hypothesized that salmonids use olfactory cues to return to natal rivers to spawn. Several studies, wherein the salmonid olfactory epithelium was altered, have concluded that salmonids without olfactory ability cannot discriminate natal streams and that functional olfactory ability is essential for their migration to spawn [3–7]. Similar conclusion was also drawn for American eels, and with the functional olfactory ability absent, anosmic eels lost the ability to migrate out of the estuary during the fall spawning migration [8]. Olfactory imprinting of dissolved amino acids in natal stream water has been reported in lacustrine sockeye salmon [9], and strong olfactory responses to natal stream water have also been found in sockeye salmon [10]. In wild anadromous Atlantic salmon, some of the olfactory receptor genes involved in the migration for reproduction have been identified [11]. These studies suggest that olfaction may be essential for the migration for reproduction in fish.

The olfactory epithelium in the nasal cavity is involved in the olfaction of fish. The olfactory functions of fish are induced by odorant elements such as steroids, bile acids and amino acids in water through the olfactory receptors in the olfactory epithelium. Subsequently, the information is processed by the central nervous system of fish to achieve the olfactory functions. To investigate the relationship between olfaction and the anadromous behavior of *C. nasus*, we sequenced the transcripts expressed in the olfactory epithelium. With this sequence information, we identified the genes and pathways involved in the migratory behavior of *C. nasus*. At present, little genomic information about *C. nasus* is available in the National Center for Biotechnology Information (NCBI) database. Therefore, the high-quality transcriptome data obtained in this study will be useful for future research on *C. nasus*.

Results and Discussion

Transcriptome sequencing and assembly

As described in the Materials and Methods, cDNA libraries for the olfactory sac of wild anadromous and non-anadromous *C. nasus* were constructed and sequenced using the Illumina platform, which produced 51,261,228 and 126,241,752 clean reads, respectively (Table 1). For anadromous and non-anadromous *C. nasus*, 117,717 and 231,219 unigenes, respectively, were obtained, and 176,510 unigenes with a mean length of 843 nucleotides were assembled from the anadromous and non-anadromous *C. nasus* unigenes (Table 1 and Figure S1). The total length of the 176,510 assembled unigenes was 148,772,175 nucleotides.

The quality of the sequence assembly result and the size distribution are shown in Figure S1. Of all the unigenes, 8,608 or over 4.8% are ≥3,000 nucleotides in length. The coding regions have been identified for 81,315 sequences (72,601 using BLASTX and 8,714 using expressed sequence tag scan; Figure S2). While it is time-consuming to obtain large cDNA collections using the traditional Sanger sequencing method, the next-generation sequencing platform has been demonstrated in this study to be useful for efficiently generating high-quality transcriptome data of *C. nasus*.

Annotation of predicted proteins and classification using COG

The putative functions of 89,456 unigenes (50.68% of all unigenes) were annotated by sequence similarity analysis with E value $\leq 1 \times 10^{-5}$ (72,127 using the NR database, 65,888 using the NT database, 61,581 using the SwissProt database, 53,575 using the KEGG database, 25,272 using the COG database, and 41,888 using gene ontology terms). However, because of the lack of genome and EST sequence data from *C. nasus*, approximately 49.32% of the unigenes could not be functionally annotated.

The E-value distribution and similarity distribution for the 72,127 unigenes (40.86% of all unigenes) that were annotated using the NR database are shown in Figure S3. The species distribution of the best BLASTX hits is also shown in Figure S3. About 66.2% of the unigenes were functionally annotated with the known fish genes. However, a small number of sequences were matched to *Paramecium tetraurelia* and *Tetrahymena thermophila* SB210 genes. These sequences may represent contaminants from sample collection or parasitic infection of *C. nasus*.

COG (clusters of orthologous groups of proteins) is a database where orthologous gene products are classified into different clusters. A total of 25,272 *C. nasus* unigenes were assigned to 25 COG categories with E value $\leq 1 \times 10^{-5}$ (Figure 1). Among these COG categories, the cluster for "general function prediction" was the largest, containing 10,278 (40.66%) of the unigenes, followed by "translation, ribosomal structure, and biogenesis" (7,169 or 28.36%), "replication, recombination, and repair" (6,315 or 24.98%), and "cell cycle control, cell division, chromosome partitioning" (6,161 or 24.37%). In addition, the "signal transduction mechanisms" cluster contained 4,092 (16.19%) unigenes.

Gene ontology assignments

To understand the functional capacity of the *C. nasus* transcriptome, 41,888 unigenes (46.8% of all unigenes) were assigned to three Gene Ontology (GO) categories: biological processes, cellular components and molecular functions (Figure 2). In the GO category of biological processes, 13,391 unigenes were involved in response to stimulus and 9,782 in signaling, both of which were enriched in this category. Of the unigenes assigned to the GO category of cellular components, 9,021 were involved in the membrane part. In addition, of the unigenes annotated with potential molecular functions, binding (27,140) and catalytic activity (16,082) were enriched in this category. GO terms of channel regulator activity (135 unigenes), electron carrier activity (256), receptor activity (1,845), and receptor regulator activity (48) were also well

Table 1. Summary of the sequences obtained from the olfactory epithelium of anadromous and non-anadromous *Coilia nasus*.

	Anadromous	Non-anadromous	
Total clean reads	51,261,228	126,241,752	
Total clean nucleotides (nt)	4,613,510,520	12,750,416,952	
Contig total number	223,325	409,459	
Unigene total number	117,717	231,219	
Contig total length (nt)	56,758,068	129,299,285	
Unigene total length (nt)	50,868,550	197,568,883	
All total number			176,510
Alltotal length (nt)			148,772,175

COG Function Classification of All-Unigene.fa Sequence

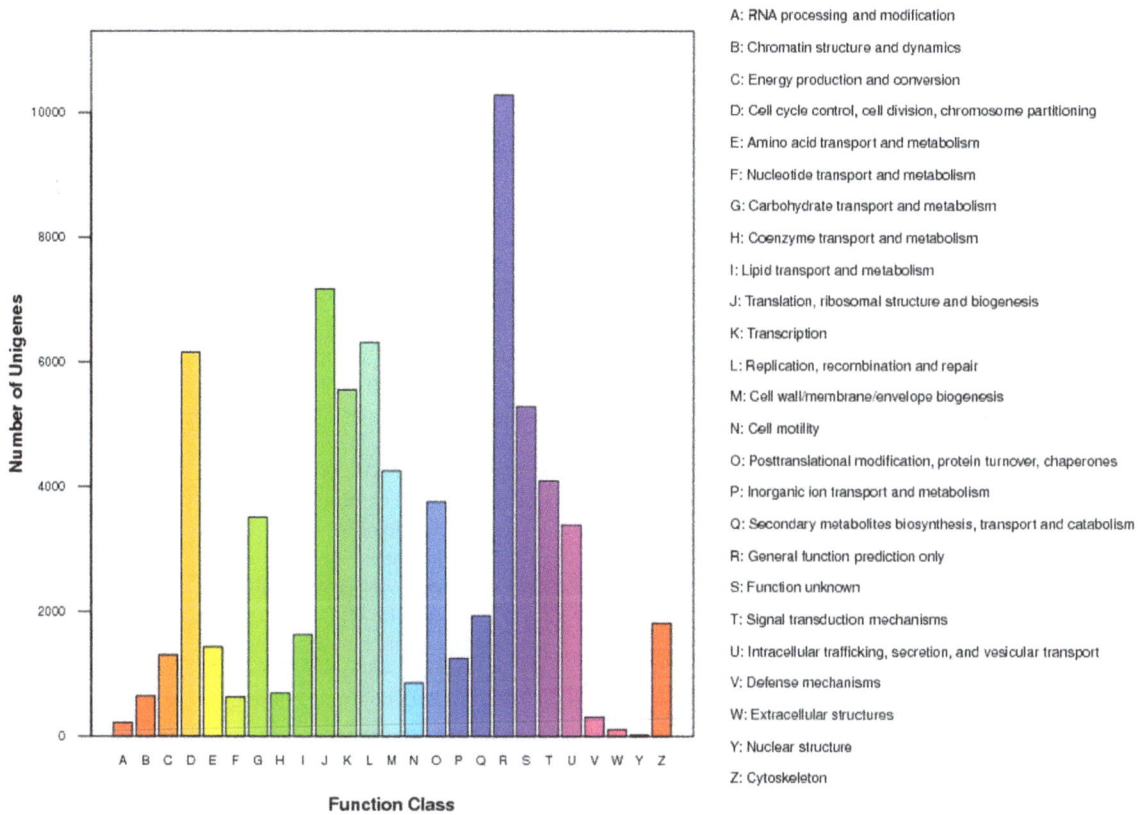

A: RNA processing and modification

B: Chromatin structure and dynamics

C: Energy production and conversion

D: Cell cycle control, cell division, chromosome partitioning

E: Amino acid transport and metabolism

F: Nucleotide transport and metabolism

G: Carbohydrate transport and metabolism

H: Coenzyme transport and metabolism

I: Lipid transport and metabolism

J: Translation, ribosomal structure and biogenesis

K: Transcription

L: Replication, recombination and repair

M: Cell wall/membrane/envelope biogenesis

N: Cell motility

O: Posttranslational modification, protein turnover, chaperones

P: Inorganic ion transport and metabolism

Q: Secondary metabolites biosynthesis, transport and catabolism

R: General function prediction only

S: Function unknown

T: Signal transduction mechanisms

U: Intracellular trafficking, secretion, and vesicular transport

V: Defense mechanisms

W: Extracellular structures

Y: Nuclear structure

Z: Cytoskeleton

Figure 1. Histogram presentation of the results from the classification using the Clusters of Orthologous Groups (COG).

represented. The large number of regulatory transcripts found in our data may indicate transcriptional plasticity in the olfactory epithelium.

Approximately 41.9% of all the transcripts of *C. nasus* did not have GO terms assigned to them. This may be because of the fact that knowledge regarding the function of *C. nasus* genes is

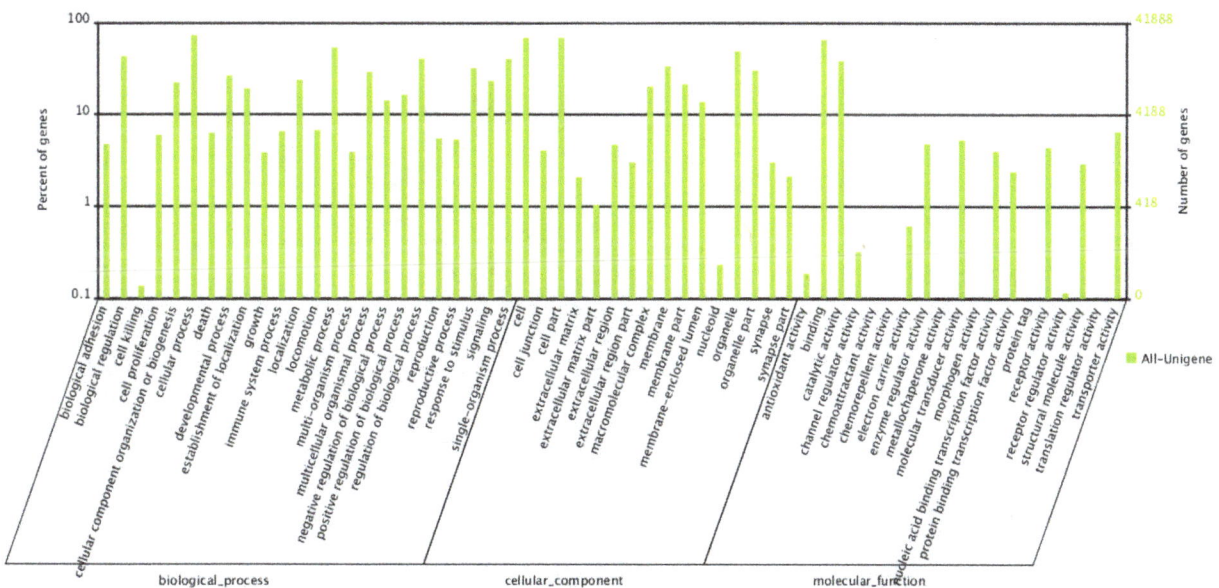

Figure 2. Histogram presentation of Gene Ontology (GO) classification. The results are divided into three GO categories: biological processes, cellular components, and molecular functions.

currently limited. It is also possible that these transcripts are from non-coding RNA genes. Nevertheless, the unannotated transcripts in the olfactory epithelium should be documented as they may be involved in the olfaction of *C. nasus*, either directly or indirectly.

Previous studies on the transcriptome of fish olfactory epithelium have been limited to the goldfish *Carassius auratus* [12]. Since this goldfish does not have the ability to migrate, comparing *C. auratus* and *C. nasus* transcriptomes may provide useful information on the molecular mechanisms of migration. We compared the GO terms of response to stimulus and binding, which may be involved in olfaction and signal transduction. *C. nasus* had a higher proportion of both terms than *C. auratus* (6.30% versus 4.40% in response to stimulus; 47.90% versus 45.70% in binding), suggesting that *C. nasus* may have higher olfaction ability than *C. auratus*.

Kyoto Encyclopedia of Genes and Genomes (KEGG) analysis

A total of 53,575 unigenes were annotated with the genes in the KEGG database. The number of unigenes in different pathways ranged from 2 to 5,243. The top 25 pathways with the highest sequence tag numbers are shown in Table 2. The top pathway (metabolic pathway) contained 5,243 unigenes. These predicted KEGG pathways may provide a useful resource for research into the spawning migration of *C. nasus* and other molecular studies in *C. nasus*.

Simple sequence repeats (SSRs) and SNPs as genetic markers

Molecular markers are a useful tool for species evolution and population differentiation studies. At present, studies of the *C. nasus* population are restricted by the lack of effective molecular markers. Through *de novo* assembly of transcriptome data, 78,852 SSRs in 54,059 sequences were detected. These SSRs include 14,998 monomers, 50,071 dimers, 9,546 trimers, 2,317 quadmers, 1,523 pentamers, and 397 hexamers (Figure S4). In addition, 224,779 single nucleotide polymorphism (SNP) sites were identified. 93,501 sites were found in anadromous *C. nasus* and 131,278 in non-anadromous *C. nasus*. There were 138,945 transition sites and 85,734 transversion sites (Table S1). The large number of putative molecular markers identified in our work may be useful for future studies on the evolution of the *C. nasus* genome, such as gene flow, genetic mapping, and genotyping.

A resource for investigation of migration genes

Previous studies on the migration of *C. nasus* have mainly focused on the behavioral and morphology aspects [1,2,13–19]. In this study, we aimed to expand this knowledge and provide new insight into the molecular mechanism of *C. nasus* migration. The transcriptome data obtained in this study provide a good resource for identifying the putative genes involved in *C. nasus* migration.

Pathway of olfactory transduction. The hypothesis of olfactory imprinting and homing for salmon assumes that some

Table 2. List of the top 25 KEGG metabolic pathways identified in the *Coilia nasus* transcriptomes.

No.	Pathway	Number (%) of ESTs	Pathway ID
1	Metabolic pathways	5,243 (9.79)	ko01100
2	Regulation of actin cytoskeleton	2,772 (5.17)	ko04810
3	Pathways in cancer	2,671 (4.99)	ko05200
4	Amoebiasis	2,288 (4.27)	ko05146
5	Focal adhesion	2,274 (4.24)	ko04510
6	Spliceosome	2,226 (4.15)	ko03040
7	MAPK signaling pathway	1,758 (3.28)	ko04010
8	RNA transport	1,651 (3.08)	ko03013
9	Endocytosis	1,602 (2.99)	ko04144
10	Tight junction	1,596 (2.98)	ko04530
11	Huntington's disease	1,581 (2.95)	ko05016
12	HTLV-I infection	1,578 (2.95)	ko05166
13	Salmonella infection	1,570 (2.93)	ko05132
14	Herpes simplex infection	1,491 (2.78)	ko05168
15	Adherens junction	1,458 (2.72)	ko04520
16	Influenza A	1,443 (2.69)	ko05164
17	Chemokine signaling pathway	1,437 (2.68)	ko04062
18	Vibrio cholerae infection	1,436 (2.68)	ko05110
19	Epstein-Barr virus infection	1,427 (2.66)	ko05169
20	Fc gamma R-mediated phagocytosis	1,378 (2.57)	ko04666
21	Vascular smooth muscle contraction	1,352 (2.52)	ko04270
22	Dilated cardiomyopathy	1,327 (2.48)	ko05414
23	Hypertrophic cardiomyopathy (HCM)	1,261 (2.35)	ko05410
24	Calcium signaling pathway	1,251 (2.34)	ko04020
25	Transcriptional misregulation in cancer	1,240 (2.31)	ko05202

odorant molecules in the natal stream are imprinted on the olfactory system of juvenile salmon during their downstream migration, and adult salmon detect the corresponding molecules to discriminate the natal stream during their homing migration [9,10,20].

In our study, the KEGG pathway of olfactory transduction (ko04740) [21–29] was used to annotate the largest number of genes (Figure 3). 547 unigenes, or 1.02% of the KEGG-annotated unigenes, were assigned to the olfactory transduction pathway.

At present, little is known about the pathway of olfactory transduction in *C. nasus*; however, relevant information can be obtained from other vertebrate species [30]. The canonical pathway of the olfactory transduction is initiated from the detection of odor molecules by odorant receptors (Rs). Binding of the odor molecules to the odorant receptors activates the $G\alpha_{olf}$-containing heterotrimeric G protein (G_{olf}), which then activates adenylyl cyclase (AC) to produce cAMP [31]. Subsequently, cAMP opens the cyclic nucleotide-gated cation channels (CNG) [32]. Ca^{2+} ions influx into the cells and depolarization occurs. Ca^{2+}-activated chloride channels (CLCA) allow an efflux of Cl^- ions, which leads to further depolarization of the cell [33–38]. The chemical signals are then converted into electronic signals that are delivered to the brain, where the signals are perceived as smells.

Elevated intracellular Ca^{2+} triggers multiple molecular events, including the down-regulation of the affinity of the CNG channel to cAMP and inhibition of the activity of AC via CAMKII (calcium/calmodulin-dependent protein kinase II)-dependent phosphorylation [24]. Longer exposure to odorants can stimulate particulate guanylyl cyclase (pGC) in cilia to produce cGMP and activate cGMP-dependent protein kinase (PKG), leading to a further increase in the amount and duration of intracellular cAMP levels, which may function to convert inactive forms of protein kinase A (PKA) to active forms [39]. PKA can also inhibit the activation of pGC as a feedback.

Termination of the response may occur at all steps of the pathway, which include receptor phosphorylation by G protein receptor kinase (GRK) or protein kinase A (PKA) and 'capping' of the phosphorylated receptor by arrestin [40–42], inhibition of adenylyl cyclase activity by CaMKII and regulation of G protein signaling 2 (RGS2) [43,44], removal of Ca^{2+} through a $Na^+–Ca^{2+}$ exchanger [45], hydrolysis of cAMP by phosphodiesterase (PDE) activity, and desensitization of the CNG channel by Ca^{2+}-calmodulin (CAM)-dependent processes [46]. However, the transcripts of arrestin, GRK and PDE involved in the response termination, and pGC are not detected in this study. This may be because *C. nasus* has a unique pathway with a lower termination ability. Since several terminators are absent in the olfactory transduction, sustained detection of odor elements in natal rivers may be possible for *C. nasus*. It is also possible that these transcripts are rare and thus undetected in this study.

Putative pheromone signaling pathway. The pheromone hypothesis was proposed based on research on Atlantic salmon *Salmon salar* and Arctic char *Salvelinus alpines* [47]. In sea lamprey, a mixture of sulfated steroids has also been demonstrated to function as a migratory pheromone [48]. Thus, the putative pheromone signaling pathway should also be considered in the study of the migration behavior of *C. nasus*.

Pheromones are secreted or excreted chemicals that can impact on the behavior of a receiving individual and trigger a social response within members of the same species. Vomeronasal type-1 receptors (V1Rs) and vomeronasal type-2 receptors (V2Rs) have been shown to function as pheromone receptors [49,50]. The binding of a pheromone to a V1R activates inhibitory adenylate cyclase G protein (Gi), and phospholipase Cβ2 (PLCβ2) is activated to produce inositol-1,4,5-trisphoshate and diacylglycerol from phosphatidylinositol-4,5-bisphoshate. This activates the transient receptor potential cation channel C2 (TRPC2). Activation of TRPC2 allows a Na^+/Ca^{2+} influx, which leads to depolarization. Recovery and adaptation of response may involve binding of CaM to TRPC2. The binding of pheromones to V2Rs activates G_o, which is a G protein involved in many signal transduction channels [30]. In V2R-expressing neurons, TRPC2 has been shown to generate depolarizing currents [30]. In this study, we identified the family of V1R and V2R, and CaM in the transcriptomes of *C. nasus*. However, TRPC2 was not detected although we identified the other members of transient receptor potential cation channels, including TRPM4, TRPV4, TRPC5, and TRPV1. It is possible that the role of TRPC2 in the pheromone signaling pathway may be superseded by the other members of the gene family.

Conclusion

By using a high-throughput sequencing approach, we obtained the high-quality *de novo* transcriptomes of *C. nasus* for the first time. Our data provide valuable information for understanding the spawning migration of *C. nasus*, and lay the foundation for future research on the genome evolution of this species, especially as the genomic sequence is still unavailable for *C. nasus*.

Materials and Methods

Ethics statement

The study was approved by the Institutional Animal Care and Use Committee of Shanghai Ocean University and performed in strict accordance with the Guidelines on the Care and Use of Animals for Scientific Purposes set by the Institutional Animal Care and Use Committee of Shanghai Ocean University.

Fish material

Three males of non-anadromous *C. nasus* were collected from Poyang Lake in Jiujiang, Jiangxi Province in China at the end of March 2012 when anadromous males of *C. nasus* had not reached Poyang Lake to spawn. The fish collection was performed with the help of fisherman Baishan Zhan with the fishing license (No. 0400051) permitted by the Jiangxi Provincial Department of Agriculture. One male of anadromous *C. nasus* was collected from the Jingjiang section of the Yangtze River in Jingjiang, Jiangsu Province in China at the beginning of April 2012 when they were migrating to spawning grounds along the Yangtze River. The fish collection was performed with the assistance of fisherman Xiping Zhou with the fishing license (No. SuChuanBu 2011 JMF254) and the special fishing license of *C. nasus* in the Yangtze River (No. SuChuanBu 2012 ZX-M032) permitted by Jiangsu Provincial Oceanic and Fishery Bureau. All fish collections were carried out in wild water, and the captured live C. nasus was immediately buried in medical ice bags ($-20°C$) until the loss of consciousness.

Before sampling, the *C. nasus* was dissected on ice and subsequently the anatomical characters of the testis gonadal development phase of *C. nasus* were rapidly checked [51]. If the individual's testis gonadal development phase was in phase III, then the olfactory capsules of *C. nasus* were collected. The operations were completed within 10 min after the loss of consciousness. After this procedure, the olfactory capsules from the non-anadromous *C. nasus* were placed into 2.0 mL tubes containing RNAlater (Ambion, US). Then the collected olfactory samples were stored at 4°C overnight and stored at $-20°C$ for 12

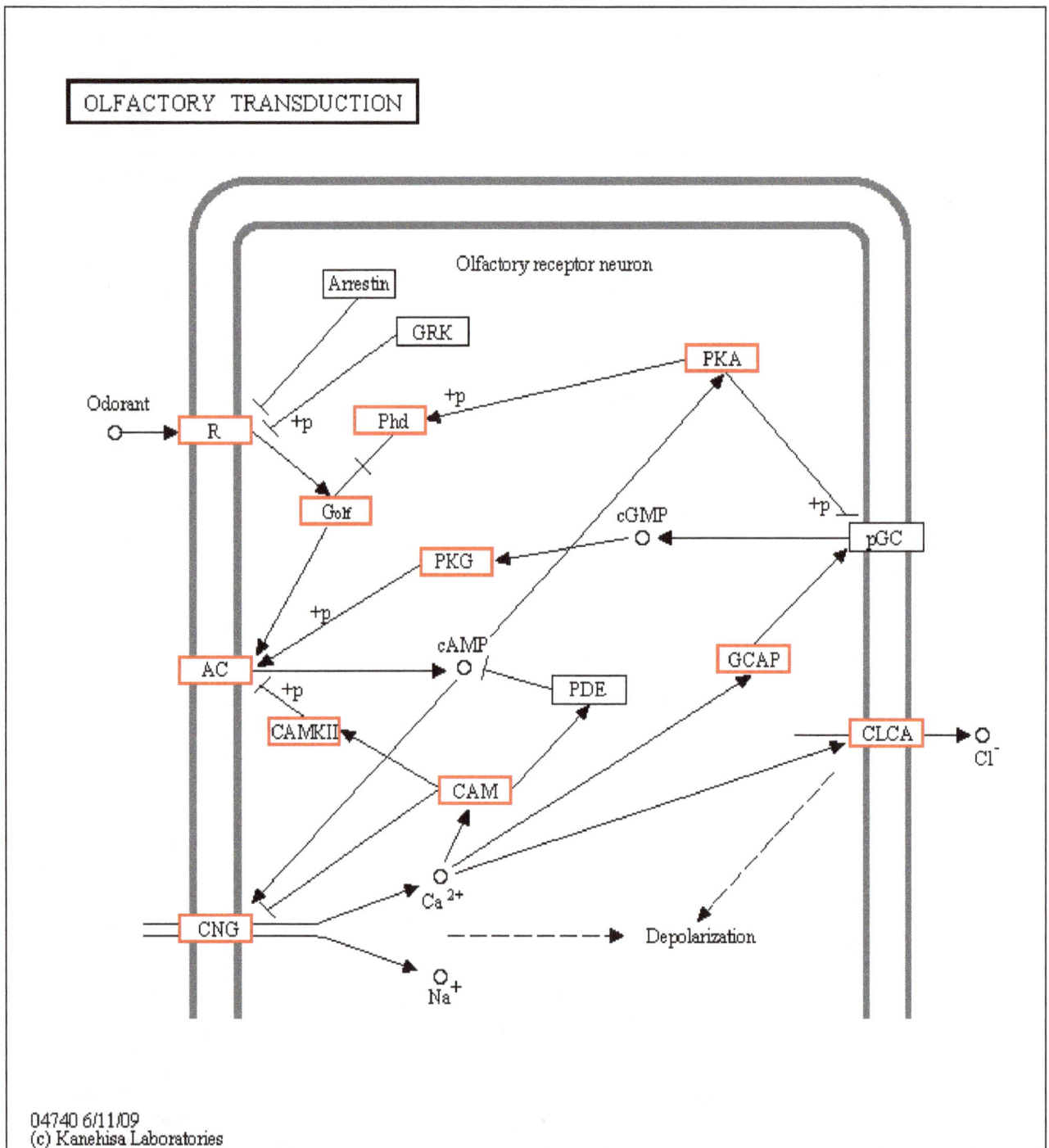

Figure 3. Functional annotation of *Coilia nasus* genes using the KEGG pathway of olfactory transduction. The genes identified in the *C. nasus* transcriptomes are shown in red boxes. R: odorant receptor; G_{olf}: $G_{\alpha olf}$-containing heterotrimeric G protein; AC: adenylate cyclase; CNG: cyclic nucleotide-gated cation channel; CLCA: calcium-activated chloride channel; GCAP: guanylyl cyclase-activating protein; Phd: phosducin; PKG: cGMP-dependent protein kinase; PKA: protein kinase A; pGC: particulate guanylyl cyclase; CAM: calmodulin; CAMKII: calcium/calmodulin-dependent protein kinase (CaM kinase) II; PDE: phosphodiesterase; Arrestin: arrestin; GRK: G protein receptor kinase.

hours during the delivery to Shanghai Ocean University, where the samples were transferred to −80°C before processing. The olfactory capsules from the anadromous *C. nasus* were immediately placed into 2.0 mL tubes and frozen in liquid nitrogen after collection and then delivered to the Shanghai Ocean University

for further processing. All the remains of above sampled fish were stored in freezer.

RNA extraction

Total RNA was isolated from samples using TRIzol reagent (Invitrogen, USA) according to the manufacturer's instructions.

The quality of purified RNA was verified on a 2100-Bioanalyzer (Agilent, USA). To prevent DNA contamination, the RNA samples were treated with DNase I. The high-quality RNA samples were then used for further experiments.

cDNA preparation and library construction

Poly(A)-containing mRNA samples were captured from total RNA with Oligo (dT)-Bead complex. The fragment mixture of the RNA fragmentation kit was added to mRNA to obtain RNA pieces with different lengths. Then single- and double-stranded cDNAs were synthesized from mRNA samples through reverse transcription using high-quality total RNA as the starting material.

The following cDNA purification was then performed. Purified cDNA fragments were suspended into End Repair Mix for end reparation and adenylate 3′ ends. Short fragments produced from the above procedures were ligated with sequencing adaptors, and then fragments with adaptors were purified and enriched with cDNA fragments through PCR. Subsequently, the purified PCR products were used to create a cDNA library. The size distribution and accurate quantification of the library were checked on a 2100-Bioanalyzer (Agilent, USA) and an ABI StepOnePlus Real-Time PCR System.

cDNA library sequencing

cDNA libraries were constructed for sequencing with Illumina Hiseq 2000. Raw sequence data were processed through the trimming of adaptor sequences, ambiguous nucleotides, and empty reads to obtain the clean data. With software Trinity and TIGR Gene Indices (TGI) Clustering tools v2.1 [52,53], the short clean reads obtained from the two types of *C. nasus* were assembled and clustered. Sequences with the fewest nucleotides that could not be extended on either end were then obtained. These sequences were called unigenes.

Unigene functional annotation and classification

The unigenes were functionally annotated by searching databases, including NR (ftp://ftp.ncbi.nih.gov/blast/db/), NT (ftp://ftp.ncbi.nih.gov/blast/db/), SwissProt (ftp://ftp.uniprot.org/pub/databases/uniprot/previous_releases/), COG (http://www.ncbi.nlm.nih.gov/COG/), gene ontology (http://www.geneontology.org/) and KEEG (http://www.genome.jp/), using BLAST with E-value $\leq 1 \times 10^{-5}$. The ESTSscan software v3.0.2 (http://www.ch.embnet.org/software/ESTScan2.html) was used to predict the coding region if a unigene had not been annotated using one of the previously mentioned databases.

Functional annotation using Gene Ontology terms (molecular functions, cellular components, and biological processes) was performed using BLAST2GO software v2.5.0 based on the NR annotation information [54]. After the gene ontology annotation, WEGO was used to obtain Gene Ontology function classification statistics of all the unigenes for understanding the species' gene function distribution [55].

The Kyoto Encyclopedia of Genes and Genomes (KEGG) database provides a systematic analysis of metabolic pathways and functions of gene products. In this study, the *C. nasus* unigenes were assigned to canonical pathways described in KEGG using BLASTX.

SSRs and SNPs analysis

Simple sequence repeats (SSRs) in the *C. nasus* unigenes were detected using the microsatellite identification tool (MISA) (http://pgrc.ipk-gatersleben.de/misa/). Detection criteria of SSRs included perfect repeat motifs of one to six base pairs and a minimum repeat number of 12 for mono-, six for di-, five for tri-, five for tetra-, four for penta-, and four for hexa-nucleotide microsatellites. SOAPsnp (http://soap.genomics.org.cn/soapsnp.html) was used to detect single nucleotide polymorphisms (SNPs) in the *C. nasus* unigenes.

Data deposition

The raw Illumina sequencing data from the olfactory epithelium of *C. nasus* were deposited in the NCBI Sequence Read Archive (SRA) Sequence Database (accession number SRP035517).

Acknowledgments

The authors gratefully acknowledge Lei Wang, Ya Zhang from our laboratory, Linhong Shen and the other staff from Administration of Fishery of Jingjiang, and the fisherman Baishan Zhan and Xiping Zhou for their assistance in sample collecting. This research was also supported by Shanghai Outstanding Undergraduate Scholarship for Interdisciplinary Training.

Author Contributions

Conceived and designed the experiments: WT DL LW. Performed the experiments: GZ DL JY. Analyzed the data: GZ LW. Contributed reagents/materials/analysis tools: WT. Wrote the paper: GZ LW WT.

References

1. Whitehead PJP, Nelson GJ, Wongratana T (1988) FAO species catalogue. Clupeoid fishes of the world (Suborder Clupeoidei). Part 2. Engraulididae. FAO Fisheries Synopsis 125(7): 460–475.
2. Yuan CM, Qin AL, Liu RH (1980) Discussion on subspecific taxonomy of the genus *Coilia* in middle and lower reaches of Yangtze River and southeast coastal China Sea. Journal of Nanjing University (Natural Sciences) 3: 67–82.
3. McBride JR, Fagerlund UHM, Smith M, Tomlinson N (1964) Olfactory preception in juvenile salmon: II. Conditioned response of juvenile sockeye salmon (*Oncorhynchus neeka*) to lake waters. Canadian Journal of Zoology 42(2): 245–248.
4. Tarrant RM (1966) Threshold of perception of eugenol in juvenile sockeye salmon. Transactions of the American Fisheries Society 95(1): 112–115.
5. Jahn LA (1967) Responses to odors by fingerling cutthroat trout from Yellowstone Lake. The Progressive Fish-Culturist 38(4): 207–210.
6. Doving KB, Westerberg H, Johnsen PB (1985) Role of olfaction in the behavior and neural responses of Atlantic salmon, *Salmon salar*, to hydrographic stratification. Canadian Journal of Fisheries and Aquatic Sciences 42(10): 1658–1667.
7. Yano K, Nakamura A (1992) Observations on the effect of visual and olfactory ablation on the swimming behavior of migrating adult chum salmon, *Oncorhynchus keta*. Ichthyological Research 39(1): 67–83.
8. Barbin GP, Parher SJ, McCleave JD (1998) Olfactory clues play a critical role in the estuarine migration of silver-phase American eels. Environmental Biology of Fishes 53: 283–291.

9. Yamamoto Y, Hino H, Ueda H (2010) Olfactory Imprinting of Amino Acids in n Lacustrine Sockeye Salmon. PLoS ONE 5(1): e8633.

10. Bandoh H, Kida I, Ueda H (2011) Olfactory Responses to Natal Stream Water in Sockeye Salmon by BOLD fMRI. PLoS ONE 6(1): e16051.

11. Johnstone KA, Lubieniecki KP, Koop BF, Davidson WS (2011) Expression of olfactory receptors in different life stages and life histories of wild Atlantic salmon (*Salmo salar*). Molecular Ecology 20(19): 4059–4069.

12. Kolmakov NN, Kube MK, Reinhardt R, Canario AV (2008) Analysis of the goldfish *Carassius auratus* olfactory epithelium transcriptome reveals the presence of numerous non-olfactory GPCR and putative receptors for progestin pheromones. BMC genomics 9: 429–445.

13. Yang JQ, Hu XL, Tang WQ, Lin HD (2008) mtDNA control region sequence variation and genetic diversity of *Coilia nasus* in Yangtze River estuary and its adjacent waters. Chinese Journal of Zoology 43(1): 8–15.

14. Guo HY, Tang WQ (2006) The relationship between sagittal otolith weight-age and its use in age determination in *Coilia nasus* from the estuary of Yangtze River. Journal of Fisheries of China 30(3): 347–352.

15. Guan WB, Chen HH, Ding HT, Xuan FJ, Dai XJ (2010) Reproductive characteristics and conditions of anadromous *Coilia ectenes* (Engraulidae) in Yangtze estuary. Marine Fisheries 32(1): 73–81.

16. Cheng WX, Tang WQ (2011) Some phenotypic varieties between different ecotypes of *Coilia nasus* in Yangtze River. Chinese Journal of Zoology 46: 33–40.

17. Zheng F, Guo HY, Tang WQ, Li HH, Liu D, et al (2012) Age structure and growth characteristics of anadromous populations of *Coilia nasus* in the Yangtze River. Chinese Journal of Zoology 47(5): 24–31.

18. Xu G, Wan J, Gu R, Zhang C, Xu P (2013) Morphological and histological studies on ovary development of *Coilia nasus* under artificial farming conditions. Journal of Fishery Sciences of China 18: 537–546.

19. Xu GC, Dong JJ, Nie ZJ, Xu P, Gu RB. (2012) Studies on lactate dehydrogenase isozymes and DNA content in different tissues of *Coilia nasus*. Journal of Shanghai Ocean University 21(4): 481–488.

20. Wisby WJ, Hasler AD (1954) Effect of olfactory occlusion on migrating silver salmon (*O. kisutch*). Journal of the Fisheries Board of Canada 11(4): 472–478.

21. Firestein S (2001) How the olfactory system makes sense of scents. Nature 413(6852): 211–218.

22. Ache BW, Young JM (2005) Olfaction: diverse species, conserved principles. Neuron 48(3): 417–430.

23. Nakamura T (2000) Cellular and molecular constituents of olfactory sensation in vertebrates. Comparative Biochemistry and Physiology Part A: Molecular & Integrative Physiology 126(1): 17–32.

24. Zufall F, Leinders-Zufall T (2000) The cellular and molecular basis of odor adaptation. Chemical senses 25(4): 473–481.

25. Moon C, Jaberi P, Otto-Bruc A, Baehr W, Palczewski K, et al. (1998) Calcium-sensitive particulate guanylyl cyclase as a modulator of cAMP in olfactory receptor neurons. The Journal of neuroscience 18(9): 3195–3205.

26. Ronnett GV, Moon C (2002) G proteins and olfactory signal transduction. Annual review of physiology 64(1): 189–222.

27. Barry PH (2003) The relative contributions of cAMP and InsP3 pathways to olfactory responses in vertebrate olfactory receptor neurons and the specificity of odorants for both pathways. The Journal of general physiology 122(3): 247–250.

28. Kohout TA, Lefkowitz RG (2003) Regulation of G protein-coupled receptor kinases and arrestins during receptor desensitization. Molecular pharmacology 63(1): 9–18.

29. Boekhoff I, Touhara K, Danner S, Inglese J, Lohse MJ, et al. (1997) Phosducin, potential role in modulation of olfactory signaling. Journal of Biological Chemistry 272(7): 4606–4612.

30. Kaupp UB (2010) Olfactory signalling in vertebrates and insects: differences andcommonalities. Nature Reviews Neuroscience 11(3): 188–200.

31. Breer H, Boekhoff I, Tareilus E (1990) Rapid kinetics of second messenger formation in olfactory transduction. Nature 345: 65–68.

32. Nakamura T, Gold GH (1987) A cyclic nucleotide-gated conductance in olfactory receptor cilia. Nature 325: 442–444.

33. Kleene SJ, Gesteland RC (1991) Calcium-activated chloride conductance in frog olfactory cilia. The Journal of neuroscience 11(11): 3624–3629.

34. Lowe G, Gold GH (1993) Non linear amplification by calcium-dependent chloride channels in olfactory receptor cells. Nature 366: 283–286.

35. Kaneko H, Putzier I, Frings S, Kaupp UB, Gensch T (2004) Chloride accumulation in mammalian olfactory sensory neurons. The Journal of neuroscience 24(36): 7931–7938.

36. Reisert J, Lai J, Yau KW, Bradley J (2005) Mechanism of the excitatory Cl response in mouse olfactory receptor neurons. Neuron 45(4): 553–561.

37. Nickell WT, Kleene NK, Kleene SJ (2007) Mechanisms of neuronal chloride accumulation in intact mouse olfactory epithelium. The Journal of physiology 583(3): 1005–1020.

38. Kurahashi T, Yau KW (1993) Co-existence of cationic and chloride components in odorant-induced current of vertebrate olfactory receptor cells. Nature 363: 71–74.

39. Moon C, Jaberi P, Otto-Bruc A, Baehr W, Palczewski K, et al. (1998) Calcium-sensitive particulate guanylyl cyclase as a modulator of cAMP in olfactory receptor neurons. The Journal of neuroscience 18(9): 3195–3205.

40. Dawson TM, Arriza JL, Jaworsky DE, Borisy FF, Attramadal H, et al. (1993) Beta-adrenergic receptor kinase-2 and beta-arrestin-2 as mediators of odorant-induced desensitization. Science 259(5096): 825–829.

41. Peppel K, Boekhoff I, McDonald P, Breer H, Caron MG, et al. (1997) G protein-coupled receptor kinase 3 (GRK3) gene disruption leads to loss of odorant receptor desensitization. Journal of Biological Chemistry 272(41): 25425–25428.

42. Mashukova A, Spehr M, Hatt H, Neuhaus EM. (2006) β-arrestin2-mediated internalization of mammalian odorant receptors. The Journal of neuroscience 26(39): 9902–9912.

43. Wei J, Zhao AZ, Chan GC, Baker LP, Impey S, et al. (1998) Phosphorylation and inhibition of olfactory adenylyl cyclase by CaM kinase II in neurons: a mechanism for attenuation of olfactory signals. Neuron 21(3): 495–504.

44. Sinnarajah S, Dessauer CW, Srikumar D, Chen J, Yuen J, et al. (2001) RGS2 regulates signal transduction in olfactory neurons by attenuating activation of adenylyl cyclase III. Nature 409 (6823): 1051–1055.

45. Reisert J, Matthews HR (1998) Na$^+$-dependent Ca^{2+} extrusion governs response recovery in frog olfactory receptor cells. The Journal of general physiology 112(5): 529–535.

46. Chen TY, Yau KW (1994) Direct modulation by Ca^{2+}–calmodulin of cyclic nucleotide-activated channel of rat olfactory receptor neurons. Nature 368(6471): 545–548.

47. Nordeng H (1971) Is the local orientation of anadromous fishes determined by pheromones ? Nature 233: 411–413.

48. Sorensen PW, Fine JM, Dvornikovs V, Jeffrey CS, Shao F, et al. (2005) Mixture of new sulfated steroids functions as a migratory pheromone in the sea lamprey. Nature Chemical Biology 1(6): 324–328.

49. Boschat C, Pélofi C, Randin O, Roppolo D, Lüscher C, et al. (2002) Pheromone detection mediated by a V1r vomeronasal receptor. Nature Neuroscience 5(12): 1261–1262.

50. Ryba NJP, Tirindelli R (1997) A new multigene family of putative pheromone receptors. Neuron 19(2): 371–379.

51. Xu GC, Nie ZJ, Zhang CX, Wei GL, Xu P, et al. (2012) Histological studies on testis development of *Coilia nasus* under artificial farming conditions. Journal of huazhong agriculture university 31(2): 247–252.

52. Grabherr MG, Haas BJ, YassourM, Levin JZ, Thompson DA, et al. (2011) Full-length transcriptome assembly from RNA-Seq data without a reference genome. Nature biotechnology 29(7): 644–652.

53. Pertea G, Huang X, Liang F, Antonescu V, Sultana R, et al. (2003). TIGR Gene Indices clustering tools (TGICL): a software system for fast clustering of large EST datasets. Bioinformatics 19(5): 651–652.

54. Conesa A, Götz S, García-Gómez JM, Terol J, Talón M, et al. (2005) Blast2GO: a universal tool for annotation, visualization and analysis in functional genomics research. Bioinformatics 21(18): 3674–3676.

55. Ye J, Fang L, Zheng H, Zhang Y, Chen J, et al. (2006) WEGO: a web tool for plotting GO annotations. Nucleic Acids Research (suppl 2): W293–297.

2

Long Distance Linkage Disequilibrium and Limited Hybridization Suggest Cryptic Speciation in Atlantic Cod

Ian R. Bradbury[1,4]*, Sharen Bowman[3], Tudor Borza[3], Paul V. R. Snelgrove[4], Jeffrey A. Hutchings[2], Paul R. Berg[5], Naiara Rodríguez-Ezpeleta[6], Jackie Lighten[2], Daniel E. Ruzzante[2], Christopher Taggart[7], Paul Bentzen[2]

1 Department of Fisheries and Oceans, St. John's, Newfoundland, Canada, 2 Marine Gene Probe Laboratory, Department of Biology, Dalhousie University, Halifax, Nova Scotia, 3 The Atlantic Genome Centre, Halifax, Nova Scotia, Canada, 4 Ocean Sciences Center, Memorial University of Newfoundland, St. John's, Newfoundland, Canada, 5 Centre for Ecological and Evolutionary Synthesis, Department of Biology, University of Oslo, Oslo, Norway, 6 AZTI-Tecnalia, Marine Research Division, Txatxarramendi ugartea z/g, Sukarrieta, Spain, 7 Department of Oceanography, Dalhousie University, Halifax, Nova Scotia, Canada

Abstract

Hybrid zones provide unprecedented opportunity for the study of the evolution of reproductive isolation, and the extent of hybridization across individuals and genomes can illuminate the degree of isolation. We examine patterns of interchromosomal linkage disequilibrium (ILD) and the presence of hybridization in Atlantic cod, *Gadus morhua*, in previously identified hybrid zones in the North Atlantic. Here, previously identified clinal loci were mapped to the cod genome with most (~70%) occurring in or associated with (<5 kb) coding regions representing a diverse array of possible functions and pathways. Despite the observation that clinal loci were distributed across three linkage groups, elevated ILD was observed among all groups of clinal loci and strongest in comparisons involving a region of low recombination along linkage group 7. Evidence of ILD supports a hypothesis of divergence hitchhiking transitioning to genome hitchhiking consistent with reproductive isolation. This hypothesis is supported by Bayesian characterization of hybrid classes present and we find evidence of common F1 hybrids in several regions consistent with frequent interbreeding, yet little evidence of F2 or backcrossed individuals. This work suggests that significant barriers to hybridization and introgression exist among these co-occurring groups of cod either through strong selection against hybrid individuals, or genetic incompatibility and intrinsic barriers to hybridization. In either case, the presence of strong clinal trends, and little gene flow despite extensive hybridization supports a hypothesis of reproductive isolation and cryptic speciation in Atlantic cod. Further work is required to test the degree and nature of reproductive isolation in this species.

Editor: Tongming Yin, Nanjing Forestry University, China

Funding: Research was supported in part by Genome Canada, Genome Atlantic and the Atlantic Canada Opportunities Agency through the Atlantic Cod Genomics and Broodstock Development Project. Supporting organizations include industrial partners (e.g., Ag West Bio Tech, SUN Microsystems, IBM, and pharmaceutical and biotechnology companies), and a complete list of supporting partners can be found at www.codgene.ca/partners.php. The funders, including commercial partners, had no role in study design, data collection and analysis, decision to publish, or preparation of the manuscript. Research funding and support was also provided by an NSERC Strategic Grant on Connectivity in Marine Fish and an NSERC Discovery Grant.

Competing Interests: Research was supported in part by Genome Canada, Genome Atlantic and the Atlantic Canada Opportunities Agency through the Atlantic Cod Genomics and Broodstock Development Project. Supporting organizations include industrial partners (e.g., Ag West Bio Tech, SUN Microsystems, IBM, and pharmaceutical and biotechnology companies), and a complete list of supporting partners can be found at www.codgene.ca/partners.php.

* Email: ibradbur@me.ca

Introduction

Understanding the complex contributions of ecological factors and genomic architecture to the formation of reproductive isolation is central to an understanding of speciation [1–3]. Evidence is accumulating that isolation and speciation can occur despite gene flow where divergent selection exists for different habitats or environments [4–6]. Study of the early stages of speciation where hybridization is common can reveal genome features involved, as well as the dominant isolating mechanisms (e.g., [5,7,8]). These processes are particularly apparent in hybrid zones and among ecological species where admixture commonly occurs and selection or barriers to hybridization maintain reproductive isolation [9–11]. Accordingly, genome wide examinations of hybrid individuals may provide unprecedented insight into the genomic architecture associated with speciation, identify sources and targets of selection [2,3,12], and clarify management and conservation objectives [13].

Recent studies continue to highlight the importance of genomic architecture to speciation and adaptation [14–16] and that patterns of linkage disequilibrium (LD) are more complex than previously thought (e.g., [17,18]). Better descriptions of both the genomic distribution of divergence and LD are necessary if architecture underlying adaptive divergence, and ultimately, the mechanisms underlying speciation are to be resolved [12,19]. How important "islands of genomic divergence" are to adaptation and speciation remains unclear, but it seems likely they play an important role during the early stages of speciation [2,12]. Speciation or isolation with gene flow is hypothesized to involve a progression of increasing isolation leading from selection at a few isolated genomic regions to genome wide divergence [1]. During this process, islands of genomic divergence may grow in size via divergence hitchhiking. Both the magnitude and size of these islands of adaptive divergence will depend on the genomic architecture and strength of selection [2,6]. Genomic island size has been reported to be between a few to hundreds of kb [20], and islands may be smaller and more isolated when gene flow is high [12,17,21,22]. As such, these genome wide patterns of divergence (i.e. frequency and size of islands of divergence, and LD among islands) may directly reflect the degree of reproductive isolation present between intraspecific groups.

Atlantic cod, *Gadus morhua*, is a commercially exploited demersal marine fish found throughout the North Atlantic characterized by high dispersal potential, large fecundities, and large population sizes [23]. Given the wide range of environments occupied, cod are likely exposed to a large range of ocean climates from temperate to high latitudes. We recently used SNP-based genome scans in Atlantic cod (*G. morhua*) in conjunction with linkage mapping to demonstrate the utility of genomic islands of adaptive divergence [16] in management and conservation efforts. These initial genomic explorations for the presence of islands of adaptive divergence identified a subset of gene-associated polymorphisms for which allele frequencies displayed parallel latitudinal clines (here designated as N (north) and S (south) types) in otherwise genetically distinct populations on both sides of the Atlantic and tested positive for signatures of selection [24]. SNPs associated with these types were non-randomly distributed across the genome and mapped to three linkage groups. Recent studies and extensive sampling have confirmed the presence of two distinct forms of two distinct groups of Atlantic cod [16] and that subsets of these SNPs are commonly identified as outliers, often associated with temperature [25–27]. The geographic co-occurrence of these forms and presence of distinct hybrid zones allows the examination of mechanisms involved in both genomic and reproductive isolation.

The goal of the present paper is to examine the nature of the trends observed previously (e.g., [16,24]) and in particular to explore patterns of interchromosomal linkage disequilibrium (ILD) and the presence of hybridization between wild Atlantic cod N and S types in the previously identified hybrid zone in the northwest Atlantic. First, we identify possible functional associations of clinal SNPs through alignment with the recently available Atlantic cod genome, and an exploration of SNP-gene associations. Second, we explore ILD among SNPs in three different linkage groups previously associated with clinal structure. Finally, we examine the degree of hybridization among these two latitudinal cod types using a Bayesian approach to assign individuals to discrete hybrid classes. We build on previous studies, which examined loci displaying evidence of environmentally associated selection in parallel on either side of the Atlantic in a subset of the loci (n = 40) and populations (n = 14) used in this study [24], and the application of these loci for fisheries management and conservation [16,28]. Here, we delve further into the mechanisms driving these patterns and present evidence of previously unknown cryptic speciation and divergence in a common, heavily exploited marine fish. The implications for the management of Atlantic cod as well as the potential for cryptic diversity in other well studied and exploited marine organisms (e.g., [13]) suggest significant gaps in our current understanding perhaps best addressed with population genomic approaches.

Materials and Methods

Sample characteristics

We re-analyzed data on 23 Atlantic cod populations all included in Bradbury *et al.* (2013) and explored the presence of LD and hybridization in the wild. These data represent cod sampled ($N = 466$) from throughout the North Atlantic in 1996–2007, and approximately 20 individuals were sampled from each location (Figure S1 in File S1). Specific details regarding all samples and locations were published elsewhere [24,29–31]. This study primarily dealt with existing DNA or tissue samples and did not involve the handling of live specimens. This study used tissue samples from fish collected during routine Government of Canada sampling of fish stocks under the supervision of the departmental animal care committee. Fish collected as part of assessment activities are exempt from animal care protocols as they are being collected under our regulatory mandate for establishing abundance estimates. This exemption is found in section 4.1.2.2 of the Canadian Council on Animal Care's Guidelines on the use of fish in research, teaching and testing. As all fish collected in this manner are deceased prior to handling, all tissue samples were collected post mortem. We isolated DNA from these samples from ethanol-preserved fin clips using a modification of a previously published glass milk procedure [32]. These samples were genotyped for 1536 SNPs (GenBankdbSNP under accession numbers ss131570222-ss131571915). Details on SNP development and genotyping were provided elsewhere [31,33]. SNPs were genotyped using the Illumina Golden Gate platform at the McGill Innovation Center. Linkage mapping information is available for most of these SNPs based on three families, including parents and F1 offspring [34].

Annotation and SNP alignment

Here we focus on linkage groups containing clinal (latitude) SNP outliers (i.e. LG2, LG7. LG12). Of SNPs previously identified as outliers, most could not be annotated using BLAST [24]. We attempted to identify associations with adjacent genes, through alignment with the recently published cod genome to identify potential ontological/functional relationships. These outlier SNPs and 80–90 bp of flanking sequence were mapped to the cod genome scaffold [35] using default parameters in BLAT [36] implemented in the Ensemble Genome Browser (http://www.ensembl.org/Gadus_morhua/Info/Index). We accepted only E values $>1.0e^{-40}$ with sequence similarity $>96\%$ as a hit to the correct position in the Atlantic cod genome. In the instance that a SNP did not map to a protein coding region of the genome, the distance (bases) to the nearest flanking genes on either side were calculated. SNPs within annotated genes and SNPs within a 5 Kb distance from an annotated gene were given a HGNC (Human Genome Nomenclature Committee) ID. We also used these annotations from the cod genome scaffold to obtain gene ontology and functionality information from the Uniprot (www.uniprot.org), or GeneCards (www.genecards.org) databases.

Linkage and divergence

In previous work, outlier tests and spatial analysis [16,24] identified a subset of these loci (n = 53, Table S1 in File S1) that displayed elevated divergence and signatures indicative of directional selection, and most were associated with latitudinal clines. These SNPs were distributed across three linkage groups. Here, we evaluated the presence of long distance ILD among all loci on the linkage groups containing outliers. LD was quantified with D' estimated using TASSEL [37]. D' represents the D statistic which measures the non-random association among alleles

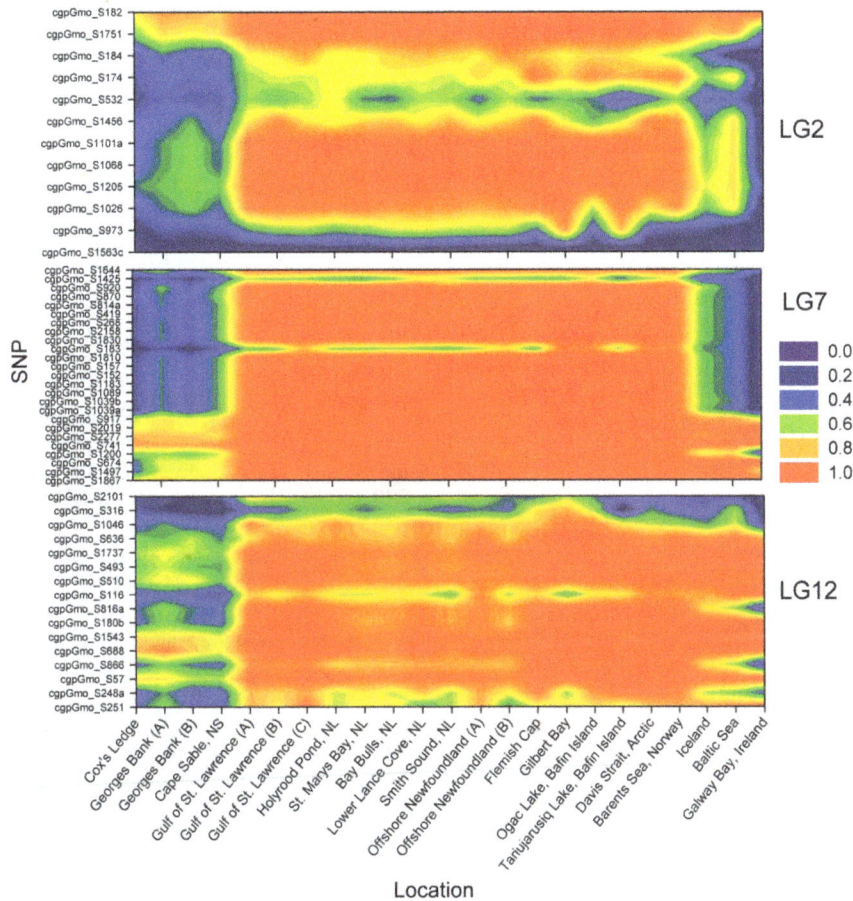

Figure 1. Clinal trends in allele frequency in Atlantic cod range-wide. Allele frequencies for outlier loci in linkage groups LG2, LG7 and LG12 from 23 locations. See Bradbury et al. [16] for specific results of outlier tests. See Figure S1 in File S1 for sample locations.

at two loci, standardized by its maximum possible value, and ranges from 0 representing no association to 1 representing complete linkage. The three linkage groups previously shown to contain outliers associated with latitudinal clinal structure were assessed for ILD among loci. D' was measured among SNPs along each combination of these groups. Similarly F_{ST} was calculated for each locus again in these three linkage groups using ARLEQUIN [38]. Both measures of D' and F_{ST} were examined in conjunction with map position using the linkage map [34].

Bayesian Tests for Hybridization

Two Bayesian approaches were used to infer the presence and frequency of various hybrid classes in the wild using only the outlier loci. Bayesian clustering was performed using STRUC-TURE v.2.2.4 [39] to estimate admixture coefficients for each individual. This approach assumes Hardy-Weinberg Equilibrium (HWE) and linkage equilibria among loci, introduces population structure, and assigns populations that are not in linkage equilibrium using a Markov chain Monte Carlo (MCMC) algorithm to estimate the number of populations (K). The algorithm was run 5 times for K = 2 to ensure convergence of values, and with a burn-in of 100 000 repetitions, 300 000 repetitions after burn-in. STRUCTURE was also run with 1 million repetitions to ensure 300 000 were sufficient. All STRUCTURE results were amalgamated among replicates using CLUMPP 1.1.2 [40] and summarized graphically using DIS-TRUCT 1.1 [41].

In addition to STRUCTURE, NEWHYBRIDS [42] was to used examine hybridization and the presence of specific hybrid classes. NEWBYBRIDS uses a Gibbs sampler and Markov chain Monte Carlo to estimate the posterior probability that genetically sampled individuals fall into each of a total of six set of hybrid or parentage categories (pure N, pure S; F1 hybrids; F2 hybrids; and both backcrosses). NEWHYBRIDS was run with a burn-in of 100 000 iterations followed by 250,000 iterations and individuals belonging to a category with probability >90% were considered correctly assigned.

To evaluate the ability of each approach to correctly identify the various hybrid classes, individuals of each class were simulated using HYBRIDLAB 1.0 [43]. All six parentage groups (N-type, S-type, F1 hybrids, F2 hybrids and reciprocal backcrosses) were simulated under random mating using 50 N-type and S-type individuals. To identify these 50 pure individuals of each group, the STRUCTURE results were filtered for individuals with admixture coefficients >0.90 and <0.10. HYBRIDLAB was then used to simulate 50 individuals of each hybrid class. Known N- and S-type as well as simulated hybrids were then used to evaluate assignment accuracy using both STRUCTURE and NEWHY-BRIDS, implementing similar parameters and methods for each as described above.

Results

Clines and SNP annotation

Outlier loci (see [16] for outlier test details) in linkage groups characterized by the presence of clinal loci [24] were re-evaluated using data on new populations. LG2 contained 12 outliers, the majority of which displayed parallel clines in both the east and west Atlantic (Figure 1). In LG7, most of 25 identified outliers displayed clear clinal trends in allele frequency in both the East and West Atlantic. Similarly LG12 displayed clinal trends in allele frequency in most outliers present. Of these three, LG7 displayed the most consistent patterns in allele frequency across locations and loci with very little variation among most individual and locus comparisons. Overall, trends among most outliers in all three linkage groups were very similar and contained 73% of all outliers identified (see [16]).

To identify genes containing or associated with (<5 kb) these SNPs, SNP sequences were mapped to cod genome scaffold [44]. Of the 53 outlier SNPs contained on these three linkage groups, 38 SNPs or ~70% were associated with known coding regions. Of these, nine mapped within gene regions, and 30 mapped within 5 kb of a known coding region in gene scaffolds (see Table S1 in File S1). Only two SNPs (cgpGMO_S182 and cgpGMO_S184) were adjacent to the same gene, MYLPF (i.e. myosin light chain, phosphorylatable, fast skeletal muscle) supporting the outlier status of this gene involved in muscle activity. In general, the identified SNP-associated genes represent a diverse array of function and possible physiological pathways. For example, cgpGmo_S917 is associated with AQP11 (Aquaporin 11) which is thought to be related to water transport (and associated with Na/K ATPase activity) across cell membranes and hence could play a key role in adaptation to different water conditions etc. Similarly, cgpGmo_S510 is associated with FASL (Fas ligand (TNF superfamily, member 6), which has been associated with signalling pathways and the immune response. Refer to Table S1 in File S1 for individual SNP results of alignment and annotation.

Interchromosomal Linkage Disequilibrium

Significant ILD was observed among all three linkage groups and in general regions of significant LD were associated with elevated divergence (i.e. F_{ST}). The largest region of consistent high divergence and D' was near the middle of LG7. This region displayed some of the highest LD values and most consistent in comparison to the other LG's. A uniform block of approximately 4 cM near the centre of LG7 was strongly associated with an approximately 3 cM section of LG2 (Figure 2). This region was also associated with high F_{ST} values between the two forms at the outlier loci on either side of the North Atlantic (Figure 2, Figure S2 in File S1). For these comparisons, D' values ranged from 0 to 1, but many loci had values >0.5 (Figure 2b). This same section of LG7 was also associated with LG12 as indicated by elevated D' values along a 15 cM section (Figure 3). Again this region was also associated with high F_{ST} values between the two forms at the outlier loci on either side of the North Atlantic (Figure 3a and Figure S2 in File S1) and D' values ranged from 0 to 0.6 with 10's of loci >0.5 (Figure 3b). LG12 also displayed lower overall divergence at outlier loci than did LG2 or LG7. Regions of LG2 and LG12 which had been associated with LG7 and elevated divergence among the two forms, displayed some evidence of ILD with each other (Figure 4a, b).

Hybridization analysis. The power to detect hybrid individuals and specific hybrid classes was evaluated using simulated hybrid individuals (see Methods). Using STRUCTURE, pure individuals were readily identified and distinguished from hybrids.

However, F1, F2, and backcrossed individuals could not easily be differentiated as both were characterized by intermediate admixture coefficients (Figure 5a). In contrast, NEWHYBRIDs correctly identified on average 99% of all individuals across all hybrid classes (Figure 5b). Only F2 individuals were not correctly identified with 100% accuracy, with success dropping to 96%. This drop is due to two F2 individuals being incorrectly identified as backcrossed individuals (Figure 5b). Given the high success rate observed for the identification of at least some pure or hybrid groups, both analyses were conducted on field collected individuals to quantify the degree of hybridization and presence of each hybrid class.

In general on both sides of the Atlantic, northern samples (Norway to Gulf of St. Lawrence) were characterized by pure N-type, with S-type and hybrid individuals dominating to the south (Figure 5). On both sides of the Atlantic, the frequency of hybrid individuals declined to the north, a trend visible in both STRUCTURE and NEWHYBRIDS analyses. The exception seems to be the sample from Ireland, which was almost entirely pure S-type, except for one individual which was identified as an F1 (Figure 5). The STRUCTURE analysis indicated significant admixture in southern populations but the presence of exact hybrid classes was difficult to discern (Figure 5). In comparison, NEWHYBRIDS identified almost all admixed individuals as F1 with little evidence of F2 and no evidence of backcrossed individuals (Figure 6). In some samples, such as one from Georges Bank, the proportion of F1s was high (>50%) (Figure 5). Frequency plots of admixture coefficients from both analyses indicate the presence of admixed individuals overall was about 15% and comprised of almost entirely F1 individuals (Figure 6).

Discussion

Hybridization along hybrid zones or between ecological species can provide rare opportunity for the study of the evolution of reproductive isolation (e.g., [3,10,45]). The degree of hybridization evident across genomes and the frequency of occurrence of hybrid classes can yield insight into the ecological and genomic processes associated with speciation [1,12,46]. The levels of linkage disequilibrium (LD) among outlier genes observed here and the lack of extensive gene flow are consistent with a hypothesis of two sympatric forms of Atlantic cod (N and S types) and support the hypothesis that these genomic regions are associated with reductions in hybrid survival and reproductive isolation. The presence of significant ILD between sympatric forms (e.g., [17,18]) has been reported elsewhere and is consistent with the presence of genome hitchhiking, but the observation that the neutral markers displayed no association with these groups [24] also suggests that its influence has been limited. Moreover the observation that genomic regions associated with these two forms represent a diverse array of possible functions and the involvement of multiple physiological pathways is consistent with the influence of a very broad selective agent such as temperature, as has been suggested previously [24]. Our results challenge current management paradigms in this species which consider both groups a single management unit in the southern portions of the range. Further study is needed to examine the geographic distributions, ecological and physiological attributes, and reproductive interactions of these two groups, and further resolve the genomic details of the linkage groups and ILD patterns identified here. Similar observations of surprisingly high levels of biocomplexity in Atlantic cod have also been made at the range margin in the waters off Greenland [27] and the eastern Atlantic [26] suggesting this may be a common phenomenon in this species. It is worth noting that patterns

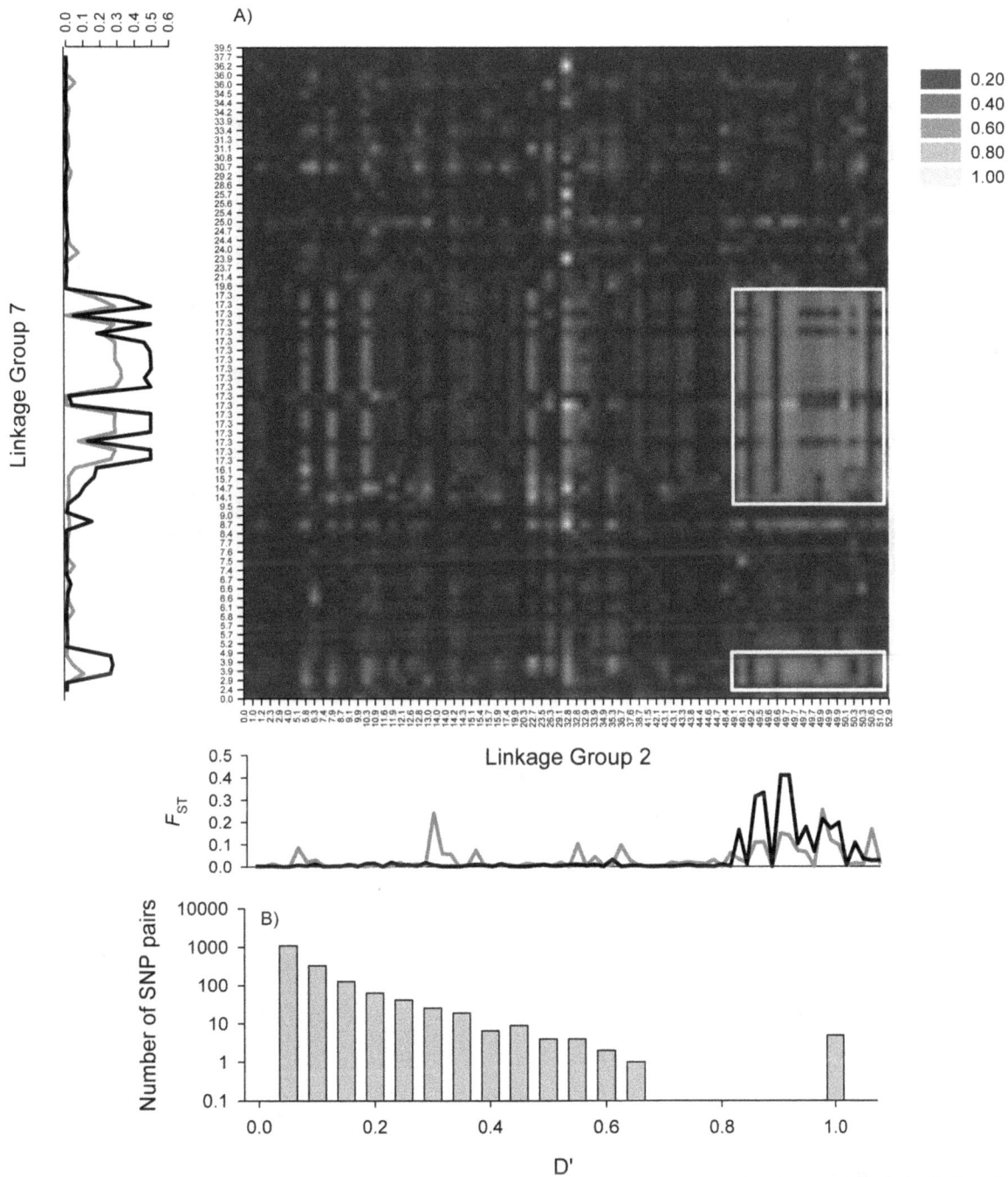

Figure 2. Interchromosomal linkage disequilibrium between LG2 and LG7. (A) Heat map of interchromosomal linkage disequilibrium between LG2 and LG7, axis labels represent linkage map positions (see [34]), note that several SNPs share a common map location. Plots of population differentiation (F_{ST}) on the left and the bottom compare divergence between N and S types, in both the west (black line) and east Atlantic (grey line). Two main regions of ILD between previously identified adaptive genomic regions are highlighted by white rectangles. (B) Histogram of D' values for all interchromosomal pairs of SNPs.

observed here seem independent of the genomic differences reported among migratory and stationary cod populations in Norway [26,47] and Iceland [48].

The parallel nature of environmental gradients in the North Atlantic provide replicated environmental and latitudinal gradients and a model for the identification of genes associated with ecological influences [49]. This work extends an earlier study identifying the clinal nature of the outlier loci associated with the

N- and S-types in this species on either side of the North Atlantic [24] by extending the geographic coverage considerably through re-analysis of samples described in Bradbury et al. [16]. With additional sampling, steep clines in allele frequency at outlier loci are consistently identified along the Scotian shelf in the west and around Iceland or southern Norway in the east. Interestingly, although most outliers in the three linkage groups examined displayed parallel clines, >50% of the loci in LG12 displayed

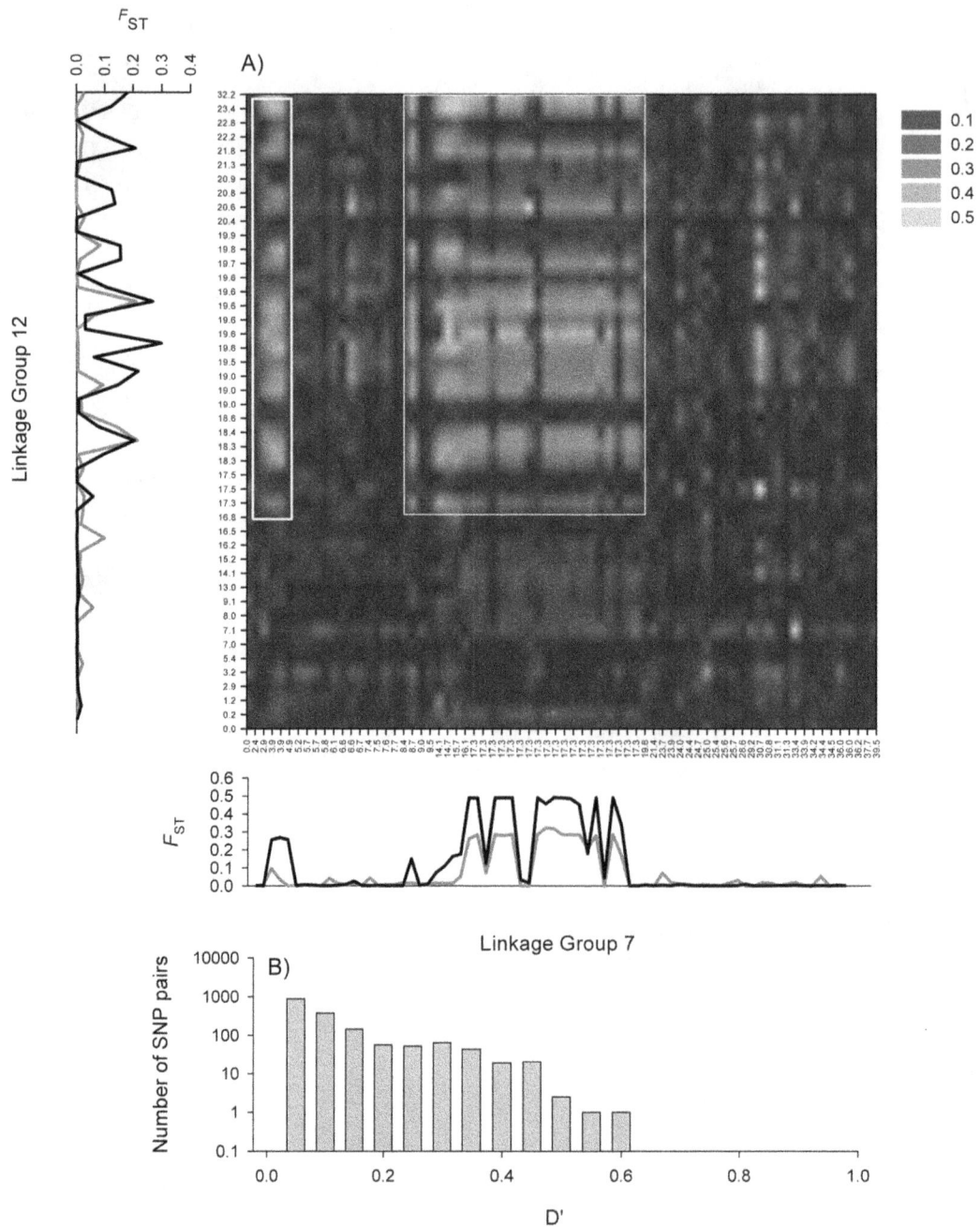

Figure 3. Interchromosomal linkage disequilibrium between LG7 and LG12. (A) Heat map of interchromosomal linkage disequilibrium between LG7 and LG12, axis labels represent linkage map positions (see [34]), note that several SNPs share a common map location. Plots of population differentiation (F_{ST}) on the left and the bottom compare divergence between N and S types, in both the west (black line) and east Atlantic (grey line). Two main regions of ILD between previously identified adaptive genomic regions are highlighted by white rectangles. (B) Histogram of D′ values for all interchromosomal pairs of SNPs.

clines only in the west and were fixed in the east. Similar latitudinal clines have been reported in other species in the Atlantic [50] as well as in the Pacific (e.g., [51]). Previously we identified temperature associations with allele frequency at 40 of these outliers supporting a hypothesis of temperature associated selection [24]. An alternate hypothesis for the formation of these clines is that these are "tension" zones formed by secondary contact of differentially adapted forms that have become trapped along environmental or ecological gradients [52]. Our observation of parallel clines requires this "trapping" of tension zones to have

occurred independently on either side of the Atlantic, and as non-outliers show no evidence of clinal structure, this explanation seems less likely. Although at present the nature of these clines and whether they represent trapped tension zones or parallel adaptive evolution remains unknown, in the context of this study, the key point of interest is that the two forms (N and S type) appear genetically distinct and appear to resist introgression despite significant interbreeding where they co-occur.

Understanding the nature of these clinal loci such as whether or not they implicate similar physiological pathways or functions is

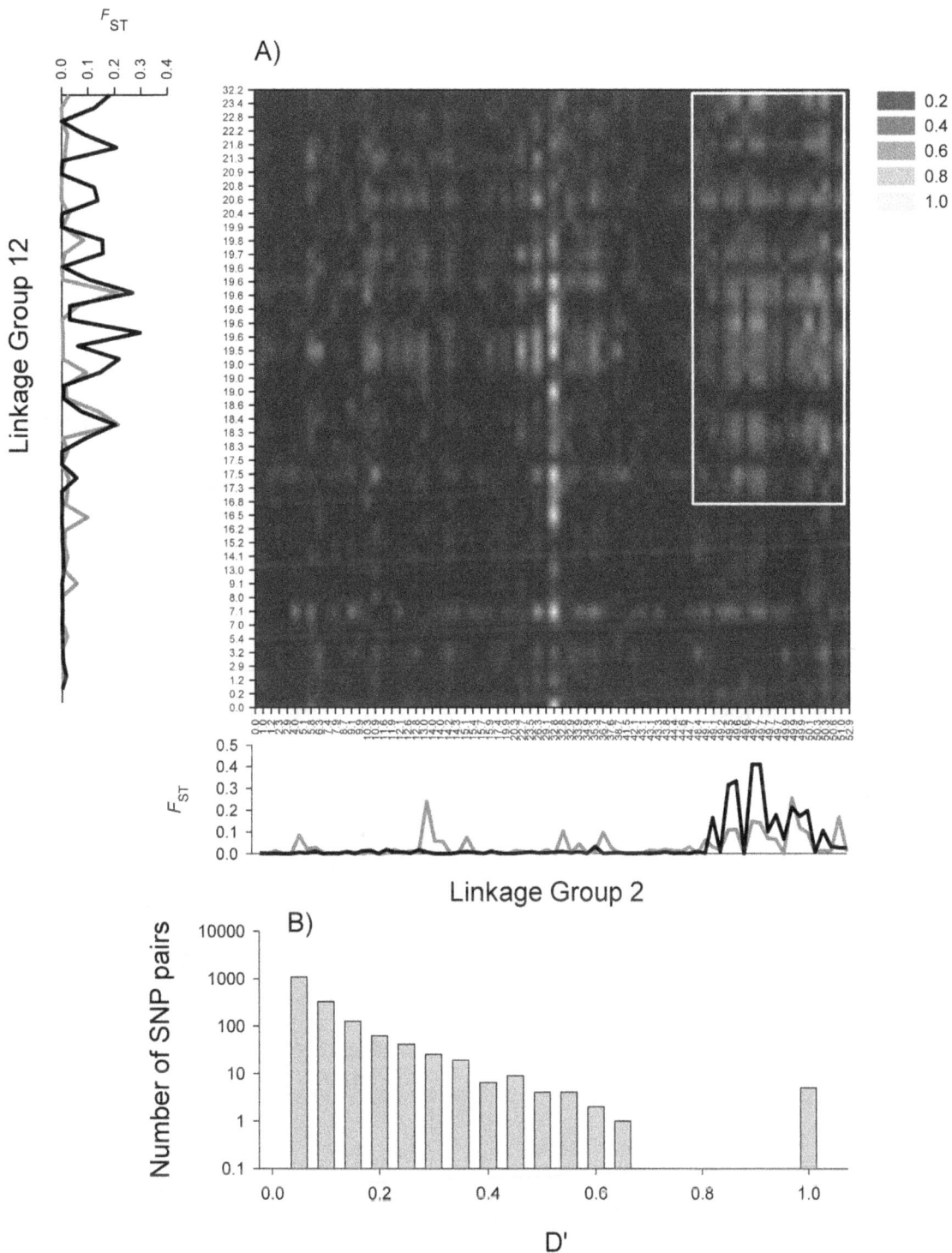

Figure 4. Interchromosomal linkage disequilibrium between LG2 and LG12. (A) Heat map of interchromosomal linkage disequilibrium between LG2 and LG12, axis labels represent linkage map positions (see [34]), note that several SNPs share a common map location. Plots of population differentiation (F_{ST}) on the left and the bottom compare divergence between N and S types, in both the west (black line) and east Atlantic (grey line). Two main regions of ILD between previously identified adaptive genomic regions are highlighted by white rectangles. (B) Histogram of D' values for all interchromosomal pairs of SNPs.

critical to the interpretation of these clines. Previous attempts to explore the gene associations of the key SNPs that distinguish the N and S types using BLAST had very limited success, producing annotations of only a few sequences [24,31,33]. Here, taking

advantage of the recently published cod genome [44], the majority of these SNPs have been located in genes or associated with genes, allowing us to explore the potential functional importance of these SNPs. Most of the SNPs associated with these N and S types are

(A) STRUCTURE

(B) NEWHYBRIDS

S-type   N-type   F1   F2   Backcross A   Backcross B

(C) STRUCTURE

(D) NEWHYBRIDS

Coxes Ledge (USA), Georges Bank A, Georges Bank B, Cape Sable (NS), Gulf of St. Lawrence A, Gulf of St. Lawrence B, Gulf of St. Lawrence C, St Marys Bay (NL), Holyrood Pond (NL), Bay Bulls (NL), Lower Lance Cove (NL), Smith Sound (NL), Newfoundland Offshore A, Newfoundland Offshore B, Flemish Cap, Gilbert Bay (LAB), Offshore Arctic, Ogac Lake, Tarijarusiq Lake, Barents Sea, Norway, Iceland, Baltic Sea, Ireland

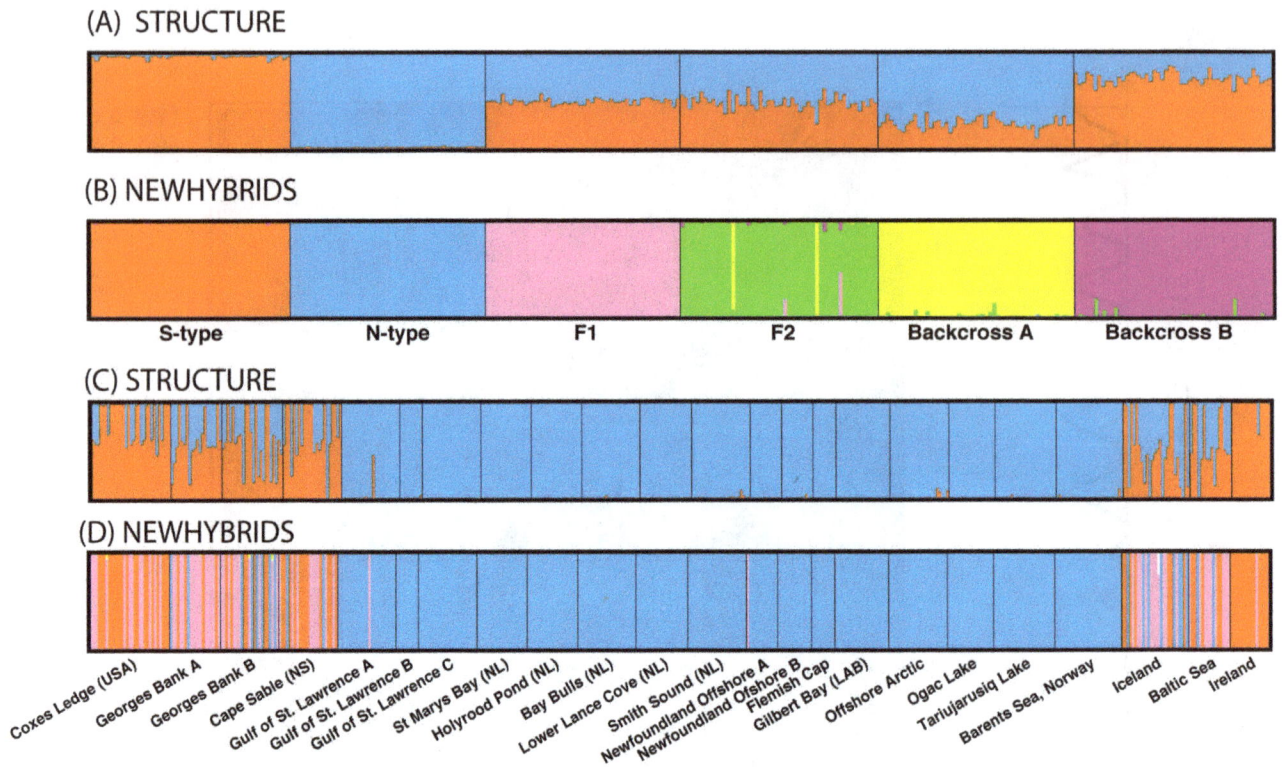

Figure 5. Bayesian assignment of simulated hybrid classes using (A) Structure with K = 2, and (B) New Hybrids with six possible hybrid classes of individuals (N type, S type, F1, F2, Backcross A, and Backcross B). Individuals simulated using HYBRIDLAB. Bayesian identification of hybrid classes of individuals present in wild Atlantic cod samples using (C) Structure, and (D) New Hybrids.

clustered in discrete regions of the genome [16], the question of how independent these types are is reasonable. The observation that all but two SNPs were associated with different genes, supports the involvement of a suite of genes in the observed clines rather than the linkage of most SNPs to a few clinal genes. These associated genes implicate a range of physiological pathways and functions with no obvious trend. This observation is consistent with the hypothesis of either a broad acting selective agent targeting numerous physiological functions (i.e. temperature), or perhaps longstanding genomic divergence and incompatibility among N and S forms. However, a hypothesis of longstanding genomic divergence does not seem compatible with observations of little to no divergence at the majority of SNPs examined. The lack of a clear and consistent link to function across many of the SNPs contrasts the outcomes of genome scans in other marine fishes, where outliers across environmental gradients are associated with genes of obvious functional importance. For example, Lamichhaney et al. [53] identify several outliers in Baltic herring populations with clear relevance to the strong salinity gradients present in the Baltic. Also, Limborg et al. [54] also identified environmentally corrected outliers in transcriptome derived SNPs in the northeast Atlantic. Or similarly, in Atlantic cod in the eastern Atlantic, Hemmer-Hansen et al. [26] identified a genomic island associated with migratory phenotype in coastal cod which included the Pantophysin gene.

The study raises the question of how the non-random association of genes on three different chromosomes can exist despite apparent high rates of interbreeding in some locations. The evidence of extensive LD among these outliers and islands of divergence on different chromosomes is consistent with either a lack of interbreeding, the suppression of recombination, or low

hybrid fitness. A lack of successful interbreeding could explain the association among islands of divergence, but the apparent common occurrence of F1 individuals suggests this is not the case. However, as F2 or backcrossed individuals were rare, it is difficult to rule out intrinsic barriers to gene flow. In theory, transmission ratio distortion (e.g., [55]) could partly explain the trends in LD observed. However this was not readily visible in mapping crosses [34] and seems unlikely. Inter-chromosome translocations could potentially play a role and should be investigated, although it seems unlikely that this could easily explain the observed LD among three chromosomes. Hohenlohe et al. [17] report significant LD among adaptive genomic regions on two different chromosomes, in threespine stickleback in marine and freshwater habitats. They concluded that population structuring in conjunction with additive and epistatic selection provided the most parsimonious explanation for ILD across these stickleback genomes. Such an explanation seems most likely here as well, though will require further study. Similar trends in ILD have been reported both in *Drosophila* [18] and sticklebacks [17] and associated with reproductive isolation or ecological speciation. Also observations of ILD have been made in situations where populations are structured either through selective breeding or strong selection such as strains of the cultivated tomato [56] or in genes associated with drug resistance in *Plasmodium falciparum* [57].

The absence of most hybrid classes supports a hypothesis of reproductive isolation and limited gene flow, although several possible explanations exist for this apparent lack of hybrid classes in southern locations where the N and S forms co-occur. First, it is possible that this analysis lacked sufficient power to detect some hybrid classes. However, the simulation and analysis of hybrid

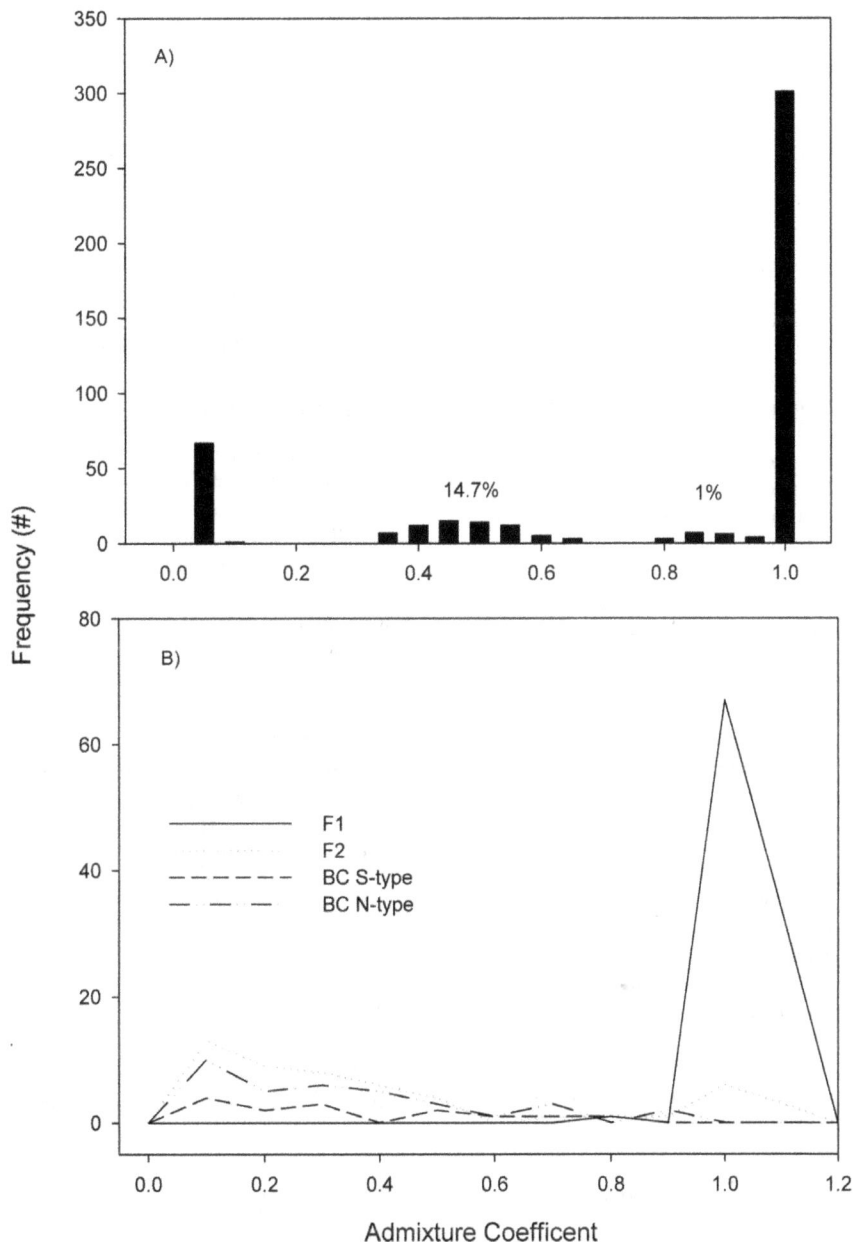

Figure 6. Frequency distributions of admixture coefficients from analysis of wild caught Atlantic cod using both (A) STRUCTURE and (B) NEWHYBRIDS.

classes supports the hypothesis that these individuals would have been correctly identified if present. A second possible explanation is that the presence of Bateson-Dobzhansky-Muller incompatibilities among islands of divergence may explain the observed linkage among chromosomes and the lack of some hybrid classes in the wild. The Bateson–Dobzhansky–Muller model predicts that postzygotic isolation evolves due to the accumulation of incompatible epistatic interactions, but few studies have quantified the relationship between genetic architecture and patterns of reproductive divergence [58]. Simulations of clines associated with BDM incompatibilities indicate that BDM epistasis with interacting locus pairs on different chromosomes resulted in the greatest long distance genomic autocorrelation in cline parameters [10]. This is largely consistent with our observations here. Finally, it is also possible that gynogenesis may play a role and explain the

relatively high numbers of F1s observed, though further work is required to explore this possibility.

In addition to the lack of some hybrid classes, the high proportion of individuals classified as F1's (>50%) in some samples was noteworthy. The number of individuals classified as F1's was highest at the mid latitudes examined (i.e. Georges Bank) and declined to the south in the east and west, with only a single F1 individual being present in the sample from Ireland. Despite the high proportion of F1's in some samples, the overall proportion among all samples was low (~15%) suggesting that some samples or regions may have been biased towards hybridization or the collection of hybrids (e.g., [59]). Admittedly it remains unclear to what degree these linked genomic islands are a product of selection and speciation ("speciation island" hypothesis) or due to an absence of gene flow and the clustering of ancestral regions of divergence in regions of low recombination ("incidental island" hypothesis) [60]. Which scenario is more likely will require further genomic and experimental work.

Regardless of which scenario is valid in this case, observation here of genetically discrete N- and S-types despite interbreeding suggests reproductive isolation and an evolutionarily significant division. There is also some evidence that the forms possess divergent phenotypes (e.g., [61,62]) and occupy different habitats (e.g., [24]). Accordingly it seems reasonable to posit that these forms represent cryptic or ecological species as has been suggested elsewhere [26]. Given the low levels of divergence across most of the genome among these forms it is reasonable to hypothesize that the ILD observed is associated with the a transition from divergence hitchhiking to genome hitchhiking in this species. A similar situation has been documented in *Anopheles* mosquitoes where two sympatric forms have been described [63] associated with islands of divergence on three chromosomes [18]. In this case, as here, identifying the role selection plays in maintaining these islands of divergence remains a challenge [18,60]. From a conservation or management perspective, the parapatric occurrence of these discrete forms pose clear challenges for the conservation of diversity in this exploited species in southern locations as has been suggested for northern locations [27]. Ultimately, a better understanding of the spatial and temporal distribution of N- and S-types will be needed if this diversity is to be reflected in the current management framework for this species.

Summary

Disentangling the contributions of ecological factors and genomic architecture to the formation of reproductive isolation is central to an understanding of speciation [1–3]. The distribution and degree of genomic divergence is expected to reflect the magnitude of reproductive isolation present among forms, and in theory can be used to evaluate where on the speciation continuum intraspecific forms reside, yet distinguishing islands of divergence due to selection and speciation from islands resulting from low regions of recombination remains a challenge. The results presented here provide convincing evidence that multiple clinal SNPs distributed across three linkage groups were associated with each other and reductions in the occurrence of hybrid classes in the wild. This work supports the hypothesis that in these sympatric N- and S-types of Atlantic cod, divergence has transitioned from divergence hitchhiking to genome hitchhiking. The results reveal cryptic divergence, possibly associated with ecological speciation in Atlantic cod on both sides of the North Atlantic. The implications of such unrecognized diversity in this extremely well studied marine fish are broad ranging, from challenging fisheries management and the conservation practices in Atlantic cod, to a broader recognition of increased biocomplexity in marine populations. Ultimately, the nature of these forms, the degree of reproductive isolation, and presence of extensive hybridization will require further field and lab study.

Supporting Information

File S1 Supporting Information file that includes Table S1, Figure S1, and Figure S2. Table S1. SNP associated genes and map position from SNP sequence alignment to cod genome using BLAT. See methods for details. Figure S1. Map of sample locations for Atlantic cod tissue samples distributed throughout the North Atlantic. Figure S2. F_{ST} between N and S types from either side of the North Atlantic (yellow – west, red – east) with linkage map distance across (A) LG2, (B) LG7 and (C) LG12.

Acknowledgments

The authors thank all who assisted with tissue collection and G.R. Carvalho and E.E. Nielsen for comments on an earlier version of this work. Research was supported in part by Genome Canada, Genome Atlantic and the Atlantic Canada Opportunities Agency through the Atlantic Cod Genomics and Broodstock Development Project. A complete list of supporting partners can be found at www.codgene.ca/partners.php. Research funding and support was also provided by a NSERC Strategic Grant on Connectivity in Marine Fish and an NSERC Discovery Grant.

Author Contributions

Conceived and designed the experiments: IRB PB. Analyzed the data: IRB PRB NR JL TB. Contributed reagents/materials/analysis tools: IRB SB PB JAH TB DER CT PVRS. Wrote the paper: IRB PB.

References

1. Feder JL, Egan SP, Nosil P (2012) The genomics of speciation-with-gene-flow. Trends Genet 28: 342–350.
2. Nosil P (2012) Ecological speciation. Oxford: Oxford University Press.
3. Feder JL, Flaxman SM, Egan SP, Comeault AA, Nosil P (2013) Geographic Mode of Speciation and Genomic Divergence. Annu Rev Ecol Evol Syst 44: 73–97.
4. Rundle H, Nosil P (2005) Ecological speciation. Ecol Lett 8: 336–352.
5. Via S (2009) Natural selection in action during speciation. Proc Natl Acad Sci U S A 106: 9939–9946.
6. Feder JL, Flaxman SM, Egan SP, Nosil P (2013) Hybridization and the build-up of genomic divergence during speciation. J Evol Biol 26: 261–266.
7. McGaugh SE, Noor MAF (2012) Genomic impacts of chromosomal inversions in parapatric *Drosophila* species. Philos Trans R Soc Lond B Biol Sci 367: 422–429.
8. Navarro A, Barton NH (2003) Chromosomal speciation and molecular divergence-accelerated evolution in rearranged chromosomes. Science 300: 321–324.
9. Barton NH, Hewitt GM (1985) Analysis of hybrid zones. Annu Rev Ecol Syst 16: 113–148.
10. Gompert Z, Parchman TL, Buerkle CA (2012) Genomics of isolation in hybrids. Philos Trans R Soc Lond B Biol Sci 367: 439–450.
11. Arnold ML (1997) Natural Hybridization and Evolution. Oxford Series in Ecology and Evolution. Oxford Oxford University Press.
12. Via S (2012) Divergence hitchhiking and the spread of genomic isolation during ecological speciation-with-gene-flow. Philos Trans R Soc Lond B Biol Sci 367: 451–460.
13. Funk WC, McKay JK, Hohenlohe PA, Allendorf FW (2012) Harnessing genomics for delineating conservation units. Trends Ecol Evol 27: 489–496.
14. Michel AP, Sim S, Powell THQ, Taylor MS, Nosil P, et al. (2010) Widespread genomic divergence during sympatric speciation. Proc Natl Acad Sci U S A 107: 9724–9729.
15. Hohenlohe PA (2010) Population genomics of parallel adaptation in threespine stickleback using sequenced RAD tags. PLoS Genet 6: e1000862.
16. Bradbury IR, Hubert S, Higgins B, Bowman S, Borza T, et al. (2013) Genomic islands of divergence and their consequences for the resolution of spatial structure in an exploited marine fish. Evol Appl 6: 450–461.
17. Hohenlohe PA, Bassham S, Currey M, Cresko WA (2012) Extensive linkage disequilibrium and parallel adaptive divergence across threespine stickleback genomes. Philos Trans R Soc Lond B Biol Sci 367: 395–408.
18. White BJ, Cheng C, Simard F, Costantini C, Besansky NJ (2010) Genetic association of physically unlinked islands of genomic divergence in incipient species of Anopheles gambiae. Mol Ecol 19: 925–939.
19. Nosil P, Feder JL (2012) Genomic divergence during speciation: causes and consequences. Philosophical Transactions of the Royal Society B: Biological Sciences 367: 332–342.
20. Nadeau NJ, Martin SH, Kozak KM, Salazar C, Dasmahapatra KK, et al. (2012) Genome-wide patterns of divergence and gene flow across a butterfly radiation. Mol Ecol 22: 814–826.
21. Feder JL, Nosil P (2010) The efficacy of divergence hitchhiking in generating genomic islands during ecological speciation. Evolution 64: 1729–1747.
22. Via S, West J (2008) The genetic mosaic suggests a new role for hitchhiking in ecological speciation. Mol Ecol 17: 4334–4345.
23. COSEWIC (2010) COSEWIC assessment and update status report on the North Atlantic cod, *Gadus moruha*, in Canada. Ottawa, ON. p. 118 p.

24. Bradbury IR, Hubert S, Higgins B, Bowman S, Paterson I, et al. (2010) Parallel adaptive evolution of Atlantic cod in the eastern and western Atlantic Ocean in response to ocean temperature. Proc R Soc Biol Sci Ser B 277: 3725–3734.

25. Nielsen EE, Hemmer-Hansen J, Poulsen N, Loeschcke V, Moen T, et al. (2009) Genomic signatures of local directional selection in a high gene flow marine organism; the Atlantic cod (*Gadus morhua*). BMC Evol Biol 9: 276.

26. Hemmer-Hansen J, Nielsen EE, Therkildsen NO, Taylor MI, Ogden R, et al. (2013) A genomic island linked to ecotype divergence in Atlantic cod. Mol Ecol 22: 2653–2667.

27. Therkildsen NO, Hemmer-Hansen J, Hedeholm RB, Wisz MS, Pampoulie C, et al. (2013) Spatiotemporal SNP analysis reveals pronounced biocomplexity at the northern range margin of Atlantic cod Gadus morhua. Evol Appl 6: 690–705.

28. Bradbury IR, Hubert S, Higgins B, Bowman S, Paterson IG, et al. (2011) Evaluating SNP ascertainment bias and its impact on population assignment in Atlantic cod, *Gadus morhua*. Molecular Ecology Resources 11: 218–225.

29. Taggart CT, Cook D (1996) Microsatellite genetic analysis of Baltic cod: a preliminary report. Manuscript Report, Marine Gene Probe Laboratory, Dalhousie University, Halifax, NS. 13.

30. Hubert S, Tarrant Bussey J, Higgins B, Curtis BA, Bowman S (2009) Development of single nucleotide polymorphism markers for Atlantic cod (Gadus morhua) using expressed sequences. Aquaculture 296: 7–14.

31. Hubert S, Higgins B, Borza T, Bowman S (2010) Development of a SNP resource and a genetic linkage map for Atlantic cod (*Gadus morhua*). BMC Genomics 11: 191.

32. Elphinstone MS, Hinten GN, Anderson MJ, Nock CJ (2003) An inexpensive and high-throughput procedure to extract and purify total genomic DNA for population studies. Mol Ecol 3: 317–320.

33. Bowman S, Hubert S, Higgins B, Stone C, Kimball J, et al. (2010) An integrated approach to gene discovery and marker development in Atlantic cod (*Gadus morhua*). Mar Biotechnol 13: 242–255.

34. Borza T, Higgins B, Simpson G, Bowman S (2010) Integrating the markers Pan I and haemoglobin with the genetic linkage map of Atlantic cod (*Gadus morhua*). BMC Research Notes 3: 261.

35. Star B, Nederbragt AJ, Jentoft S, Grimholt U, Malmstrom M, et al. (2011) The genome sequence of Atlantic cod reveals a unique immune system. Nature 477: 207–210.

36. Kent WJ (2002) BLAT–the BLAST-like alignment tool. Genome Res 12: 656–664.

37. Bradbury PJ, Zhang Z, Kroon DE, Casstevens TM, Ramdoss Y, et al. (2007) TASSEL: software for association mapping of complex traits in diverse samples. Bioinformatics 23: 2633–2635.

38. Excoffier L, Lischer HEL (2010) Arlequin suite ver 3.5: a new series of programs to perform population genetics analyses under Linux and Windows. Molecular Ecology Resources 10: 564–567.

39. Pritchard JK, Stephens M, Donnelly P (2000) Inference of population structure using multilocus genotype data. Genetics 155: 945.

40. Jakobsson M, Rosenberg NA (2007) CLUMPP: a cluster matching and permutation program for dealing with label switching and multimodality in analysis of population structure. Bioinformatics 23: 1801–1806.

41. Rosenberg NA (2004) Distruct: a program for the graphical display of population structure. Mol Ecol Notes 4: 137–138.

42. Anderson EC, Thompson EA (2002) A model-based method for identifying species hybrids using multilocus genetic data. Genetics 160: 1217–1229.

43. Nielsen EE, Bach LA, Kotlicki P (2006) hybridlab (version 1.0): a program for generating simulated hybrids from population samples. Mol Ecol Notes 6: 971–973.

44. Star B, Nederbragt AJ, Jentoft S, Grimholt U, Malmstrom M, et al. (2011) The genome sequence of Atlantic cod reveals a unique immune system. Nature 477: 207–210.

45. Barton NH, Gale KS (1993) Genetic analysis of hybrid zones. In: Harrison RG, editor. Hybrid zones and the evolutionary process. Oxford: Oxford University Press. pp. 13–45.

46. Nosil P, Feder JL (2012) Genomic divergence during speciation: causes and consequences. Philos Trans R Soc Lond B Biol Sci 367: 332–342.

47. Knutsen H, Jorde PE, Andre C, Stenseth NC (2003) Fine-scaled geographical population structuring in highly mobile marine species: the Atlantic cod. Mol Ecol 12: 385–394.

48. Pampoulie C, Ruzzante DE, Chosson V, Jorundsdottir TD, Taylor L, et al. (2006) The genetic structure of Atlantic cod (Gadus morhua) around Iceland: insight from microsatellites, the *Pan* I locus, and tagging experiments. Can J Fish Aquat Sci 63: 2660–2674.

49. Schmidt PS, Serrao E, Pearson G, Riginos C, Rawson P, et al. (2008) Ecological genetics in the North Atlantic: environmental gradients and adaptation at specific loci. Ecology 89: S91–S107.

50. Crawford DL, Place AR, Powers DA (1990) Clinal variation in the specific activity of lactate dehydrogenase-B. J Exp Zool 255: 110–113.

51. Sotka EE, Wares JP, Barth JAB, Grosenberg RK, Palumbi SR (2004) Strong genetic clines and geographical variation in gene flow in the rocky intertidal barnacle *Balanus glandula*. Mol Ecol 13: 2143–2156.

52. Bierne N, Welch J, Loire E, Bonhomme F, David P (2011) The coupling hypothesis: why genome scans may fail to map local adaptation genes. Mol Ecol 20: 2044–2072.

53. Lamichhaney S, Martinez Barrio A, Rafati N, Sundstrom G, Rubin CJ, et al. (2012) Population-scale sequencing reveals genetic differentiation due to local adaptation in Atlantic herring. Proc Natl Acad Sci U S A 109: 19345–19350.

54. Limborg MT, Helyar SJ, De Bruyn M, Taylor MI, Nielsen EE, et al. (2012) Environmental selection on transcriptome-derived SNPs in a high gene flow marine fish, the Atlantic herring (*Clupea harengus*). Mol Ecol 21: 3686–3703.

55. Hahn MW, White BJ, Muir CD, Besansky NJ (2012) No evidence for biased co-transmission of speciation islands in *Anopheles gambiae*. Philos Trans R Soc Lond B Biol Sci 367: 374–384.

56. Robbins MD, Sim S-C, Yang W, Van Deynze A, van der Knaap E, et al. (2011) Mapping and linkage disequilibrium analysis with a genome-wide collection of SNPs that detect polymorphism in cultivated tomato. J Exp Bot 62: 1831–1845.

57. Adagu IS, Warhurst DC (2001) Plasmodium falciparum: linkage disequilibrium between loci in chromosomes 7 and 5 and chloroquine selective pressure in Northern Nigeria. Parasitology 123: 219–224.

58. Fierst JL, Hansen TF (2010) Genetic architecture and postzygotic reproductive isolation: evolution of Bateson-Dobzhansky-Muller incompatibilities in a polygenic model. Evolution 64: 675–693.

59. Oliveira E, Salgueiro P, Palsson K, Vicente JL, Arez AP, et al. (2008) High levels of hybridization between molecular forms of Anopheles gambiae from Guinea Bissau. J Med Entomol 45: 1057–1063.

60. Turner TL, Hahn MW (2010) Genomic islands of speciation or genomic islands and speciation? Mol Ecol 19: 848–850.

61. Purchase CF, Brown JA (2000) Interpopulation differences in growth rates and food conversion efficiencies of young Grand Banks and Gulf of Maine Atlantic cod (Gadus morhua). Can J Fish Aquat Sci 57: 2223–2229.

62. Hutchings JA, Swain DP, Rowe S, Eddington JD, Puvanendran V, et al. (2007) Genetic variation in life-history reaction norms in a marine fish. Proc R Soc Lond B Biol Sci 274: 1693–1699.

63. Turner TL, Hahn MW, Nuzhdin SV (2005) Genomic islands of speciation in *Anopheles gambiae*. PLoS Biol 3: e285.

3

Small Changes in Gene Expression of Targeted Osmoregulatory Genes When Exposing Marine and Freshwater Threespine Stickleback (*Gasterosteus aculeatus*) to Abrupt Salinity Transfers

Annette Taugbøl[1]*, **Tina Arntsen**[1], **Kjartan Østbye**[1,2], **Leif Asbjørn Vøllestad**[1]

1 Centre for Ecological and Evolutionary Synthesis (CEES), Department of Biosciences, University of Oslo, Blindern, Norway, **2** Hedmark University College, Department of Forestry and Wildlife Management, Campus Evenstad, Elverum, Norway

Abstract

Salinity is one of the key factors that affects metabolism, survival and distribution of fish species, as all fish osmoregulate and euryhaline fish maintain osmotic differences between their extracellular fluid and either freshwater or seawater. The threespine stickleback (*Gasterosteus aculeatus*) is a euryhaline species with populations in both marine and freshwater environments, where the physiological and genomic basis for salinity tolerance adaptation is not fully understood. Therefore, our main objective in this study was to investigate gene expression of three targeted osmoregulatory genes (Na^+/K^+-ATPase (ATPA13), cystic fibrosis transmembrane regulator (CFTR) and a voltage gated potassium channel gene (KCNH4) and one stress related heat shock protein gene (HSP70)) in gill tissue from marine and freshwater populations when exposed to non-native salinity for periods ranging from five minutes to three weeks. Overall, the targeted genes showed highly plastic expression profiles, in addition the expression of ATP1A3 was slightly higher in saltwater adapted fish and KCNH4 and HSP70 had slightly higher expression in freshwater. As no pronounced changes were observed in the expression profiles of the targeted genes, this indicates that the osmoregulatory apparatuses of both the marine and landlocked freshwater stickleback population have not been environmentally canalized, but are able to respond plastically to abrupt salinity challenges.

Editor: Vincent Laudet, Ecole Normale Supérieure de Lyon, France

Funding: The study was supported by the Norwegian Research Council. The funders had no role in study design, data collection and analysis, decision to publish, or preparation of the manuscript.

Competing Interests: The authors have declared that no competing interests exist.

* Email: annette.taugbol@ibv.uio.no

Introduction

The ability to respond rapidly to environmental change is beneficial in variable environments, making phenotypically plastic organisms better adapted in unstable and unpredictable environments [1]. The capacity of an organism to respond to its environment is facilitated by the environmentally induced alteration of gene and protein expression. While the evolution of plasticity depends on the trait(s) in question and the source of environmental variation, there is a general acceptance that the ability to be plastic may be constrained by a variety of costs underlying the plastic responses [2]. As such, evolutionary theory predicts loss of plasticity after periods of environmental stability, when environmental constancy eliminates or weakens the source of selection that was formerly important for its maintenance, given that the cost for the trait is high [3], or through environmentally induced genetic assimilation [4] which reduces the environmental influence on trait expression.

Phenotypic plasticity of a trait is generally assumed to be under selection when a single organism is exposed to several environments during its lifetime which each select for different trait values. Most fish species are stenohaline, living either in fresh or salt water [5,6] where they are exposed to the same type of osmoregulatory challenge during their lifetime. For fish living in marine waters, the concentration of ions is much higher in the water compared to the environment inside the cell, and surrounding ions diffuse into the cell while water is lost. The situation is reversed for a freshwater fish, where the surroundings are ion depleted, making the fish passively loose ions and gain water. In order to maintain a relatively stable internal osmotic environment, fish counteract these effects by a variety of specialized physiological mechanisms, mainly in the gills [7] and the kidney [8], and these genetic adaptations can limit movement between salinities. Only a very limited number of species are truly euryhaline [6], being able to osmoregulate in a wide variety of salinity environments. Even fewer can tolerate extreme changes in osmolality over short time scales, such as the killifishes (*Fundulus spp.*) [9] and the threespine stickleback (*Gasterosteus aculeatus*) [10,11].

The threespine stickleback (order, Gasterosteiformes, family Gasterosteidae; hereafter stickleback) is a small fish that was originally a marine species [12]. However, since the last glaciation, sticklebacks have colonized a large number of brackish and freshwater systems throughout the northern hemisphere and are now occupying an extremely wide haloniche [11,13]. Many of the newly formed freshwater populations have become landlocked due to the isostatic uplifting of the land following deglaciation, and the stickleback in these habitats have consequently been separated from the sea for thousands of years. If the costs of having a plastic osmoregulatory machinery is high, it is expected that these landlocked freshwater stickleback populations should have lost the ability to osmoregulate efficiently in saltwater. However, studies suggest that freshwater populations of stickleback still possess the osmoregulatory machinery enabling them to handle abrupt changes in salinity [10,14], despite having been separated from the marine environment for up to 10–18 000 years [12]. This indicates that during adaptation to freshwater environments, the osmoregulatory physiology of landlocked sticklebacks has not been environmentally assimilated, or alternatively, the functionality of the osmoregulatory apparatus and its genomic architecture may not be open for selective change due to pleiotropic gene-interactions and is thus expected to remain similar in freshwater and marine populations.

One way to test if fish are adapted to a particular haloniche is to expose individuals to salinity challenges by transferring individuals from the original salinity to a test-salinity, tracking the expression of relevant genes through time. Earlier experiments show that stickleback easily tolerate transfers from freshwater to fully marine salinity, as well as the reverse [10,14]. However, it is not clear if the same osmoregulatory machinery is functioning at all times. The aim of this study was to assess the effect of experimental manipulation of salinity on the expression of genes important for osmoregulation. For this purpose, we collected adult fish from one marine and one freshwater site and exposed fish from each population to either 0 or 30 PSU (practical salinity units) for periods varying from 5 minutes to 3 weeks. The expression of four candidate genes was then followed through time (Fig. 1); three of the included genes are related to ion-pumps recognized to be under selection in marine-freshwater gradients (Na$^+$/K$^+$-ATPase (ATP1A3), cystic fibrosis transmembrane regulator (CFTR), voltage gated potassium channel gene (KCNH4)), and one is a stress related heat shock protein gene gene (HSP70), also associated with osmotic stress.

The objective of the study was to i) assess how target genes were expressed in the native salinity (assuming adaptation), ii) assess how target genes were affected when freshwater adapted fish were exposed to saltwater and iii) assess how target genes were affected when saltwater adapted fish were exposed to freshwater. As the osmoregulatory challenges are opposite in freshwater and marine environments, with ion secretion needed in saltwater and ion uptake needed in freshwater, we expect that osmoregulatory genes upregulated in freshwater will be downregulated in saltwater, and vice versa. We further expect the stress-related gene to have elevated expression levels in the beginning of the exposure for both groups due to handling and the physiological challenges associated with changing gene expression.

Materials and Methods

Fish and maintenance conditions

Adult stickleback were captured at two locations near Oslo, Norway (Figure 1a, b), during May and June 2010 and 2011. Fish from the marine population are known to breed there, and are not migratory, as many populations are known to be elsewhere [15,16,17]. The marine site (Sandspollen; 59° 39′ 58″N; 10° 35′ 11″ E) has a salinity that fluctuates between 22–29 PSU, while the freshwater pond (Glitredammen; 59° 55′ 53″N; 10° 29′ 55″ E; elevation 82.8 m above sea level) is stable at 0 PSU. The marine fish is comprised only of the completely plated morph (having a full row of lateral plates along its body flank), while the freshwater population only has the low plated morph (with lateral plates in the front region only). The two locations are geographically isolated by approximately 35 km by shortest distance through water, where about 8.5 km is through the river Sandvikselva that contains several steep waterfalls and dams. The age of the lake has been estimated at 7800 years before present using the program Sealevel32 [18]. The program uses information on postglacial land uplift and water level rise to estimate lake age. Downstream movement of fish from Glitredammen is possible, but upstream movement from the sea is impossible.

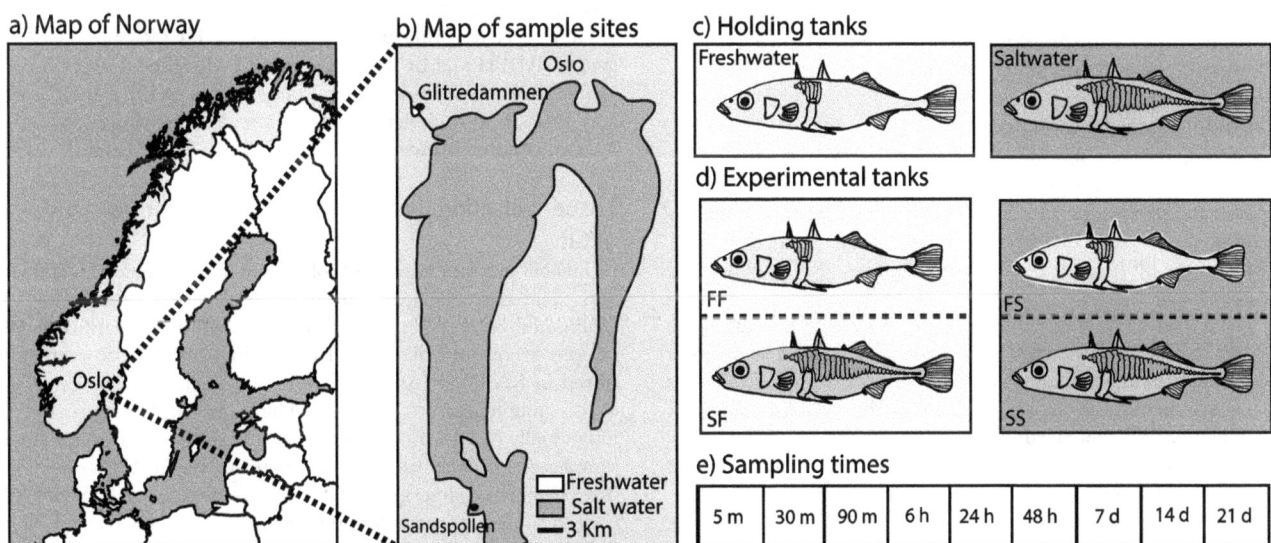

Figure 1. Study area and experimental design. a) Map of Norway showing the position of the sampling sites b) Locations of the two sampling sites, Glitredammen (freshwater) and Sandspollen (marine) c) Wild caught fish from both sampling locations were taken into the lab and placed in holding tanks of their native salinity for a minimum of three weeks. After acclimation, two groups of eight fish from both populations were exposed to either saltwater (30 PSU) or freshwater (0 PSU). The exposure tanks were divided in two by a perforated wall, so that both populations could be exposed to the same water quality at the same time d) Exposure times before the fish was collected and gill tissue was sampled.

After capture, the fish were transported to the aquarium facility at the University of Oslo and acclimated to holding conditions in their native salinity for minimum three weeks prior to the experiment. Two glass holding-tanks (500L) with either salt (30 PSU) or fresh water (0 PSU) were used for acclimation (Figure 1c), using biologically activated canister filters (EHEIM professional 3600), and UV-filtration. The acclimation tanks were covered with black plastic in front and on the sides to reduce visual stress. Further, to reduce potential male nesting behavior, the tanks were not equipped with any environmental enrichment, leaving the tanks free of sand and vegetation. The temperature in the tanks was maintained at room temperature (approx. 20°C) and the light regime was set at a 12:12 light:dark cycle. The fish were fed two times a day with frozen red bloodworms throughout the acclimation and exposure period.

Experimental design and protocol

The experimental setup consisted of 80 L tanks; the tanks were either filled with 0 PSU water or with 30 PSU water (Figure 1c), and covered with black sheets to reduce visual stress. A grey plastic wall divided each experimental tank into two 40 L compartments, where perforation ensured water movement between compartments (Figure 1d).

At the start of an experiment, 8 fish that appeared healthy were collected from each holding tank and placed directly in either the 30 or the 0 PSU experimental tank. Saltwater fish were therefore tested in either saltwater (SS; control) or in freshwater (SF). The freshwater fish were also tested in saltwater (FS) or in freshwater (FF; control) (Figure 1d).

The fish were exposed for different time periods, lasting between 5 minutes and 3 weeks (Figure 1e). The time periods were selected to cover short-term effects as well as long-term changes in gene expression. After each experiment, the fish were quickly netted out of the experimental tanks, immediately killed by a swift blow to the head and was directly processed for tissue collection.

Ethics statement

The experiment was approved by the Norwegian Animal experimentation and care committee (permit no ID 2705) and all efforts were made to minimize suffering.

Candidate gene expression

Candidate genes for osmoregulation were selected based on published studies on divergence in gene expression between marine and freshwater sticklebacks [19], studies identifying outlier regions in DNA sequences between marine and freshwater sticklebacks [20], and preliminary Illumina RNA-sequencing results (Table 1).

The targeted Na^+/K^+-ATPase gene, ATP1A3, has displayed salinity dependent regulation in fish when acclimated to different salinities, including killifish, *Fundulus heteroclitus* [21]. ATP1A3 is a plasma membrane protein that helps the establishment and maintenance of the electrochemical gradients of sodium and potassium ions across the plasma membrane by coupling the exchange of two extracellular K+ ions for three intracellular Na+ ions to the hydrolysis of one molecule of ATP [22], thereby ensuring a relatively constant osmolarity of cells and blood plasma. The protein is powering salt secretion in saltwater fish and salt absorption in freshwater fish [23].

The cystic fibrosis transmembrane regulator, CFTR, is an apical membrane anion channel involved in chloride secretion, and establishes an electrical driving force for trans-epithelial sodium secretion that generate the osmotic driving force for water

flow, yielding an isotonic secretory product. As a candidate gene for saltwater adaptation [24], previous studies have also shown an upregulation of chloride cells and CFTR expression in a Hawaiian goby (*Stenogobius hawaiiensis*) [25] and in killifish [26] exposed to salt water.

While able to tolerate a wide range of salinity, whole genome sequencing of marine and freshwater sticklebacks have identified several chromosomal regions that have undergone parallel selection after freshwater invasion, indicating adaptive divergence and evolutionary change across the marine-freshwater boundary [20]. One identified region differing between marine and freshwater sticklebacks was an inversion with alternative functional exons of the voltage gated potassium channel gene, KCNH4, on either side, suggesting marine and freshwater specific isoforms [20]. However, although small, parallel changes in the sequences of genes may result from similar selection pressure across environments [27,28,29], it is the functional gene products and its regulation through expression that gives rise to the phenotype. Therefore, when candidate regions or loci linked to adaptive divergence have been identified, the regions should be tested in function, such as their role in gene regulation in a relevant ecological setting. The primer pairs in this study does not distinguish between marine and freshwater isoforms as the spanning ends of the inversion are identical down to a few base-pairs and the mutations are seemingly located within introns of the gene.

Both the physical handling when fish are transferred between tanks and the changes in water salinity are stressful, thus a stress-related heat-shock protein, HSP70, known to be affected by osmotic stress [30] was also included in the study.

To control for variation in expression levels not due to the experimental treatment, we used two reference genes, Elongation factor 1 alpha (EF1α) and Gluceraldehyde-3-phosphate dehydrogenase (GADPH). EF1α has been used successfully as reference gene in a similar study on sticklebacks [19] as well as in other gene expression studies on fish [26]. GADPH is a commonly used reference gene and has been stably expressed in a wide array of studies spanning predator cues [31] to exposure to offshore produced water [32].

Gene specific primers for target genes CFTR and reference genes GADPH and EF1α were previously designed and optimized (Table 1). Primers for additional target genes (ATP1A3, KCNH4 and HSP70) were designed based on genetic sequences from the Enseml genome browser and NCBI Primer-Blast (Table 1).

Tissue collection, RNA isolation, cDNA synthesis and qPCR

The gill plays an important role in the maintenance of blood ion and acid–base balance in both freshwater- and seawater acclimated fish [7,33,34]. After each fish was sacrificed, gill samples were immediately collected using sterilized tweezers and stored in RNA*later* (Ambion RNA, Life Technologies) according to the manufacturers protocol. The sampled fish were stored individually in 70% EtOH. The mRNA was isolated from the gill samples from each fish separately, using the mRNA direct kit (Invitrogen) as described by the manufacturer. The mRNA concentration and purity was quantified using Bioanalyzer (Agilent 2100 Bioanalyzer) and the Agilent RNA 6000 Pico Kit (Agilent Technologies) according to the protocol, and all samples were diluted down to 0.125 µg/µL before cDNA synthesis. The cDNA was prepared using the Superscript VILO cDNA synthesis kit (Invitrogen by Life Technologies) as described by the manufacturer, and the concentration was checked spectrophotometrically using Nanodrop (NanoDrop Teqnologies INC).

Table 1. Primers used for qPCR expression analysis of threespine stickleback genes.

Target	Gene name	Ensembl gene ID	Primer sequences	°C	E (%)	Reference
ATP1A3b	ATPase, Na+/K+ transporting, alpha 3b polypeptide	ENSGACG00000009524	F: AGCCGAGATCCCCTTCAACTCCA	60	99.07	*This study*
			R: GCTCCTTCCCCTGCACCAGGA			
CFTR	Cystic fibrosis transmembrane conductance regulator	ENSGACG00000009039	F: GCAGGCCTCTTCTTCACCAA	58	98.51	McCairns et al. (2009)
			R: TCCAGATAGAGGCTGATGTTCTTG			
KCNH4	Potassium voltage-gated channel subfamily H member 4	ENSGACG00000008648	F: CACAGTGACCTCTCTGGTGC	60	99.29	*This study*
			R: AGACATGAGCAGGGTCAGGA			
HSP70	Heat shock protein 70	ENSGACG00000013048	F: ATCGGTATTGACCTGGGCAC	60	99.20	*This study*
			R: GGTATCGGTGAACGCCACAT			
Reference						
EF1α	Elongation factor 1	ENSGACG00000002182	F: CATTGTCACTTACCTGAATCACATGA	60	99.26	McCairns et al. (2009)
			R: TGTGGCATTTAACAACATTTCCA			
GADPH	Glyceraldehyde-3-phosphate dehydrogenase	ENSGACG00000005864	F: CAAACCGTTGGTGACAGTATTTG	60	99.9	Sanago et al. (2011)
			R: GCACTGAGCATAAGGACACATCTAA			

The cDNA concentration was diluted down to 15 ng/μL (±1.5 ng/μL) prior to qPCR amplification after testing for optimization (standard curves, two-fold serial dilutions on pooled cDNA) and association curves for each primer pair (all primer pairs tested on dilution curves at 58 and 60°C). Primer efficiencies were calculated using the formula $E = (10^{-1/slope})-1$. All primer pairs had efficiencies between 95–100% and presented a single product, confirmed with a melting curve.

The qPCR reaction was performed on a Lightcycler 480 (Roche) using SYBR Green PCR Master Mix (Roche). Each 20 μL reaction contained 5.0 μL of the optimized concentrated cDNA, 1.0 μL of each primer, 10.0 μL of SYBR Green and 3.0 μL H2O. The thermocycle program included an enzyme activation step for 5 min, followed by 45 cycles of 95°C for 10 s, 58/60°C for 20 s and 72°C for 20 s. After the amplification phase, a dissociation curve was generated to confirm the presence of a single amplicon. The individual samples were run on duplicated plates, along with three negative reverse-transcriptase controls and an eightfold serial dilution to calibrate plate variation between runs. The obtained cycle threshold (C_q) values for the individuals were adjusted for plate efficiency, and duplicated reactions that differed by more than 0.5 C_q values were checked manually and removed from the analysis.

Statistical analysis

The target gene C_q values were normalized using the mean of the two control genes. Both the EF1α and GADPH were relatively stably expressed across the various time points but did differ somewhat between treatments. The C_q values for GADPH were slightly higher in the SS and SF treatments than in the other treatments ($F_{3, 237} = 6.64$, P<0.001), but did not differ between treatments for EF1α ($F_{3, 235} = 1.43$, P=0.236). The relative expression levels were expressed as the individuals normalized C_q-values of the target transcript, and expressed relative to the mean values of a control group, here set to the 5 min exposure, for each treatment group (SS, SF, FF, FS). This method gives the fold change in expression relative to the control [35].

Variation in fold change of the expression of the different target genes was tested using general linear models. Each treatment group was tested at 9 different time points, where time can be classified both as a continuous variable (in minutes) as well as an ordinal factor (1–9). Preliminary analyses indicated that using continuous time was the better modeling approach and was therefore used in the model, expressed on a log-scale. To account for non-linear effects we included a squared term for time. The general model structure was thus:

$$Y = \text{Treatment} + \text{Time} + \text{Time}^2 + \text{Treatment} * \text{Time} + \text{Treatment} * \text{Time}^2$$

where treatment is the four different treatment types (SS, SF, FF, FS). The best (most parsimonious) model was selected using backward selection, using the Bayesian information criterion (BIC). BIC puts a heavier penalty on parameter number than the more commonly used Akaike information criterion (AIC).

Results

In general all experimental fish handled the transfer to the experimental water qualities well, both when transferred to the control water quality (groups SS and FF) and to the treatment salinity (SF, FS). A total of 9 fish from the freshwater population died across treatments during the experiment (6 in FF and 3 in FS), whereas no marine fish died. A total of 288 sticklebacks were used throughout this study.

We used the gene expression levels at 5 minutes of exposure as the control against which all fold level changes in expression was compared. Overall there were only minor differences in C_q – levels among the various treatment groups for the target genes (Table 2, Figure 2, Appendix S1).

The model that best explained variation in fold change in ATP-expression contained the ln-time and treatment factor ($F_{4, 204} = 14.636$; P<0.0001). Overall there was a tendency for the relative expression to increase with time, and the time-adjusted

Table 2. Gene expression of four osmoregulatory genes (see Table 1) (C_q-values; mean $\pm$ se) for marine and freshwater stickleback after 5 min exposure to either salt or freshwater.

Treatment	CFTR	ATP	KCNH4	HSP70
SS	1.15 3±0.009 (7)	1.111±0.009 (7)	1.415±0.013 (7)	1.272±0.020 (7)
SF	1.133±0.013 (4)	1.121±0.010 (6)	1.429±0.014 (6)	1.257±0.022 (6)
FF	1.158±0.009 (8)	1.126±0.008 (8)	1.413±0.012	1.196±0.019 (8)
FS	1.148±0.011 (6)	1.121±0.010 (6)	1.374±0.014 (6)	1.215±0.022 (6)
	$F_{3, 21} = 0.836$, P=0.489	$F_{3, 23} = 1.383$, P=0.273	$F_{3, 23} = 2.836$, P=0.061	$F_{3, 23} = 3.028$, P=0.050

Summary statistics from an analysis of variance for each gene is also given. Treatments are: SS (saltwater fish in saltwater), SF (saltwater fish in freshwater), FF (freshwater fish in freshwater) and FS (freshwater fish in saltwater).

mean fold change was slightly larger for the SS and FS (SS: 1.547±0.093; FS: 1.555±0.105) than for the FF and SF treatments (SF: 1.283±0.094; FF: 1.193±0.105). To further examine how the fold-change varied with treatment we re-ran the analyses grouping the fish into those tested in freshwater (FF, SF) and those tested in saltwater (SS, FS). Overall, the best model using this model-structure fit better to the data than the best model using treatment group (BIC treatment group = 478.0; BIC treatment water type = 466.0). The best model contained the interaction with time and water treatment type.

The model that best explained variation in fold change in CFTR-expression contained the ln-time and treatment factor ($F_{4, 206} = 13.041$; P<0.0001). Overall there was a tendency for the relative expression to increase with time, and the time-adjusted mean fold change was larger for the two control treatments (SS: 1.592±0.087; FF: 1.427±0.099) than for the transfer treatments (SF: 0.903±0.088; FS: 1.155±0.094). The transfer treatments and the control treatments differed significantly (Tukey HSD post hoc test, P<0.05).

The model that best explained variation in fold change in KCNH-expression contained the ln-time ($F_{1,207} = 14.937$; P = 0.0001) and treatment factor ($F_{3,207} = 14.593$; P<0.0001). Overall there was a tendency for the relative expression to increase with time (0.087±0.023). To further examine how the fold-change varied with treatment we reran the analyses grouping the fish into those tested in freshwater (FF, SF) and those tested in saltwater (SS, FS). Overall the best model using this model structure fit better to the data than the best model using treatment group (BIC treatment group = 527.1; BIC treatment water type = 537.6). The best model contained time ($F_{1, 209} = 14.944$; P<0.0001 and water treatment type ($F_{1, 209} = 43.995$; P<0.0001). The KCNH-expression was significantly elevated in freshwater (1.488±0.079), whereas it was significantly decreased in saltwater (0.756±0.077).

The model that best explained variation in fold change in HSP70-expression only contained the treatment factor ($F_{3, 203} = 2.688$; P = 0.048). Overall, mean fold change was larger than one for fish originating from saltwater (SS: 1.286±0.181; SF: 1.526±0.182) whereas if was smaller than one for the fish originating from freshwater (FF: 0.865±0.209; FS: 0.902±0.205). The fish from freshwater and saltwater differed significantly in expression level (Tukey HSD post hoc test, P<0.05). However, despite the expression levels being quite different between the ecotypes, the explanatory power of the model was low ($R^2 = 0.038$).

Discussion

Teleost fishes maintain nearly constant internal osmotic concentration and have osmoregulatory machinery fine-tuned to the external salinity in either salt or fresh water. However, some species can tolerate a wide range of environmental salinity, also with only short or no acclimation. Our main objective in this study was to investigate the expression of three relevant osmoregulatory genes and one stress related gene in marine and freshwater threespine stickleback, when exposed to non-native salinity for various periods. The study showed that both populations were capable of handling a direct transfer to a new and very different salinity. Survival was high in all treatments and the variation in gene expression was relatively small. This suggests that the capacity for osmoregulation in a wide range of salinity regimes has not been lost in either population. And yet, interestingly, no pronounced changes were observed in the expression profiles of the genes targeted in this study. This suggests that the ability of these fish to reverse the direction of their osmoregulation has not been canalized or lost.

Expression of ATP1A3 was elevated in saltwater

In this study, the ATP1A3 expression was lower in the FF treatment group at all time points, with a mean difference of 0.03 on an exponential scale compared to SS. Further, the SF group did show a weak down-regulation of ATP1A3 compared to SS and the expression levels stabilized around the FF values after approximately 24 hours. Comparing the FF and the FS, the FF had a consistently lower expression of ATP1A3, equally as the SS group.

As the Na^+/K^+-ATPase transporters both secrete and absorb salt in order to obtain a nearly constant internal osmotic concentration when in marine- and freshwater, respectively [7], it might be expected that the long-term expression of the protein stabilize on equal levels. Nevertheless, when a fish experiences a change in external osmolarity, the expression level is expected to shift in order to handle the novel osmotic and ionic stress: it is therefore surprising that we see so little change in the overall expression on the shorter time-scale in this study. Overall, the main finding of ATP1A3 being less expressed in freshwater is in accordance with most research on salmonid fish, where gill Na^+/K^+-ATPase activity is higher in seawater acclimated fish and decreases following migration into freshwater [36,37]; other fish species also show this pattern, including sea bass, *Dicentrarchus labrax* [38] and flounder, *Platichthys flesus* [39]. However, yet other studies have shown that the expression levels for the equivalent ATP isoform in Atlantic salmon (*Salmo salar*) did not change as a result of freshwater exposure [40], or as in killifish,

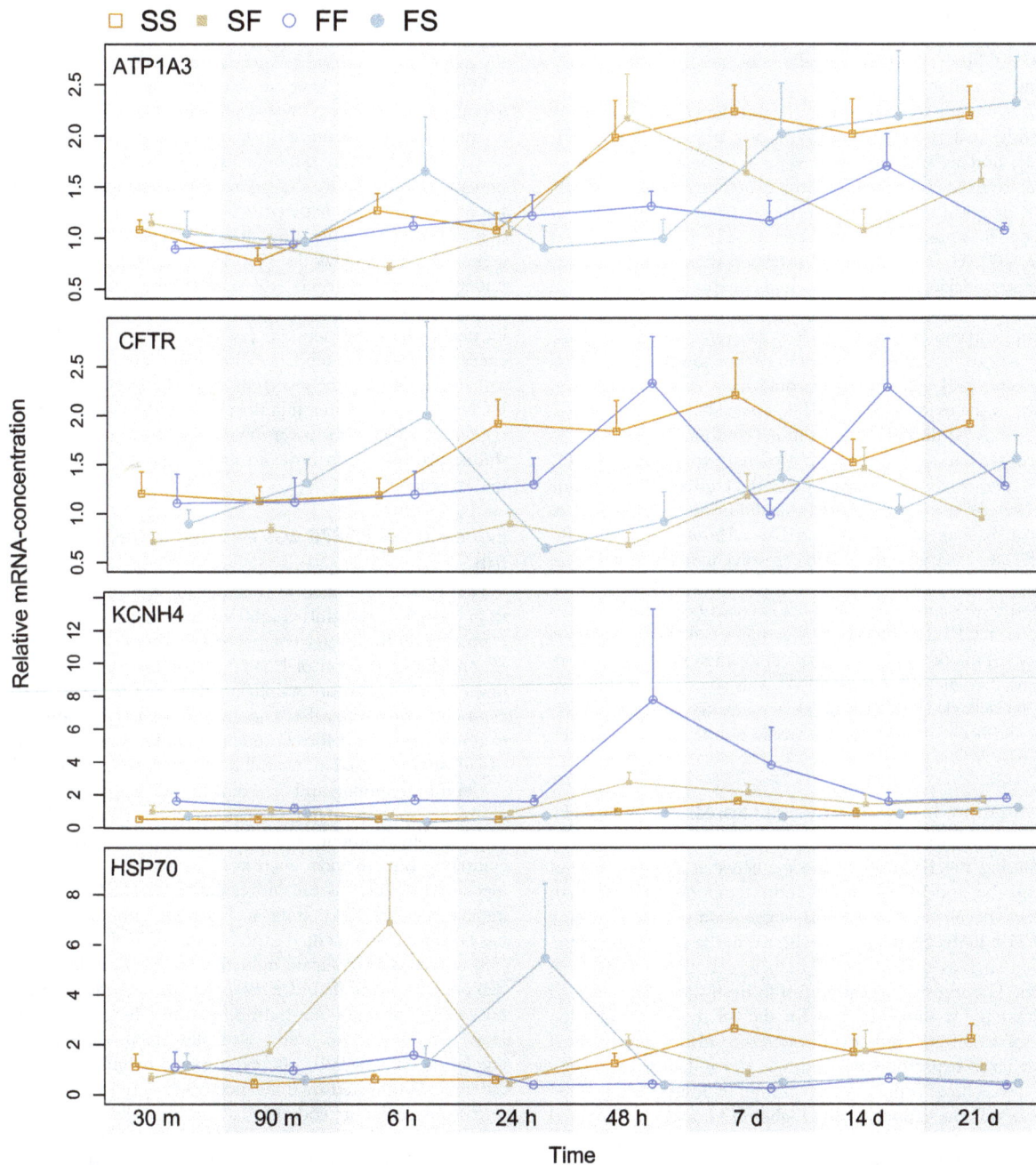

Figure 2. Relative mRNA-expression for the four targeted genes, ATP1A3, CFTR, KCNH4 and HSP70 in saltwater control (SS, dark orange), saltwater fish exposed to freshwater (SF, light orange), freshwater control (FF, dark blue) and freshwater fish exposed to saltwater (FS, light blue) for exposure periods relative to the 5 minute exposure. Values between 0–1 indicate lower expression, and values over 1 indicate higher expression relative to the expression at 5 minutes.

where ATP1A3 was up-regulated in freshwater [21]. Overall, previous reviews on fish osmoregulation state that the role of gill Na^+/K^+-ATPase is uncertain [41], unclear [7], or that the energy required for sodium uptake can be generated only by Na^+/K^+-ATPase [42]. It should however be noted that the cellular localization of the Na^+/K^+-ATPase transporter in this study is not known, and a histological analysis of the gene localization during a salinity challenge could provide additional information on the osmoregulatory function in stickleback.

No major changes in CFTR expression between treatments

Overall, only small changes in CFTR expression were observed across the treatments in the present study. CFTR had a slightly higher expression level in SS compared to FF, but only in the early time-periods (5 min to 24 hours), where the overall expression was reduced in both treatments compared to T = 5 min. After the first day, the expression stabilizes for both groups. Transferring marine fish into freshwater (SF) lead to an overall reduced expression of CFTR compared to the SS treatment, but this trend was not consistent across all time points. The freshwater fish transferred to saltwater (FS) had a higher expression of CFTR compared to FF for the first 6 hours, but was generally stable across all time points, indicating no major change in expression.

CFTR is thought to be central in the ion excretion at the gills of marine fish, as it is involved in the passive transport of chloride ions [24], and as such the expression is expected to decrease following freshwater acclimation. Anticipated increased expression of CFTR in apical membranes in response to transfer to saltwater have been illustrated in Atlantic salmon [43] and eel, *Anguilla anguilla*, [44]. However, contrary to expectations, another study on lab-reared stickleback observed a higher expression of CFTR in freshwater compared to saltwater [19]. Further, when comparing long-term expression levels of CFTR in saltwater exposed Atlantic salmon, the expression levels tended to decline towards the control after 30 days [43]. Although we would have expected a higher difference in the expression of CFTR, especially between the two control groups (SS and FF), there is much confounding evidence of CFTR-expression across taxa [45]; reported expression of CFTR in freshwater spans from not expressed at all in Mosambique tilapia, *Oreochromis mossambicus* [46], diffuse in Killifish [47] to no change in expression in striped bass, *Morone saxatilis* [48], indicating an overall complicated involvement of CFTR in freshwater osmoregulation [47]. An alternative to differential expression of the ion-transporter CFTR could be a redistribution and reuse of CFTR proteins, as Marshall et al. [47] illustrated movement of the protein from an apical location in SW to a more diffuse and basolateral location in FW. This could also be the case for the stickleback in this study, as a rearrangement of CFTR-proteins and hence its activity state would not be picked up by the qPCR analysis.

Much variation, but higher overall expression of KCNH4 in freshwater fish

Comparing the KCNH4-expression between the two control treatments (SS and FF), there was more temporal variation than would have been expected: KCNH4-expression decreased during the first 24 h in the SS treatment, whereas in the FF treatment the expression increased the first 48 h, before both seemed to be stabilized. The transfer of marine fish to freshwater (SF) followed approximately the same curve as for the SS treatment, and the transfer of freshwater fish to saltwater (FS) also demonstrate a down-regulated expression during the first 24 h, before expression increased reaching control levels (5 min exposure) after three weeks. Overall, the expression was higher in FF and FS compared to SS and SF. However, as there was no overall trend in the expression of KCNH4 for any of the groups in this study, this indicates that the gene is involved in other processes than osmoregulation, or that the osmoregulatory function of this transcript is located in other organs than the gill.

KCNH4 is a voltage-gated ion channel protein that is sensitive to voltage changes in the cell membrane and is known to have several functions, including regulation of cell volume, maintaining resting currents and affecting cardiac contractility [49]. Recent genome re-sequencing of 21 individual stickleback from marine and freshwater habitats across their global distribution revealed 81 loci underlying repeated parallel divergence in marine and freshwater ecotypes, including three chromosomal inversions [20]. One of the inversion sites has marine and freshwater specific 3' exons of the KCNH4-gene, indicating parallel ecological selection on the breaking sites of the inversion, and possibly also directly on KCNH4 [20]. Although it is clear that chromosomal rearrangements can contribute to speciation [50,51,52], it is less evident how they do so and which mechanisms are involved. It is

therefore important to identify whether inversion events have led to profound changes in the expression pattern of genes involved in the inversion, in relevant experimental setups.

While we do not have any information on the chromosomal arrangements in these two populations, it is still interesting to quantify the expression of the gene across different environments, as one would expect different expression profiles if the gene is under osmoregulatory selection. Chromosomal inversions are known to alter gene activity, either by causing non-functionalization of the gene, generating alternative splice sites or by altering gene regulatory networks [53,54]. In a study on development under different thermal selection regimes, one population of *Drosophila subobscura* had different expression patterns for loci located within and between inversions, where significant differences in expression tended to be more commonly found inside rather than outside the inversion [55]. The increased expression of KCNH4 in freshwater found in this study indicates a potential regulatory effect of the inversion on chromosome 1, however more studies are needed in order to disentangle the complete effect of the inversion.

Expression of HSP70 was elevated in freshwater adapted fish

HSP70 was identified as a candidate gene for detecting short-term osmotic stressful conditions in stickleback, with higher expression in freshwater compared to saltwater. When comparing SS and FF expression of HSP70, the marine fish had an overall lower expression before stabilizing after the first 24 h. Marine fish exposed to freshwater (SF) had an increased expression compared to SS the first 6 h before normalizing and freshwater fish exposed to saltwater (FS) had an overall lower expression compared to FF.

Capture, handling and crowding are all factors that can initiate a stress response in fish, as can short-term fluctuations in the physical environment. Physiological responses to stressors are complex, but include increased activity of cellular defense mechanisms, such as the up-regulation of HSP-genes [56]. The involvement of HSP70 in the acclimation of fish to salinity changes has been well documented experimentally [57,58,59]. Larsen et al. [59] observed a significant induction of HSP70 in kidney tissue of two populations of flounder, *Platichtys flesus*, when introduced to non-native salinities for both short- and long term exposures. However, the same study also illustrated tissue-specific up-regulation of HSP70, as expression in gill and liver was differentially affected by differences in salinity [59]. Other similar studies on Atlantic cod, *Gadus morhua*, illustrated expression differentiations in both gills and kidney after salinity transplantations during the first 24 hours [60], similar to the result in this study.

Conclusions and perspectives

In our study system, sticklebacks from the marine population are genetically (Østbye et al., unpublished data) and morphologically [61] differentiated from the freshwater population living in the river below the waterfall. However, as the populations are genetically differentiated, a surprising result of this study is how little variation in gene expression was observed when the fish were directly transferred to the contrasting salinity treatments, and additionally, how little it differed between the two populations in their native salinities.

Movement between environments of different salinity is physiologically costly [62], and resident populations experiencing different salinity levels are predicted to be locally adapted to their native habitat as traits promoting euryhalinity are expected to be rapidly lost if they are not under selection [63]. It is likely that adaption to the local environment takes time, but the freshwater fish in this study have been separated from the marine populations for more than 7000 years (between 3500 to 7000 generations assuming a two year or one year life cycle), likely sufficient time for local adaptation given reasonable selection [64,65]. Further, population genetic studies on stickleback have recognized salinity to be a major factor in the distribution of genotypes in systems that exchange migrants [66,67], also across high gene-flow environments such as in the Baltic ocean [68], indicating that adaption to salinity is under selection.

Based on survival alone, the results from this study suggest that over the >3500 generations of adaptions to freshwater environments, the osmoregulatory physiology of landlocked stickleback has not been significantly canalized or experienced strong selection as they have retained their capacity for osmoregulation in saltwater. Additionally the locally adapted marine stickleback can osmoregulate in freshwater, despite originating as a marine species. This indicates that for stickleback, the cost of retaining osmoregulatory plasticity is small, or that the traits promoting euryhalinity in stickleback are under strong selection or pleitropically linked to other traits under strong selection. However, for fish living in marine and freshwater environments, the selection pressure for osmoregulation is still opposite, indicating that the stickleback must have alternative cell-regulating mechanisms for survival in unfamiliar salinities. In this study we only targeted gene expression values by quantifying the amount of mRNA extracted from gill tissue, however, the amount of mRNA does not necessarily imply equal concentrations of the functional proteins as a number of mechanisms can limit or increase the production. Protein expression profiles, also from kidney tissue, could have revealed different results. Other alternative strategies the stickleback may be utilizing could include changes in activity state of the ion-transport proteins by movement within the cell (activation/down-regulation), or by re-using the proteins by reversing the orientation in the cell-membrane, or by modulations following the interaction of other proteins [69,70,71].

Whatever method it is that the stickleback seems to be employing to osmoregulate so effectively, it has created a species that is incredibly well able to colonize new habitats regardless of the salinity they find themselves in, which has been a huge asset to this species in its spread throughout the northern hemisphere. Additional studies targeting the exact genetic and physiological mechanisms for the wide salinity tolerance in marine and freshwater stickleback are needed to understand the stickleback's incredible capacity for ion secretion and absorption. Their ability to adapt immediately to the environmental demands, with no apparent increase in physiological stress, is as unusual as it is intriguing, especially as this has been so evolutionarily important for this widespread species.

Acknowledgments

We thank Anders Herland and Haaken Hveding Christensen for assistance with fish maintenance, Monica Solbakken, Martin Malmstrøm, Nanna Winger Steen, Ave Tooming-Klunderud and Mari Espelund for help in the lab, Kjetill S. Jakobsen, Alexander J. Nederbragt and Bastian Staar for interesting discussions and ideas and Anna Mazzarella for constructive comments to the manuscript.

Author Contributions

Conceived and designed the experiments: AT LAV KØ. Performed the experiments: AT TA KØ. Analyzed the data: AT TA LAV. Wrote the paper: AT LAV.

References

1. West-Eberhard MJ (2005) Phenotypic accommodation: adaptive innovation due to developmental plasticity. Journal of Experimental Zoology Part B-Molecular and Developmental Evolution 304B: 610–618.
2. DeWitt TJ, Sih A, Wilson DS (1998) Costs and limits of phenotypic plasticity. Trends in Ecology & Evolution 13: 77–81.
3. Masel J, King OD, Maughan H (2007) The loss of adaptive plasticity during long periods of environmental stasis. American Naturalist 169: 38–46.
4. Lande R (2009) Adaptation to an extraordinary environment by evolution of phenotypic plasticity and genetic assimilation. Journal of Evolutionary Biology 22: 1435–1446.
5. Schultz ET, McCormick SD (2013) Euryhalinity in an evolutionary context. In: McCormick SD, Farrell AP, Brauner CJ, editors. Fish physiology: Euryhaline fishes. Oxford: Elsevier Science. pp. 477–529.
6. Edwards SL, Marshall WS (2012) Principles and patterns of osmoregulation and euryhalinity in fishes. In: Stephen D. McCormick APF, Colin JB, editors. Fish Physiology: Academic Press. pp. 1–44.
7. Evans DH, Piermarini PM, Choe KP (2005) The multifunctional fish gill: dominant site of gas exchange, osmoregulation, acid-base regulation, and excretion of nitrogenous waste. Physiological Reviews 85: 97–177.
8. Varsamos S, Nebel C, Charmantier G (2005) Ontogeny of osmoregulation in postembryonic fish: A review. Comparative Biochemistry and Physiology a-Molecular & Integrative Physiology 141: 401–429.
9. Griffith RW (1974) Environment and salinity tolerance in the genus *Fundulus*. Copeia 1974: 319–331.
10. Heuts MJ (1947) Experimental studies on adaptive evolution in *Gaserosteus-aculeatus L*. Evolution 1: 89–102.
11. Wootton RJ (1976) The biology of the sticklebacks. New York: Academic Press.
12. Bell MA (1977) Late Miocene marine Threespine stickleback, *Gasterosteus aculeatus*, and its zoogeographic and evolutionary significance. Copeia: 277–282.
13. Bell MA, Foster SA (1994) The evolutionary biology of the threespine stickleback. New York: Oxford University Press.

14. Grøtan K, Østbye K, Taugbøl A, Vøllestad LA (2012) No short-term effect of salinity on oxygen consumption in threespine stickleback (*Gasterosteus aculeatus*) from fresh, brackish, and salt water. Canadian Journal of Zoology 90: 1386–1393.
15. Raeymaekers JAM, Maes GE, Audenaert E, Volckaert FAM (2005) Detecting Holocene divergence in the anadromous-freshwater three-spined stickleback (*Gasterosteus aculeatus*) system. Molecular Ecology 14: 1001–1014.
16. Von Hippel FA, Weigner H (2004) Sympatric anadromous-resident pairs of threespine stickleback species in young lakes and streams at Bering Glacier, Alaska. Behaviour 141: 1441–1464.
17. McPhail JD (1994) Speciation and the evolution of reproductive isolation in the sticklebacks (*Gasterosteus*) of south-western British Columbia. In: Bell AM, Foster JR, editors. The evolutionary biology of the threespine stickleback. Oxford: Oxford University Press. pp. 399–437.
18. Møller JJ (2003) Relative sea level change in Fennoscandia. Net version 3.00. University of Tromsø: Department of Geology, TMU.
19. McCairns RJS, Bernatchez L (2010) Adaptive divergence between freshwater and marine sticklebacks: Insights into the role of phenotypic plasticity from an intergrated analysis of candidate gene expression. Evolution 64: 1029–1047.
20. Jones FC, Grabherr MG, Chan YF, Russell P, Mauceli E, et al. (2012) The genomic basis of adaptive evolution in threespine sticklebacks. Nature 484: 55–61.
21. Whitehead A, Roach JL, Zhang SJ, Galvez F (2012) Salinity- and population-dependent genome regulatory response during osmotic acclimation in the killifish (*Fundulus heteroclitus*) gill. Journal of Experimental Biology 215: 1293–1305.
22. Mobasheri A, Avila J, Cozar-Castellano I, Brownleader MD, Trevan M, et al. (2000) Na+, K+-ATPase isozyme diversity; Comparative biochemistry and physiological implications of novel functional interactions. Bioscience Reports 20: 51–91.

23. Bonting SL (1970) Sodium-pottassium activated adenosine triphosphatase and cation transport. In: Bittar EE, editor. Membranes and ion transport. New York: John Wiley & Sons. pp. 257–363.

24. Silva P, Solomon R, Spokes K, Epstein FH (1977) Ouabain inhibition of gill Na-K ATPase: relationship to active chloride transport. Journal of Experimental Zoology 199: 419–426.

25. McCormick SD, Sundell K, Bjornsson BT, Brown CL, Hiroi J (2003) Influence of salinity on the localization of Na+/K+-ATPase, Na+/K+/2Cl(−)cotransporter (NKCC) and CFTR anion channel in chloride cells of the Hawaiian goby (Stenogobius hawaiiensis). Journal of Experimental Biology 206: 4575–4583.

26. Scott GR, Richards JG, Forbush B, Isenring P, Schulte PM (2004) Changes in gene expression in gills of the euryhaline killifish Fundulus heteroclitus after abrupt salinity transfer. American Journal of Physiology-Cell Physiology 287: C300–C309.

27. Hoekstra HE (2006) Genetics, development and evolution of adaptive pigmentation in vertebrates. Heredity 97: 222–234.

28. Chan YF, Marks ME, Jones FC, Villarreal G, Shapiro MD, et al. (2010) Adaptive evolution of pelvic reduction in Sticklebacks by recurrent deletion of a Pitx1 enhancer. Science 327: 302–305.

29. Rosenblum EB, Roempler H, Schoeneberg T, Hoekstra HE (2010) Molecular and functional basis of phenotypic convergence in white lizards at White Sands. Proceedings of the National Academy of Sciences of the United States of America 107: 2113–2117.

30. Sørensen J, Kristensen T, Loeschcke V (2003) The evolutionary and ecological role of heat shock proteins. Ecology Letters 6: 1025–1037.

31. Sanogo YO, Hankison S, Band M, Obregon A, Bell AM (2011) Brain transcriptomic response of threespine sticklebacks to cues of a predator. Brain Behavior and Evolution 77: 270–285.

32. Knag AC, Taugbøl A (2013) Acute exposure to offshore produced water has an effect on stress- and secondary stress responses in three-spined stickleback Gasterosteus aculeatus. Comparative Biochemistry and Physiology C-Toxicology & Pharmacology 158: 173–180.

33. Krogh A (1937) Osmotic regulation in fresh water fishes by active absorption of chloride ions. Zeitschrift vergleichende Physiologie 24: 656–666.

34. Evans DH (2008) Teleost fish osmoregulation: what have we learned since August Krogh, Homer Smith, and Ancel Keys. American Journal of Physiology-Regulatory Integrative and Comparative Physiology 295: R704–R713.

35. Livak KJ, Schmittgen TD (2001) Analysis of relative gene expression data using real-time quantitative PCR and the 2(T)(-Delta Delta C) method. Methods 25: 402–408.

36. Bystriansky JS, Schulte PM (2011) Changes in gill H+-ATPase and Na+/K+-ATPase expression and activity during freshwater acclimation of Atlantic salmon (Salmo salar). Journal of Experimental Biology 214: 2435–2442.

37. Shrimpton JM, Patterson DA, Richards JG, Cooke SJ, Schulte PM, et al. (2005) Ionoregulatory changes in different populations of maturing sockeye salmon Oncorhynchus nerka during ocean and river migration. Journal of Experimental Biology 208: 4069–4078.

38. Jensen MK, Madsen SS, Kristiansen K (1998) Osmoregulation and salinity effects on the expression and activity of Na+, K+-ATPase in the gills of European sea bass, Dicentrarchus labrax (L.). Journal of Experimental Zoology 282: 290–300.

39. Stagg RM, Shuttleworth TJ (1982) Na+, K+ ATPase, quabain binding and quabain-sensitive oxygen consumption on gills from Platichthys flesus adapted to seawater and freshwater. Journal of comparative physiology 147: 93–99.

40. Folmar LC, Dickhoff WW (1980) The parr-smolt transformation (smoltification) and seawater adaptation in salmonids: A review of selected literature. Aquaculture 21: 1–37.

41. Perry SF (1997) The chloride cell: Structure and function in the gills of freshwater fishes. Annual Review of Physiology 59: 325–347.

42. Kirschner LB (2004) The mechanism of sodium chloride uptake in hyperregulating aquatic animals. Journal of Experimental Biology 207: 1439–1452.

43. Singer TD, Clements KM, Semple JW, Schulte PM, Bystriansky JS, et al. (2002) Seawater tolerance and gene expression in two strains of Atlantic salmon smolts. Canadian Journal of Fisheries and Aquatic Sciences 59: 125–135.

44. Wilson JM, Antunes JC, Bouca PD, Coimbra J (2004) Osmoregulatory plasticity of the glass eel of Anguilla anguilla: freshwater entry and changes in branchial ion-transport protein expression. Canadian Journal of Fisheries and Aquatic Sciences 61: 432–442.

45. Havird JC, Henry RP, Wilson AE (2013) Altered expression of Na+/K+-ATPase and other osmoregulatory genes in the gills of euryhaline animals in response to salinity transfer: A meta-analysis of 59 quantitative PCR studies over 10 years. Comparative Biochemistry and Physiology D-Genomics & Proteomics 8: 131–140.

46. Hiroi J, McCormick SD, Ohtani-Kaneko R, Kaneko T (2005) Functional classification of mitochondrion-rich cells in euryhaline Mozambique tilapia (Oreochromis mossambicus) embryos, by means of triple immunofluorescence staining for Na/K+-ATPase, Na+/K+/2Cl(−) cotransporter and CFTR anion channel. Journal of Experimental Biology 208: 2023–2036.

47. Marshall WS, Lynch EA, Cozzi RRF (2002) Redistribution of immunofluorescence of CFTR anion channel and NKCC cotransporter in chloride cells during adaptation of the killifish Fundulus heteroclitus to sea water. Journal of Experimental Biology 205: 1265–1273.

48. Madsen SS, Jensen LN, Tipsmark CK, Kiilerich P, Borski RJ (2007) Differential regulation of cystic fibrosis transmembrane conductance regulator and Na+, K+-ATPase in gills of striped bass, Morone saxatilis: effect of salinity and hormones. Journal of Endocrinology 192: 249–260.

49. Gutman GA, Chandy KG, Grissmer S, Lazdunski M, McKinnon D, et al. (2005) International union of pharmacology. LIII. Nomenclature and molecular relationships of voltage-gated potassium channels. Pharmacological Reviews 57: 473–508.

50. Ellegren H, Smeds L, Burri R, Olason PI, Backstrom N, et al. (2012) The genomic landscape of species divergence in Ficedula flycatchers. Nature 491: 756–760.

51. Rieseberg LH (2001) Chromosomal rearrangements and speciation. Trends in Ecology & Evolution 16: 351–358.

52. Noor MAF, Grams KL, Bertucci LA, Reiland J (2001) Chromosomal inversions and the reproductive isolation of species. Proceedings of the National Academy of Sciences of the United States of America 98: 12084–12088.

53. Kirkpatrick M, Barton N (2006) Chromosome inversions, local adaptation and speciation. Genetics 173: 419–434.

54. Matzkin LM, Merritt TJS, Zhu CT, Eanes WF (2005) The structure and population genetics of the breakpoints associated with the cosmopolitan chromosomal inversion In(3R)Payne in Drosophila melanogaster. Genetics 170: 1143–1152.

55. Laayouni H, Garcia-Franco F, Chavez-Sandoval BE, Trotta V, Beltran S, et al. (2007) Thermal evolution of gene expression profiles in Drosophila subobscura. Bmc Evolutionary Biology 7.

56. Moseley P (2000) Stress proteins and the immune response. Immunopharmacology 48: 299–302.

57. Deane E, Kelly S, Luk J, Woo N (2002) Chronic salinity adaptation modulates hepatic heat shock protein and insulin-like growth factor I expression in black sea bream. Marine Biotechnology 4: 193–205.

58. Fangue NA, Hofmeister M, Schulte PM (2006) Intraspecific variation in thermal tolerance and heat shock protein gene expression in common killifish, Fundulus heteroclitus. Journal of Experimental Biology 209: 2859–2872.

59. Larsen PF, Nielsen EE, Williams TD, Loeschcke V (2008) Intraspecific variation in expression of candidate genes for osmoregulation, heme biosynthesis and stress resistance suggests local adaptation in European flounder (Platichthys flesus). Heredity 101: 247–259.

60. Larsen PF, Nielsen EE, Meier K, Olsvik PA, Hansen MM, et al. (2012) Differences in salinity tolerance and gene expression between two populations of Atlantic Cod (Gadus morhua) in response to salinity stress. Biochemical Genetics 50: 454–466.

61. Bjærke O, Østbye K, Lampe HM, Vøllestad LA (2010) Covariation in shape and foraging behaviour in lateral plate morphs in the three-spined stickleback. Ecology of Freshwater Fish 19: 249–256.

62. Moyle PB, Cech JJ (1996) Fishes. An introduction to ichthyology. 3rd edition. Upper Saddle River, N.J.: Prentice Hall.

63. Schultz ET, McCormick SD (2012) Euryhalinity in an evolutionary context. In: Stephen D. McCormick APF, Colin JB, editors. Fish physiology: Academic Press. pp. 477–533.

64. Kinnison MT, Hendry AP (2001) The pace of modern life II: from rates of contemporary microevolution to pattern and process. Genetica 112: 145–164.

65. Hendry AP, Kinnison MT (1999) Perspective: The pace of modern life: Measuring rates of contemporary microevolution. Evolution 53: 1637–1653.

66. McCairns RJS, Bernatchez L (2008) Landscape genetic analyses reveal cryptic population structure and putative selection gradients in a large-scale estuarine environment. Molecular Ecology 17: 3901–3916.

67. Taugbøl A, Junge C, Quinn TP, Herland A, Vøllestad LA (2014) Genetic and morphometric divergence in threespine stickleback in the Chignik catchment, Alaska. Ecology and Evolution 4: 144–156.

68. DeFaveri J, Jonsson PR, Merilä J (2013) Heterogeneous genomic differentiation in marine threespine stickleback: Adaptation along an environmental gradient. Evolution 67: 2530–2546.

69. Pertl H, Pockl M, Blaschke C, Obermeyer G (2010) Osmoregulation in Lilium pollen grains occurs via activation of the plasma membrane H+ ATPase activity by 14-3-3 proteins. Plant Physiology 154: 1921–1928.

70. Szczesnaskorupa E, Browne N, Mead D, Kemper B (1988) Positive charges at the NH2 terminus convert the membrane-anchor signal peptide of cytochrome P-450 to a secretory signal peptide. Proceedings of the National Academy of Sciences of the United States of America 85: 738–742.

71. Hartmann E, Rapoport TA, Lodish HF (1989) Predicting the orientation of eukaryotic membrane-spanning proteins. Proceedings of the National Academy of Sciences of the United States of America 86: 5786–5790.

4

Comparison of Mercury Contamination in Live and Dead Dolphins from a Newly Described Species, *Tursiops australis*

Alissa Monk[1,2,3]*, **Kate Charlton-Robb**[2,4], **Saman Buddhadasa**[5], **Ross M. Thompson**[1,2]

1 Institute for Applied Ecology, University of Canberra, Bruce, Australian Capital Territory, Australia, 2 School of Biological Sciences, Monash University, Clayton, Victoria, Australia, 3 Dolphin Research Institute, Hastings, Victoria, Australia, 4 Australian Marine Mammal Conservation Foundation, Hampton East, Victoria, Australia, 5 National Measurement Institute, Commonwealth Government, Port Melbourne, Victoria, Australia

Abstract

Globally it is estimated that up to 37% of all marine mammals are at a risk of extinction, due in particular to human impacts, including coastal pollution. Dolphins are known to be at risk from anthropogenic contaminants due to their longevity and high trophic position. While it is known that beach-cast animals are often high in contaminants, it has not been possible to determine whether levels may also be high in live animals from the same populations. In this paper we quantitatively assess mercury contamination in the two main populations of a newly described dolphin species from south eastern Australia, *Tursiops australis*. This species appear to be limited to coastal waters in close proximity to a major urban centre, and as such is likely to be vulnerable to anthropogenic pollution. For the first time, we were able to compare blubber mercury concentrations from biopsy samples of live individuals and necropsies of beach-cast animals and show that beach-cast animals were highly contaminated with mercury, at almost three times the levels found in live animals. Levels in live animals were also high, and are attributable to chronic low dose exposure to mercury from the dolphin's diet. Measurable levels of mercury were found in a number of important prey fish species. This illustrates the potential for low dose toxins in the environment to pass through marine food webs and potentially contribute to marine mammal deaths. This study demonstrates the potential use of blubber from biopsy samples to make inferences about the health of dolphins exposed to mercury.

Editor: Christopher D. Marshall, Texas A&M University, United States of America

Funding: Funded by contributions from the West Gippsland Catchment Management Authority (http://www.wgcma.vic.gov.au/), Gippsland Lakes Task Force and Coast Action/Coastcare (http://www.coastcare.com.au/). The funders had no role in study design, data collection and analysis, decision to publish, or preparation of the manuscript. Ross Thompson was funded in part by an ARC Future Fellowship (FT110100957).

Competing Interests: The authors have declared that no competing interests exist.

* Email: alissa.monk@canberra.edu.au

Introduction

Marine ecosystems are increasingly subject to a wide range of pressures, including over-exploitation, acidification, climate change, invasion and pollution [1]. Marine organisms in many areas are subject to high extinction risk as a result of human activities. Up to a third of coral reef fishes in northern Australia are considered vulnerable to extinction [2], and widespread extinctions are being observed across a range of marine taxa [1]. Marine mammals are considered especially vulnerable, and the International Union for the Conservation of Nature Red Book lists 25% of marine mammal species as threatened [3]. Recently a major analysis of the threatening processes for marine mammals identified those occurring in productive coastal waters as being particularly at risk, but also noted the lack of data on a number of rare species globally [4].

Coastal marine areas which are adjacent to major human population centres are subject to major threats as a result of human activities, including anthropogenic inputs of pollutants, particularly heavy metals [4,5,6]. The relative longevity of marine mammals, coupled with their high trophic position in food chains, results in potential for bioaccumulation of contaminants [7,8].

One contaminant of particular concern is mercury. Mercury is highly toxic and has detrimental health effects in mammals including neurological disorders, immunosuppression and reproductive disorders that can all lead to death [9,10]. Mercury is a naturally occurring element however centuries of human activities, including mining and coal burning, have led to increases in levels where it has become a major concern to both human and environmental health. In marine systems, mercury is retained in sediment where it can be taken up into the food web and biomagnified to high concentrations in the upper trophic levels and bioaccumulates in higher trophic organisms such as dolphins [11]. Mercury contamination in coastal waters therefore represents a potential major health risk to marine mammal populations.

Analysis of the factors contributing to marine mammal deaths is particularly problematic. Contaminant analysis of beach-cast dead animals at a number of locations internationally has found high levels of heavy metals in a range of taxa, including beluga whales [12], dolphins [13] and seals [14] (Table 1). However directly implicating anthropogenic contaminants in the deaths of marine mammals is difficult, because it is usually not possible to determine whether levels in dead animals are any higher than those in apparently healthy members of the population. It has been difficult

Table 1. Comparison of concentrations of mean total mercury (mg/kg wet weight) found in beach-cast dead marine mammal populations worldwide.

Species	Area	Mercury		References
		Blubber	Liver	
Tursiops australis	Victoria, Australia	3.64	420.00	This Study
Tursiops aduncus	South Australia	-	475.78	[34]
Tursiops truncatus	Eastern Australia	-	16.36	[10]
Delphinus delphis	New Zealand	-	71.00	[8]
Tursiops truncatus	Israel coast, Mediterranean	1.50	97.00	[35]
Stenella coeruleoalba	Israel coast, Mediterranean	1.60	181.00	[35]
Stenella coeruleoalba	French coast, Mediterranean	0.86	217.73	[26] #
Stenella coeruleoalba	Apulian coast, Mediterranean	1.38	189.16	[36]
Sousa chinensis	Hong Kong	-	42.94	[9] #
Phocoena phocoena	England and Wales	-	16.15	[23]
Stenella longirostris	Gulf of California, USA	-	21.32	[37] #
Tursiops truncatus	South Carolina Coast, USA	-	17.8	[7]

#converted to wet weight using conversion factor of 1:3 for liver and 1:2 for blubber.

therefore to assess the potential for heavy metals in the environment to directly impact on marine mammal mortality. This has contributed to the more general lack of data on the endangering processes which are impacting marine mammals [4].

In 2011 a new species of dolphin, the Burrunan dolphin (*Tursiops australis*), was described from coastal waters of south-eastern Australia using a combination of genetic and morphological traits [15]. More recent phylogenetic analyses, using whole mitochondrial genome sequencing, further validates *T. australis* as a new species and a sister group to all other *Tursiops* lineages [16]. The species appears to be restricted to inshore waters of southern Australia (Victoria, Tasmania and South Australia [15,17,18,19]. Only two small resident populations are known, from Port Phillip Bay and the Gippsland Lakes (approximately 100 and 50 individuals respectively. Both populations live in shallow semi-enclosed coastal water bodies which receive inflows from watersheds with extensive urban and agricultural development. In the case of the larger population (Port Phillip Bay) the animals live adjacent to Melbourne, a city of 4.5 million people. This has led to concern that the species may be subject to contamination from coastal pollution. This risk is intensified by evidence from population studies which have shown that population sizes of this species are small relative to other bottlenose populations in Australia and from around the world [15]. This species has also been proposed as a putative ancestral node for *Tursiops* diversification in this region [16] and as such is a highly significant species.

This study compares contaminant levels in beach cast and live *Tursiops australis* individuals to determine whether there is potential for contaminants to be contributing to dolphin mortality.

Materials and Methods

Ethics Statement

Collection of samples was conducted under the Wildlife Act 1975 Research Permit #10003250, issued by Victorian Department of Environment and Primary Industries (DEPI; Victorian State Government) and was approved by the Biological Sciences Animal Ethics Committee (Monash University) BSCI/2008/21.

We were able to assay contaminant levels (arsenic, lead, total mercury, selenium and summed polychlorinated biphenyls (PCBs)) in tissue from both beach-cast ('dead') and live *T. australis*. Ten blubber and six liver samples were obtained on necropsy from *T. australis* individuals found dead on beaches between 2004 and 2009. Tissue samples were collected during necropsies from fresh dead (code 2) or early moderate decomposition (code 3) carcasses (Table 2). Skin and blubber samples were taken from twenty live individuals (14 male and 6 female) using a dart biopsy approach [20] in 2007. Samples were stored at −20°C until analysed. The age of five of the dead animals was previously determined [19](Table 2). For live animals we limited our sampling to mature individuals to eliminate age class as a variable. Mature individuals were defined as being of a length of approximately 2.5 m.In addition, samples were taken from fish species in the region which are known to form part of the dolphins' diet [19], in order to determine likely pathways for mercury ingestion. Mercury levels were measured in the main food web compartments through sampling of 5–10 g of muscle tissue from fish. Fish were sourced from local fishermen who operate in the same areas as the dolphins feed.

Sample preparation and subsequent toxicological analysis of both dolphin and fish tissue was carried out at the National Measurement Institute (NMI) (Melbourne, Victoria; Commonwealth Government). Contaminant levels were determined on fresh tissue using a standard Nitric Acid digestion and detection with both ICP-MS and ICP-OES. Certified Reference Materials were analysed together with the samples for quality assurance. To validate whether using biopsy sized blubber portions (weight range 2–3 g) was comparable to larger samples (50–100 g) taken from necropsied animals, biopsy-sized sub-samples were taken from the larger samples and mercury results compared between the two methods (all other contaminants occurred at very low levels, Table 3). No significant differences were found ($F_{(1,14)} = 0.17$, $p = 0.69$). We also tested whether there was any difference in mercury levels based on gender and there were no significant sex differences in mercury concentration for either live ($F_{(1,18)} = 0.62$, $p = 0.40$) or beach cast dolphins ($F_{(1,8)} = 0.32$, $p = 0.81$), therefore sex was pooled for all further analyses. We investigated whether

Table 2. Biological characteristics of beach cast individuals of *Tursiops australis* from Victoria, Australia.

Location	Date collected	Sex	Age (yrs)	Length (m)	Decomposition code	Blubber layer at dorsal fin (cm)
Altona	02/10/2004	F	-	-	3	.5
Geelong	19/09/2005	F	-	2.62	2	2.5
Port Fairy	27/10/2006	M	8	2.66	3	1.5
San Remo	23/04/2007	M	-	2.27	3	1.5
Poddy Bay	25/10/2007	M	11	2.7	3	-
Mitchell River	01/11/2007	F	20	2.78	3	1.8
Paynesville	04/11/2007	M	-	2.73	2	1.5
Beaumaris	21/01/2008	M	21	2.55	3	1.2
Point Henry	23/01/2008	M	13	2.36	2	1.4
Clifton Springs	14/11/2008	M	-	2.20	2	1.4

there was a temporal effect on mercury levels in beach cast dolphins and no effect was found ($F_{(1,8)} = 0.107$, p = 0.752).

As most published studies have reported mercury levels in liver tissue, which we could not sample in live dolphins, we combined data from this study and data from previous published studies (see studies with both blubber and liver mercury concentrations Table 1) to form a regression between liver and blubber levels, (n = 30 R^2 = 0.59, p<0.001, ln(liver level) = 1.1107 × ln(blubber level)+4.5851) and converted values from live dolphin blubber samples to liver concentrations to facilitate comparison with previously published studies.

Results and Discussion

Arsenic, lead, selenium and PCBs were detected but were lower than levels known to cause health effects [21] (Table 3). We acknowledge that it is likely a combination of contaminants that burden the animals, however given that no other contaminant tested is at high levels it is likely that mercury is contributing the majority of the toxicological burden on these animals.

The total mercury in blubber of dead adult *T. australis* from coastal Victoria was 2.7 times higher on average than the values for the live animals (Figure 1) (one factor ANOVA for a comparison between live and dead animals; $F_{(1,28)} = 36.04$, p< 0.001). For the beach-cast animals, liver values ranged from 100 to 840 mg/kg, while in live animals the range was estimated to be between 28 and 483 mg/kg. Placed in an international context, *T. australis* has a higher average concentration of mercury than has been reported for small cetaceans in East Australia, New Zealand, America, England, Hong Kong and the Mediterranean (Table 1).

There is little information on actual physiological tolerances to contaminants in free-living animals, however it has been noted that there is a limit to the concentration of mercury that an animal can tolerate. For mammalian liver tissues, this limit appears to be within the range of 100–400 mg/kg, wet weight, of liver tissue [21]. In study by Ronald *et al.*, [22], seals were fed different daily doses of mercury. Compared to the control group, the low dose group showed a decline in appetite and body weight and a reduction in activity after approximately 60 days. The high dose group had to be force fed after 4 days, became lethargic, suffered from continuous weight loss and had died by day 26. This shows that even at lower doses, mercury is having behavioural effects and at high doses it can cause death. The average total mercury concentration in the liver of the high dose animals after death was 138 mg/kg. Nearly all dead dolphins from this study had higher

Table 3. Concentrations of contaminants (mg/kg wet weight) in blubber from individuals of *Tursiops australis* from Victoria, Australia.

Contaminant	Dead(n = 10)	Live (n = 20)
Arsenic	0.39±0.07	0.23±0.02
	(0.13–0.80)	(<0.10–0.38)
Lead	0.061±0.011	2.91±0.615
	(0.05–0.14)	(0.76–12)
Total Mercury	3.64±0.68	1.32±0.20
	(1.40–7.20)	(0.32–4.20)
Selenium	2.88±0.74	1.69±0.2
	(0.80–6.50)	(0.52–3.9)
∑PCB	3275.54±9.46.28	-
	(258.80–8055.3)	

Values shown are means with standard errors and range in brackets underneath.

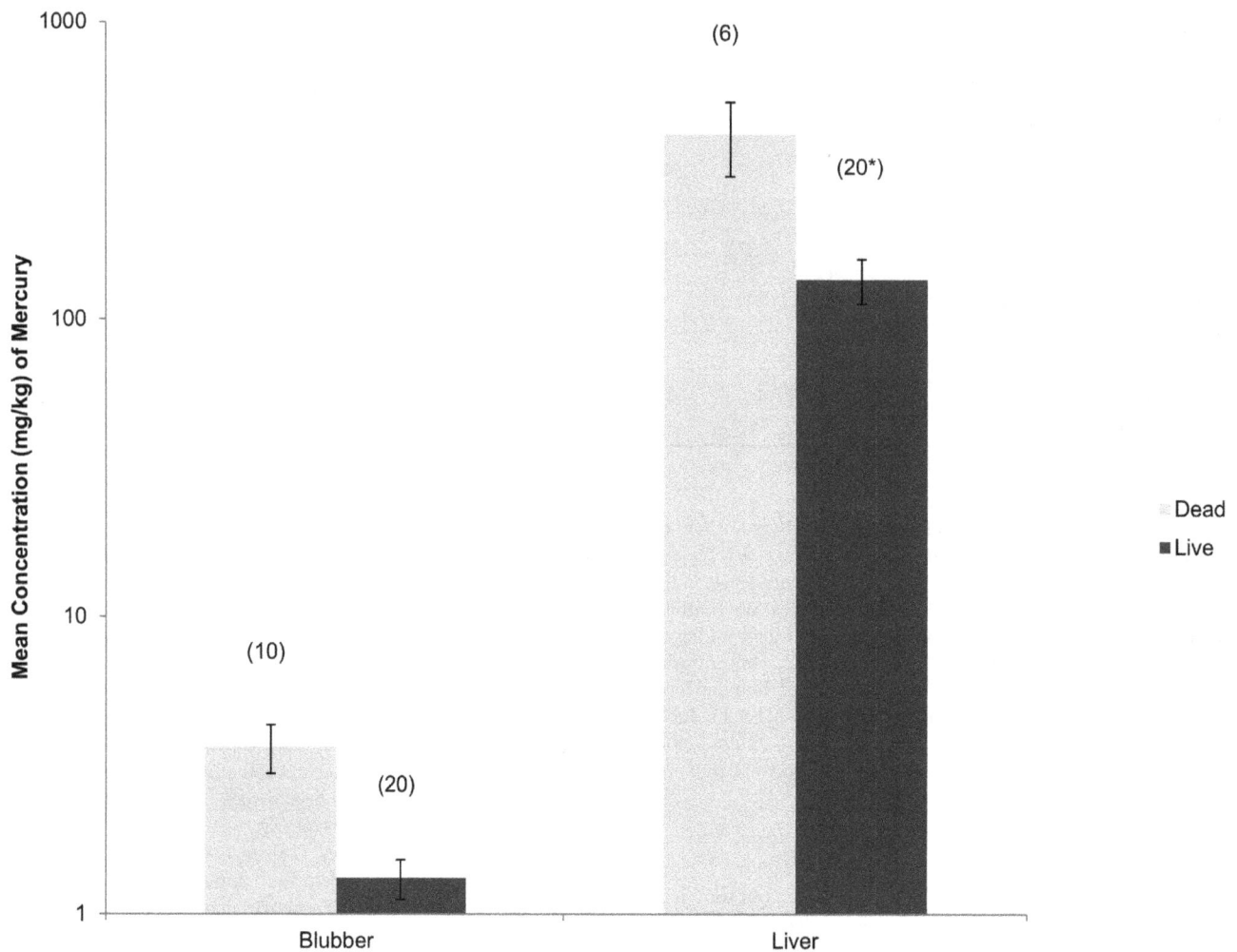

Figure 1. Concentration of total mercury (mg/kg wet weight) in blubber and liver from live and dead individuals of *Tursiops australis* from Victoria, Australia. Values shown are means with standard errors, sample sizes are shown in brackets above each bar. Values marked * were estimated from a regression between liver and blubber levels from worldwide levels.

mercury levels in their livers than the dead seals from the Ronald *et al.*, [22] study. It is also important to note that some of the predicted liver concentrations of live (assumed healthy) dolphins from this study also exceeded this concentration, emphasising the importance of knowing the mercury concentrations in the live populations before making any conclusions about the possible effects in dead animals. Bennett *et al.*, [23] found that harbor porpoises that died from infectious disease had significantly higher mercury levels than healthy animals that died from physical trauma, suggesting that mercury causes immunosuppression at levels much lower than found in the current study (Table 1). Full histopathology studies were not carried out in this study and therefore a relationship between the presence of infectious disease and mercury concentrations could not be determined. Studies into the neurological effects of mercury in seals [24] and polar bears [25] have found that low levels of mercury affects neurochemical pathways that have essential roles in multiple aspects of animal health, behaviour, reproduction, and survival. Based on these studies and given the high concentration of mercury found in *T. australis*, it is highly likely that mercury is affecting animal health, potentially through immunosuppression and influences on neurochemical pathways.

Ratios of mercury: selenium in liver tissue are indicative of toxicological stress in mammals as physiological processes act to bind methyl mercury and store it as the insoluble compound tiemmannite (HgSe) [26,27]. This detoxification process results in a 1:1 molar ratio of mercury and selenium. An average ratio of 1:1.06±0.03 was found in the beach-cast Victorian dolphins, suggesting these dolphins are under toxicological stress from mercury.

This study represents a significant advance by being able to compare beach-cast to live animals within the same population, potentially implicating mercury in the morbidity of animals within a region. The ability to biopsy live animals and compare blubber mercury concentrations to those from beach-cast individuals has considerable potential in marine mammal studies. Whilst a previous study by Stavros *et al.*, [28] has found a slightly stronger relationship between mercury levels found in liver and skin, and used the method to make inferences about the health of free ranging animals, there are some limitations to using their method. Their study involved capture and release of animals and as such they are able to obtain larger samples of skin (approximately 0.5 g dry weight) than is able to be collected from the dart biopsy approach (0.03 g wet weight). Skin samples from our study were small and used for genetic analyses preferentially over toxicological

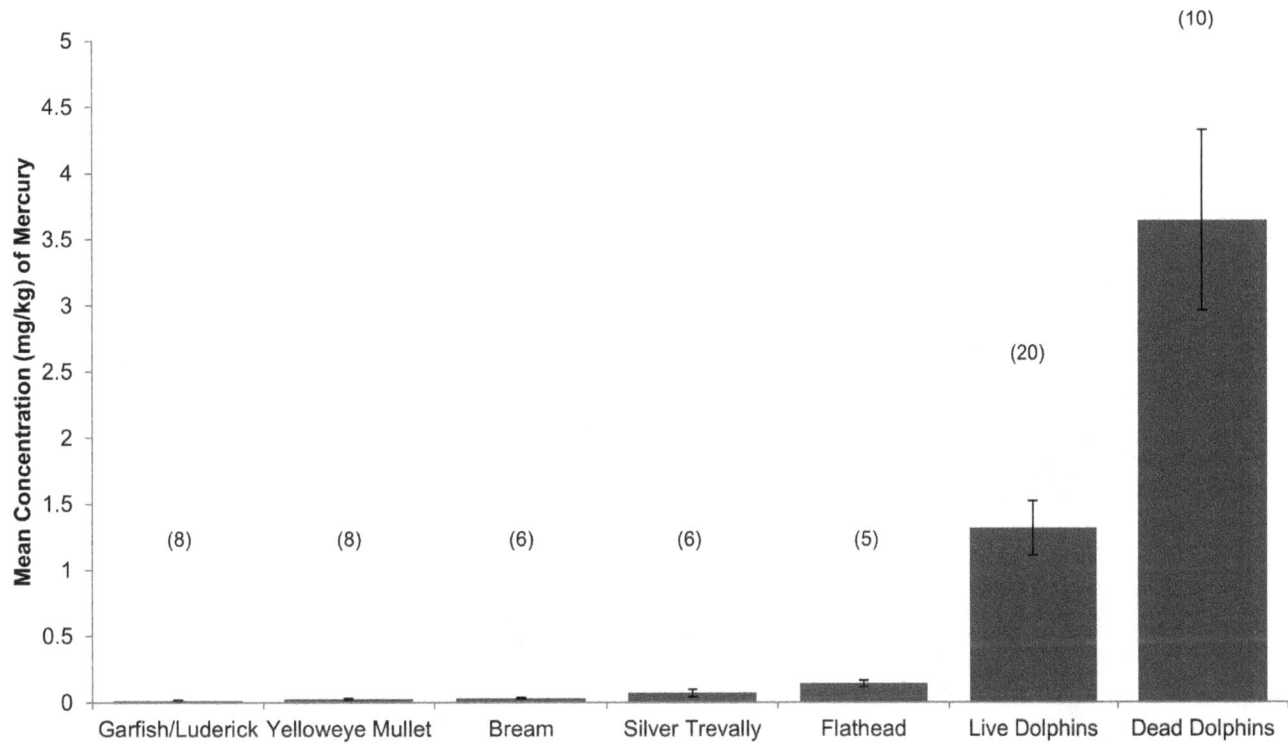

Figure 2. Concentration of total mercury (mg/kg wet weight) in potential dolphin prey muscle and blubber from live and dead *Tursiops australis* from coastal Victoria. Values shown are means with standard errors, sample sizes are shown in brackets above each bar.

analyses, meaning that the use of the blubber component was the only practical approach.

Age has been found to be an important factor in mercury concentrations in marine mammals. As an animal gets older, the level of mercury is predicted to increase due to bioaccumulation [29]. Given only five animals in this study were aged it was not realistic to determine whether there was any association between mercury contamination and age. We can however state that the individuals sampled were not all older animals and, importantly, that levels observed were not highest in the oldest animals. We can conclude that the animals dying are not just old age animals that have accumulated high levels of mercury over their long lifetime, as three out of the four animals aged were considered middle aged (Table 2).

We were not able to clearly identify a single point of origin for the mercury. Of note, the region was extensively mined for gold in the late 19th and early 20th Centuries, activities which included wide spread use of mercury to extract gold ore [30]. It is likely that mercury is entering the semi-enclosed waters in which *T. australis* is found via major river catchments and becoming incorporated into the marine food web. Using a previous stable isotope study of trophic relationships for this species [19] we were able to identify the main prey species of the dolphins and assess their mercury levels. Analysis of mercury loads in dolphin prey found no clear single source of contamination, although all prey items contained measurable amounts of mercury (Figure 2). Mercury is readily absorbed across the gastro-intestinal tract from prey items, and the volumes of food ingested over the lifetime of an animal mean that even low doses in prey items may result in high levels of exposure [29]. The levels of mercury found in prey tissue were much lower than those in the dolphins, consistent with a number of previous studies of marine mammals [29,31–33]. However mercury is

strongly biomagnified in marine food chains [31–33], and bioaccumulates in muscle tissue, blubber and in the liver of marine mammals [29]. The values that we found indicate that even relatively low doses of contaminants in marine environments can result in high loadings to long-lived, high trophic level predators. A comprehensive food web study is needed to be able to calculate the amount of mercury *T. australis* is consuming.

Our comparison of levels from beach-cast and live animals shows the potential for a role for mercury contamination in the mortality of these animals. This, along with the fact that mercury concentrations were within a range known to have multiple health effects, would suggest that mercury is negatively impacting on these dolphins. This is of grave concern, considering that this species is newly described, appears to occur over a very limited range and has a small known population size. However a limitation of this study is that we only analysed for total mercury. We acknowledge that to better understand the effects of mercury, speciation of mercury should be completed to identify what form the mercury is in. Another limitation of the study is that we didn't do a screen for all possible contaminants, in particular dioxins, PAHs and organic pesticides that are known to have detrimental effects at low concentrations. Pilot work on dolphins from this area has suggested that concentrations of these contaminants are low (Monk, unpublished data) but we cannot state with certainty that they are not contributing to mortality in some cases.

To better understand and interpret the levels present in dead dolphins from a population, it is important to compare it to levels found in the live population. Future toxicological research should also include parallel studies to investigate the actual effects of mercury on the dolphins including histopathology, measurement of neurochemical biomarkers and measurement of genomic and genetic biomarkers.

Author Contributions

Conceived and designed the experiments: AM KCR RT. Performed the experiments: AM KCR SB. Analyzed the data: AM KCR RT.

Contributed reagents/materials/analysis tools: SB. Wrote the paper: AM KCR SB RT.

References

1. Worm B, Sandow M, Oschlies A, Lotze HK, Myers RA (2005) Global patterns of predator diversity in the open oceans. Science 309: 1365–1369.
2. Graham NAJ, Chabanet P, Evans RD, Jennings S, Letourneur Y, et al. (2011) Extinction vulnerability of coral reef fishes. Ecology Letters 14: 341–348.
3. IUCN 2013. The IUCN Red List of Threatened Species. Version 2013.2. < http://www.iucnredlist.org>. Downloaded on 8 January, 2014.
4. Davidson AD, Boyer AG, Kim H, Pompa-Mansilla S, Hamilton MJ, et al. (2012) Drivers and hotspots of extinction risk in marine mammals. Proceedings of the National Academy of Sciences of the United States of America 109: 3395–3400.
5. Burke L, Kura Y, Kassem K, Revenga C, Spalding M, et al. (2001) Pilot Analysis of Global Ecosystems: Coastal Ecosystems. World Resources Institute, Washington D.C.
6. Pulster EL, Smalling KL, Maruya KA (2005) Polychlorinated biphenyls and toxaphene in preferred prey fish of coastal southeastern U.S. bottlenose dolphins (*Tursiops truncatus*). Environmental Toxicology and Chemistry 24: 3128–3136
7. Beck KM, Fair P, McFee W, Wolf D (1997) Heavy metals in livers of bottlenose dolphins stranded along the south Carolina coast. Marine Pollution Bulletin 34: 734–739.
8. Stockin KA, Law RJ, Duignan PJ, Jones GW, Porter L, et al. (2007) Trace elements, PCBs and organochlorine pesticides in New Zealand common dolphins (*Delphinus* sp.). Science of The Total Environment 387: 333–345.
9. Parsons EC (1998) Trace metal pollution in Hong Kong: Implications for the health of Hong Kong's Indo-Pacific hump-backed dolphins (*Sousa chinensis*). The Science of The Total Environment 214: 175–184.
10. Law RJ, Morris RJ, Allchin CR, Jones BR, Nicholson MD (2003) Metals and organochlorines in small cetaceans stranded on the east coast of Australia. Marine Pollution Bulletin 46: 1206–1211.
11. Das K, Debacker V, Pillet S, Bouquegneau JM (2003) Heavy metals in marine mammals. pp. 135–167 in: Vos JG, Bossart GD, Fournier M, O'Shea TJ, editors. Toxicology of marine mammals. New Perspectives: Toxicology and the Environment. Taylor & Francis: London. 643 pp.
12. Becker PR, Krahn MM, Mackey EA, Demiralp R, Schantz MM, et al. (2000) Concentrations of polychlorinated biphenyls (PCB's), chlorinated pesticides, and heavy metals and other elements in tissues of belugas, *Delphinapterus leucas*, from Cook Inlet, Alaska. Marine Fisheries Review 62: 81–98.
13. Muir DCG, Wagemann R, Grift NP, Norstrom RJ, Simon M, et al. (1988) Organochlorine chemical and heavy metal contaminants in white-beaked dolphins (*Lagenorhynchus albirostris*) and pilot whales (*Globicephala melaena*) from the coast of Newfoundland, Canada. Archives of Environmental Contamination and Toxicology 17: 613–629.
14. Olsson M, Karlsson B, Ahnland E (1994) Diseases and environmental contaminants in seals from the Baltic and the Swedish west coast. The Science of The Total Environment 154: 217–227.
15. Charlton-Robb K, Gershwin L-a, Thompson R, Austin J, Owen K, et al. (2011) A New Dolphin Species, the Burrunan Dolphin *Tursiops australis* sp. nov., Endemic to Southern Australian Coastal Waters. PLoS ONE 6: e24047.
16. Moura AE, Nielsen SCA, Vilstrup JT, Moreno-Mayar JV, Gilbert MTP, et al. (2013) Recent Diversification of a Marine Genus (*Tursiops* spp.) Tracks Habitat Preference and Environmental Change. Systematic Biology 62: 865–877.
17. Charlton K, Taylor AC, McKechnie SW (2006) A note on divergent mtDNA lineages of bottlenose dolphins from coastal waters of southern Australia. Journal of Cetacean Research and Management 8: 173–179.
18. Möller LM, Bilgmann K, Charlton-Robb K, Beheregaray L (2008) Multi-gene evidence for a new bottlenose dolphin species in southern Australia. Molecular Phylogenetics and Evolution 49: 674–681.
19. Owen K, Charlton-Robb K, Thompson R (2011) Resolving the Trophic Relations of Cryptic Species: An Example Using Stable Isotope Analysis of Dolphin Teeth. PLoS ONE 6.
20. Krutzen M, Barre LM, Moller LM, Heithaus MR, Simms C, et al. (2002) A biopsy system for small cetaceans: darting success and wound healing in *Tursiops* spp. Marine Mammal Science 18: 863–878.
21. Piotrowski JK, Coleman DO (1980) Environmental hazards of heavy metals: Summary evaluation of lead, cadmium and mercury. Monitoring and Assessment Research Centre Report 20. University College of London.
22. Ronald K, Tessaro SV, Uthe JF, Freeman HC, Frank R (1977) Methylmercury poisoning in the harp seal (*Pagophilus groenlandicus*). Science of The Total Environment 8: 1–11.
23. Bennett PM, Jepson PD, Law RJ, Jones BR, Kuiken T, et al. (2001) Exposure to heavy metals and infectious disease mortality in harbour porpoises from England and Wales. Environmental Pollution 112: 33–40.
24. Basu N, Kwan M, Chan HM (2006) Mercury but not Organochlorines Inhibits Muscarinic Cholinergic Receptor Binding in the Cerebrum of Ringed Seals (*Phoca hispida*). Journal of Toxicology and Environmental Health, Part A 69: 1133–1143.
25. Basu N, Scheuhammer AM, Sonne C, Letcher RJ, Born EW, et al. (2009) Is Dietary Mercury of Neurotoxicological Concern to Wild Polar Bears (*Ursus maritimus*)? Environmental Toxicology and Chemistry 28: 133–140.
26. Augier H, Park WK, Ronneau C (1993) Mercury contamination of the striped dolphin *Stenella coeruleoalba* Meyen from the French Mediterranean coasts. Marine Pollution Bulletin 26: 598–604.
27. Chen M-H, Shih C-C, Chou CL, Chou L-S (2002) Mercury, organic-mercury and selenium in small cetaceans in Taiwanese waters. Marine Pollution Bulletin 45: 1–12.
28. Stavros H-CW, Stolen M, Durden WN, McFee W, Bossart GD, et al. (2011) Correlation and toxicological inference of trace elements in tissues from stranded and free-ranging bottlenose dolphins (Tursiops truncatus). Chemosphere 82: 1649–1661.
29. Wagemann R, Trebacz E, Boila G, Lockhart WL (1998) Methylmercury and total mercury in tissues of arctic marine mammals. The Science of The Total Environment 218: 19–31.
30. Fabris G, Theodoropoulos T, Sheehan A, Abbott B (1999) Mercury and Organochlorines in Black Bream, *Acanthopagrus butcheri*, from the Gippsland Lakes, Victoria, Australia: Evidence for Temporal Increases in Mercury levels. Marine Pollution Bulletin 38: 970–976.
31. Das K, Beans C, Holsbeek L, Mauger G, Berrow SD, et al. (2003) Marine mammals from Northeast Atlantic: relationship between their trophic status as determined by delta C-13 and delta N-15 measurements and their trace metal concentrations. Marine Environmental Research 56: 349–365.
32. Bisi TL, Lepoint G, Azevedo A, Dorneles P, Flach L, et al. (2012) Trophic relationships and mercury biomagnification in Brazilian tropical coastal food webs. Ecological Indicators 18: 291–302.
33. Atwell L, Hobson KA, Welch HE (1998) Biomagnification and bioaccumulation of mercury in an arctic marine food web: insights from stable nitrogen isotope analysis. Canadian Journal of Fisheries and Aquatic Sciences 55: 1114–1121.
34. Lavery TJ, Butterfield N, Kemper CM, Reid RJ, Sanderson K (2008) Metals and selenium in the liver and bone of three dolphin species from South Australia, 1988–2004. Science of The Total Environment 390: 77–85.
35. Roditi-Elasar M, Kerem D, Hornung H, Kress N, Shoham-Frider E, et al. (2003) Heavy metal levels in bottlenose and striped dolphins off the Mediterranean coast of Israel. Marine Pollution Bulletin 46: 503–512.
36. Cardellicchio N, Giandomenico S, Ragone P, Di Leo A (2000) Tissue distribution of metals in striped dolphins (*Stenella coeruleoalba*) from the Apulian coasts, Southern Italy. Marine Environmental Research 49: 55–66.
37. Ruelas JR, Paez-Osuna F, Perez-Cortes H (2000) Distribution of Mercury in Muscle, Liver and Kidney of the Spinner Dolphin (*Stenella longirostris*) Stranded in the Southern Gulf of California. Marine Pollution Bulletin 40: 1063–1066.

5

Retention of Habitat Complexity Minimizes Disassembly of Reef Fish Communities following Disturbance: A Large-Scale Natural Experiment

Michael J. Emslie*, Alistair J. Cheal, Kerryn A. Johns

Australian Institute of Marine Science, Townsville, Queensland, Australia

Abstract

High biodiversity ecosystems are commonly associated with complex habitats. Coral reefs are highly diverse ecosystems, but are under increasing pressure from numerous stressors, many of which reduce live coral cover and habitat complexity with concomitant effects on other organisms such as reef fishes. While previous studies have highlighted the importance of habitat complexity in structuring reef fish communities, they employed gradient or meta-analyses which lacked a controlled experimental design over broad spatial scales to explicitly separate the influence of live coral cover from overall habitat complexity. Here a natural experiment using a long term (20 year), spatially extensive ($\sim$115,000 kms^2) dataset from the Great Barrier Reef revealed the fundamental importance of overall habitat complexity for reef fishes. Reductions of both live coral cover and habitat complexity had substantial impacts on fish communities compared to relatively minor impacts after major reductions in coral cover but not habitat complexity. Where habitat complexity was substantially reduced, species abundances broadly declined and a far greater number of fish species were locally extirpated, including economically important fishes. This resulted in decreased species richness and a loss of diversity within functional groups. Our results suggest that the retention of habitat complexity following disturbances can ameliorate the impacts of coral declines on reef fishes, so preserving their capacity to perform important functional roles essential to reef resilience. These results add to a growing body of evidence about the importance of habitat complexity for reef fishes, and represent the first large-scale examination of this question on the Great Barrier Reef.

Editor: Maura (Gee) Geraldine Chapman, University of Sydney, Australia

Funding: The study was supported by Australian Institute of Marine Science and the Australian Government's Marine and Tropical Sciences Research Facility and National Environment Research Program (Tropical Ecosystems Hub). The funders had no role in study design, data collection and analysis, decision to publish, or preparation of the manuscript.

Competing Interests: The authors have declared that no competing interests exist.

* Email: m.emslie@aims.gov.au

Introduction

Habitat complexity is fundamentally important for the maintenance of high biodiversity across a range of ecosystems [1–5]. Coral reef ecosystems are among the most diverse on the planet with reefs with higher habitat complexity often housing more species than less complex reefs due to the greater variety of niches and shelter [6–8]. Habitat complexity on coral reefs has two major components; the underlying substrate rugosity and the skeletal structure provided by live and dead hard corals. Coral reefs are subject to many types of disturbance that can have negligible to severe impacts on coral cover and habitat complexity. For example, disturbances such as *Acanthaster planci* (crown-of-thorns starfish) outbreaks and coral bleaching cause coral mortality but leave skeletons intact [9–11], so habitat complexity remains largely unchanged in the short term. Subsequently, coral skeletons may erode due to natural processes causing longer term declines in habitat complexity. Conversely, waves from storms can obliterate entire coral colonies removing the habitat complexity previously afforded by their skeletons [11,12]. However, loss of coral structures due to storms or skeletal erosion will not necessarily lead to low habitat complexity if substrate rugosity is high. Indeed, reefs with high substrate rugosity should maintain a greater diversity of organisms than reefs with low substrate rugosity once hard corals are removed, with the exception of those organisms fundamentally dependent on intact coral skeletons or living coral tissue for survival.

Disturbances on coral reefs can dramatically impact the diversity, abundance and community structure of reef fishes, because many fish species are closely associated with live corals and their structures [6–8,13–15]. To date, many studies have attributed changes in fish communities to loss of hard coral cover [9,13,16–20]. Numerous reef fishes rely on hard corals for food and/or shelter and many of these species decline in abundance following hard coral decline [9,16–22]. However, numerous fish species with seemingly limited reliance on hard corals *per se* (e.g. non-corallivorous butterflyfishes, large predators, some herbivorous fishes) have also declined in abundance following disturbances, and in these cases the role of habitat complexity has been implicated [8,9,11,13,20,23]. Declines in abundance and diversity of reef fishes following disturbances can be detrimental to ecosystem functioning and reef resilience due to a reduction in

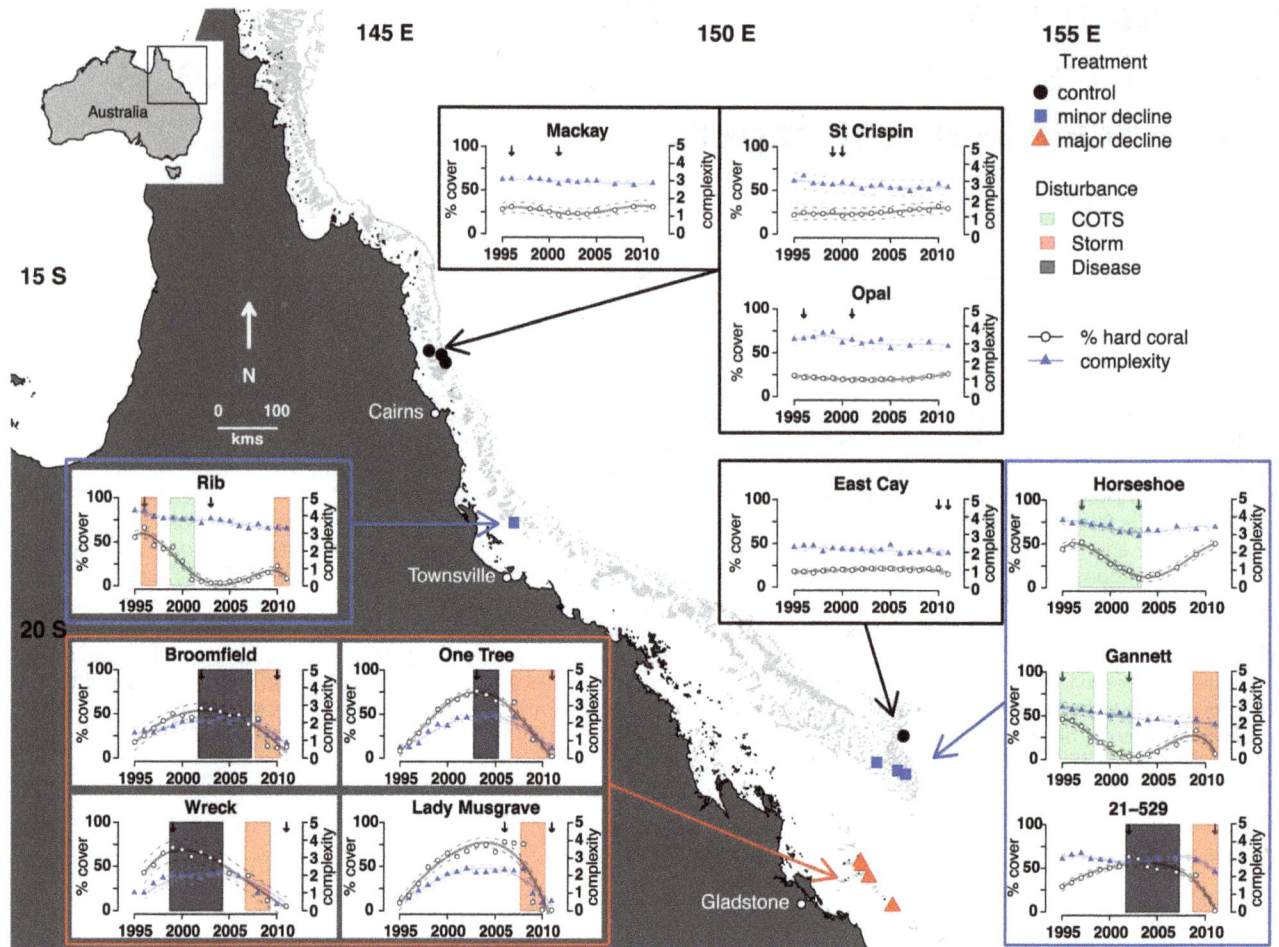

Figure 1. Location of the study reefs in each of the three treatments (Major Decline, Minor Decline and Control). Small panels display trends in hard coral cover and habitat complexity, along with shaded periods of time when disturbances (COTS = *Acanthaster planci* outbreaks, storms & coral disease) occurred. Points are raw data means, while solid lines indicate modelled average trends and dotted lines show 2 x standard errors from a linear mixed effects model fitted separately to hard coral cover and habitat complexity. Arrows mark the years of greatest and least hard coral cover.

the capacity of reef fishes to perform trophic functions. For example, a reduction in the number and diversity of herbivorous fishes decreases their capacity to prevent proliferation of macro-algae that may limit recovery of corals following disturbances [24–32]. Clearly, declines in both live corals and habitat complexity must be important to reef fishes, and disentangling the relative influence of each will provide clues to the relative threat to reef fishes of disturbances which do and do not alter habitat complexity.

It has previously been demonstrated through experimentation [7,33,34] and longer term datasets [12–14,16–22] that reductions in habitat complexity and live coral cover adversely affect reef fish communities. Manipulative experiments have generally been conducted at restricted spatio- temporal scales, typically small ($\sim$10 s of m^2) patch reefs surveyed over several months [7,33,34], and results are difficult to scale up to ecosystem levels. Projects conducted over larger spatio-temporal scales have generally employed gradient/regression type analyses (e.g. [13]) or meta-analyses (e.g. [11]), which are useful approaches for highlighting relationships among variables, changes in variables along a gradient and for integrating many disparate datasets, but lack rigorous experimental designs with which to definitively attribute

causation. Here we use data collected from reefs spread over 115,000 km^2 of the Great Barrier Reef (GBR), gathered over 20 years and employ a natural experiment to formally test how the loss of live coral versus loss of habitat complexity influences reef fish community structure, the diversity of reef fish families and functional groups, and the abundance of individual species.

Methods

Sampling

Data were gathered as part of the Long Term Monitoring Program at the Australian Institute of Marine Science (GBRMPA permit number G13/36390.1); in which fish and benthic communities have been surveyed on 47 reefs of the GBR since 1995. Large-scale disturbances, such as storms and *A. planci* outbreaks that have occurred over the last two decades on the GBR [22,35–36], facilitate opportunities to test macro-ecological hypotheses that due to their scope, require manipulations of a scale (100 s kilometres) that are logistically impossible for researchers to attempt using traditional experimental frameworks [37].We were able to perform a natural experiment to investigate the effects of reductions in live coral cover versus habitat complexity on reef fish communities, by retrospectively assigning replicate reefs into three treatments based on the effects of disturbances. Eight reefs were chosen based on comparable levels of live coral cover (>50%) and subsequent similar and very large relative declines in cover (~90%) due to disturbances. These reefs were separated into two equal treatments based on relative reductions in habitat complexity: 1. a major decline in habitat complexity from high/moderate to very low levels (hereafter "Major Decline"), and 2. a minor decline in habitat complexity from high to moderate levels (hereafter "Minor Decline"). A further four reefs had minimal declines in hard coral cover and no change in habitat complexity (hereafter "Control"; Fig. 1). Even though reefs in each treatment were unevenly distributed geographically (Fig. 1), 77% of fish species were common to all reefs in the study thus enabling valid comparisons of changes to fish communities. Furthermore, our analysis determined the magnitude of change in individual species abundance and community structure, plus the proportion of the community affected (irrespective of identity) before and after disturbances. Thus species identity *per se* was not important but rather the magnitude of changes and the proportion of the community affected.

Three sites of five permanently marked 50 m transects were situated in comparable reef slope habitats (n = 15 transects per reef) and were surveyed on SCUBA annually from 1995 until 2006 and then biennially thereafter. From 1995 until 2005, the benthic community was described using a 30-cm video swathe along the transects. Forty frames from each video transect were sampled and the benthic organisms beneath five points projected on to each frame in a quincunx pattern were identified to the finest taxonomic resolution possible, yielding 200 samples per transect. After 2006, a digital still image was taken every metre along each transect, and forty images were selected and analysed as before [38]. These data were then converted to percent cover of total hard coral for use in univariate analyses. For multivariate analyses, data were converted to percent cover of finer taxonomic groupings that included different growth forms of the most abundant coral family Acroporidae and other hard corals (including all other non-Acroporidae hard coral families), fire coral (genus *Millepora*), soft corals, coralline, turf and macro-algae, rubble, dead coral, sand, abiotic, sponges and other (rare benthic organisms of very low abundance e.g., ascidians, anemones). Fish communities were surveyed concurrently on the same transects using underwater visual census. The abundance and number of species of fishes recorded during surveys were taken from a list of 215 mobile, diurnally active species (including the families Acanthuridae, Chaetodontidae, Labridae, Lethrinidae, Lutjanidae, Pomacentridae, Scaridae, Siganidae, Zanclidae and the commercially important *Plectropomus* spp., hereafter "coral trout"). While parrotfishes are now considered as a tribe Scarinae within the family Labridae, we use the term "Scaridae" to distinguish this group of fishes from other Labridae. We define "species richness" as the number of species recorded and use this term hereafter. Cryptic species such as gobies and blennies were not included. Two transect widths were used: 50×1 m belts for the Pomacentridae and 50×5 m belts for the remaining families [39]. Habitat complexity was independently estimated retrospectively by two observers using a scale of zero (least complex - minimal vertical relief, few holes, crevices and overhangs) to five (most complex - high vertical relief, many holes, crevices and overhangs) from $360°$ video panoramas taken at the start of each transect. This 0 to 5 scale correlates strongly with a range of other rugosity metrics and has been found to be a good predictor of reef fish diversity and abundance [40].

Analyses

To provide the clearest picture of absolute changes in fish communities under varying degrees of change in habitat complexity, we compared metrics of reef fish communities at times of greatest (hereafter "Before") and least (hereafter "After") percent coral cover (indicated by arrows in Fig 1). All analyses were conducted in R [41]. To visualise the changes in fish and benthic communities before and after disturbances, we performed a non-metric Multi-Dimensional Scaling (nMDS) based on the Bray-Curtis similarity co-efficient using the iso-MDS package. To reduce the influence of highly abundant taxa, benthic cover data were row centred and square-root transformed. Similarly, to visualise changes to the whole community rather than a few highly abundant species, fish abundances were row centred and fourth root transformed prior to analysis. To examine the magnitude of change in fish and benthic communities before and after disturbances, we conducted a permutational multivariate analysis of variance using distance matrices and assessed the sums of squares for each Treatment and used the ADONIS function from the VEGAN package in R [41]. As the Treatment by Time interaction was significant, we re-ran the analysis separately for each Treatment (Major Decline, Minor Decline, Control).

Changes in fish and benthic communities were further investigated using Bayesian hierarchical models [42], fitted separately for hard coral cover, habitat complexity, total fish species richness and the species richness of eight reef fish families surveyed (Acanthuridae, Chaetodontidae, Labridae, Lethrinidae, Lutjanidae, Pomacentridae, Scaridae, Siganidae), plus the commercially important coral trout (*Plectropomus* spp.). In order to assess the effects of loss of habitat complexity and live coral on functional roles performed by reef fishes, we examined changes to the species richness of broad functional groups including corallivorous and generalist butterflyfishes, herbivores, planktivores and predators. Models had the fixed factors of Time (Before or After) and Treatment (Major Decline, Minor Decline, Control), and random factors of reef, site and transect. Most variables were modelled against a gaussian distribution in the MCMCglmm package [43]; however some were modelled against negative binomial distributions (log link) to account for zero-inflation and over-dispersion inherent in ecological count data [44] (Table S1). Negative-binomial models were fitted through Just Another Gibbs Sampler (JAGS) via the R2JAGS package in R and used non-informative, flat gaussian priors and the posterior distributions were derived from three Markov chain Monte Carlo (MCMC) (see Table S1 for further model details including number of iterations, burn in and thinning). Model convergence and mixing of Markov chains was assessed visually from trace plots and autocorrelation of the chains was always less than 0.2. Inferences about temporal changes were based on 95% Bayesian Higher Posterior Density (HPD) intervals of cell means predicted from posterior distributions of model parameters. Specific post-hoc contrasts were examined including differences in Time (before and after disturbance) among Treatments and differences among Treatments.

We assessed changes in the abundance of individual reef fish species by plotting a comparable metric to account for differences in initial coral cover [45], calculated as the percent change in abundance from before to after disturbance;

$$\%\text{difference} = \ln[(A_{a,i} - A_{b,i})/A_{b,i}] \times 100$$

Where A_b and A_a were mean values at before and after disturbance respectively. Fish species were only included in these analyses if their summed abundance was ≥ 10 per reef (= 15 transects) in one of the two years. Changes in individual species abundance were then averaged across the four reefs within each Treatment.

Results

Benthic and fish community structure changed from times of greatest to least coral cover, but the magnitude of change varied among habitat complexity treatments (Fig 2). On reefs with a major decline in complexity, there were substantial shifts in the structure of both fish communities (ADONIS Time: $F = 19.134$, d. f. $= 1$, $Pr(>F) = 0.001$) and benthic communities (ADONIS Time: $F = 85.902$, d. f. $= 1$, $Pr(>F) = 0.001$) (Fig 2). Similarly, a large shift occurred in the benthic communities on reefs with minor declines in habitat complexity, (ADONIS Time: $F = 32.429$, d. f. $= 1$, $Pr(>F) = 0.001$), but a much smaller shift was evident for the fish communities (ADONIS Time: $F = 2.1751$, d. f. $= 1$, $Pr(>F) = 0.059$) on these reefs compared to those in the Major Decline treatment (Fig 2). Very little change occurred in either the fish communities (ADONIS Time: $F = 0.3885$, d. f. $= 1$, $Pr(>F) = 0.909$) or benthic communities (ADONIS Time: $F = 1.0507$, d. f. $= 1$, $Pr(>F) = 0.304$) on Control reefs (Fig 2).

Hard coral cover declined in all treatments but the decline was negligible on Control reefs. Habitat complexity only declined substantially on Major Decline reefs; reductions were minimal on reefs in the Minor Decline treatment and were similar to changes at Control reefs (Fig 3). Reductions in fish total species richness and the species richness of the Chaetodontidae and Labridae occurred on reefs in both complexity decline treatments, though the loss was greatest on in the Major Decline reefs (Fig 3). Also, species richness of Acanthuridae, Lutjanidae, Pomacentridae, Scaridae and coral trout declined on reefs in the Major Decline treatment, but not on those in the Minor Decline or Control treatments (Fig 3). There were large declines of species richness of all functional groups of fishes on Major Decline reefs (Fig 3). However, the species richness of only two functional groups, corallivorous butterflyfishes and predators, declined on Minor Decline reefs and these reductions were substantially smaller than those on reefs in the Major Decline treatment. There was no substantial decline in species richness of any functional group on Control reefs (Fig. 3).

Changes in the abundance of individual species varied substantially among the three habitat complexity treatments (Fig 4), with major declines in habitat complexity impacting a greater number of species than minor declines. On Major Decline reefs, 75% of species declined in abundance, 56% of species lost half their abundance and 18% were locally extirpated (declined to zero) (Fig 4). In comparison, the abundance of less than half (48%) of the fish species declined on reefs in the Minor Decline treatment, 24% declined in abundance by half and only 3% of species were locally extirpated (Fig 4). Fish species on Control reefs were far less affected; 22% of species declined in abundance, with only 3% declining by half and no species being locally extirpated (Fig 4).

The major loss of habitat complexity also greatly reduced the capacity of reef fishes to perform their functional roles. Among the functionally important herbivorous fishes, fourteen species declined in abundance by 50% or more on reefs that underwent major declines in habitat complexity, compared to four species on reefs with a minor decline and only one species on Control reefs. Additionally, abundances of some commercially important fishery species such as coral trout, were reduced to zero on Major Decline reefs, but declined by less than 5% on Minor Decline reefs. In addition, obligate corallivores accounted for a large proportion of

the species that declined in abundance in the Minor Decline treatment, but accounted for a much smaller proportion of the substantially greater number of species that declined on reefs with major declines in complexity.

Discussion

Using long-term data at ecologically meaningful scales on the GBR, this study has demonstrated the fundamental importance of habitat complexity for the maintenance of diverse fish communities, which is critical for maintaining healthy ecosystem function. Among reefs which underwent large declines in live coral cover, it was only on those reefs where habitat complexity also declined markedly that reef fish communities underwent wholesale reductions in diversity, species abundances and functional capacity. Previously small scale manipulative experiments [7,33,34], gradient/regression type analyses [13,20], or meta-analyses [11] had proposed the importance of habitat complexity for reef fishes, but whether these results reflected a broad-scale truth had not been rigorously tested. Our large-scale, natural experiment was able to demonstrate the generality of habitat complexity as a fundamental driver of reef fish community structure on the GBR, supporting findings in other regions [11,13,20,46,47]. We showed that major loss of habitat complexity affected a broad array of reef fishes from all trophic/functional groups. Additionally, although major loss of hard coral but not habitat complexity caused declines in some fish species, mostly those intimately associated with hard corals, the role of corals was not as important if overall habitat complexity remained moderate to high. Such results suggest that reefs which undergo major reductions in overall habitat complexity following disturbances will support depauperate reef fish communities, with a reduced ability to perform critical functional roles that contribute to the resilience of coral reefs.

While decreases in abundance of coral dependent species following loss of live coral were expected irrespective of changes in habitat complexity [20,22], the sweeping reductions in abundance of most reef fish species following major reduction in habitat complexity was more surprising (but see [11,13]). Large predatory fishes, planktivorous damselfishes and various herbivores were included in these decreases despite most having no obvious dependence on corals, implying that these fishes are dependent on habitat complexity for their survival, most likely through the provision of shelter and food sources. Clearly, habitat complexity affords shelter not only through live corals, but also through dead coral skeletons and by caves, cracks and fissures in the substrate. Where fish abundance declined due to lack of shelter, it was uncertain whether this resulted from migration to more suitable habitat, either around the reef or into deeper water, or from increased mortality resulting from the lack of refugia from predation. Whatever the mechanism of these declines, such dramatic shifts in reef fish community structure have implications for the ecological functioning of coral reef communities.

The extirpation of numerous species of fishes following major declines in habitat complexity contributed to a major reduction in fish diversity, with species from a range of trophic affiliations lost. High fish diversity usually equates to increased functional diversity (the number of functional groups at a site) and functional redundancy (the number of species within a functional group), both key components of reef resilience [48–53]. Higher functional diversity should enhance the capacity of a reef to deal with disturbances while functional redundancy provides a form of ecological insurance for the maintenance of a functional role despite losses of some species due to disturbances. Thus it seems

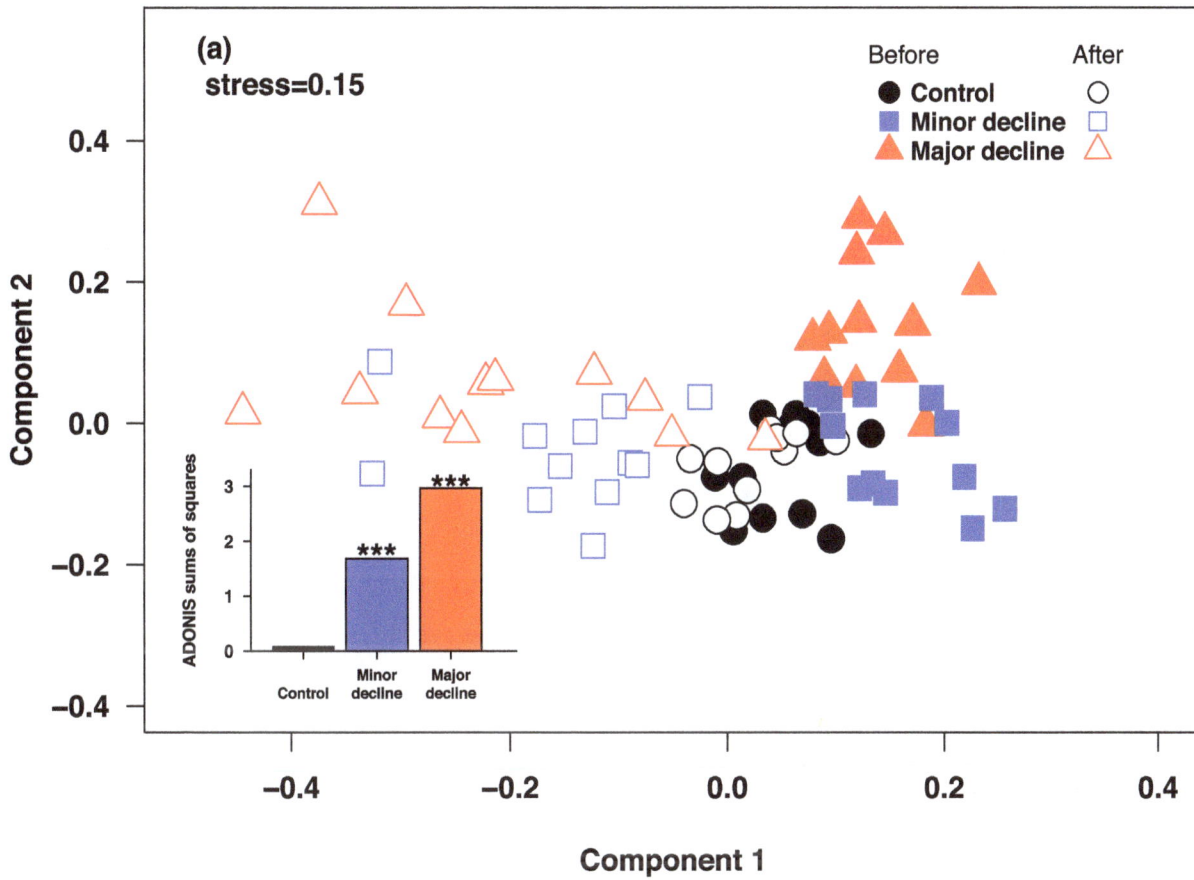

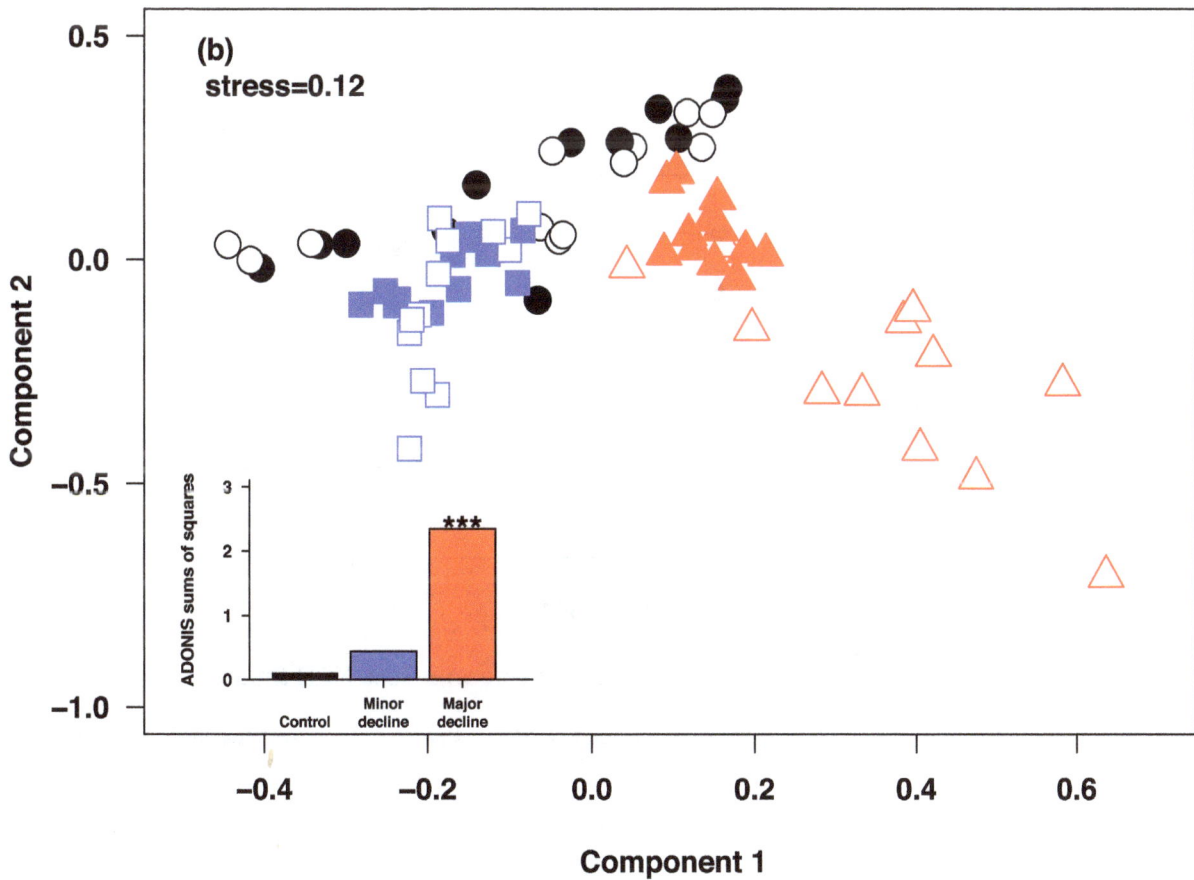

Figure 2. Multi-dimensional plot based on Bray-Curtis similarity coefficients of (a) square-root transformed percent benthic cover and (b) fourth-root transformed fish species abundances. Each panel presents changes to communities following disturbances for the three treatments (Major Decline, Minor Decline and Control). A full model ADONIS analysis revealed a significant interaction for both benthic communities (ADONIS Treatment*Time: F = 14.293, d. f. = 2, Pr(>F) = 0.001) and fish communities (ADONIS Treatment*Time: F = 4.9225, d. f. = 2, Pr(>F) = 0.001). Changes from times of greatest to least coral cover were further examined by separate ADONIS for each individual Treatment (Major Decline, Minor Decline and Control), and the small inset bar graphs display the effect sizes (Sums of Squares) from these individual analyses. ***: Pr(>F) = <0.001

highly likely that resilience will be diminished following major losses of habitat complexity. For example, the functional contribution of herbivorous fishes to reef resilience has been well established. Many species of herbivorous reef fishes have the capacity to prevent algal overgrowth and aid coral recovery through their grazing activities, thereby preventing undesirable shifts to a macro-algal dominated state [24,32,54]. In this study, the disappearance of fourteen species of herbivorous fishes on reefs where there were major declines of habitat complexity is likely to result in increased vulnerability to such phase shifts (but see [55]).

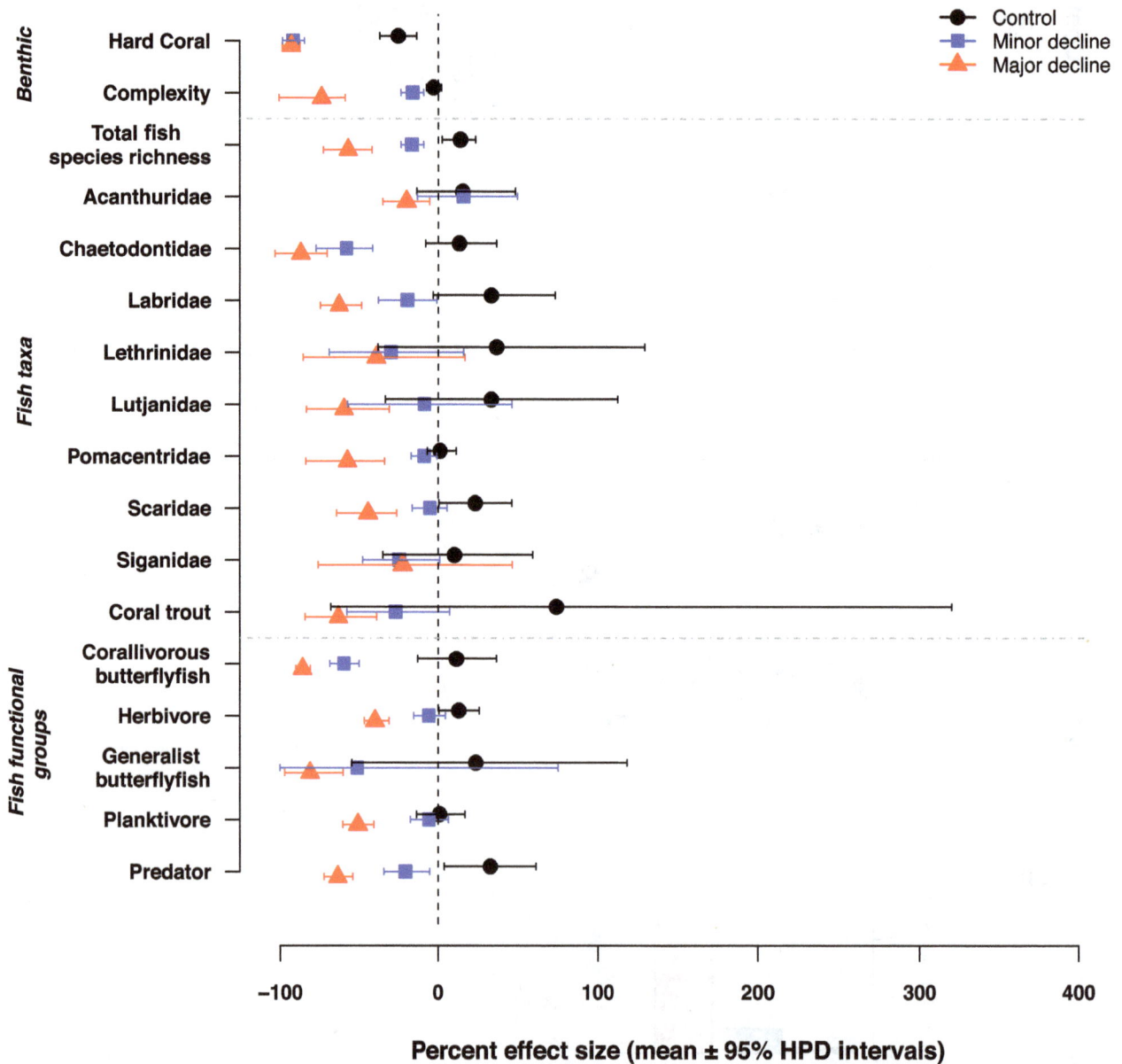

Figure 3. Differences in hard coral cover, habitat complexity, total species richness of fishes and species richness of eight fish families and five broad functional groups for each of the three treatments (Major Decline, Minor Decline and Control). Data are average effect sizes from generalized linear mixed effects model expressed as a per cent change from the time of greatest to least coral cover. Inferences about temporal changes were based on 95% Bayesian Highest Posterior Density (HPD) intervals of cell means predicted from posterior distributions of model parameters derived via Markov-chain Monte Carlo (MCMC) sampling. Effects are considered significant if the HPD intervals do not intersect zero.

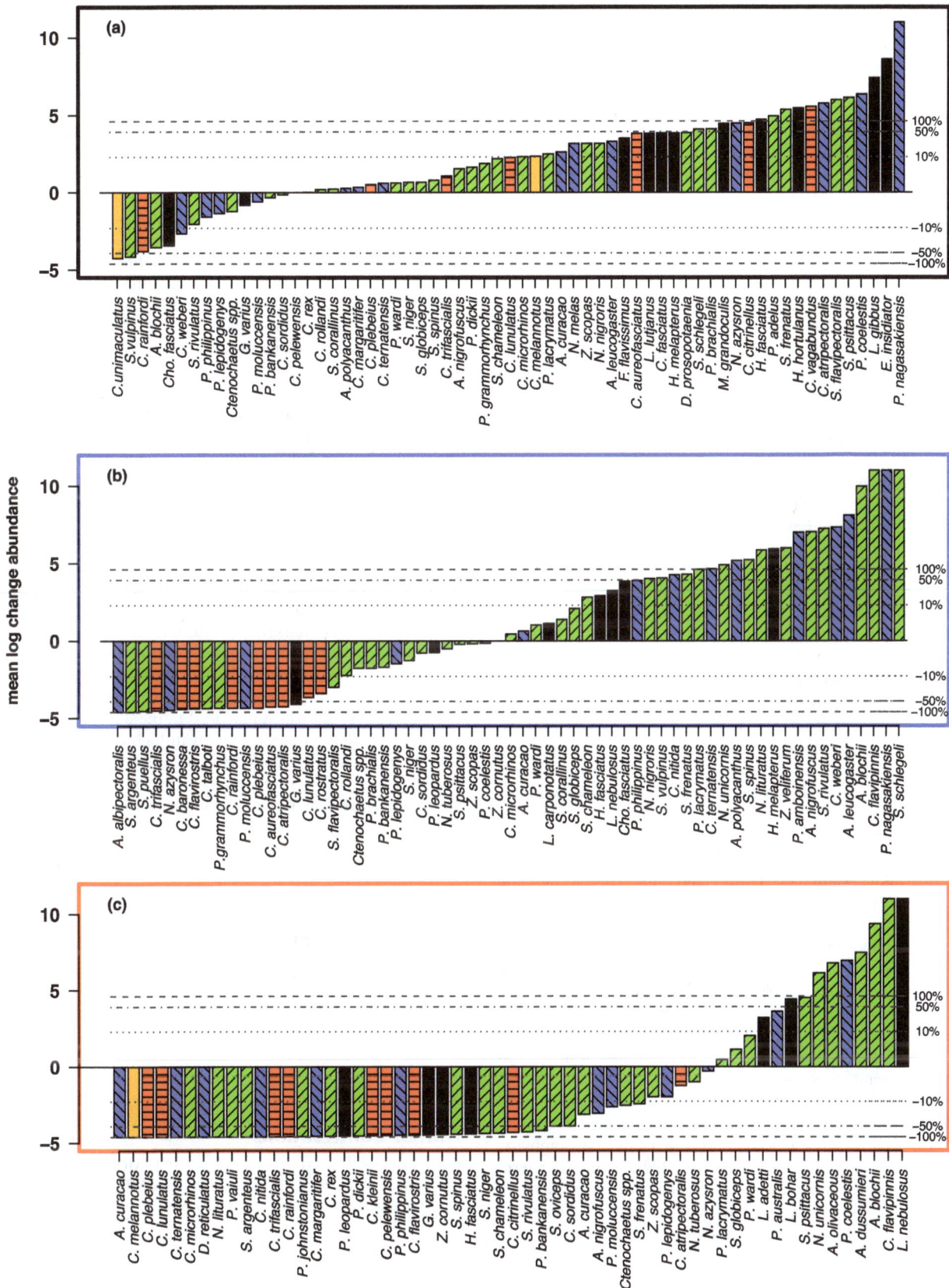

Figure 4. Average percentage change in abundance of individual fish species between times of greatest and least hard coral cover for (a) control reefs (b) reefs that underwent minor declines in complexity (c) reefs that underwent major decline in complexity. Fish species were only included in analyses if their reef wide abundance was ≥10 in one of the two years. Changes in individual species abundance at each reef were then averaged across the four reefs in each Treatment (Major Decline, Minor Decline and Control). Note that the y axis scale is in

natural log units and dotted horizontal lines represent 10, 50 and 100% changes in abundance and that error bars were not included to improve clarity. Coloured bars represent trophic affiliations: green with right diagonal hatching = herbivores, blue with left diagonal hatching = planktivores, red with horizontal hatching = corallivorous butterflyfishes, orange solid bars = generalist (non-coral feeding) butterflyfish, black solid bars = predators. A list of species abbreviations on the x-axis and their corresponding species names are found in Table S2.

While the role of herbivorous fishes in reef resilience has been well established, the contributions of many other reef fishes to reef resilience and healthy ecosystem functioning is less clear. However, what is certain is that the loss of a range of coral reef species performing many functional roles will likely have unknown consequences for ecosystem functioning. For example, reductions in the diversity and abundance of corallivorous fishes (e.g. butterflyfishes) will lower coral mortality [56], because corallivorous butterflyfishes can consume between 9 and 13% of the available tissue biomass of coral, representing 50 to 80% of the total annual productivity [57]. The loss of corallivorous fishes following disturbances will therefore remove substantial predation pressures from newly recruited corals and may ultimately aid recovery. Conversely, the loss of corallivorous fishes may deleteriously affect recovery as high diversity and abundance of corallivorous butterflyfishes has been demonstrated to slow or halt the transmission of coral disease [58]. Future research focused on the role played by corallivorous butterflyfishes in coral dynamics shortly following disturbances could aid our understanding of what impact, if any, the loss of corallivorous fishes plays in reef resilience and ecosystem functioning.

It appears that the short term loss following disturbances of adult fishes not directly dependent on live coral relates more closely to the lack of available shelter rather than to loss of living corals *per se*. Similarly, findings of diverse coral reef fish assemblages on artificial structures largely devoid of corals supports the idea that shelter provided by habitat complexity is fundamentally important to coral reef fish communities [59–61]. However, many reef fishes use live coral as a cue for settlement, including taxa that do not utilise live coral as adults [18]. Although fish communities may be relatively unaffected by coral mortality when habitat complexity is retained, shifts in community structure may lag behind disturbances if fish recruitment is suppressed by limited availability of living coral, while natural mortality of surviving fishes continues. Furthermore, the erosion of coral skeletons after some disturbances such as *A. planci* outbreaks, coral bleaching and coral disease slowly decreases habitat complexity, and may also produce lagged declines in fishes [20,62]. However, in this study adult fish populations were not depleted while habitat complexity remained, providing a buffer to fish population declines while coral is recovering in those cases. Thus in normal circumstances, lagged effects are likely to be balanced by coral recovery and new fish recruitment as long as complexity remains following disturbance. Nevertheless, lagged effects in reef fishes may potentially become more important in future decades, especially if predictions of increased coral bleaching and ocean acidification are correct [63]. In summary, while the retention of habitat complexity reduces the short term impact of disturbances on fish communities, the regeneration of live coral is essential for the maintenance of complex habitats and therefore, to the recovery and long term persistence of diverse reef fish communities.

While previous studies have identified the link between habitat complexity and reef fishes, many of these studies have focused on subsets of the fish community (e.g. [20,21,64], but see [46]). We were able to tease apart the roles of reductions in coral cover versus habitat complexity on a large proportion of diurnally active and conspicuous reef fish communities over ecologically meaningful scales. To our knowledge, this is the first large-scale natural experiment conducted on the GBR to investigate the fundamental contribution of habitat complexity in driving reef fish community change. These results illustrated that reef fish communities are more adversely affected by disturbances which degrade both live coral cover and habitat complexity (i.e. storms), than those which reduce cover of live corals only (i.e. coral bleaching and outbreaks of *A. planci*). Such results should be of interest to reef managers, particularly given our finding that the major fishery target species, the coral trout (*Plectropomus* spp.) disappeared from sites of major complexity decline, with socio-economic ramifications for fishers utilising this resource. In addition, the impact of storms on reef fish communities at sites where coral skeletons account for most of the habitat complexity will be equally devastating irrespective of any zoning to protect target species from fishing. In effect, the benefits afforded by reserve zoning can be reversed almost instantaneously. Conversely, protection of fish communities at sites where complexity of the underlying substrate is high would better preserve important functional processes performed by reef fishes, encouraging rapid recovery in the event that coral cover is removed. Given the prospect of increases in storm intensity with climate change [65] which may lead to the architectural collapse of coral reefs [66], protecting sites with high underlying substrate complexity should be considered to alleviate vulnerability to disassembly of reef fish communities, reductions in the functional roles they perform and much diminished reef resilience.

Acknowledgments

We thank all colleagues, past and present, who have contributed to gathering the data used in this study. The insightful comments of Nick Graham, Aaron MacNeil and Murray Logan are also appreciated.

Author Contributions

Conceived and designed the experiments: MJE AJC KAJ. Performed the experiments: MJE AJC KAJ. Analyzed the data: MJE KAJ. Contributed to the writing of the manuscript: MJE AJC KAJ.

References

1. MacArthur RH, MacArthur JW (1961) On bird species diversity. Ecology 42: 594–598.
2. Heck KL Jr, Wetstone GS (1977) Habitat complexity and invertebrate species and abundance in tropical seagrass. J Biogeogr 4: 135–142.
3. Russell BC (1977) Population and standing crop estimates for rocky reef fishes of North-East New Zealand, New Zealand. J Mar Freshw Res 11: 23–36.
4. Crowder LB, Cooper WE (1982) Habitat structural complexity and the interactions between bluegills and their prey. Ecology 63: 1802–1813.

5. Spies TA (1998) Forest structure: a key to the ecosystem. Northwest Sci 72: 34–39.
6. Risk MJ (1972) Fish diversity on a coral reef in the Virgin Islands. Atoll Res Bull 153: 1–7.
7. Syms C, Jones GP (2000) Disturbance, habitat structure and the dynamics of a coral reef fish community. Ecology 81: 2714–2729.
8. Gratwicke B, Speight MR (2005) Effects of habitat complexity on Caribbean marine fish assemblages. Mar Ecol Prog Ser 292: 301–310.
9. Sano M, Shimizu M, Nose Y (1987) Long-term effects of destruction of hermatypic corals by *Acanthaster planci* infestation on reef fish communities at Iriomote Island, Japan. Mar Ecol Prog Ser 37: 191–199.
10. Sano M (2004) Short-term effects of a mass coral bleaching event on a reef fish assemblage at Iriomote Island, Japan. Fish Sci 70: 41–46.
11. Wilson SK, Graham NAJ, Pratchett MS, Jones GP, Polunin NVC (2006) Multiple disturbances and the global degradation of coral reefs: are reef fishes at risk or resilient? Glob Change Biol 12: 2220–2234. (doi: 10.1111/j.1365-2486.2006.01252.x)
12. Halford AR, Cheal AJ, Ryan D, Williams DMcB (2004) Resilience to large-scale disturbance in coral and fish assemblages on the Great Barrier Reef. Ecology 85: 1892–1905.
13. Graham NAJ, Wilson SK, Jennings S, Polunin NVC, Bijoux JP, et al. (2006) Dynamic fragility of oceanic coral reef ecosystems. Proc Natl Acad Sci USA 103: 8425–8429.
14. Emslie MJ, Cheal AJ, Sweatman H, Delean S (2008) Recovery from disturbance of coral and reef fish communities on the Great Barrier Reef, Australia. Mar Ecol Prog Ser 371: 177–190.
15. Pratchett MS, Munday PL, Wilson SK, Graham NAJ, Cinner JE, et al. (2008) Effects of climate induced coral bleaching on coral-reef fishes – ecological and economic consequences. Ocean Mar Biol Ann Rev 46: 251–296.
16. Woodley JD, Chornesky EA, Clifford PA, Jackson JBC (1981) Hurricane Allen's impact on Jamaican coral reefs. Science 214: 749–755.
17. Bell JD, Galzin R (1984) Influence of live coral cover on coral reef fish communities. Mar Ecol Prog Ser 15: 265–274.
18. Jones GP, McCormick MI, Srinivasan M, Eagle J (2004) Coral declines threaten fish biodiversity in marine reserves. Proc Natl Acad Sci USA 101: 8251–8253.
19. Cheal AJ, Wilson SK, Emslie MJ, Dolman AM, Sweatman H (2008) Response of reef fish communities to coral declines on the Great Barrier Reef. Mar Ecol Prog Ser 372: 211–223.
20. Graham NAJ, Wilson SK, Pratchett MS, Polunin NVC, Spalding MD (2009) Coral mortality versus structural collapse as drivers of corallivorous butterflyfish decline. Biodivers Conserv 18: 3325–3336.
21. Luckhurst BE, Luckhurst K (1978) Analysis of influence of substrate variables on coral-reef fish communities. Mar Biol 49: 317–323.
22. Emslie MJ, Pratchett MS, Cheal AJ (2011) Effects of different disturbance types on butterflyfish communities of Australia's Great Barrier Reef. Coral Reefs 30: 461–471. (doi 10.1007/s00338-011-0730-x)
23. Wilson SK, Dolman AM, Cheal AJ, Emslie MJ, Pratchett MS, et al. (2009) Maintenance of fish diversity on disturbed coral reefs. Coral Reefs 28: 3–14.
24. Hughes TP (1994) Catastrophes, phase shifts and large-scale degradation of a Caribbean coral reef. Science 265: 1547–1551.
25. Hughes TP, Rodrigues MJ, Bellwood DR, Ceccarelli D, Hoegh-Guldberg O, et al. (2007) Phase shifts, herbivory and the resilience of coral reefs to climate change. Curr Biol 17: 360–365.
26. Lewis SM (1986) The role of herbivorous fishes in the organisation of a Caribbean reef community. Ecol Monogr 56:183–200.
27. Mantyka CS, Bellwood DR (2007) Direct evaluation of macroalgal removal by herbivorous coral reef fishes. Coral Reefs 26:435–442.
28. Burkepile DE, Hay ME (2008) Herbivore species richness and feeding complementarity affect community structure and function on a coral reef. Proc Natl Acad Sci USA 105:16201–16206.
29. Burkepile DE, Hay ME (2010) Impact of herbivore identity on algal succession and coral growth on a Caribbean reef. PLoS ONE 5(1):e8963.
30. Mumby PJ (2006) The impact of exploiting grazers (Scaridae) on the dynamics of Caribbean coral reefs. Ecol Appl 16:747–769.
31. Mumby PJ, Hastings A, Edwards HJ (2007) Thresholds and resilience of Caribbean coral reefs. Nature 450:98–101.
32. Cheal AJ, MacNeil MA, Cripps E, Emslie MJ, Jonker M, et al. (2010) Coral-macroalgal phase shifts or reef resilience: links with diversity and functional roles of herbivorous fishes on the Great Barrier Reef. Coral Reefs 29:1005–1015.
33. Lewis AR (1997) Effects of experimental coral disturbance on the structure of reef fish communities on large patch reefs. Mar Ecol Prog Ser 161: 37–50.
34. Coker DJ, Graham NAJ, Pratchett MS (2012) Interactive effects of live coral and structural complexity on the recruitment of reef fishes. Coral Reefs 31: 919–927. (doi 10.1007/s00338-012-0920-1)
35. Osborne K, Dolman AM, Burgess SC, Johns KA (2011) Disturbance and the dynamics of coral cover on the Great Barrier Reef (1995-2009). PLoS ONE 6(3): e17516. doi:10.1371/journal.pone.0017516
36. De'ath G, Fabricius KE, Sweatman H, Puotinen M (2012) The 27-year decline of coral cover on the Great Barrier Reef and its causes. Proc Natl Acad Sci USA 109 (44): 17995–17999. (doi/10.1073/pnas.1208909109)

37. Wellington GM, Victor BC (1985) El Nino mass coral mortality: a test of resource limitation in a coral reef damselfish population. Oecologia 68: 15–19.
38. Jonker M, Johns K, Osborne K (2013) Surveys of benthic reef communities using underwater digital photography and counts of juvenile corals. Long-term Monitoring of the Great Barrier Reef Standard Operational Procedure Number 10. Australian Institute of Marine Science, Townsville, Queensland, Australia.
39. Halford AR, Thompson AA (1994) Visual census surveys of reef fish. Long-term Monitoring of the Great Barrier Reef Standard Operational Procedure Number 3. Australian Institute of Marine Science, Townsville, Queensland, Australia.
40. Wilson SK, Graham NAJ, Polunin NVC (2007) Appraisal of visual assessments of habitat complexity and benthic composition on coral reefs. Mar Biol 151: 1069–1076. (doi 10.1007/s00227-006-0538-3)
41. R Core Team (2013) R: A language and environment for statistical computing. R Foundation for Statistical Computing, Vienna, Austria. ISBN 3-900051-07-0. Available: http://www.R-project.org. Accessed 2014 Jul 20.
42. Gelman A, Hill J (2007) Data Analysis Using Regression and Multilevel/Hierarchical Models. Cambridge University Press: New York.
43. Hadfield JD (2010) MCMC methods for multi-response generalized linear mixed models: the MCMCglmm R package. J Stat Softw 33:1–22.
44. Zuur AF, Saveliev AA, Ieno EN (2012) Zero inflated models and generalized linear mixed models with R. Highland Statistics Limited, Newburgh, UK.
45. Graham NAJ, McClannahan TR, MacNeil MA, Wilson SK, Polunin NVC, et al. (2008) Climate warming, Marine Protected areas and the ocean-scale integrity of coral reef ecosystems. PLoS ONE 3(8): e3039. (doi:10.1371/journal.pone.0003039)
46. Friedlander AM, Parrish JD (1998) Habitat characteristics affecting reef fish assemblages on a Hawaiian coral reef. J Exp Mar Biol Ecol 224: 1–30.
47. Graham NAJ, Nash KL (2012) The importance of structural complexity in coral reef ecosystems. Coral Reefs 32: 315–326.
48. Elmqvist T, Folke C, Nyström M, Peterson G, Bengtsson J, et al. (2003) Response diversity, ecosystem change, and resilience. Front Ecol Environ 1: 488–494.
49. Folke C, Carpenter S, Walker B, Scheffer M, Elmqvist T, et al. (2004) Regime shifts, resilience and biodiversity in ecosystem management. Annu Rev Ecol Evol Syst 35: 557–581.
50. Walker B (1992) Biodiversity and ecological redundancy. Conserv Biol 6: 18–23.
51. Petersen GC, Allen CR, Holling CS (1998) Ecological resilience, biodiversity and scale. Ecosystems 1: 6–18.
52. Hooper DU, Chapin FS, Ewel JJ, Hector A, Inchausti P, et al. (2005) Effects of biodiversity on ecosystem functioning: a consensus of current knowledge. Ecol Monogr 75: 3–35.
53. Folke C (2006) The re-emergence of a perspective for social-ecological systems analyses. Glob Environ Change 16: 253–267.
54. Bellwood DR, Hughes TP, Folke C, Nyström M (2004) Confronting the coral reef crisis. Nature 429: 827–833.
55. Cheal AJ, Emslie MJ, MacNeil MA, Miller I, Sweatman H (2013) Spatial variation in the functional characteristics of herbivorous fish communities and the resilience of coral reefs. Ecol Appl 23: 174–188.
56. Rotjan RD, Lewis SM (2008) Impact of coral predators on tropical reefs. Mar Ecol Prog Ser 367: 73–91.
57. Cole AJ, Lawton RJ, Wilson SK, Pratchett MS (2012) Consumption of tabular acroporid corals by reef fishes: a comparison with plant-herbivore interactions. Funct Ecol 26(2): 307–316.
58. Cole AJ, Chong-Seng KM, Pratchett MS, Jones GP (2009). Coral feeding fishes slow progression of black band disease. Coral Reefs 28: 965.
59. Alevizon WS, Gorham JC (1989) Effects of artificial reef deployment on nearby resident fishes. Bull Mar Sci 44:646–661.
60. Burt JA, Feary DA, Cavalcante G, Bauman AG, Usseglio P (2013) Urban breakwaters as reef fish habitat in the Persian Gulf. Mar Poll Bull 72: 342–350.
61. Pradella N, Fowler AM, Booth DJ, Macreadie PI (2014) Fish assemblages associated with oil industry structures on the continental shelf of north-western Australia. J Fish Biol 84: 247–255. (doi:10.1111/jfb.12274)
62. Graham NA, Wilson SK, Jennings S, Polunin NVC, Robinson J, et al. (2007) Lag effects in the impacts of mass coral bleaching on coral reef fish, fisheries and ecosystems. Cons Biol 21:1291–1300. (doi 10.1111/j.1523-1739.2007.00754.x)
63. Hoegh-Guldberg O, Mumby PJ, Hooten AJ, Steneck RS, Greenfield P, et al. (2007) Coral reefs under rapid climate change and ocean acidification. Science 318:1737–1742. (doi 10.1126/science.1152509)
64. Noonan SHC, Jones GP, Pratchett MS (2012) Coral size, health and structural complexity: effects on the ecology of a coral reef damselfish. Mar Ecol Prog Ser 456: 127–137.
65. Webster PJ, Holland GJ, Curry JA, Chang HR (2005) Changes in tropical cyclone number, duration and intensity in a warming environment. Science 309: 1844–1846.
66. Alvarez-Filip L, Dulvy NK, Gill JA, Cote IM, Watkinson AR (2009) Flattening of Caribbean coral reefs: region-wide declines in architectural complexity. Proc R Soc Lond B Biol Sci 276: 3019–3025.

6

Genetic Variations in Two Seahorse Species (*Hippocampus mohnikei* and *Hippocampus trimaculatus*): Evidence for Middle Pleistocene Population Expansion

Yanhong Zhang[1], Nancy Kim Pham[2], Huixian Zhang[1], Junda Lin[2], Qiang Lin[1]*

1 Key Laboratory of Tropical Marine Bio-resources and Ecology, South China Sea Institute of Oceanology, Chinese Academy of Sciences, Guangzhou, China, **2** Vero Beach Marine Laboratory, Florida Institute of Technology, Vero Beach, Florida, United States of America

Abstract

Population genetic of seahorses is confidently influenced by their species-specific ecological requirements and life-history traits. In the present study, partial sequences of mitochondrial cytochrome b (*cytb*) and control region (CR) were obtained from 50 *Hippocampus mohnikei* and 92 *H. trimaculatus* from four zoogeographical zones. A total of 780 base pairs of *cytb* gene were sequenced to characterize mitochondrial DNA (mtDNA) diversity. The mtDNA marker revealed high haplotype diversity, low nucleotide diversity, and a lack of population structure across both populations of *H. mohnikei* and *H. trimaculatus*. A neighbour-joining (NJ) tree of *cytb* gene sequences showed that *H. mohnikei* haplotypes formed one cluster. A maximum likelihood (ML) tree of *cytb* gene sequences showed that *H. trimaculatus* belonged to one lineage. The star-like pattern median-joining network of *cytb* and CR markers indicated a previous demographic expansion of *H. mohnikei* and *H. trimaculatus*. The *cytb* and CR data sets exhibited a unimodal mismatch distribution, which may have resulted from population expansion. Mismatch analysis suggested that the expansion was initiated about 276,000 years ago for *H. mohnikei* and about 230,000 years ago for *H. trimaculatus* during the middle Pleistocene period. This study indicates a possible signature of genetic variation and population expansion in two seahorses under complex marine environments.

Editor: Patrick Callaerts, VIB & Katholieke Universiteit Leuven, Belgium

Funding: This study was funded by the Outstanding Youth Foundation in Guangdong Province (S2013050014802), the National Natural Science Foundation of China (41176146, 41306148), and the US National Science Foundation East Asia and Pacific Summer Institute Fellowship (EAPSI,OISE-1209841). The funders had no role in study design, data collection and analysis, decision to publish, or preparation of the manuscript.

Competing Interests: The authors have declared that no competing interests exist.

* Email: linqiangzsu@163.com

Introduction

Examining patterns of genetic diversity, population structure, and expansion has become an important part in the management plans of endangered populations, and population size is the major determinant of population well-being and extinction risk [1]. In marine environments, population genetics are often impacted by species-specific ecological requirements and life-history traits [2]. The complex and dynamic interactions between the physical and biological environment and the physiology, behaviour, and life histories of individual taxa can apparently lead to the differentiation of marine populations [3].

For marine species, climatic events can undoubtedly impact their historical biogeography; however, marine patterns are relatively poorly known because of the high geological complexity and biological diversity [4]. Based on the endemism of the marine biota in the Northwest Pacific Ocean, three zoogeographical zones have been identified, i.e., the Oriental Zone, Japan Warm-Temperate Zone, and the Tropical Zone [5]. These three zones are defined largely by ecological rather than by historical factors [5]. Sea surface temperature has been postulated to be the primary factor, which governs the formation of these zoogeographical zones, rather than other environmental factors [5].

In marine environments, some species that have long-lived, free-swimming, and feeding (planktotrophic) larval phases probably have relatively high dispersal abilities, and this promotes genetic exchange between populations [2]. However, seahorses are at the lower end of the marine fish dispersal continuum and retain historical patterns [6]. All seahorses are vulnerable to habitat damage because of their feeble swimming ability and small home range behaviour [7]. Many seahorse species undergo a planktonic newborn stage between two and six weeks, after which they settle down into sessile habitats [7,8]. A long planktonic period is likely to create widespread gene flow across geographically disconnected populations, resulting in genetic homogeneity [9].

The three-spot seahorse *Hippocampus trimaculatus* and Japanese seahorse *H. mohnikei* are the most abundant and economically important seahorse species along China's coast. *H. trimaculatus* has a wide distribution range throughout the tropical and sub-tropical regions in Southeast Asia and is about 8–15 cm in body length [10,11]. *H. mohnikei* is a small body size (about 5–8 cm in body length) and inshore-water species, and is generally found in seagrass areas less than 10 m deep [12,13]. The distribution of *H. mohnikei* is limited in the Northeast Asian Sea [13]. It is important to know the population structure in order to conserve these and other seahorse species because of heavy exploitation and environmental changes. Here, we demonstrated the mtDNA diversity among *H. trimaculatus* and *H. mohnikei* along China's coast by sequence analyses of cytochrome *b* (*cytb*) and control region (CR) haplotypes. We then compared the characteristic modes of their genetic variations and the evidence for population expansion across past climatic events.

Materials and Methods

Sample collection

A total of 50 *H. mohnikei* were sampled from Yangmadao and Laizhouwan along North China's coast, which belonged to the Oriental Zone (OZ) [3], and a total of 92 *H. trimaculatus* individuals were sampled from nine localities along Southeast China's coast, and all samples were pooled into three groups from three zoogeographical zones, i.e., Warm-Temperate Zone (WTZ), sub-Tropical Zone (sTZ), and Tropical Zone (TZ) [3] (Table 1 and Fig. 1). Most specimens of *H. mohnikei* and *H. trimaculatus* were collected by researchers on board trawl boats, and a few were obtained with the help of local fishermen and buyers; all seahorses were immediately preserved until DNA isolation. Seahorse project is a key study in the South China Sea Institute of Oceanology, and also a key research focus in the Chinese Academy of Sciences. Seahorses used in this experiment have been absolutely approved for the use of research work, and sampling areas in our study are public, and there is no special policy to protect the seahorses. Some research work in our laboratory aims to obtain detailed information about wild seahorses and then provide data that may lead to the protection of seahorses in some areas. All seahorse samples utilized in this study have received animal ethics approval for experimentation by the Chinese Academy of Sciences. We have provided a scanned certificate for our investigation on seahorses.

Molecular analysis

A small amount of tissue from the tail of each seahorse was removed and macerated using phosphate buffered saline (PBS) buffer for extraction. The macerating tissue and muscle from the fresh seahorses were frozen in liquid nitrogen and then ground into powder. Genomic DNA was extracted using the AxyPrep Multisource Genomic DNA Miniprep Kit (Axygen Biosciences, USA) following the manufacturer's protocol with minor modifications: the tissue homogenate was incubated at 56°C for 2 hours during cell lysis with Proteinase K. All DNA samples were stored at -80°C until polymerase chain reaction (PCR) amplification. A part of the mitochondrial *cytb* gene (895 base pairs) was amplified employing the seahorse specific primers: forward shf 5′-CTACCTGCACCATCAAATATTTC-3′ and reverse shr2 5′-CGGAAGGTGAGTCCTCGTTG 3′ [6]. DNA amplification of CR sequences followed the methodology published previously [27]. All PCR reactions were carried out in a total volume of 50 μl, utilizing 3 μl (10-100 ng) DNA, 0.25 μl Taq DNA polymerase (5 U/μl, TaKaRa, China-Japan Joint Company, Dalian, China), 1 μl of each primer (10 μM), 4 μl dNTP Mixture (2.5 mM), 5 μl Ex Taq Buffer (10×), and 35.75 μl ddH$_2$O. The thermocycling sequence was conducted as follows: an initial step of 94°C (3 min); a second step of 35 cycles of 94°C (30 s), 50°C (30 s), and 72°C (75 s); and a final step of 72°C (10 min). The primers and amplification conditions used for CR were as described in Teske et al. [27]. Amplified PCR products were checked on 1.5% agarose gels and purified for sequencing using the E. Z. N. A. Gel Extraction Kit (Omega, USA). *Cytb* genes and CR were commercially sequenced using PCR purified products

from both forward and reverse primers (BGI, China). Sequences were assembled and edited using Bioedit 7.0.9.0 [14], and subsequently aligned utilizing ClustalW [15]. Sequences were submitted to GenBank (accession numbers: *cytb* for *H. trimaculatus* KC519325-KC519363 and *H. mohnikei* KC527556-KC527584; CR for *H. trimaculatus* KJ158359-KJ158392 and *H. mohnikei* K J158393-KJ158419).

Population genetic analyses

Genetic diversity indices (based on composition and transition/transversion bias) were calculated using MEGA5 [16]. The numbers of individuals (*n*), number of variable sites (*ns*), number of haplotypes (*np*), haplotype diversity (*h*), nucleotide diversity (*π*), and average number of nucleotide differences (*k*) for each species' population were estimated using the software DnaSP 5.10.00 [17].

Demographic reconstruction

Pairwise mismatch distributions, sum of square deviations (SSD), and raggedness index (R) were performed using Arlequin 3.1 [18] for all sampling locations combined to find evidence of past demographic expansion. According to coalescent theory, a population at demographic equilibrium usually exhibits a multi-modal mismatch distribution, but is usually unimodal following a recent population demographic or range expansion [19]. If the test statistics show no significant SSD value and low R value, it means that the population has experienced sudden expansion [20]. We also tested the neutral theory in Arlequin 3.1 employing Tajima's *D* [21] and Fu's *Fs* [22]. Tajima's test is the most conservative test of neutrality; whereas, Fu's *Fs* is the most powerful test for population growth. Expectations of Fu's *Fs* and Tajima's *D* are significantly negative values ($P<0.05$) in a sudden expansion population [22]. The relationship Tau $= 2ukt/g$ was used to estimate the time of expansion (t), where k is the number of nucleotides assayed; u is the mutation rate per nucleotide; and g is the generation interval. An average mutation rate of 6.3×10^{-9} per site per year for the seahorse *cytb* gene was assumed based on a generation time of approximately 1 year [23]. To analyze population structure, an analysis of molecular variance (AMOVA) was utilized in the Arlequin 3.1 software. We also used the Bayesian skyline plot (BSP) [24] implemented in BEAST [25] to assess historical changes in effective population size. An uncorrelated lognormal relaxed clock was employed for *cytb* alignment. Divergence time was estimated to be 2% per million years (My) based on the entire mtDNA molecule, which was widely used for bony fish [26,27].

Phylogenetic analyses

The *cytb* sequences were subjected to the phylogenetic analysis based on a neighbor-joining (NJ) tree and the maximum likelihood (ML) method in PAUP 4.0b10 [28]. We selected the best-fit nucleotide substitution model for each locus using the Akaike information criterion (AIC) in the program Modeltest 3.0 [29]. *H. kuda* (JX217831.1) was used as an outgroup for *H. mohnikei*. *H. trimaculatus* sequences (AY322461, AY322457, AY322452, AY322437, AY322435, AY322454, AY322462, AY322471, AY322473, AY322467, AY322475, AY322468, AY322464, and AY322434) were obtained from GenBank as additional reference sequences for phylogeography analysis. Polymorphic sites were determined using DnaSP 5.10.00, and these sites were utilized to construct an unrooted median-joining haplotype network using software package Network 4.6.1.0 [30].

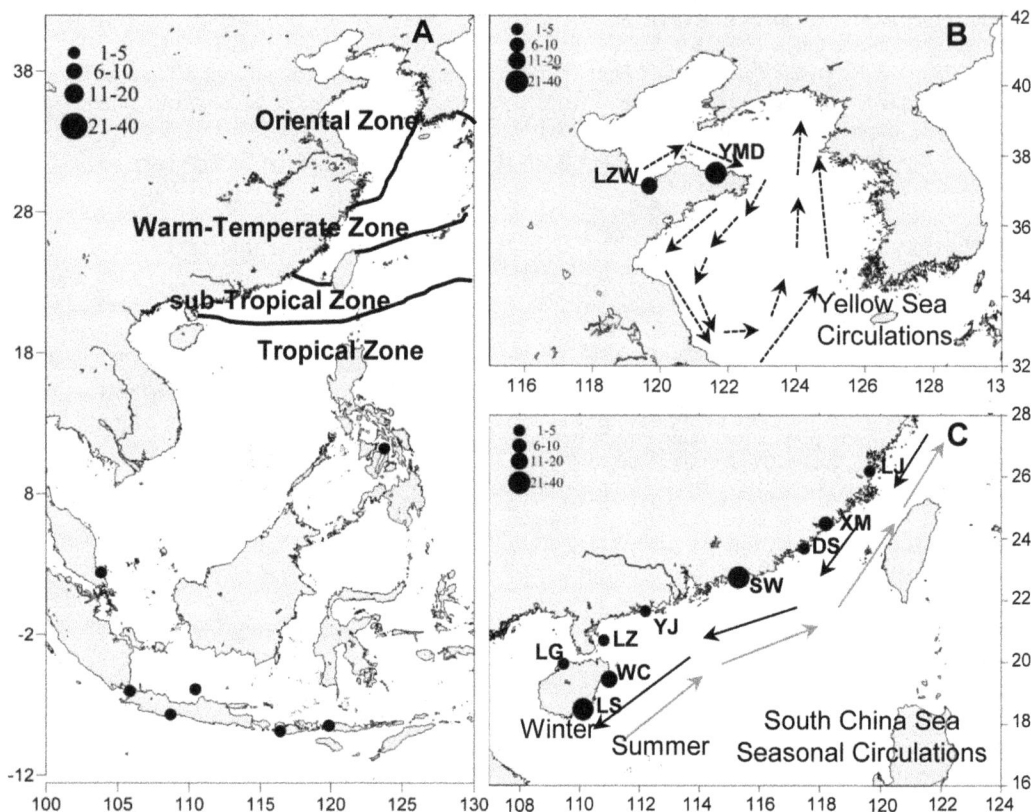

Figure 1. Sampling localities and sizes of *H. mohnikei* **and** *H. trimaculatus.* The zoogeographical zones are modified by the previous report of Briggs [7] (names in the figure A, and zones are delineated by black lines). Alphabetical letters indicate the seahorse groups: (A) outgroup for *H. trimaculatus*; (B) *H. mohnikei* group (under the influence of the Yellow Sea circulations (dotted arrows) [48]); and (C) *H. trimaculatus* group (under the influence of overall seasonal circulation in the South China Sea [black arrows in winter and grey arrows in summer] [44,45]).

Table 1. Sampling location and sample size (n) of *H. mohnikei* and *H. trimaculatus.*

	Location	Latitude	Longitude	n
H. mohnikei				
LZW	Laizhouwan	37.13°N	119.68°E	20
YMD	Yangmadao	37.50°N	121.65°E	30
H. trimaculatus				
Warm-Temperature Zone (WTZ)	—	—	—	15
LJ	Lianjiang	26.18°N	119.64°E	5
XM	Xiamen	24.48°N	118.20°E	6
DS	Dongshan	23.69°N	117.48°E	4
sub-Tropical Zone (sTZ)	—	—	—	35
SW	Shanwei	22.75°N	115.31°E	33
YJ	Yangjiang	21.65°N	112.20°E	1
LZ	Leizhou	20.71°N	110.81°E	1
Tropical Zone (TZ)	—	—	—	42
LG	Lingao	19.95°N	109.45°E	2
WC	Wenchang	19.45°N	110.99°E	15
LS	Lingshui	18.48°N	110.11°E	25

Results

DNA sequence variability and genetic diversity

A total of 780 bp of the *cytb* fragment were unambiguously sequenced in 50 specimens of *H. mohnikei* and 92 specimens of *H. trimaculatus*. The mean number of nucleotide composition in *H. mohnikei* was T = 27.1%, C = 15.5%, A = 34.0%, and G = 23.4%; in *H. trimaculatus*, it was T = 28.1%, C = 15.4%, A = 30.4%, and G = 26.1%. The transition and transversion rates were 1.77 and 8.82, respectively. Of the 780 bp nucleotide sequence of *H. mohnikei*, there were 25 variable sites, accounting for 3.2% of the length of the sequence. Of these variable sites, there were 12 parsimony informative sites, accounting for 48.0% of the length of the sequence. In *H. trimaculatus*, the number of variable sites was 45, accounting for 5.7% of the length of the sequence, and the number of parsimony informative sites was 10, accounting for 22.2% of the variable sites.

Of the 50 *H. mohnikei cytb* sequenced, there were 29 unique haplotypes (Hj1-Hj29), which were distributed across the two sampled populations as follows: 22 haplotypes from Yangmadao and 11 from Laizhouwan. There were four haplotypes shared by the two populations. The overall haplotype (*h*) and nucleotide diversity (π) in *H. mohnikei* was 0.946 and 0.00353, respectively (Table 2). Of the 92 *H. trimaculatus cytb* sequenced, there were 39 unique haplotypes (Ht1-Ht39), which distributed across the three sampled population pools as follows: seven haplotypes from WTZ, 17 from sTZ, and 24 haplotypes from TZ. The overall haplotype (*h*) and nucleotide diversity (π) in *H. trimaculatus* was 0.902 and 0.00281, respectively (Table 2). Of the *H. mohnikei* and *H. trimaculatus* CR sequenced, there were 27 haplotypes (JC1-JC27) and 34 haplotypes (TC1-TC34), respectively.

Phylogenetic analyses

The relationships of haplotypes based on *cytb* was determined using a neighbor-joining (NJ) algorithm for *H. mohnikei* (Fig. 2A) and maximum likelihood (ML) method for *H. trimaculatus* (Fig. 2B) with bootstrap values indicated above each branch. Partitions with <50% support were not shown. Sequences representing the major clusters of mtDNA haplotypes detected in a recent survey of a diverse sample of lineage A and lineage B (west and east of Wallace's Line, respectively) for the three-spot seahorses from Southeast Asia [6] are included in Fig. 2 and labeled according to their GenBank accession numbers. The haplotypes A4, A12, A22, and A30 (from Malaysia), A20, A25, and A29 (from Java) were almost clustered together with all of the haplotypes in this study. However, haplotypes B1, B2, B6, B7, B10, B12, and B14 were clustered into another branch, which was unlike A lineage. All of the *H. trimaculatus* haplotypes grouped with the A lineage, suggesting an origin on the Asiatic side of Wallace's Line.

The median-joining network among haplotypes of *H. mohnikei* populations presented a star-like distribution trend. For the *cytb* haplotype network, the highest frequency haplotype was Hj12, occupying a central position in the network (Fig. 3A). Other haplotypes were associated with Hj12 by spur. Based on the coalescent theory, the *H. mohnikei* populations experienced a significant population expansion [31]. Because haplotype Hj12 is in the basal position of the network, it was the most widely distributed haplotype and may be the ancestor haplotype. For the CR haplotype network, the highest frequency haplotype was JC1. The *H. mohnikei* haplotype networks were consistent with the NJ phylogenetic tree. For *H. trimaculatus*, the network suggested little or no association between haplotypes geographically, which was consistent with the ML phylogenetic tree. For *cytb* haplotype

Table 2. Genetic diversity of mitochondrial cytochrome *b* for *H. mohnikei* and *H. trimaculatus*.

		n	ns	np	h	π	k
H. mohnikei	YMD	30	19	22	0.961±0.023	0.00340±0.00036	2.65517
	LZW	20	16	11	0.912±0.056	0.00368±0.00080	2.87500
	All	50	25	29	0.946±0.023	0.00353±0.00039	2.75763
H.trimaculatus	WTZ	15	7	7	0.872±0.067	0.00217±0.00038	1.69231
	sTZ	35	21	17	0.899±0.034	0.00273±0.00037	2.12773
	TZ	42	29	24	0.919±0.028	0.00307±0.00035	2.39489
	All	92	45	39	0.902±0.020	0.00281±0.00023	2.19401

Numbers of individuals (*n*), number of segregating sites (*ns*), number of haplotypes (*np*), haplotype diversity (*h*), nucleotide diversity (π), and average number of nucleotide differences (*k*).

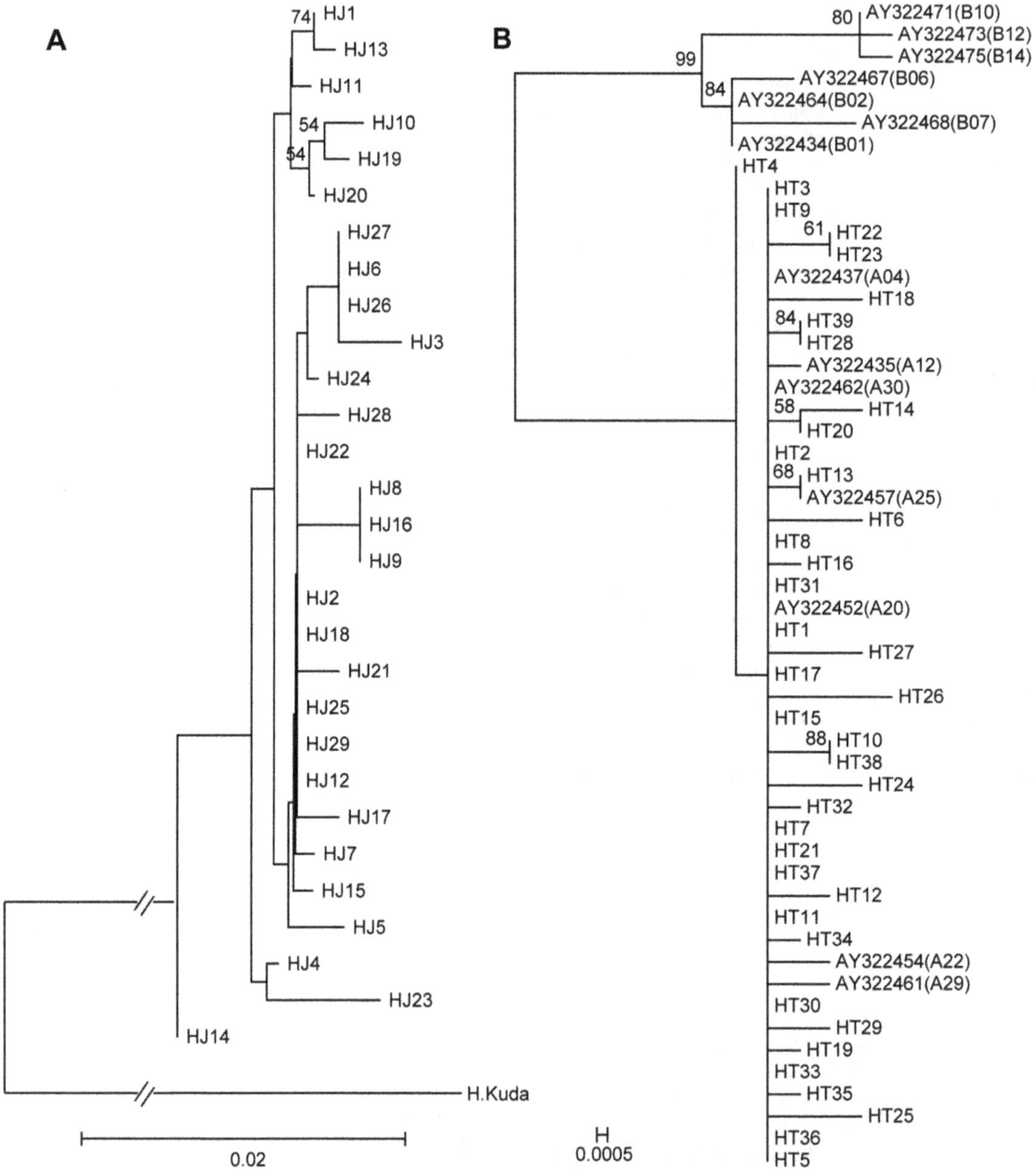

Figure 2. Phylogenetic relationships among *H. mohnikei* and *H. trimaculatus* based on *cytb* genes. NJ tree generated from *H.mohnikei* haplotypes, plus one outgroup (*H. kuda*) (A) and ML tree generated from *H. trimaculatus* haplotypes (B). Cytochrome b sequences are compared to GenBank sequences from reference samples. Figures on branches indicate the degree of bootstrap support (only values above 50% bootstrap support are illustrated).

network, the highest frequency haplotype was Ht1 and Ht2, followed by Ht5, which consisted of 19, 19, and 10 individuals, respectively, and they occupied a central position in the network (Fig. 3B). For the CR haplotype network, the highest frequency haplotype was TC2 and TC9, which both consisted of 17 individuals (Fig. 3B).

Population structure

AMOVA analyses identified that there was not a strong geographic subdivision between the two populations of *H. mohnikei* sampled. Only 1.85% of the total variance was attributed to differences among populations. However, 98.15% of the total variance was attributed to differences within populations. Fixation

index further supports a rise in gene flow with $\Phi_{ST} = 0.01849$ (P> 0.05) indicating no genetic structure. AMOVA analyses also showed that there was not a strong geographic subdivision between the three populations of *H. trimaculatus* sampled ($\Phi_{ST} = 0.00032$, P>0.05).

Historical demography

Demographic history changes were analyzed for *H. mohnikei* and *H. trimaculatus* populations using neutrality tests and mismatch distributions. Tajima's *D* and Fu's *Fs* values in *H. mohnikei* were significantly negative in all populations. For *H. trimaculatus*, Tajima's *D* and Fu's *Fs* values were significantly negative (P<0.05) in all populations, except that Tajima's *D* was

A   *H. mohnikei*

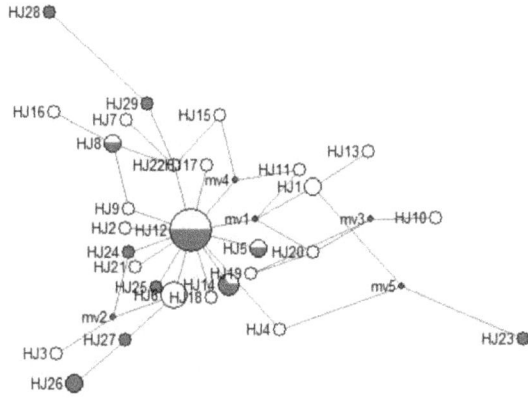
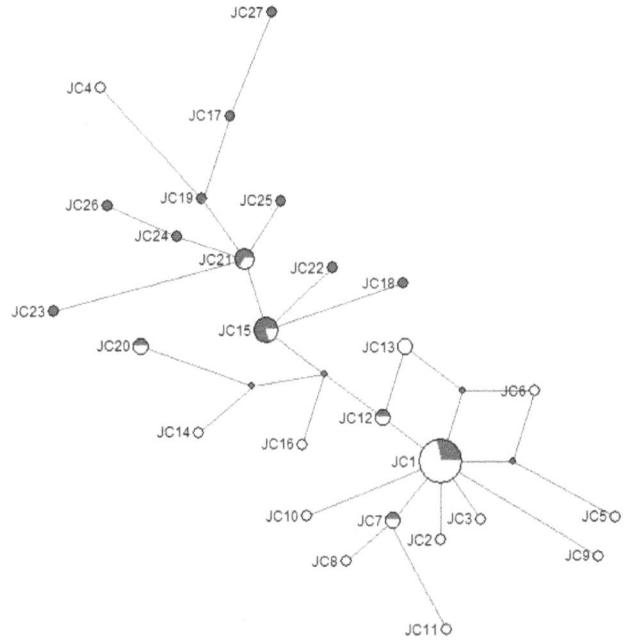

B   *H. trimaculatus*

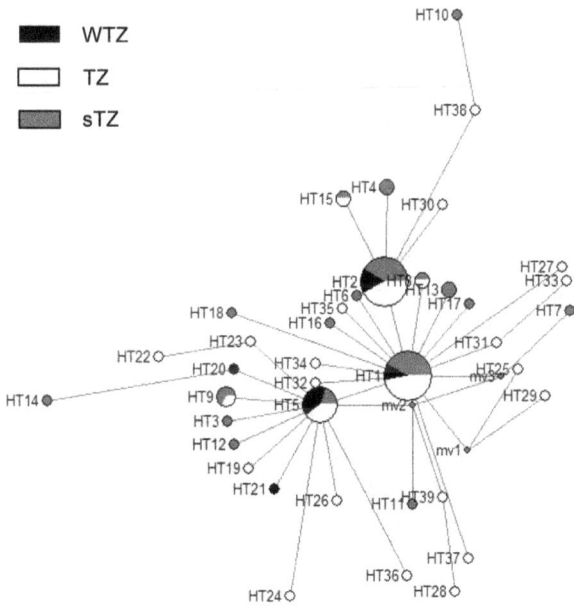
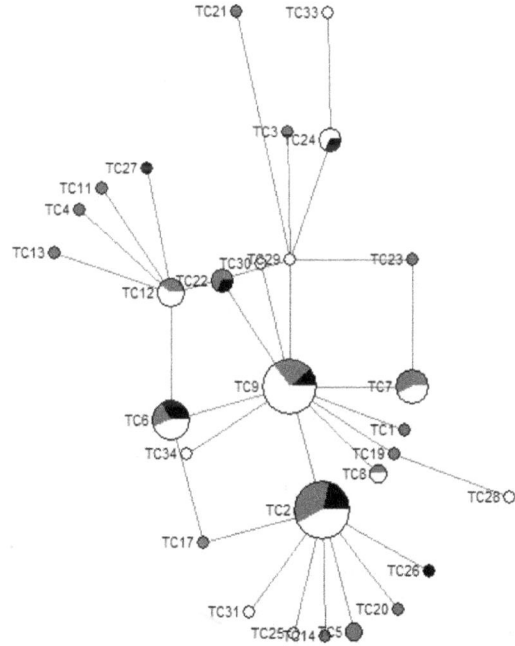

Cytochrome b   Control region

Figure 3. Statistical parsimony network showing phylogenetic relationships among *H. mohnikei* **(A) and** *H. trimaculatus* **(B) haplotypes.** The area of each circle is proportional to the number of specimens sharing that haplotype. Small open squares represent hypothesized intermediate haplotypes not observed in our sample.

not significant in the WTZ group (Table 3). Due to significantly negative Tajima's *D* and Fu's *Fs* values for all seahorse populations combined in the present study (except for negative, but not significant Tajima's *D* values for the WTZ population), we speculate that the populations may have experienced population expansion in the past. The *cytb* and CR data sets exhibited a unimodal mismatch distribution (Fig. 4), which indicated that both seahorse species may have undergone a recent population demographic or range expansion. At the same time, a small and

not statistically significant SSD and R-value showed population expansion (Table 3). Estimates of the time since the start of population expansion of the *H. mohnikei* ranged from about 276,000 years ago, while *H. trimaculatus* populations in China started expanding approximate 230,000 years ago; the oldest dates were for TZ, and the youngest were for WTZ (Table 3). The results of BSP (Fig. 5) also rejected population stability for both species. BSP estimates of *H. mohnikei* group suggested that the population has expanded about 11-fold, from about 0.55 to about

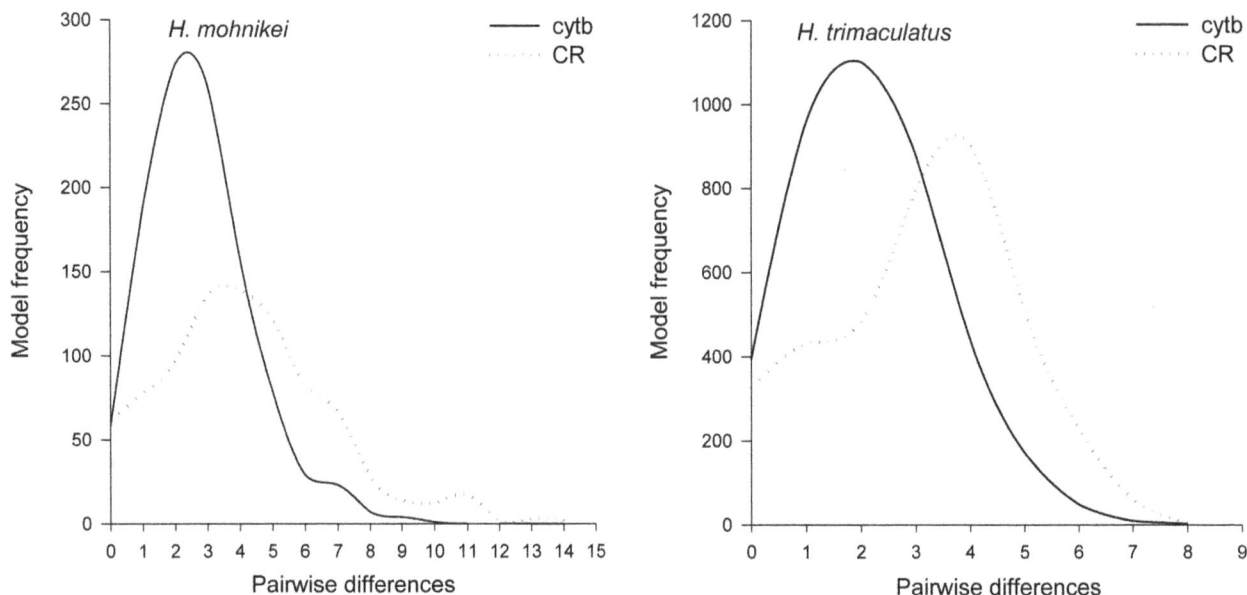

Figure 4. Mismatch distributions for *cytb* and CR sequences found in *H. mohnikei* and *H. trimaculatus*. In each case, the curve represents the observed frequency of pairwise differences among haplotypes.

6. BSP estimates of *H. trimaculatus* group suggested that the population has expanded about 7-fold, from about 1 to about 7.

Discussion

Genetic diversity and structure

In the present study, the haplotype diversity at a high level and the nucleotide diversity in the lower-middle-level indicated that *H. mohnikei* and *H. trimaculatus* populations may have experienced a long period of stable evolution, or there were different lineages along China's coast. This pattern of high haplotype diversity is common in marine fish and consistent with previous studies of *Sardina pilchardus* [32], *Schizothorax prenanti* [33], *H. trimaculatus* [6,34] *H. ingens* [35], and *Hoplostethus atlanticus* [36]. High haplotype diversity at a gene locus within populations may have also been caused by other factors, such as large population size, environmental heterogeneity, life-history traits, origin, as well as ages of the species [37]. The pattern of genetic variability with high haplotype diversity, but relatively low nucleotide diversity, suggests that the population has undergone population expansion [38]. Genetic variability is considered to be the foundation of evolution and can be affected by many factors, such as mutation rates, effective population size, and gene flow [39]. Gene flow is a constraint on local genetic differentiation, and the adaptation between populations and low gene flow between populations can lead to genetic subdivision of populations [40,41].

An advantage of using *cytb* over nuclear genes is that an mtDNA gene tree can yield insights into population history that may be lost due to recombination in a nuclear gene tree. That an NJ tree separated the *cytb* haplotypes with high bootstrap support indicates distinct genetic structuring between the east and west coast populations of seahorses along India's coasts [34]. As shown in Fig. 2, due to the low levels of genetic variation present between our sampled populations, the NJ tree generated from *H. mohnikei* haplotypes and ML tree generated from *H. trimaculatus* haplotypes had low support values, implicating no obvious genetic structure in *H. mohnikei* and *H. trimaculatus* populations. Many

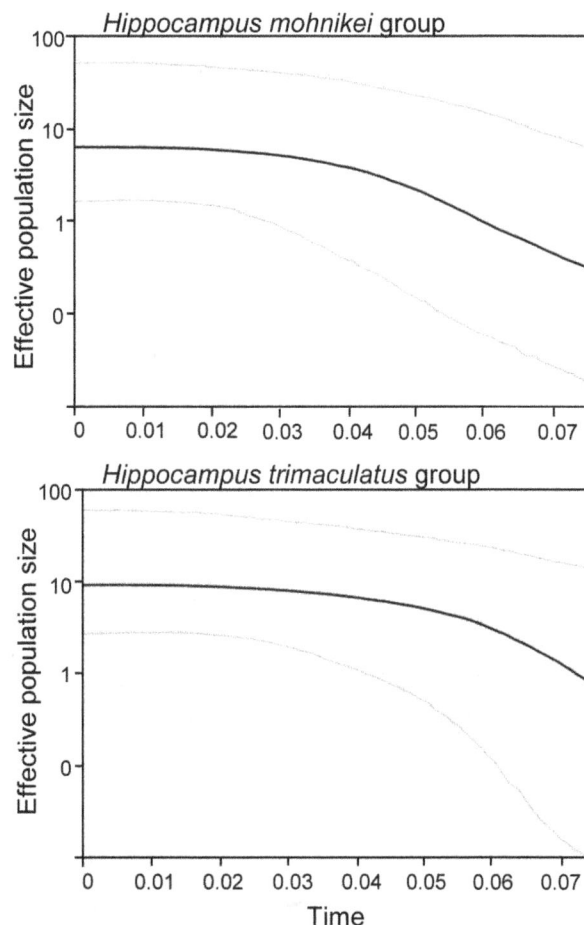

Figure 5. Bayesian skyline plots representing historical demographic trends of *H. mohnikei* and *H. trimaculatus* in China. The mean estimate is enclosed within the 95% highest posterior densities.

Table 3. Demographic statistics for mitochondrial cytochrome *b* of *H. mohnikei* and *H. trimaculatus*.

	Tajima'sD		Fu's Fs		Mismatch Distribution		Tau
	D	P	D	P	SSD	Raggedness index	
H. mohnikei							
YMD	−1.54678	0.04500	−21.79502	0.00000	0.00786	0.05820	2.76562
LZW	−1.52178	0.04500	−5.04335	0.00300	0.00322	0.02595	2.78125
TOTAL	−1.69266	0.02500	−26.46206	0.00000	0.00093	0.03759	2.71484
H.trimaculatus							
WTZ	−0.94729	0.18800	−2.83361	0.02200	0.00631	0.08087	1.85156
sTZ	−1.98345	0.00800	−11.83494	0.00000	0.00074	0.04849	2.12500
TZ	−2.19551	0.00200	−22.30991	0.00000	0.00062	0.03645	2.46875
TOTAL	−2.40148	0.00000	−27.17594	0.00000	0.00064	0.04186	2.25586

Neutrality tests: Tajima's D, Fu's Fs, and expansion (coalescence) time under the sudden expansion assumption in mutation-generations (τ); D: values of Tajima' D and Fu's Fs; P: p-values of Tajima' D and Fu's Fs; SSD: sum of the square deviations.

bootstrap values for nodes were low, indicating that the substructure within the major clusters is uncertain.

AMOVA analyses indicated the absence of significant population genetic differentiation across *H. mohnikei* and *H. trimaculatus* populations. *H. trimaculatus* has the most widespread distribution range, indicating potentially high dispersal capabilities; whereas, *H. mohnikei* is confined to China's Bohai Sea and Yellow Sea, indicating potentially low dispersal capabilities.

Population structure is affected by genetic drift, local adaptation, and gene flow. In a marine environment, the development of population structure is confidently influenced by factors that affect dispersal, such as ocean currents, historical variance, and geographic distance coupled with differences in dispersal ability and habitat discontinuity [35]. The possible explanation for the homogeneity of populations of *H. mohnikei* and *H. trimaculatus* was the high level of gene flow. Although the mobility of seahorses is feeble, marine currents make passive dispersal possible. Overall seasonal circulation in the South China Sea is cyclonic in winter and anticyclonic in summer, with a few stable eddies [42,43]. The seasonal circulation is mostly driven by monsoon winds, and also related to water exchange between the South China Sea and the East China Sea through the Taiwan Strait, and between the South China Sea and the Kuroshio Current through the Luzon Strait (Fig. 1) [42,43]. Several other fish species in the area, which have pelagic larval and/or juvenile stages, show genetic homogeneity among populations and could be passively transported by ocean currents [44,45].

The Yellow Sea circulations play an important role in the passive dispersal of *H. mohnikei*. The eastward Lubei coastal current flows along the northern part of the Shandong Peninsula, and then turns south in Chengshanjiao; however, northeastward currents in the Lunan coast flows from southwest to northeast all year round (Fig. 1). At the same time, there is an offshore mesoscale anticyclonic in Qingdao-Shidao [46]. Therefore, the coastal currents of the Yellow Sea might have limited the dispersal range of *H. mohnikei*. On the other hand, even if there are a few number of *H. mohnikei* migrating to the South China Sea with the ocean currents, the environment may not be suitable for inhabitation of the population, such as the water temperature. These ecological differences often result in varying dispersal, which plays an important role in determining the phylogeographical structure of marine species. The Sea of Japan, East China Sea, and South China Sea have been isolated during the glacial periods [47]. Recent molecular studies indicated that some widespread marine species exhibited phylogeographical patterns corresponding to these three glacial refugia [48]. Geographic boundaries during the Pleistocene may also have played an important role in these species dispersion.

Population expansion

The population of *H. mohnikei* displayed a genetic pattern typical of a population that has undergone a recent population expansion due to its one common haplotype present across the range, most haplotypes unique to single sites, and a pattern of a shallow star-shaped haplotype network. As shown in Fig. 3B, the distribution of the central and abundant haplotype Ht1 and Ht2, extended from WTZ to TZ, which supports that *H. trimaculatus* in this region has undergone range expansion. The range expansion was a recent phenomenon and may not have obtained the migration-drift equilibrium, as shown by the lack of phylogeographical structure [38]. A similar star-like pattern of genetic relatedness among haplotypes was seen in other seahorses, such as *H. hippocampus* (L. 1758) [49].

Due to significantly negative Tajima's D and Fu's Fs values for all seahorse populations combined in the present study (except for negative, but not significant Tajima's D values for the WTZ population), we speculate that the populations may have undergone population expansion in the past. Furthermore, mismatch distributions were calculated for *H. mohnikei* and *H. trimaculatus* to investigate the hypothesis of a population expansion. Previous studies have revealed that population bottlenecks and population expansions have a significant effect on the pattern of genetic polymorphism among haplotypes in the population [19]. These theoretical studies demonstrate that populations in stable demographic equilibrium have a multimodal mismatch distribution (ragged and chaotic); whereas, the distribution appears unimodal after recent demographic expansions [19,50]. The mismatch distributions for *H. mohnikei* and *H. trimaculatus* populations were unimodal and fully consistent with a population expansion. The expansion was initiated 276,000 years ago for *H. mohnikei* and 230,000 years ago for *H. trimaculatus* during the middle Pleistocene period. The Pleistocene, which spans from about 1.6 Myr to 10,000 years before the present, was punctuated by a series of large glacial-interglacial changes [51]. It was probably a result of a high dispersal potential, which was particularly advantageous during the rising of sea water temperatures and levels [52,53,54]. Responding to the climatic events, marine ecosystems make corresponding changes in species distributions, and abundances and productivity [55]. Glaciation-interglaciation events and associated changes in the marine environment probably have had great effects in the demographic history of many marine and coastal fish, such as *Beryx decadactylus* [45], *Hoplostethus atlanticus* [36], and *Glyptocephalus stelleri* [56].

Conclusions

This study demonstrates the genetic variation and population expansion for two seahorses, *H. mohnikei* and *H. trimaculatus*, which are small feeble swimming fish and confidently influenced by their species-specific ecological requirements and life-history traits. Both seahorses have experienced population expansions since the mid-Pleistocene, and the population span of the expansion of *H. mohnikei* is larger and occurred earlier than that of *H. trimaculatus*. The observed lack of population differentiation can be explained by this past population expansion and present-day juvenile or sub-adult dispersal. Our study detected the absence of significant genetic divergence across the South China Sea in *H. trimaculatus*, suggesting that broad-scale conservation management strategies may be appropriate for this species. As a connective study on seahorses, future work will aim to assess the stability of the genetic variation and population expansion in the near future with the possible impact from heavy exploitation of seahorses and environmental change along China's coast.

Acknowledgments

We are grateful to Dr. Zexia Gao of Central China Agricultural University and Dr. Adeljean L.F.C. Ho of the Florida Institute of Technology for their valuable comments on the manuscript, and Geng Qin for assistance in collecting the seahorse samples.

Author Contributions

Conceived and designed the experiments: QL YHZ. Performed the experiments: YHZ NKP HXZ. Analyzed the data: YHZ. Contributed reagents/materials/analysis tools: YHZ QL. Wrote the paper: YHZ QL JDL.

References

1. Reed DH, O'Grady JJ, Brook BW, Ballou JD, Frankham R (2003) Estimates of minimum viable population sizes for vertebrates and factors influencing those estimates. Biological Conservation 113: 23–34.
2. Palumbi SR (1994) Genetic divergence, reproductive isolation, and marine speciation. Annual Review of Ecology and Systematics: 547–572.
3. Silva SE, Silva IC, Madeira C, Sallema R, Paulo OS, et al. (2013) Genetic and morphological variation in two littorinid gastropods: evidence for recent population expansions along the East African coast. Biological Journal of the Linnean Society 108: 494–508.
4. McKenzie K (1991) Implications of shallow Tethys and the origin of modern oceans. Australian Systematic Botany 4: 37–40.
5. Briggs JC (1995) Global biogeography: Elsevier Science.
6. Lourie SA, Vincent ACJ (2004) A marine fish follows Wallace's Line: the phylogeography of the three-spot seahorse (*Hippocampus trimaculatus*, Syngnathidae, Teleostei) in Southeast Asia. Journal of Biogeography 31: 1975–1985.
7. Foster S, Vincent A (2004) Life history and ecology of seahorses: implications for conservation and management. Journal of Fish Biology 65: 1–61.
8. Scales H (2010) Advances in the ecology, biogeography and conservation of seahorses (genus Hippocampus). Progress in Physical Geography 34: 443–458.
9. Bohonak AJ (1999) Dispersal, gene flow, and population structure. Quarterly Review of Biology: 21–45.
10. Lourie S, Foster S, Cooper E, Vincent A (2004) A guide to the identification of seahorses. Washington, D.C.: University of British Columbia and World Wildlife Fund.
11. Kim IS, Lee WO (1995) First record of the seahorse fish, *Hippocampus trimaculatus* (Pisces: Syngnathidae) from Korea. Korean Journal of Zoology (Korea Republic) 38: 74–77.
12. Masuda H, Muzik KM (1984) The fishes of the Japanese Archipelago: Tokai University Press Tokyo.
13. Lourie SA, Vincent AC, Hall HJ (1999) Seahorses: an identification guide to the world's species and their conservation.
14. Hall TA (1999) BioEdit: a user-friendly biological sequence alignment editor and analysis program for Windows 95/98/NT; pp. 95–98.
15. Larkin M, Blackshields G, Brown N, Chenna R, McGettigan P, et al. (2007) Clustal W and Clustal X version 2.0. Bioinformatics 23: 2947–2948.
16. Tamura K, Peterson D, Peterson N, Stecher G, Nei M, et al. (2011) MEGA5: molecular evolutionary genetics analysis using maximum likelihood, evolutionary distance, and maximum parsimony methods. Molecular Biology and Evolution 28: 2731–2739.
17. Librado P, Rozas J (2009) DnaSP v5: a software for comprehensive analysis of DNA polymorphism data. Bioinformatics 25: 1451–1452.
18. Excoffier L, Laval G, Schneider S (2005) Arlequin (version 3.0): an integrated software package for population genetics data analysis. Evolutionary bioinformatics online 1: 47.
19. Rogers AR, Harpending H (1992) Population growth makes waves in the distribution of pairwise genetic differences. Molecular Biology and Evolution 9: 552–569.
20. Dsouli-Aymes N, Michaux J, De Stordeur E, Couloux A, Veuille M, et al. (2011) Global population structure of the stable fly (Stomoxys calcitrans) inferred by mitochondrial and nuclear sequence data. Infection, Genetics and Evolution 11: 334–342.
21. Tajima F (1989) Statistical method for testing the neutral mutation hypothesis by DNA polymorphism. Genetics 123: 585–595.
22. Fu YX (1997) Statistical tests of neutrality of mutations against population growth, hitchhiking and background selection. Genetics 147: 915–925.
23. Curtis J, Vincent A (2006) Life history of an unusual marine fish: survival, growth and movement patterns of Hippocampus guttulatus Cuvier 1829. Journal of Fish Biology 68: 707–733.
24. Heled J, Drummond A (2008) Bayesian inference of population size history from multiple loci. BMC Evolutionary Biology 8: 289.
25. Drummond AJ, Rambaut A (2007) BEAST: Bayesian evolutionary analysis by sampling trees. BMC Evolutionary Biology 7: 214.
26. Grewe PM, Krueger CC, Aquadro CF, Bermingham E, Kincaid HL, et al. (1993) Mitochondrial DNA variation among lake trout (Salvelinus namaycush) strains stocked into Lake Ontario. Canadian Journal of Fisheries and Aquatic Sciences 50: 2397–2403.
27. Teske P, Cherry M, Matthee C (2003) Population genetics of the endangered Knysna seahorse, *Hippocampus capensis*. Molecular Ecology 12: 1703–1715.
28. Swofford D (2002) PAUP 4.0 b10: Phylogenetic analysis using parsimony. Sinauer Associates, Sunderland, MA, USA.
29. Posada D, Crandall KA (1998) Modeltest: testing the model of DNA substitution. Bioinformatics 14: 817–818.
30. Bandelt HJ, Forster P, Röhl A (1999) Median-joining networks for inferring intraspecific phylogenies. Molecular Biology and Evolution 16: 37–48.
31. Slatkin M, Hudson RR (1991) Pairwise comparisons of mitochondrial DNA sequences in stable and exponentially growing populations. Genetics 129: 555–562.
32. Tinti F, Di Nunno C, Guarniero I, Talenti M, Tommasini S, et al. (2002) Mitochondrial DNA sequence variation suggests the lack of genetic heteroge-

neity in the Adriatic and Ionian stocks of Sardina pilchardus. Marine Biotechnology 4: 163–172.

33. Song Z, Song J, Yue B (2008) Population genetic diversity of Prenant's schizothoracin, Schizothorax prenanti, inferred from the mitochondrial DNA control region. Environmental Biology of Fishes 81: 247–252.

34. Goswami M, Thangaraj K, Chaudhary BK, Bhaskar LVSK, Gopalakrishnan A, et al. (2009) Genetic heterogeneity in the Indian stocks of seahorse (*Hippocampus kuda* and *Hippocampus trimaculatus*) inferred from mtDNA cytochrome b gene. Hydrobiologia 621: 213–221.

35. Saarman NP, Louie KD, Hamilton H (2010) Genetic differentiation across eastern Pacific oceanographic barriers in the threatened seahorse *Hippocampus ingens*. Conservation Genetics 11: 1989–2000.

36. Varela AI, Ritchie PA, Smith PJ (2012) Low levels of global genetic differentiation and population expansion in the deep-sea teleost *Hoplostethus atlanticus* revealed by mitochondrial DNA sequences. Marine biology 159: 1049–1060.

37. Nei M (1987) Molecular evolutionary genetics: Columbia University Press.

38. Slatkin M (1993) Isolation by distance in equilibrium and non-equilibrium populations. Evolution: 264–279.

39. Amos W, Harwood J (1998) Factors affecting levels of genetic diversity in natural populations. Philosophical Transactions of the Royal Society B: Biological Sciences 353: 177.

40. Slatkin M (1987) Gene flow and the geographic structure of natural. Science 3576198: 236.

41. Wei DD, Yuan ML, Wang BJ, Zhou AW, Dou W, et al. (2012) Population Genetics of Two Asexually and Sexually Reproducing Psocids Species Inferred by the Analysis of Mitochondrial and Nuclear DNA Sequences. PloS one 7: e33883.

42. Hu J, Kawamura H, Hong H, Qi Y (2000) A review on the currents in the South China Sea: seasonal circulation, South China Sea warm current and Kuroshio intrusion. Journal of Oceanography 56: 607–624.

43. Xu X, Qiu Z, Chen H (1982) The general descriptions of the horizontal circulation in the South China Sea; pp. 137–145.

44. Hoarau G, Borsa P (2000) Extensive gene flow within sibling species in the deep-sea fish *Beryx splendens*. Comptes Rendus de l'Académie des Sciences-Series III-Sciences de la Vie 323: 315–325.

45. Friess C, Sedberry G (2011) Genetic evidence for a single stock of the deep-sea teleost Beryx decadactylus in the North Atlantic Ocean as inferred from mtDNA control region analysis. Journal of Fish Biology 78: 466–478.

46. Fagao Z, Hanli M, Yangui L (1987) Analysis of drift bottle and drift card experiments in Bohai Sea and Huanghai Sea (1975–80). Chinese Journal of Oceanology and Limnology 5: 67–72.

47. Wang P (1999) Response of Western Pacific marginal seas to glacial cycles: paleoceanographic and sedimentological features. Marine Geology 156: 5–39.

48. Liu JX, Gao TX, Wu SF, Zhang YP (2007) Pleistocene isolation in the Northwestern Pacific marginal seas and limited dispersal in a marine fish, *Chelon haematocheilus* (Temminck & Schlegel, 1845). Molecular Ecology 16: 275–288.

49. Woodall L, Koldewey H, Shaw P (2011) Historical and contemporary population genetic connectivity of the European short-snouted seahorse *Hippocampus hippocampus* and implications for management. Journal of Fish Biology 78: 1738–1756.

50. Harpending H (1994) Signature of ancient population growth in a low-resolution mitochondrial DNA mismatch distribution. Human Biology 66: 591.

51. Imbrie J, Boyle E, Clemens S, Duffy A, Howard W, et al. (1992) On the structure and origin of major glaciation cycles 1. Linear responses to Milankovitch forcing. Paleoceanography 7: 701–738.

52. Lee HJ, Boulding EG (2007) Mitochondrial DNA variation in space and time in the northeastern Pacific gastropod, Littorina keenae. Molecular Ecology 16: 3084–3103.

53. Teske PR, Papadopoulos I, McQuaid CD, Newman BK, Barker NP (2007) Climate change, genetics or human choice: why were the shells of mankind's earliest ornament larger in the pleistocene than in the holocene? PLoS ONE 2: e614.

54. Crandall ED, Frey MA, Grosberg RK, Barber PH (2008) Contrasting demographic history and phylogeographical patterns in two Indo-Pacific gastropods. Molecular Ecology 17: 611–626.

55. Webb T, Bartlein P (1992) Global changes during the last 3 million years: climatic controls and biotic responses. Annual Review of Ecology and Systematics 23: 141–173.

56. Xiao Y, Gao T, Zhang Y, Yanagimoto T (2010) Demographic history and population structure of blackfin flounder (*Glyptocephalus stelleri*) in Japan revealed by mitochondrial control region sequences. Biochemical Genetics 48: 402–417.

7

Disruption of the Thyroid System by the Thyroid-Disrupting Compound Aroclor 1254 in Juvenile Japanese Flounder (*Paralichthys olivaceus*)

Yifei Dong, Hua Tian, Wei Wang, Xiaona Zhang, Jinxiang Liu, Shaoguo Ru*

Marine Life Science College, Ocean University of China, Qingdao, Shandong Province, The People's Republic of China

Abstract

Polychlorinated biphenyls (PCBs) are a group of persistent organochlorine compounds that have the potential to disrupt the homeostasis of thyroid hormones (THs) in fish, particularly juveniles. In this study, thyroid histology, plasma TH levels, and iodothyronine deiodinase (IDs, including ID_1, ID_2, and ID_3) gene expression patterns were examined in juvenile Japanese flounder (*Paralichthys olivaceus*) following 25- and 50- day waterborne exposure to environmentally relevant concentrations of a commercial PCB mixture, Aroclor 1254 (10, 100, and 1000 ng/L) with two-thirds of the test solutions renewed daily. The results showed that exposure to Aroclor 1254 for 50 d increased follicular cell height, colloid depletion, and hyperplasia. In particular, hypothyroidism, which was induced by the administration of 1000 ng/L Aroclor 1254, significantly decreased plasma TT_4, TT_3, and FT_3 levels. Profiles of the changes in mRNA expression levels of IDs were observed in the liver and kidney after 25 and 50 d PCB exposure, which might be associated with a reduction in plasma THs levels. The expression level of ID_2 mRNA in the liver exhibited a dose-dependent increase, indicating that this ID isotype might serve as sensitive and stable indicator for thyroid-disrupting chemical (TDC) exposure. Overall, our study confirmed that environmentally relevant concentrations of Aroclor 1254 cause significant thyroid disruption, with juvenile Japanese flounder being suitable candidates for use in TDC studies.

Editor: Cheryl S. Rosenfeld, University of Missouri, United States of America

Funding: This work was supported by the National Natural Science Foundation of China (31202001) www.nsfc.gov.cn, Natural Science Foundation of Shandong Province (ZR2012CQ010) www.sdnsf.gov.cn, and Marine Public Scientific Research Funding Project (2012418012) www.soa.gov.cn. The funders had no role in study design, data collection and analysis, decision to publish, or preparation of the manuscript.

Competing Interests: The authors have declared that no competing interests exist.

* Email: rusg@ouc.edu.cn

Introduction

Polychlorinated biphenyls (PCBs) have been listed as one of 21 persistent organic pollutants (POPs) under the Stockholm Convention, due to their recalcitrance to degradation and tendency to biomagnify up the food chain. PCBs are widely studied TDCs that potentially cause various abnormalities in the thyroid system of vertebrates [1–3], especially in amphibians and mammals [4]. Recently, the disturbance of fish thyroid systems by PCBs has received increasing research focus; however, the thyroidal responses of fish to PCBs has shown variable results in different studies [5]. For example, Aroclor 1242 and 1254 (commercial PCBs mixtures) lowered plasma 3,5,3' -triiodothyronine (T_3) levels without altering plasma thyroxine (T_4) levels when fed to adult coho salmon (*Oncorhynchus kisutch*) [6]. The injection of Aroclor 1254 has been shown to increase plasma T_3 levels and delay the plasma T_4 surge commonly associated with smoltification [7]. These variable effects on thyroid hormone (THs) levels may be related to the physiological stage or age of fish used in different laboratory studies. Most of these studies preferentially used adult fish of sufficient size/age to either obtain adequate blood samples for THs measurement or the assessment of other thyroid indices, while only a few studies have used juvenile fish to assess the thyroid disrupting effects of PCBs [5].

Some researchers recommended that young developing fish should be the focus of future studies on thyroid disruption, because juvenile fish are more dependent on the regulation of THs and more sensitive to TDCs compared to adult fish [5,8]. THs have been linked to a multitude of important functions in early development of fish, such as growth, tissue differentiation, and metamorphosis [8,9]. In Japanese flounder (*Paralichthys olivaceus*), the exogenous administration of THs or elevation of endogenous T_4 levels by thyroid stimulating hormone (TSH) induces advanced metamorphosis, while thiourea (TU, an anti-thyroid drug) treatment delays the metamorphosis process [10,11]. Exogenous THs also induce the transition of muscle proteins, replacement of erythrocytes, skin pigmentation, and development of the gastric glands in fish [12,13]. These findings indicate that THs are fundamental for the early development and growth of fish, and that TH disruption in juvenile fish may cause growth retardation or abnormal development. Therefore, juvenile fish are assumed to be particularly susceptible to thyroid disruption.

A series of endpoints have been proposed to assess the effects of PCBs on the fish thyroid cascade, and mainly include central controlled effects and peripheral controlled effects [8]. Measurement of the central control of the thyroid cascade may be accomplished *via* thyroid histopathological analysis, in addition to

Table 1. Nucleotide sequences of primers used for real-time PCR and product sizes.

Gene	GenBank Accession No.	Primer sequence (5′–3′)	Amplicon size (bp)
ID$_1$	AB362421	GGTGGTGGACGAAATGAATG	147
		TCCAGTAACGAACGCACCTCT	
ID$_2$	AB362422	GCACCAGAACTTGGAGGAGAG	142
		GCACACTCGTTCGTTAGACACA	
ID$_3$	AB362423	TGGCTGGAGCAGTACAGGAG	103
		TGAGGCAGAATGGGCAGA	
5S-rRNA	AB154836	CCATACCACCCTGAACAC	102
		CGGTCTCCCATCCAAGTA	

measurement of plasma total and free THs levels [14–18]. The peripheral control of the conversion of T_4 to T_3 may be assessed *via* a suite of iodothyronine deiodinase (IDs) activities in the liver or other extra-thyroid tissues [19,20]. Three ID isotypes are mainly expressed in teleosts, with these enzymes presenting different catalytic properties [21]. In particular, ID$_1$ exhibits both outer ring-deiodination (ORD) and inner ring-deiodination (IRD) activities; however, when combined with its preferred substrate, 3,3′,5′-triiodo-L-thyronine (rT$_3$), this enzyme is considered to become even more involved in the degradation of THs, particularly the inactivation of rT$_3$ to 3,3′-diiodo-L-thyronine (3,3′-T$_2$). ID$_2$ activates the ORD pathway, by converting T_4 into T_3. ID$_3$ catalyses the IRD pathway, which converts T_4 and T_3 into the inactive metabolites rT$_3$ and 3,3′-T$_2$, respectively [22–24].

The Japanese flounder is an economically important species that is considered to be an ideal model organism for the study of thyroid disruption. The important roles of THs during the early stage of the development of this flatfish have been extensively demonstrated, particularly during metamorphosis [25–28]. To date, effects of PCBs on the thyroid system of the Japanese flounder remain unclear. This study aimed to obtain an integrated insight into the effects of environmentally relevant concentrations of Aroclor 1254 on the thyroid system of juvenile Japanese flounder. Changes in the development and growth of this fish species were examined, and the tissue levels of PCB congeners were measured. We anticipate that these analyses will indicate the potential suitability of using juvenile Japanese flounder as candidates for use in TDC studies.

Materials and Methods

Ethics statement

The fish were handled according to the National Institute of Health guidelines for the handling and care of experimental animals. The animal utilization protocol was approved by the Institutional Animal Care and Use Committee of the Ocean University of China. All surgery was performed under MS-222 anesthesia, and all efforts were made to minimize suffering.

Animals

Experimental trials were conducted in the marine life science college of Ocean University of China. A total of 360 juvenile Japanese flounder (80 days post hatching) were purchased from a commercial fish farm in China. The fish were raised in 240-L tanks containing 200 L of sand-filtered natural seawater (pH 8.0±0.1; 33 ppt salinity) at an ambient temperature (23±3°C). To minimize the aggressive behavior of juvenile fish, a 24-h dark photoperiod (light/dark cycle, 0/24 h) was maintained, with the tanks only being lit up 10 min before each feeding. Fish were fed a commercial flounder feed (Marubeni Nisshin feed, Chuo-Ku, Japan) 4 times a day (2% total fish weight per tank per day) between 08:00 and 20:00. Fish were allowed to acclimate to experimental conditions for 2 weeks prior to the initiation of experiments. The average wet body weight (W$_T$) of the fish used in the experiment was 6.21±1.77 g, and the total body length (L$_T$) was 8.04±1.54 cm.

Table 2. The contents of 7 tracer PCB congeners in juvenile Japanese flounder.

Test item	Control	10 ng/L	100 ng/L	1000 ng/L
PCB28	N/D	N/D	N/D	N/D
PCB52	6.90	10.44	28.53	156.15
PCB101	9.21	20.95	52.33	266.03
PCB118	4.04	12.70	40.98	209.51
PCB153	1.14	7.89	18.48	94.39
PCB138	1.77	6.99	30.18	156.30
PCB180	N/D	0.99	2.078	7.78
Total (ng/kg ww)	21.07	59.99	173.48	890.18

N/D: not detected.

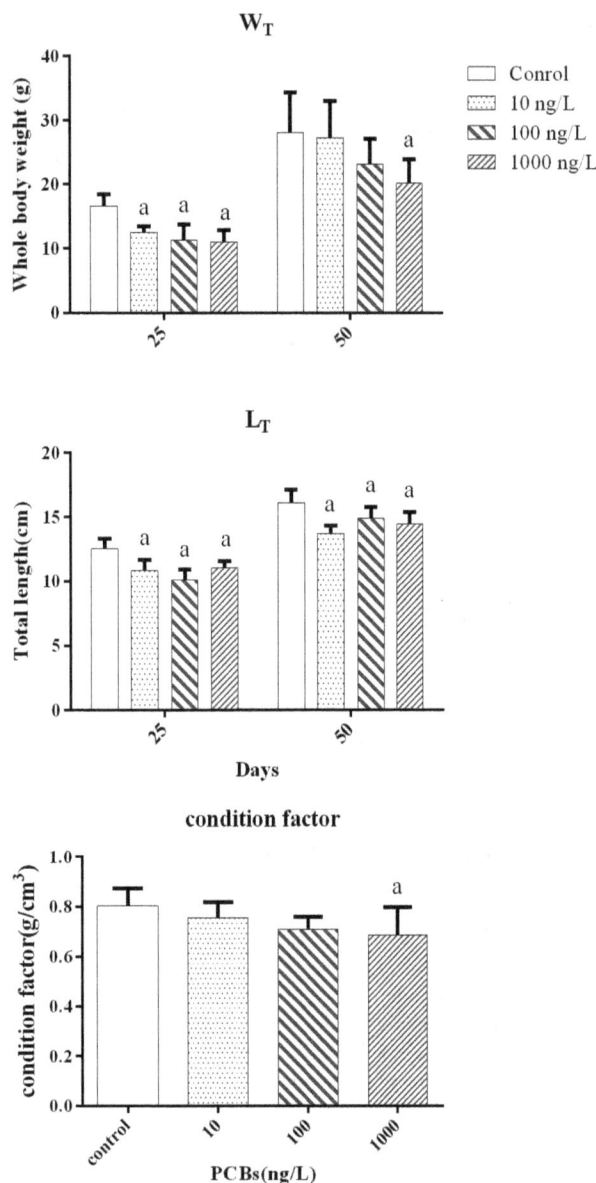

Figure 1. Effect on the total body length, body weight, and condition factor in juvenile Japanese flounder exposed to 0, 10, 100, and 1000 ng/L Aroclor 1254 for 25 and 50 d. The condition factor was calculated at the end of 50 d exposure. [a] $P<0.05$ indicates significant differences between the exposure groups and corresponding control.

Experimental design and fish sampling

Fish were randomly assigned to a control group and 3 treatment groups (size of each group $n = 90$ in each case). Juvenile Japanese flounder were exposed to Aroclor 1254 (AccuStandard Inc, NH, USA, CAS 11097-69-1) at 0 (control), 10, 100, and 1000 ng/L. Aroclor 1254 stock concentrate (1 mg/mL) was made up in ethanol (50 mg Aroclor 1254 was dissolved in 50 mL ethanol). During exposure, two-thirds of the test solutions were changed once per day, and the appropriate amount of seawater and stock solution was added to maintain the specified chemical concentrations.

Fish were deprived of food on the last day of exposure. After 25 and 50 days of exposure, all fish were anesthetised in MS-222 (Sigma, St. Louis, MO, USA), and rinsed with distilled water. The L_T and W_T of the fish in each tank ($n = 9$) were measured to calculate the condition factor (CF = $100 \times W_T$ (g)/L_T (cm)^3). Blood was collected in heparinised tubes by puncturing the caudal vein within 3 min of netting the fish. After centrifugation, plasma was collected and stored at $-80°C$ until RIA. In particular, at the 25-day sampling point, the plasma of 2–3 fish was pooled ($n = 9$). The liver and kidney tissues ($n = 9$) were isolated, frozen in liquid nitrogen, and stored at $-80°C$ until further processing. For the histology analysis, the thyroid tissues enclosed in the subpharyngeal area that were sampled at the end of 50 d exposure ($n = 9$) were fixed in formalin fixative for 24 h at $4°C$, and stained with hematoxylin-eosin.

PCB contaminant analysis

The real concentrations for the 7 tracer PCB congeners in whole fish was measured with GC-MS as described in [29]. At the end of 50 d exposure, 2–3 fish (approximately 50 g in total weight) from each group were randomly sampled, lyophilized, and homogenised in 20 g anhydrous sodium sulphate, and were then placed into a Soxhlet extractor. Samples with 7 types of ^{13}C recovery internal standards (PCB 28, 52, 101, 118, 153, 138, and 180) were extracted by 350 mL hexane/dichloromethane = 1:1 (v/v) for 24 h. After primary purification by gel permeation chromatography, the extracts were then placed in acid silica (44% sulphuric acid, w/w) for further purification and component separation. Hexane (20 mL) was used for the complete elution of PCBs. The final eluate was concentrated to 100 μL under nitrogen and then transferred to a GC vial with ^{13}C-PCB 202 inlet internal standards. The PCB congeners were analysed using an Agilent 6890N/5973i GC-MS system (Agilent Technologies Inc., Palp Alto, USA). The GC-MS analytical parameters have been conducted by referring to Environmental Protection Series: Reference method for the analysis of polychlorinated biphenyls (EPS 1/RM/31, Canada).

Thyroid histological processing

All histopathological endpoints were assayed, as described in [16], with minor modifications. Serial sections were examined under a light microscope until 6–14 follicles/fish were found and photographed. Follicular cell height on the pictures was quantified using Image-Pro plus (version 6.0.0.260). Follicular cell height was determined by obtaining 6 measurements at regular intervals along the follicle perimeter (i.e. 36–84 follicular cell height measurements for each fish and 324–756 measurements for each treatment). A grading system was applied for the hyperplasia evaluation: Grade 1, focal hyperplasia; Grade 2, thyroid follicular cell with less than 50% hyperplasia; and Grade 3, thyroid follicular cell with more than 50% hyperplasia. The average score of 9 fish from each treatment was used, which was based on the sum of the grade of each fish. The sum of the number of colloid deletions (per 10 follicles) was calculated (colloid deletion follicle/10 follicles).

RNA isolation and quantitative RT-PCR

The procedures for RNA extraction and gene expression analysis were performed as previously described by [31]. In brief, total RNA was isolated from the liver and kidney using TRIzol reagent (Invitrogen, Carlsbad, CA, USA) following the manufacturer's instructions. Equal amounts of RNA (1 μg) were reverse-transcribed into cDNA using a PrimeScript RT reagent kit (Takara Bio Inc., Shiga, Japan). Primers were designed for the specific amplification of ID_1, ID_2, ID_3, and 5S-rRNA (an internal control) according to the sequences published in GenBank (Table 1).

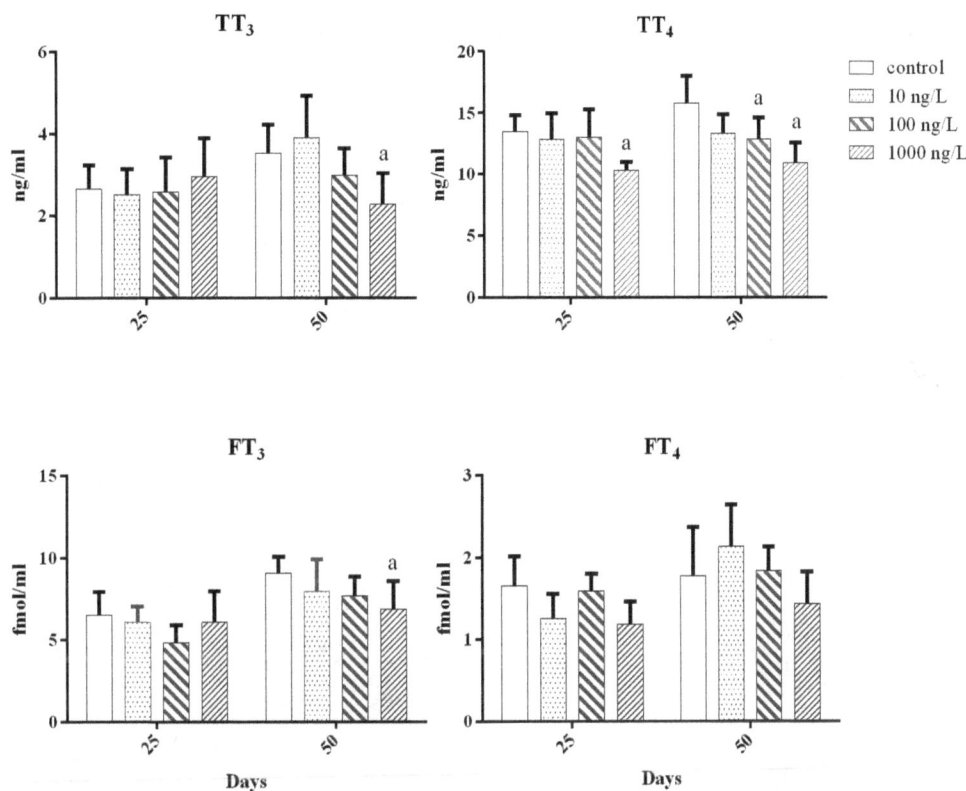

Figure 2. Quantification of plasma TT$_3$, TT$_4$, FT$_3$, and FT$_4$ contents in Japanese flounder exposed to 0, 10, 100, and 1000 ng/L of Aroclor 1254 for 25 and 50 d. [a] $P<0.05$ indicates significant differences between the exposure groups and corresponding control.

All reactions were run on a Eppendorf MasterCycler ep *RealPlex*[4] (Eppendorf, Wesseling-Berz-dorf, Germany). Parallel PCR reactions were conducted to amplify the target gene and 5S-rRNA. Real-time PCR was performed in 20 μL reaction mixtures containing 1× SYBR *Premix Ex Taq* (Takara Bio Inc., Shiga, Japan), 0.4 μM for each primer, 0.4 μL of ROX Reference Dye (Takara Bio Inc., Shiga, Japan), and 4 μL of first-strand cDNA (template). The thermal profile was 95°C for 30 s followed by 40 cycles of 95°C for 5 s and 60°C for 30 s. To ensure that a single product was amplified, melting curve analysis was performed on the PCR products at the end of each PCR run. In addition, 2% agarose gel electrophoresis of the PCR products was performed to confirm the presence of single amplicons of the correct predicted size (not shown). 5S-rRNA transcripts were used as housekeeping genes to standardize the results and to eliminate variations in mRNA and cDNA quantity and quality. 5S-rRNA levels were not affected by any of the experimental conditions in the study. The target gene mRNA abundance in each sample, relative to the abundance of 5S-rRNA, was calculated by the formula $2^{-\Delta\Delta Ct}$ and plotted on a logarithmic scale [31].

Hormone assay

Muscular TT$_3$, TT$_4$, FT$_3$, and FT$_4$ concentrations were measured by radio immunoassay (RIA) (Beijing North Institute of Biological Technology, Beijing, China) according to the manufacturer's instructions. The assay detection limits were 0.05 ng/mL for TT$_3$, 2 ng/mL for TT$_4$, 0.5 fmol/mL for FT$_3$, and 1 fmol/mL for FT$_4$. The inter- and intra-assay coefficients of variation for all the stated hormones were <10% and <15%, respectively.

Statistics

All data are presented as the mean ± standard deviation. Data normality was verified using the Kolmogorov-Smirnov test [32], and homogeneity of variance was checked by Levene's test. If the data failed to pass the test, a logarithmic transformation of the data was performed and retested. Significant differences were assessed between each treatment and the control using one-way analysis of variance (ANOVA), followed by Tukey's multiple comparisons test. $P<0.05$ was considered to be statistically significantly different. All statistical tests were conducted using GraphPad PRISM (Version 6.00) software.

Results

PCB concentrations in Japanese flounder juvenile

The concentrations of 7 tracer PCB congeners in juvenile Japanese flounder are shown in Table 2. A concentration-dependent bioconcentration of Aroclor 1254 was measured in the whole body of all exposure groups. In the 1000 ng/L treatment, the total concentration of measured PCB congeners (including PCB28, PCB52, PCB101, PCB118, PCB153, PCB138, and PCB180) reached 890.18 ng/g ww.

Effects of Aroclor 1254 on the growth of juvenile Japanese flounder

During exposure, mortality rates were below 10% in all groups. As shown in Fig. 1, after 25 days of exposure, Aroclor 1254 significantly reduced W$_T$ and L$_T$ in all treatments. After 50 days of exposure, 10 ng/L and 100 ng/L Aroclor 1254 did not affect W$_T$, but

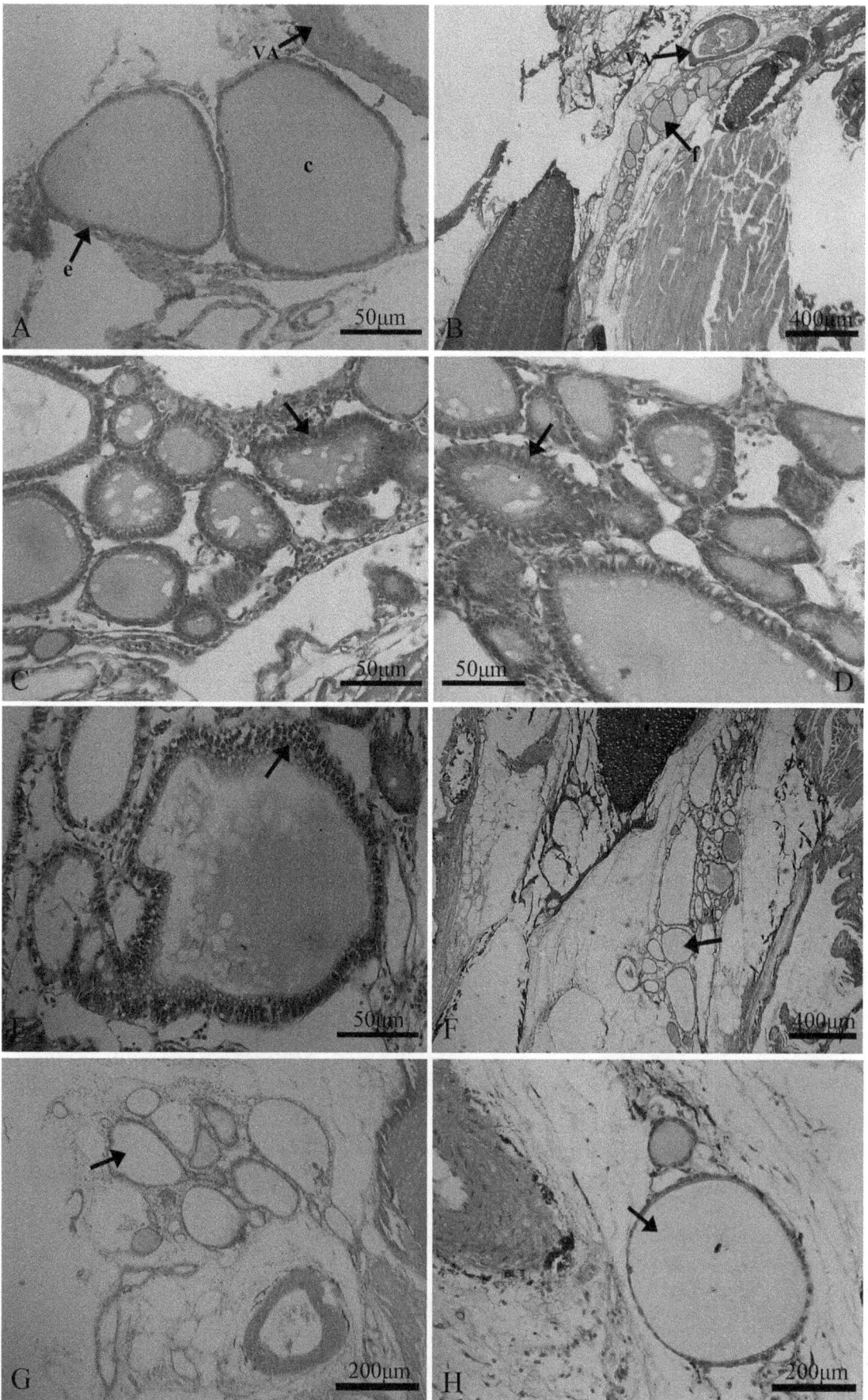

Figure 3. Histological structure of thyroid follicles in juvenile Japanese flounder exposed to 0, 10, 100, and 1000 ng/L Aroclor 1254 for 50 d. (A) and (B) control fish presenting ovoid follicles of variable sizes filled with colloid and lined with squamous follicle cells; (C) and (D) significantly increased epithelial cell height with a little colloid depletion in the lumen after exposure to 100 ng/L Aroclor 1254. (E) Focal hyperplasia in fish exposed to 1000 ng/L. (F) and (G) colloid depletion in fish exposed to 1000 ng/L. (H) Dispersed and reticular colloid in fish exposed to 1000 ng/L. VA = ventral aorta, f = thyroid follicle, c = colloid, and e = thyroid follicle epithelial cell.

significantly inhibited L_T. Exposure to 1000 ng/L Aroclor 1254 for 50 days significantly reduced W_T, L_T, and CF, relative to the control.

Effects of Aroclor 1254 on plasma TT_4, TT_3, FT_4, and FT_3 levels

The effects of Aroclor 1254 on plasma THs levels are shown in Fig. 2. In flounder exposed to different concentrations of Aroclor 1254 for 25 days, the TT_3, FT_3, and FT_4 levels in the plasma were not significantly altered by any of the treatments, whereas plasma TT_4 levels significantly decreased in the 1000 ng/L group. After 50 days of Aroclor 1254 exposure, both plasma TT_3 and FT_3 levels significantly decreased in the 1000 ng/L group, with plasma TT_4 levels showing a dose-dependent decrease, which was significant at concentrations of 100 and 1000 ng/L, while plasma FT_4 levels remained unaltered.

Effects of Aroclor 1254 on thyroid histopathology

The control fish presented oval thyroid follicles of variable sizes that were filled with colloid. In addition, the follicles were line with a single layer of cuboidal to flat follicle epithelial cells (Fig. 3A, B). Representative histopathological abnormalities in Japanese flounder exposed to different concentrations of Aroclor 1254 for 50 days are shown in Fig. 3C–H, including increased epithelial cell height (Fig. 3C, D), hyperplasia (Fig. 3E), and colloid depletion (Fig. 3F, G). Compared to the control group, the colloid observed in the 100 ng/L and 1000 ng/L groups was foamy in appearance, and colloid density decreased (Fig. 3H). For the quantitative analyses, significantly increased levels of follicular epithelial cell height, hyperplasia, and colloid depletion were observed in the 100 and 1000 ng/L Aroclor 1254 treatments (Fig. 4).

Effects of Aroclor 1254 on ID_1, ID_2 and ID_3 mRNA expression in the liver and kidney

As shown in Fig. 5, after 25 days of exposure to Aroclor 1254, ID_1 mRNA levels in the kidney were significantly higher in the 10 and 100 ng/L grouts; however, no significant difference was observed for the liver in any of the treatments. The significant up-regulation of ID_2 and ID_3 mRNA levels was observed in both the kidney and liver of all treatments. In juvenile Japanese flounder exposed to Aroclor 1254 for 50 days, significantly higher ID_1 mRNA levels were obtained in the kidney and liver of the100 ng/L and 10 ng/L groups, respectively. The transcription of ID_2 mRNA in the kidney was significantly stimulated on exposure to 100 and 1000 ng/L Aroclor 1254, which were significantly upregulated in the liver for all treatments. The transcription levels of ID_3 in the kidney and the liver were not significantly altered by any Aroclor 1254 treatment.

Discussion

Our results showed that exposure to Aroclor 1254 significantly decreased plasma TT_4 and TT_3 levels (Fig. 2). However, interpretation of PCBs on the fish thyroid system is exceedingly complex, and does not appear to elicit consistent, detectable plasma TH responses (Table 3). At least three categories of factors have to be considered: 1) test-species variable, 2) the variable

composition of PCB mixtures, and 3) the distinction between exposure and effect due in part to thyroid compensation [5].

In this study, Aroclor 1254 exposure inhibited the L_T, W_T, and CF of juvenile Japanese flounder, which probably led to growth retardation. Crane et al. [15] found that ammonium perchlorate reduces plasma T_4 levels, which inhibited the development of fathead minnow (*Pimephales promelas*) larvae. Schmidt et al. [18] reported that exposure of zebrafish larvae to potassium-perchlorate caused a significant decrease in both plasma T_4 levels and CF. The current study also found that Aroclor 1254 exposure causes plasma T_3 and T_4 levels to decline. Because THs are important in the development and growth of teleosts, particularly during the early life stages, this type of thyroid disruption might inhibit the growth of juvenile Japanese flounder.

However, exposure to PCBs produced different results in adult and juvenile fish. For instance, the study by Schnitzler et al. [29] showed that one PCB mixture induced muscle T_4 levels to decrease in adult sea bass (*Dicentrarchus labrax*), without affecting body length, body weight, or specific growth rates. Iwanowicz et al. [33] reported that the intraperitoneal (*i. p.*) injection of 5 mg/kg Aroclor 1248 caused plasma T_3 levels to decrease in the brown bullhead (*Ameiurus nebulosus*), but had no significant effects on plasma T_4 levels or CF. Following exposure to PCB 126 by *i. p.* injection lower plasma T_4 concentrations was observed in adult lake trout (*Salvelinus namaycush*), whereas it had no effect on fish growth or condition [34]. In adult fish, abundant stores of THs have been found in muscles and other tissues, in addition to the plasma pool, thyroid tissues [35]. These TH stores in extra-thyroidal tissues might be released into the bloodstream or peripheral tissues to compensate thyroid disruption induced by exposure to exogenous compounds. Brown et al. [36] found that muscle T_3 and T_4 contents rapidly reduced in rainbow trout exposed to the PCB 126, with few changes in the histology of thyroid follicles and growth rate. This finding indicates that adult fish have a mechanism to compensate for the thyroid system, enabling them to balance available TH content in peripheral tissues, which does not affect growth. In contrast, the peripheral tissues of juveniles contained relatively low TH levels; therefore, TH deficiency in juveniles might be more likely to trigger a negative feedback regulation compared to adult fish, inducing a series of cascading effects that involve the hypothalamus-pituitary-thyroid (HPT) axis to maintain TH homeostasis. Thus, thyroid tissue might stimulate TH synthesis in juvenile Japanese flounder exposed to Aroclor 1254, based on the observed increase in epithelial cell height, hyperplasia of thyroid follicular epithelial cells, and colloid deletion in the current study. This phenomenon might, to some extent, be attributed to the feedback response to Aroclor 1254 within the thyroid cascade.

The severity of colloid depletion and epithelial cell height are routinely employed markers for identifying thyroid disruption [16]. Crane et al. [15] pointed out that colloid depletion indicates serious injuries, close to the collapse of follicles. In the present study, juvenile Japanese flounder exposed to 100 and 1000 ng/L Aroclor 1254 had significantly greater thyroid follicular epithelial cell height, which reduced colloid area. Many irregularly shaped follicles, some without colloids, were observed, particularly in the highest exposure group. These degenerative changes of the thyroid

Epithelial cell heigh

Hyperplasia

Colloid deletion

PCBs (ng/L)

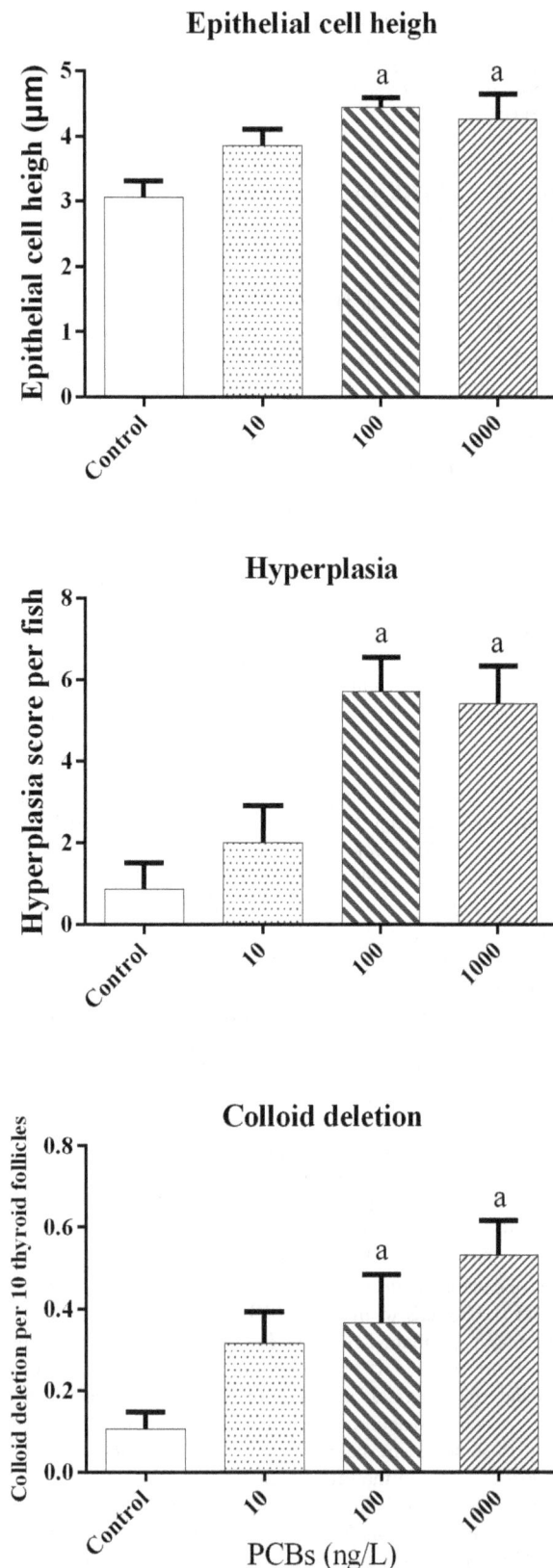

Figure 4. Measurement of epithelial cell height, colloid deletion, and hyperplasia of thyroid follicle in juvenile Japanese flounder exposed to Aroclor 1254 for 50 d. [a] $P<0.05$ indicates significant differences between the exposure groups and corresponding control.

tissues might cause hypothyroidism in juvenile Japanese flounder, preventing them from balance the decrease in TT_4 baselines in the 2 Aroclor groups with the highest concentrations (100 and 1000 ng/L) after 50 d exposure.

In particular, changes in thyroid tissue histology caused by Aroclor 1254 exposure were similar to those induced by perchlorate. Perchlorate blocks the iodine uptake of thyroid follicles by competitively inhibiting iodide and sodium/iodine transport proteins from combining; thereby, hindering the synthesis of THs [37]. Consequently, the decline in TH levels might stimulate TSH secretion from the pituitary through the feedback pathway, and eventually cause compensatory hypertrophy, hyperplasia, and colloid reduction of thyroid follicular cells [14,16,18]. In contrast, some inorganic chemicals, like Cd^{2+}, directly damage thyroid follicles by inducing lipid peroxidation; thus, affecting TH synthesis. Therefore, the toxicity mechanism of Aroclor 1254 on thyroid follicles might be similar to that of perchlorate, rather than the direct effect of heavy metals, such as Cd^{2+}. In other words, Aroclor 1254 probably causes plasma TH levels to decrease in juvenile Japanese flounder; thereby, inducing the compensatory hypertrophy and hyperplasia of thyroid follicular cells through negative feedback pathways, to promote TH synthesis.

Previous studies have shown that deiodinase in fish is sensitive to environmental contaminants, such as metals, polychlorinated biphenyls, and pesticides [38–40]. Van der Geyten et al. [41] demonstrated that changes in hepatic ID_1 and ID_2 activities tend to be consistent with that of their mRNA levels, indicating pre-translated regulation, by which deiodinase mRNA levels coincide with deiodinase enzyme activities. In addition, Picard-Aitken et al. [42] suggested that deiodination gene expression could be used as sensitive biomarkers to indicate thyroid disruption in fish on exposure to environmental chemicals. In the present study, the gene expression of IDs in juvenile Japanese flounder was sensitive to exposure to Aroclor 1254. After 25 and 50 d exposure, Aroclor 1254 stimulated the transcription of ID_2 mRNA in the kidney and liver, which would result in more T_4 being converted into T_3. Another study also found that exposure of sea bass to a mixture of Aroclor 1254 and 1248 led to a significant increase in ID_2 activities [29]. ID_2 mRNA expression tended to be the most sensitive and stable indicator for thyroid disruption in the present study, because it showed a dose-dependent increase in all treatment groups after both 25 and 50 days exposure, especially in the liver. However, it is difficult to distinguish whether Aroclor 1254 has a direct or indirect disrupting effect on the thyroid system by triggering compensatory mechanisms within the thyroid system; consequently, it is difficult to explain how the thyroid status of juvenile Japanese flounder exposed to Aroclor 1254 is altered by only a few indicators. Blanton and Specker [8] suggested that the actions of certain xenobiotics at different levels of the fish thyroid cascade could not be independently monitored by any biomarker. However, ID_2 represents one important indicator for interpreting disruption to the thyroid cascade in fish exposed to environmental contaminants.

After 25 d exposure, 10 and 100 ng/L Aroclor 1254 caused ID_1 and ID_3 mRNA expression levels to increase, especially in the kidney. This response would accelerate the metabolism of T_3, which helps maintain plasma THs homeostasis. At the highest dose, the mRNA expression of ID_2 in the kidney and liver was significantly upregulated, while the expression of ID_1 showed no significant change. This result also indicates the presence of a compensatory response to decreased plasma TT_4 levels, to maintain stable plasma TT_3 levels; otherwise, the increased mRNA expression of ID_3 in the kidney and liver might aggravate

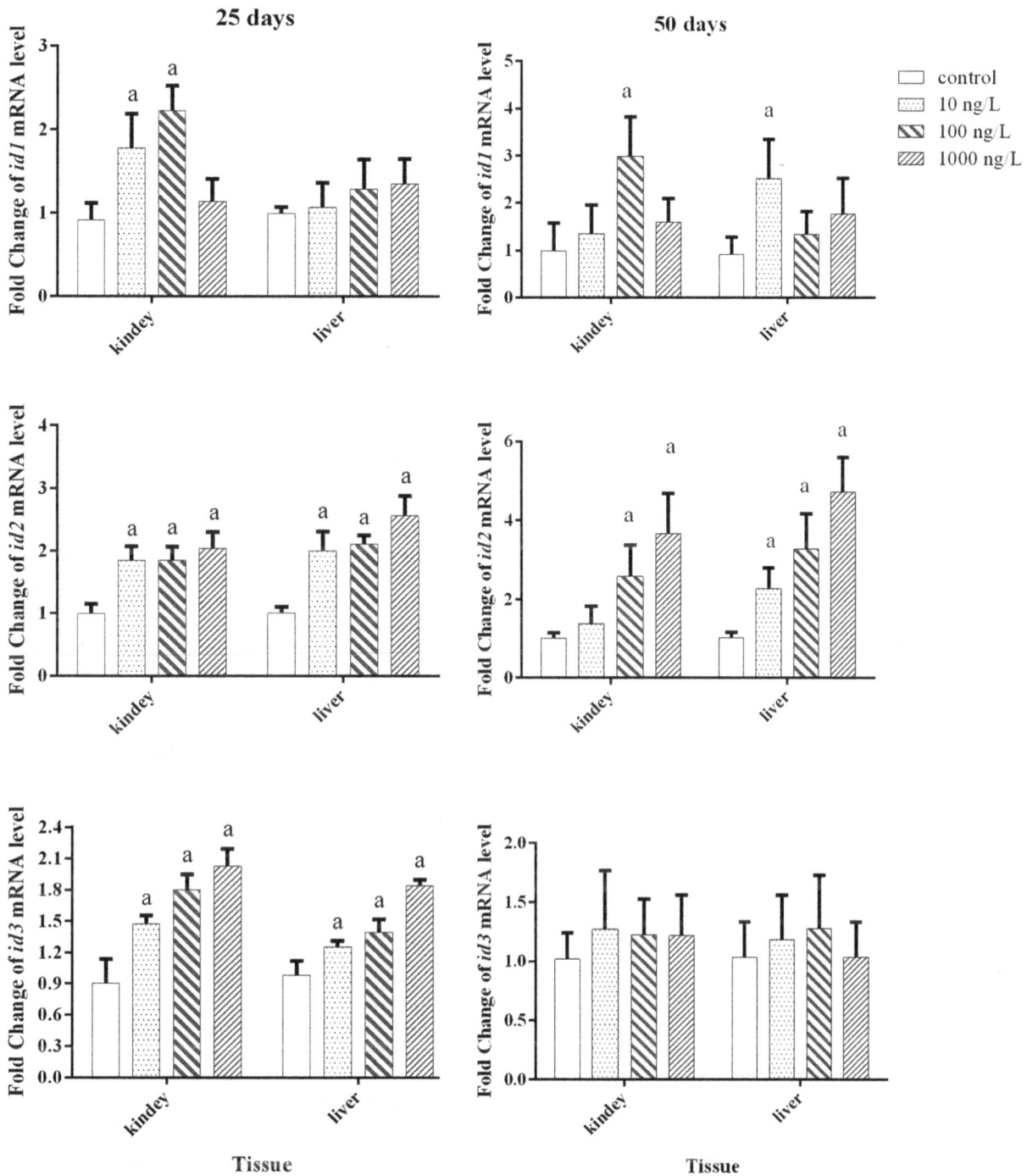

Figure 5. Relative mRNA expression levels of ID$_1$, ID$_2$, and ID$_3$ in the kidney and liver of juvenile Japanese flounder exposed to 0, 10, 100, and 1000 ng/L of Aroclor 1254 for 25 and 50 d. Fold change (y-axis) represents the expression of the target gene mRNA relative to that of the control group (equals 1 by definition). [a] $P<0.05$ indicates significant differences between the exposure groups and corresponding control.

the reduction in plasma TT$_4$. In studies of tilapia, van der Geyten et al. [43] found that ID$_2$ activity in the liver and ID$_3$ activity in the gill decreased with declining T$_3$ concentrations, which is responsible for balancing the reduction in T$_3$. Schnitzler et al. [29] suggested that PCB-induced changes in deiodinase activity offset the decline in plasma T$_3$ levels. Adams et al. [44] suggested that elevated T$_4$ ORD activity serves as a homeostatic adjustment to offset increased systemic T$_3$ clearance. After 50 d exposure, the decrease in plasma TT$_3$ levels at the highest dose was mostly due

to hypothyroidism, which caused a drop in thyroidal T$_4$ production and secretion; thus, exceeding the regulation ability of IDs, and also resulting in lower FT$_3$ levels.

Some authors have found that a change in plasma TH levels alters the ID$_3$ expression. For example, Higgs and Eales [45] found that a decrease in fish T$_4$ levels leads to a decrease in the metabolic clearance level of T$_4$; in other words, a decrease in the ID$_3$ level. A study by Van der Geyten et al. [41] showed that a decrease in the TT$_4$ and TT$_3$ levels of tilapia with thyroid

Table 3. The effects of PCBs on fish plasma thyroid hormone homostasis.

Type of PCBs	Dose	Exposure days	Species	Ages of fish	T_4	T_3	Reference
PCB 126	25 µg/kg	210	Salvelinus namaycush	Adult	↑	–	[34]
PCB 126	500 µg/kg	7	Hippoglossoides platessoides	Adult	–	–	[44]
PCB 77	1000 µg/kg	90	Thymallus arcticus	Adult	–	–	[49]
PCB 77	500 µg/kg	7	Hippoglossoides platessoides	Adult	↑	–	[44]
Clophen A50	500 mg/kg	10	Platichthys flesus	Adult	–	–	[50]
Aroclor 1254	150 µg/kg	42	Oncorhynchus kisutch	Adult	→	↑	[6]
Aroclor 1254	1 mg/kg	30	Micropogonias undulatus	Adult	↑	→	[51]
Aroclor 1254	0.5 µg/g	35	Oreochromis niloticus	Adult	–	–	[38]
Aroclor 1248	5 mg/kg	21	Ameiurus nebulosus	Adult	–	→	[33]
1254:1248	50 µg/g	84	Oncorhynchus kisutch	Adult	–	→	[5]
1254:1248	10 µg/g	120	Dicentrarchus labrax	Adult	→	→	[30]

↑, increase; ↓, decrease; –, no effect.

dysfunction caused by methimazole exposure caused ID_1 and ID_2 levels to increase and ID_3 levels to decrease. However, the current study found that a decrease in plasma TH levels did not influence the mRNA expression of ID_3 after 50 days of exposure. Coimbra et al. [38] found that at 21 and 35 days after tilapia were exposed to Aroclor 1254, TT_3 and TT_4 levels showed no significant changes, whereas ID_3 activity levels in the liver significantly increased, while the activity of ID3 increased in the gill after 21 days of exposure. The exact reason for this phenomenon requires further study.

At present, higher exposure concentrations of PCBs are often used to investigate their thyroid-disrupting effects on adult fish (Table 3). Of note, PCB concentrations detected in the environment are far lower than those used in these exposure experiments. For example, the PCBs content of the surface water and sediment in the Minjiang River Estuary, China, are 985 ng/L and 34.39 µg/kg on average, respectively [46]. The total concentration of PCBs ranged from 2.33 µg/kg to 44 µg/kg in the marine sediments in Barcelona, Spain [47], and 10 µg/kg to 899 µg/kg in the surface sediments of Naples Harbour, Italy [48]. Adult sea bass fed with the equivalent of actual environmental concentrations of the mixture of Aroclor 1260 and 1254 only showed reduced muscle T_3 levels, with no significant changes in muscle T_4 levels and thyroid histology; however, exposure to the same contaminants at concentrations 10 times above actual environmental concentrations led to a decrease in both T_3 and T_4 levels in muscles, and caused follicular degeneration [29]. This study found that even environmentally relevant concentrations of Aroclor 1254 caused significant disruption to the thyroid system of flounder juveniles, including changes in thyroid histopathology, altered plasma TH levels, and modulation in the expression levels of IDs mRNA in the liver and kidney. This result supported the hypothesis that juvenile fish are more sensitive to PCBs compared to adult fish, making them suitable candidate animal models for studying TDCs.

Many TDCs, such as sodium perchlorate, have been reported to affect the early growth and development of teleosts [15,16,18]. Mechanisms underlying the effects of PCBs on the early life stages of fish development *via* their thyroid disrupting abilities should be investigated in future studies, not only to delineate the disrupting effects of PCBs at individual and ecological levels, but also to establish some links between the macroscopic effects and the microscopic mechanisms for a more comprehensive ecological risk assessment of these pollutants. In particular, flatfish species, including Japanese flounder, experience a unique and critical process of metamorphosis during development, when the larvae shift from a planktonic to a benthic mode of life, with this process being primarily controlled by the thyroid system. Therefore, the larvae of Japanese flounder may represent an excellent model organism for investigating the effects of PCBs on the thyroid system and fish development in future studies.

Acknowledgments

The authors are grateful to all members in the lab for their help.

Author Contributions

Conceived and designed the experiments: YD SR HT. Performed the experiments: YD JL. Analyzed the data: YD XZ. Contributed reagents/materials/analysis tools: JL. Contributed to the writing of the manuscript: HT XZ WW SR. Obtained permission for use of fish fertilized eggs: JL.

References

1. Hansen LG (1998) Stepping backward to improve assessment of PCB congener toxicities. Environmental Health Perspectives 106: 171–189.

2. Safe SH (1994) Polychlorinated biphenyls (PCBs): environmental impact, biochemical and toxic responses, and implications for risk assessment. CRC Critical Reviews in Toxicology 24: 87–149.

3. UNEP website. Available: http://www.chem.unep.ch/Legal/ECOSOC/UNEP%20Consolidated%20List%2010%20May%202010.pdf. Accessed 2014 July 15.

4. Jugan M-L, Levi Y, Blondeau J-P (2010) Endocrine disruptors and thyroid hormone physiology. Biochemical pharmacology 79: 939–947.

5. Brown SB, Adams BA, Cyr DG, Eales JG (2004) Contaminant effects on the teleost fish thyroid. Environmental Toxicology and Chemistry 23: 1680–1701.

6. Leatherland J, Sonstegard R (1978) Lowering of serum thyroxine and triiodothyronine levels in yearling coho salmon, Oncorhynchus kisutch, by dietary mirex and PCBs. Journal of the Fisheries Board of Canada 35: 1285–1289.

7. Folmar LC, Dickhoff WW, Zaugg WS, Hodgins HO (1982) The effects of aroclor 1254 and no. 2 fuel oil on smoltification and sea-water adaptation of coho salmon (Oncorhynchus kisutch). Aquatic toxicology 2: 291–299.

8. Blanton ML, Specker JL (2007) The hypothalamic-pituitary-thyroid (HPT) axis in fish and its role in fish development and reproduction. Crit Rev Toxicol 37: 97–115.

9. Cyr DG, Eales J (1996) Interrelationships between thyroidal and reproductive endocrine systems in fish. Reviews in Fish Biology and Fisheries 6: 165–200.

10. Inui Y, Tagawa M, Miwa S, Hirano T (1989) Effects of bovine TSH on the tissue thyroxine level and metamorphosis in prometamorphic flounder larvae. General and comparative endocrinology 74: 406–410.

11. Okada N, Morita T, Tanaka M, Tagawa M (2005) Thyroid hormone deficiency in abnormal larvae of the Japanese flounder Paralichthys olivaceus. Fisheries Science 71: 107–114.

12. Power D, Llewellyn L, Faustino M, Nowell M, Björnsson BT, et al. (2001) Thyroid hormones in growth and development of fish. Comparative Biochemistry and Physiology Part C: Toxicology & Pharmacology 130: 447–459.

13. Yamano K (2005) The role of thyroid hormone in fish development with reference to aquaculture. Japan Agricultural Research Quarterly 39: 161.

14. Bradford CM, Rinchard J, Carr JA, Theodorakis C (2005) Perchlorate affects thyroid function in eastern mosquitofish (Gambusia holbrooki) at environmentally relevant concentrations. Environmental science & technology 39: 5190–5195.

15. Crane HM, Pickford DB, Hutchinson TH, Brown JA (2005) Effects of ammonium perchlorate on thyroid function in developing fathead minnows, Pimephales promelas. Environmental health perspectives 113: 396.

16. Liu FJ, Wang JS, Theodorakis CW (2006) Thyrotoxicity of sodium arsenate, sodium perchlorate, and their mixture in zebrafish Danio rerio. Environmental science & technology 40: 3429–3436.

17. Mukhi S, Patiño R (2007) Effects of prolonged exposure to perchlorate on thyroid and reproductive function in zebrafish. Toxicological sciences 96: 246–254.

18. Schmidt F, Schnurr S, Wolf R, Braunbeck T (2012) Effects of the anti-thyroidal compound potassium-perchlorate on the thyroid system of the zebrafish. Aquat Toxicol 109: 47–58.

19. Eales J, Brown S (1993) Measurement and regulation of thyroidal status in teleost fish. Reviews in Fish Biology and Fisheries 3: 299–347.

20. Eales JG, Brown SB, Cyr DG, Adams BA, Finnson KR (1999) Deiodination as an index of chemical disruption of thyroid hormone homeostasis and thyroidal status in fish. ASTM SPECIAL TECHNICAL PUBLICATION 1364: 136–164.

21. Orozco A, Valverde-R C (2005) Thyroid hormone deiodination in fish. Thyroid 15: 799–813.

22. Köhrle J (1999) Local activation and inactivation of thyroid hormones: the deiodinase family. Molecular and cellular endocrinology 151: 103–119.

23. Moreno M, Berry MJ, Horst C, Thoma R, Goglia F, et al. (1994) Activation and inactivation of thyroid hormone by type I iodothyronine deiodinase. FEBS letters 344: 143–146.

24. Van der Geyten S, Byamungu N, Reyns G, Kühn E, Darras V (2005) Iodothyronine deiodinases and the control of plasma and tissue thyroid hormone levels in hyperthyroid tilapia (Oreochromis niloticus). Journal of endocrinology 184: 467–479.

25. Inui Y, Miwa S (1985) Thyroid hormone induces metamorphosis of flounder larvae. General and comparative endocrinology 60: 450–454.

26. Miwa S, Inui Y (1991) Thyroid hormone stimulates the shift of erythrocyte populations during metamorphosis of the flounder. Journal of Experimental Zoology 259: 222–228.

27. Miwa S, Yamano K, Inui Y (1992) Thyroid hormone stimulates gastric development in flounder larvae during metamorphosis. Journal of Experimental Zoology 261: 424–430.

28. Yamano K, Miwa S, Obinata T, Inui Y (1991) Thyroid hormone regulates developmental changes in muscle during flounder metamorphosis. General and comparative endocrinology 81: 464–472.

29. Schnitzler JG, Celis N, Klaren PH, Blust R, Dirtu AC, et al. (2011) Thyroid dysfunction in sea bass (Dicentrarchus labrax): underlying mechanisms and effects of polychlorinated biphenyls on thyroid hormone physiology and metabolism. Aquat Toxicol 105: 438–447.

30. Tian H, Ru S, Bing X, Wang W (2010) Effects of monocrotophos on the reproductive axis in the male goldfish (Carassius auratus): Potential mechanisms underlying vitellogenin induction. Aquatic toxicology 98: 67–73.

31. Livak KJ, Schmittgen TD (2001) Analysis of Relative Gene Expression Data Using Real-Time Quantitative PCR and the $2^{-\Delta\Delta Ct}$ Method. methods 25: 402–408.

32. Drezner Z, Turel O, Zerom D (2010) A Modified Kolmogorov-Smirnov Test for Normality. Taylor & Francis 39: 693–704.

33. Iwanowicz LR, Blazer VS, McCormick SD, Vanveld PA, Ottinger CA (2009) Aroclor 1248 exposure leads to immunomodulation, decreased disease resistance and endocrine disruption in the brown bullhead, Ameiurus nebulosus. Aquat Toxicol 93: 70–82.

34. Brown SB, Evans RE, Vandenbyllardt L, Finnson KW, Palace VP, et al. (2004) Altered thyroid status in lake trout (Salvelinus namaycush) exposed to co-planar 3,3',4,4',5-pentachlorobiphenyl. Aquat Toxicol 67: 75–85.

35. Fok P, Eales J, Brown S (1990) Determination of 3, 5, 3'-triiodo-L-thyronine (T₃) levels in tissues of rainbow trout (Salmo gairdneri) and the effect of low ambient pH and aluminum. Fish physiology and biochemistry 8: 281–290.

36. Brown SB, Fisk AT, Brown M, Villella M, Muir DC, et al. (2002) Dietary accumulation and biochemical responses of juvenile rainbow trout (Oncorhynchus mykiss) to 3, 3', 4, 4', 5-pentachlorobiphenyl (PCB 126). Aquatic toxicology 59: 139–152.

37. Leung AM, Pearce EN, Braverman LE (2010) Perchlorate, iodine and the thyroid. Best Practice & Research Clinical Endocrinology & Metabolism 24: 133–141.

38. Coimbra AM, Reis-Henriques MA, Darras VM (2005) Circulating thyroid hormone levels and iodothyronine deiodinase activities in Nile tilapia (Oreochromis niloticus) following dietary exposure to Endosulfan and Aroclor 1254. Comp Biochem Physiol C Toxicol Pharmacol 141: 8–14.

39. Li W, Zha J, Li Z, Yang L, Wang Z (2009) Effects of exposure to acetochlor on the expression of thyroid hormone related genes in larval and adult rare minnow (Gobiocypris rarus). Aquatic Toxicology 94: 87–93.

40. Zhang X, Tian H, Wang W, Ru S (2013) Exposure to monocrotophos pesticide causes disruption of the hypothalamic-pituitary-thyroid axis in adult male goldfish (Carassius auratus). Gen Comp Endocrinol 193: 158–166.

41. Van der Geyten S, Toguyeni A, Baroiller JF, Fauconneau B, Fostier A, et al. (2001) Hypothyroidism induces type I iodothyronine deiodinase expression in tilapia liver. Gen Comp Endocrinol 124: 333–342.

42. Picard-Aitken M, Fournier H, Pariseau R, Marcogliese DJ, Cyr DG (2007) Thyroid disruption in walleye (Sander vitreus) exposed to environmental contaminants: Cloning and use of iodothyronine deiodinases as molecular biomarkers. Aquatic toxicology 83: 200–211.

43. Van der Geyten S, Mol K, Pluymers W, Kühn E, Darras V (1998) Changes in plasma T3 during fasting/refeeding in tilapia (Oreochromis niloticus) are mainly regulated through changes in hepatic type II iodothyronine deiodinase. Fish Physiology and Biochemistry 19: 135–143.

44. Adams BA, Cyr DG, Eales JG (2000) Thyroid hormone deiodination in tissues of American plaice, Hippoglossoides platessoides: characterization and short-term responses to polychlorinated biphenyls (PCBs) 77 and 126. Comparative Biochemistry and Physiology Part C: Pharmacology, Toxicology and Endocrinology 127: 367–378.

45. Higgs DA, Eales J (1977) Influence of food deprivation on radioiodothyronine and radioiodide kinetics in yearling brook trout, Salvelinus fontinalis (Mitchill), with a consideration of the extent of l-thyroxine conversion to 3, 5, 3'-triiodo-L-thyronine. General and comparative endocrinology 32: 29–40.

46. Zhang Z, Hong H, Zhou J, Huang J, Yu G (2003) Fate and assessment of persistent organic pollutants in water and sediment from Minjiang River Estuary, Southeast China. Chemosphere 52: 1423–1430.

47. Castells P, Parera J, Santos F, Galceran M (2008) Occurrence of polychlorinated naphthalenes, polychlorinated biphenyls and short-chain chlorinated paraffins in marine sediments from Barcelona (Spain). Chemosphere 70: 1552–1562.

48. Sprovieri M, Feo ML, Prevedello L, Manta DS, Sammartino S, et al. (2007) Heavy metals, polycyclic aromatic hydrocarbons and polychlorinated biphenyls in surface sediments of the Naples harbour (southern Italy). Chemosphere 67: 998–1009.

49. Palace VP, Allen-Gil SM, Brown SB, Evans RE, Metner DA, et al. (2001) Vitamin and thyroid status in arctic grayling (Thymallus arcticus) exposed to doses of 3, 3', 4, 4'-tetrachlorobiphenyl that induce the phase I enzyme system. Chemosphere 45: 185–193.

50. Besselink H, Van Beusekom S, Roex E, Vethaak A, Koeman J, et al. (1996) Low hepatic 7-ethoxyresorufin-O-deethylase (EROD) activity and minor alterations in retinoid and thyroid hormone levels in flounder (Platichthys flesus) exposed to the polychlorinated biphenyl (PCB) mixture, Clophen A50. Environmental Pollution 92: 267–274.

51. LeRoy KD, Thomas P, Khan IA (2006) Thyroid hormone status of Atlantic croaker exposed to Aroclor 1254 and selected PCB congeners. Comp Biochem Phyiol C Toxicol Pharmacol 144: 263–271.

8

Examining the Prey Mass of Terrestrial and Aquatic Carnivorous Mammals: Minimum, Maximum and Range

Marlee A. Tucker*, Tracey L. Rogers

Evolution and Ecology Research Centre, School of Biological, Earth and Environmental Sciences, The University of New South Wales, Sydney, New South Wales, Australia

Abstract

Predator-prey body mass relationships are a vital part of food webs across ecosystems and provide key information for predicting the susceptibility of carnivore populations to extinction. Despite this, there has been limited research on the minimum and maximum prey size of mammalian carnivores. Without information on large-scale patterns of prey mass, we limit our understanding of predation pressure, trophic cascades and susceptibility of carnivores to decreasing prey populations. The majority of studies that examine predator-prey body mass relationships focus on either a single or a subset of mammalian species, which limits the strength of our models as well as their broader application. We examine the relationship between predator body mass and the minimum, maximum and range of their prey's body mass across 108 mammalian carnivores, from weasels to baleen whales (Carnivora and Cetacea). We test whether mammals show a positive relationship between prey and predator body mass, as in reptiles and birds, as well as examine how environment (aquatic and terrestrial) and phylogenetic relatedness play a role in this relationship. We found that phylogenetic relatedness is a strong driver of predator-prey mass patterns in carnivorous mammals and accounts for a higher proportion of variance compared with the biological drivers of body mass and environment. We show a positive predator-prey body mass pattern for terrestrial mammals as found in reptiles and birds, but no relationship for aquatic mammals. Our results will benefit our understanding of trophic interactions, the susceptibility of carnivores to population declines and the role of carnivores within ecosystems.

Editor: Kornelius Kupczik, Max Planck Institute for Evolutionary Anthropology, Germany

Funding: This work is supported by the Australian Research Council grant # LP0989933 to TLR. The funders had no role in study design, data collection and analysis, decision to publish, or preparation of the manuscript.

Competing Interests: The authors have declared that no competing interests exist.

* Email: marlee.tucker@unsw.edu.au

Introduction

Examining patterns in predator-prey relationships provides information on predation pressure (e.g. on specific size guilds) [1,2], the impact of decreasing prey species on predators [3] and the potential for trophic cascades and the collapse of prey populations [4–6]. However, previous research on predator-prey body mass relationships in mammalian carnivores has focused upon the mean mass of prey, largely ignoring the minimum and maximum body mass of prey consumed by predators. It is important to include the minimum, maximum and range of prey mass consumed as it allows the examination of the upper and lower limits of carnivore prey selection. In addition, prey selection provides information such as energetic requirements (e.g. intake rates), which is often used for predicting the susceptibility of carnivores to population declines, the role of carnivores within ecosystems and community structure [13].

Larger-sized predators can utilise a wide variety of prey types because they have large home ranges [7] that provide access to a diversity prey species [8], as well as a wide gape size that allows them to feed on prey of a variety of sizes. Despite this, large predators tend to eat larger-sized prey [13]. It is not always profitable for large species to feed on small-sized prey due to capture inefficiency as it is costly to pursue small-sized prey in relation to the small energetic benefit gained [9]. The minimum and maximum size of prey should scale positively with predator body mass, resulting in there being no relationship between predator body mass and diversity of prey size (i.e. dietary niche breadth - DNB) [10,11]. However, if maximum prey size scales positively with predator mass and minimum prey size does not this will result in a larger diversity of prey size for larger predators (i.e. wider DNB).

Our knowledge of mammalian broad-scale patterns of prey-size range is limited. There has been limited work investigating the prey mass of African predators and its effect on the system [12]. However, this work largely focuses upon predation pressure on prey species, particularly herbivorous mammals. The remaining predator-prey body mass research is based the mean prey mass of predators [13,14]. Investigations into other animal groups include predatory fish [15,16], reptiles [11,17] and birds [10], where there is a general consensus that there is a positive relationship between predator body mass and prey minimum, maximum and range in mass, except for fish where the evidence is conflicting (positive or no relationship between predator mass and minimum prey mass).

Using minimum, maximum and range of prey mass for 108 carnivorous mammals from the orders Carnivora and Cetacea, we investigated the nature of the relationship between carnivore body mass and prey body mass and how living in either the marine or terrestrial environment has impacted this relationship. This study has two objectives: first to examine the influence of physical environment on minimum, maximum and range of prey mass; and second to investigate the influence physical environment has had on the distribution of minimum, maximum and range of prey mass. Based on previous research [11,13], we predict that prey mass (minimum, maximum and range) will be positively correlated with predator body mass for terrestrial carnivores. However, for marine carnivores there could be two possible outcomes: first, prey mass (minimum, maximum and range) could be positively correlated with body mass similar to terrestrial carnivores and other marine non-mammalian predators [16]; or second, there could be no relationship between prey mass (minimum, maximum and range) and predator body mass. No relationship between predator mass and prey mass is a possibility due to the high abundance of small species that form dense aggregations in aquatic environments (e.g. krill or fish), which lead to an increase in the encounter rates between aquatic predators and these small prey species. With both small and large predators exposed to these abundant food resources, this would result in both small and large predators feeding upon small prey species and therefore suggest no relationship between predator mass and prey mass in aquatic systems.

By examining the moments (e.g. mean, mode, skewness etc.) of the prey mass distributions, we can gather information on how the mass of the prey consumed by carnivorous mammals differs or is similar across different environments. This information is important for building our knowledge of predator-prey relationships and the drivers behind these relationships.

Materials and Methods

Ethics statement

All data in this study were extracted from published sources; hence no permission or approval for obtaining the data was required.

Database

Data were collated on the minimum and maximum prey mass (kg) consumed by 108 carnivorous mammal species (Appendix S1). Table 1 provides a summary of the orders and families sampled. Prey mass range was calculated by subtracting the minimum prey mass from the maximum prey mass. Mean body mass (kg) was also collected for these 108 predator species using the database PanTHERIA [18]. All carnivores were classified as terrestrial or aquatic, where aquatic species forage in water to survive (e.g. foraging) and terrestrial species forage on land to survive. All values including carnivore mass and prey mass were $\log_{10}$ transformed prior to all analyses.

Phylogeny

We required a single phylogenetic tree to examine minimum prey mass, maximum prey mass and prey mass range in carnivorous mammals. Phylogenetic information was obtained from the Fritz et al. [19] mammal supertree containing 5,020 species and branch lengths proportional to time since divergence. This tree was pruned using Mesquite ver 2.74 [20] to create the tree (n = 108) based on the data in our database. Sotalia guianensis (Guiana dolphin) was positioned within the pruned tree based on the topologies of Caballero et al. [21]. Due to insufficient phylogenetic information, the Fritz et al. [19] tree included soft polytomies where more than two species diverge at a single point in time. To resolve the polytomies, we used a semi-automated polytomy resolver for dated phylogenies [22]. The polytomy

resolution involved two steps; 1) R 3.0.2 [23] was used to create an XML input file containing topology constraints and input commands for BEAST, and 2) the XML input file was run through the program BEAST 1.8 [24] which uses a Bayesian Markov chain Monte Carlo (MCMC) algorithm to permute the unresolved relationships within the tree based on the birth-death model. This produced 1,000 alternative phylogenetic trees to be used for the phylogenetic comparative analyses and the ancestral state reconstructions.

Analyses

We applied a model selection approach to test the level of support for alternative models of prey mass evolution. The best model was selected using second-order Akaike's information criterion with a correction for sample size (AICc; [25]). The model with the lowest AICc value reflects the model with the highest support, although any other models within two units of the lowest model were also considered to be likely candidates. We selected the cut-off of <2.0 ΔAIC based on previous studies who have identified that models below this threshold are generally equally supported, models between 4–7 ΔAIC have some support and models >10 ΔAIC have no support [26–28]. To compute AICc values, we applied each model as a phylogenetic generalized least squares (PGLS) regression using the CAPER package in R [29] to each of the 1000 trees (see previous section). PGLS regression also computes a λ parameter using maximum likelihood that estimates whether the extent of phenotypic variation among species (e.g., mean body mass and associated home range size) is correlated to phylogeny. When λ is close to 1, phenotypic differentiation among present-day taxa reflects the phylogenetic relationships among those species and is the product of Brownian evolution. When λ is close to 0 phenotypic differentiation is unrelated to phylogeny and might be the outcome of adaptive evolution [30].

We performed three separate PGLS analyses for minimum prey mass, maximum prey mass and range of prey mass respectively. For each we examined the level of support of the relationship between prey mass (minimum, maximum or range), carnivore body mass and environment across 108 species. The models were formulated as; (a) $\beta_0 + \beta_{mass} * \beta_{environment}$, where environment was coded as "terrestrial" or "aquatic" and included an interaction term between carnivore mass and environment; (b) $\beta_0 + \beta_{mass}$, which assumed body mass was the only variable predicting prey mass; (c) β_0, the evolutionary null model in which no predictor variable was included and subsequently modelled variance in species prey mass as the outcome of Brownian evolution (i.e. under Brownian motion, trait evolution proceeds as a random walk through trait space and Brownian motion has been proposed as a null model of evolution for testing hypotheses of trait evolution [31]).

To examine the effect of phylogeny and ecology on the minimum and maximum prey mass of carnivores, we ran variance component analyses [32]. Variance was examined between species, focusing on the contribution of order, family, genus, mass and environment (aquatic or terrestrial). Variance components analysis was performed using the lme4 package [33] in R version 2.13.2.

To gain an understanding on the shape of the prey mass distributions across species and environments, we extracted the descriptive statistics including the mean, median, mode, range, minimum, maximum, standard deviation (S.D.), skewness and kurtosis. Skewness measures the degree of asymmetry of a distribution. If the skewness value is positive the data has a right skewed distribution and a negative value suggests left skewed data.

Table 1. Summary of the orders and families included in our study sample.

Order	Family
Carnivora	Canidae
	Felidae
	Herpestidae
	Hyaenidae
	Mephitidae
	Mustelidae
	Otariidae
	Phocidae
	Procyonidae
	Ursidae
	Viverridae
Cetacea	Balaenidae
	Balaenopteridae
	Cetotheriidae
	Delphindae
	Delphinidae
	Eschrichtiidae
	Kogiidae
	Monodontidae
	Phocoenidae
	Physeteridae
	Pontoporiidae
	Ziphiidae

Kurtosis measures height of the curve relative to its standard deviations. Data with a peaked distribution with values around zero (i.e. normal distribution) have a positive kurtosis value, whereas negative values between 0 and -1 implies that the data has a flat distribution and values lower than -1.5 suggest a bimodal distribution.

Results

Phylogenetic Generalized Least Squares Regression

The model including an interaction between body mass and environment (mass-environment) was the best supported model for minimum prey mass, maximum prey mass and prey mass range (Table 2). However, for the maximum prey mass and prey mass range there was less than 2 ΔAIC units between the mass model and the mass-environment model, suggesting that these models are equally supported. In both cases, the addition of environment explained a limited amount of additional variance (3% for both maximum prey and prey range) however for minimum prey mass, environment explained an additional 7% of variance. The phylogenetic signal (λ) was consistently high for the mass model for minimum, maximum and prey size range (>0.7). Lambda however, was mixed for the mass-environment models, where it was low (<0.5) for the maximum prey and prey range size, it was higher (0.62) for the minimum prey mass.

We found there was no significant difference (confidence intervals overlap 0) in intercept between terrestrial and aquatic carnivores for minimum (CI -2.22, 1.54; Fig. 1A), maximum

(CI -2.48, 1.47; Fig. 1B) and range of prey mass (CI -2.70, 1.00; Fig. 1C). There was a significant difference (confidence intervals do not overlap 0) in slope between terrestrial and aquatic carnivores for minimum (CI 0.37, 1.93), maximum (CI 0.24, 1.94) and range of prey mass (CI 0.49, 2.16). Despite the negative slopes of the aquatic regression lines, these values were not significantly different from 0 for minimum, maximum or prey mass range. The terrestrial regression slopes were positive and significantly different from 0 for minimum, maximum or prey mass range (Fig. 1A, B and C).

Variance Components Analysis

Order, family and genus explained the maximum proportion of variance in prey mass of carnivores (58–73%; Table 3), providing additional support for the strong influence of phylogenetic relatedness on the prey mass consumed by carnivorous mammals. Body mass of the carnivore explained a relatively large degree of variance (32–39%), where environment had little influence over the mass of prey consumed (<0.01–3%).

Descriptive Statistics

Examining the descriptive statistics, the minimum prey mass distribution across all mammals is positively skewed (skew = 0.11), while maximum and range of prey mass is negatively skewed (−0.40 to −0.37; Table S1). The prey mass data has a normal distribution for minimum, maximum and range of prey mass with kurtosis values between 0.62 and 1.18 (Table S1).

When examining the aquatic and terrestrial prey mass distributions, we find that aquatic carnivores tend to feed on smaller-sized prey, with mean prey mass for minimum, maximum and the range being lower than terrestrial carnivores (Fig. 2, Table S2 and S3). Aquatic carnivore prey mass distributions are all negatively skewed (−0.39 to −0.63), but in terrestrial carnivores negatively-skewed distribution are only seen for maximum and range prey mass distributions (−0.25 and −0.35). The prey mass distribution for terrestrial carnivores is relatively flat as suggested by the negative kurtosis values (−0.25 to −0.75). For aquatic species, the prey mass distributions are normally distributed with kurtosis value between 0.88 and 3.17. Additionally, aquatic carnivores feed on prey spanning 12 900 kg, compared with 2 700 kg for terrestrial carnivores.

Discussion

The best model of prey mass evolution includes both carnivore mass and environment, although environment explains a small percentage (~8%) of variance in prey size. In spite of this, aquatic and terrestrial mammalian carnivores have different relationships suggesting different optimal foraging strategies. Aquatic mammalian carnivores have no relationship between prey (neither minimum nor maximum) and predator body mass, unlike terrestrial mammalian carnivores where there is a positive prey-predator (both minimum and maximum) body mass relationship. In contrast to terrestrial predators, larger marine carnivores do not have to actively pursue prey with large body mass to meet their energetic requirements [13]. The abundance of small-sized prey in aquatic and marine environments (Fig. 2) is likely to have driven these patterns in marine carnivores. As well as prey availability, it is also important to note the effect of dimensionality on predator-prey relationships and consumption rates. In 3D environments, it has been demonstrated that consumption rates are higher, not only the baseline rates but also the scaling exponent [34]. This not only has an impact on predator-prey relationships (e.g. larger consumer-resource body mass ratios) but also the strength of

Table 2. Level of support for explanatory models of prey mass evolution in carnivorous mammals.

Prey mass	Model	ΔAICc	ΔAICc 95% CI (upper, lower)	Lambda	Effect size (r)
Minimum	$\beta_0 + \beta_{mass} \times \beta_{environment}$	0.0	NA	0.62	0.28
	$\beta_0 + \beta_{mass}$	5.8	4.99, 6.58	0.71	0.10
	β_0	5.5	4.78, 6.28	0.64	NA
Maximum	$\beta_0 + \beta_{mass} \times \beta_{environment}$	0.0	NA	0.48	0.28
	$\beta_0 + \beta_{mass}$	0.9	0.04, 2.07	0.74	0.23
	β_0	4.8	4.23, 5.29	0.57	NA
Range	$\beta_0 + \beta_{mass} \times \beta_{environment}$	0.0	NA	0.30	0.28
	$\beta_0 + \beta_{mass}$	1.6	0.23, 3.1	0.73	0.23
	β_0	5.3	4.54, 5.84	0.57	NA

Results are from phylogenetic least squares (PGLS) regression analyses computed for 1000 alternative resolutions of the mammalian phylogeny. Model terms include carnivore body mass (β_{mass}), environment either aquatic or terrestrial ($\beta_{environment}$) and the intercept (β_0).

interactions between trophic levels and the stability of the community.

In addition to the effect of differences in prey availability and dimensionality, there are differences in body size patterns across aquatic and terrestrial carnivores. Marine mammals tend to have larger body sizes compared to terrestrial mammals because of the relaxation of biomechanical constraints and increased thermoregulatory constraints [35]. This has an effect on the predator-prey relationships, which tend to be more accentuated in the marine environment due to this disparity in body size between predators and their prey. This has been illustrated by the presence of larger predator-prey mass ratios in aquatic environments [14,36].

The lack of relationship between prey mass and predator mass could be because predator morphology is a large driver behind the prey choice of aquatic mammalian carnivores. Morphology, such as gape size, is a limiting factor for aquatic carnivores physically unable to capture larger-sized prey (e.g. altered morphology of appendages to aid with swimming rather than grasping). Additionally, gape limitation in aquatic systems is believed to impact the number of trophic levels that predators can feed from [37]. This is due to aquatic ecosystems being size structured, where body size generally increases with trophic level [38]. Dentition is also likely to be important as carnivores with highly overlapping ranges often have different tooth morphology driven by competition for resources [39]. In the aquatic system mammals have specialised feeding morphology including keratinized baleen plates for filter feeding (mysticete whales), multi-cuspidate interlocking teeth for krill sieving (e.g. crabeater and leopard seals), simple teeth with reduced serration for catching fish (piscivory e.g. dolphins), reduced teeth (e.g. Ross seal), rounded teeth (e.g. walrus) for eating hard-shelled molluscs or even the loss of teeth (e.g. sperm whale and beaked whales) for eating soft-bodied molluscs. Having specialised dentition can minimise resource competition; however it can leave these specialist carnivores vulnerable to higher extinction pressure if their prey populations were to collapse or become extinct [40,41].

The addition of environment into the PGLS models explains greater variance and has the highest phylogenetic signal only for minimum prey mass and not for models of maximum or range in prey mass. This suggests there are differences between aquatic and terrestrial mammalian carnivores in the patterns of minimum prey body mass only. There are a great number of large aquatic mammalian carnivores feeding on small-sized prey whereas all large terrestrial mammalian carnivores are tied to feeding upon

large-sized prey to maximise their energetic intake while minimising their expenditure [13]. In aquatic environments, particularly the marine system, the combination of the high abundance of prey below 500 g, and the schooling nature of these prey, makes it efficient for large carnivores to switch to feeding on smaller prey.

The combined results from the PGLS (phylogenetic signal; λ) and the variance components analysis suggest that phylogenetic relatedness is a major influence of prey mass distribution patterns across carnivorous mammals. The driving factor behind this result are the baleen whales, as they represent closely related species that share a common feeding strategy. This group represents some of the largest species today (up to 200 tonnes) and they feed on some of the smallest prey species (e.g. zooplankton). Baleen whales consist of species from Balaenopteridae and Balaenidae, and all use filter feeding to capture their prey. There are also other feeding strategies, such as pack hunting, that are generally shared across taxonomic levels (i.e. at the family and genus level).

The most likely reason behind the scatter present in the relationship between body mass and prey mass of terrestrial carnivores (Fig. 1A–C), is related to carnivore feeding strategy. Terrestrial carnivores following two feeding strategies: large prey consumers or small prey consumers [13,42]. Terrestrial carnivore species weighing 21.5 kg or less feed on invertebrates and small vertebrate prey species (<10 kg) [42]. Above the 21.5 kg threshold, carnivores must shift to feeding on large vertebrate prey to meet their energetic requirements [42]. Several carnivore species that feed upon large vertebrate prey have evolved cooperative or pack hunting strategies, which confer several benefits. One benefit of hunting in a pack is the minimisation of the energy expended whilst hunting, but also maximising the size of the prey captured and prey capture efficiency [43,44]. An increase in the size of the prey captured, energetic intake and hunting success also follows an increase in the number of individuals within the group [43]. Another advantage of hunting cooperatively is that individuals may gain other benefits including increased body size or reproductive success. For example, individual male fosa (*Cryptoprocta ferox*) that forage in groups tend to have larger body size, which also enables increased competition and mating success [45].

There are various drivers influencing the patterns of prey mass consumed by carnivorous mammals. The size of prey chosen by carnivores is driven by the trade-off between energy acquisition and expenditure, the available ecological niches and the dimen-

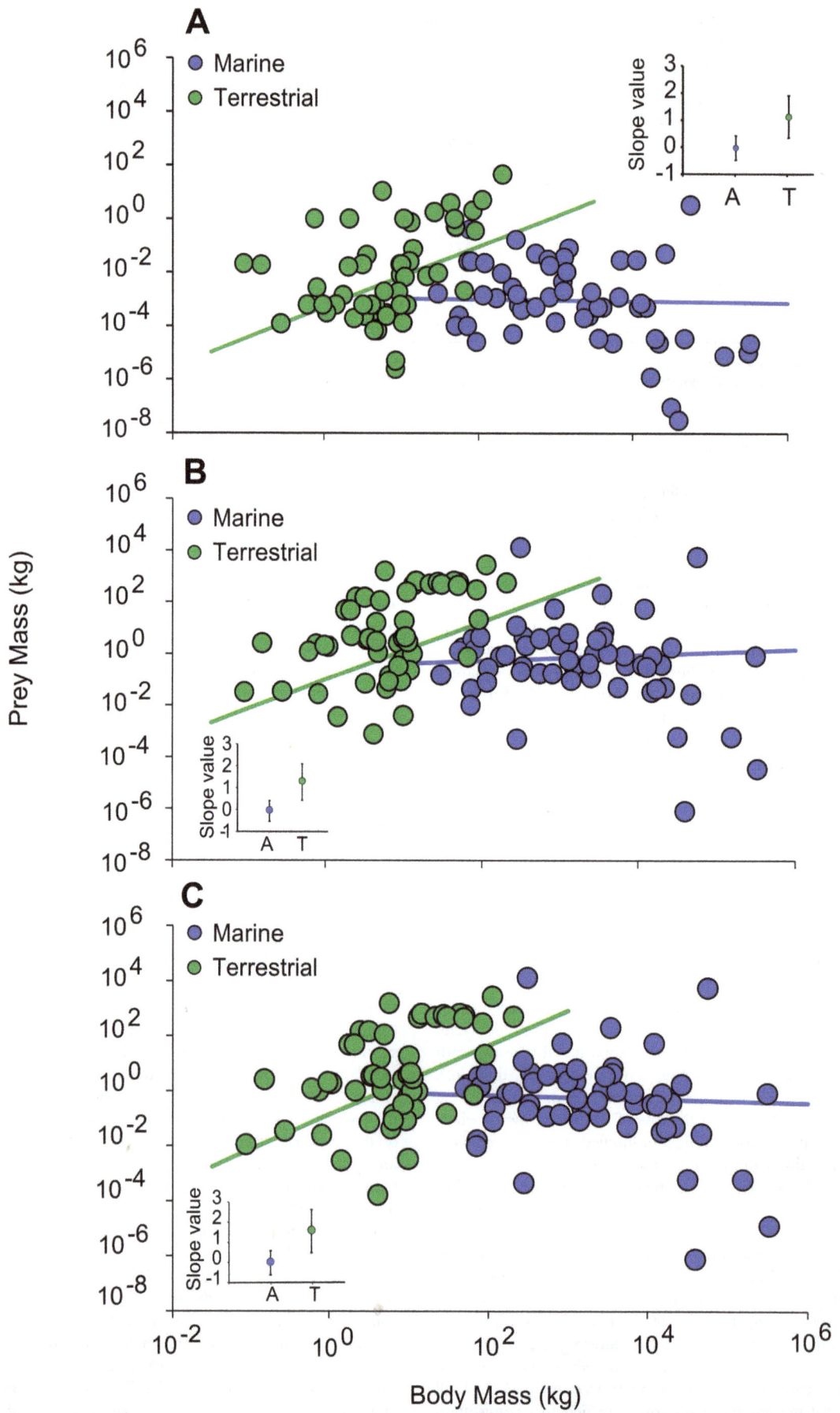

Figure 1. Minimum prey mass (A), maximum prey mass (B) and prey mass range (C) as a function of carnivore body mass compared for terrestrial (green circles) and aquatic (blue circles) species. Each datum represents a species mean value. The solid green line is the phylogenetic regression of terrestrial mammals: (A) log(Y) = 1.13log(X)-3.3, (B) log(Y) = 1.12log(X)-1.01 and (C) log(Y) = 1.26log(X)-0.87. The solid blue line is the phylogenetic regression of aquatic mammals: (A) log(Y) = −0.03log(X)-2.96, (B) log(Y) = 0.11log(X)+-0.50 and (C) log(Y) = −0.07log(X)-0.02. Insert: intercept values and confidence intervals (CI) for aquatic (A) and terrestrial (T) species. Values were calculated from phylogenetic least squares (PGLS) regression analyses applied to 1000 alternative resolutions of the mammalian phylogeny. Error bars represent CI's for the intercept values and are calculated using the standard error (SE) multiplied by 1.96.

sionality of the environment [34]. Foraging animals minimise their energetic costs while maximising energetic gains by moving towards an optimal foraging strategy. As different species evolved different foraging strategies, driven by their evolutionary history and environmental influences, this shapes the patterns of prey mass that are consumed by carnivores. Additionally, the mass of prey utilised by carnivores is influenced by the ecological niches available to them and the resource encounter rate, both of which have an effect on the foraging strategy and the prey availability [46–48].

Carnivore energetic costs and prey density both influence the minimum prey mass consumed. Carnivores tend to feed on prey above a certain mass due to the increasing inefficiency of feeding on small prey, because a low capture rate will arise when carnivores forage on prey much smaller than themselves [9]. However, this can be overcome in instances where prey species are in high densities, as illustrated by marine carnivores (e.g. baleen whales) who can survive feeding upon prey less than 1 g (e.g. krill and other invertebrates). This is further highlighted by the lower minimum prey mass of aquatic carnivores compared to that of terrestrial carnivores (Fig. 2).

On the other end of the scale, maximum prey mass is driven by morphological constraints (i.e. gape and locomotion) and energetic costs. A carnivore feeding on prey considerably larger than their

own mass will result in a mismatch of reaction time, where the carnivore will respond at a slower rate than that of the prey and will end with the prey escaping and the carnivore in an energetic deficit [9]. Additionally, while it would be ideal for all carnivores to feed on large-sized prey species (providing all carnivores could successfully capture large prey), it would be considerably costly from an energetic perspective, not to mention the increased amount of time spent processing and ingesting the prey [13,49].

Modal prey mass is influenced by environmental drivers. Based on the data used in the study, the minimum prey mass range (0.0001 to 0.01 kg) is the most commonly utilised by carnivorous species from the aquatic and terrestrial environments. The type of prey included within this weight range are invertebrates (aquatic and terrestrial), small mammals, fish and squid. The most common maximum prey range differs between environments, with terrestrial carnivores predominantly feeding on prey between 1 to 100 kg (i.e. small and large mammal prey), compared with 0.01 to 1 kg for aquatic carnivores (i.e. invertebrate, squid and fish prey). With the prey-weight categories of 0.0001–0.01 kg, 1–100 kg and 0.01–1 kg being the most abundant, this suggests that feeding within these ranges is a common foraging strategy across carnivores. Additionally, modal prey mass will also be driven by the characteristics of the environment, where primary productivity and trophic interactions will shape the size of prey available and

Table 3. Variance components analysis of prey mass across 108 carnivorous mammal species.

Prey Mass	Variance Source	Variance Component	Total Variance Explained (%)
Minimum	Total	3.17	100
	Order	0.18	5.65
	Family	1.13	35.65
	Genus	0.52	16.65
	Mass	1.23	38.83
	Environment	0.11	3.43
Maximum	Total	3.31	100
	Order	0.43	13.79
	Family	0.19	6.26
	Genus	1.64	52.60
	Mass	1.05	33.70
	Environment	<0.01	<0.01
Range	Total	3.50	100
	Order	0.40	11.33
	Family	0.17	4.88
	Genus	1.81	51.69
	Mass	1.12	31.97
	Environment	<0.01	<0.01

Categories include minimum prey mass (smallest prey size consumed), maximum prey size (largest prey size consumed) and range of prey mass (maximum minus minimum prey mass).

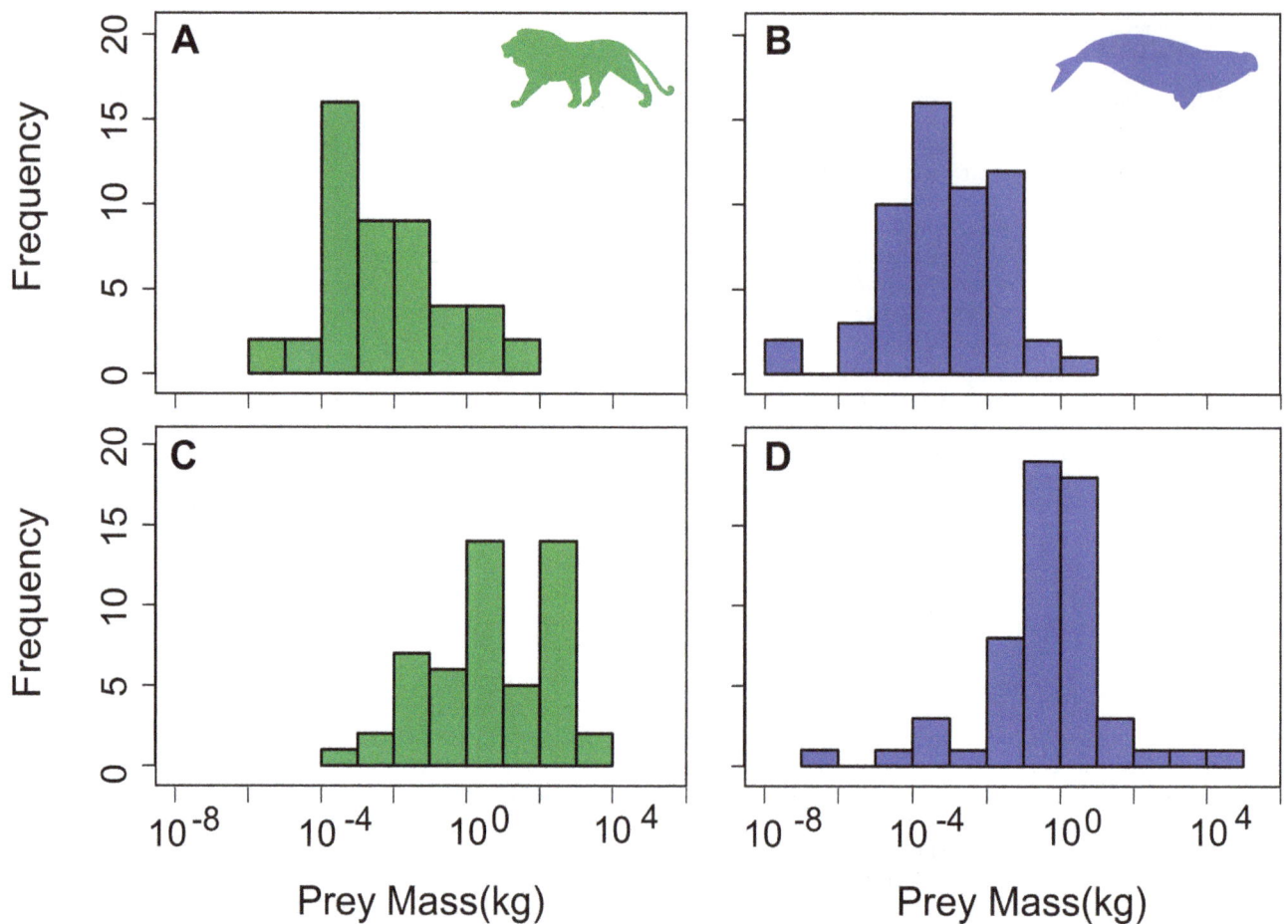

Figure 2. Distributions of the minimum prey mass for (A) terrestrial carnivorous mammals (green bars) and (B) aquatic carnivorous mammals (blue bars), and the maximum prey mass for (C) for terrestrial carnivorous mammals (green bars) and (D) for aquatic carnivorous mammals (blue bars). Silhouettes by uncredited and Chris Huh were downloaded from http://phylopic.org.

the abundance of these prey species. For example, productivity within the marine environment is driven by small, single-celled organisms, allowing higher availability of productivity to consumers and higher predator-prey body mass ratios [14].

In summary, carnivorous mammals within our sample differ in the size of prey they consume and this is influenced by a suite of factors including phylogenetic relatedness, carnivore body mass, the characteristics of the environment in which the carnivore resides, and has evolved within, as well as carnivore energetics. Whilst phylogenetic relatedness and carnivore body mass are the dominant drivers of the prey mass consumed by mammals, the physical environment has a role in the minimum-size prey that can be consumed. Previous research has shown that there is a positive relationship between carnivore mass and the mass of their prey. However, we have demonstrated that this is not the case for aquatic mammalian carnivores. Differences in environmental characteristics including primary productivity and food-web structure are driving the differences in prey mass consumed across aquatic and terrestrial carnivores.

Optimal foraging strategies in mammalian carnivores differ not only across species but also physical environments, which needs to be accounted for when thinking about carnivore behaviour. Gaining a better understanding of the relationship between mammalian carnivores and their prey, predator strategies and the factors driving these patterns, aids with predictions of the

susceptibility of carnivores to population declines and the role of carnivores within ecosystems.

Supporting Information

Table S1 Descriptive statistics for the prey mass distributions across 108 carnivorous mammals.

Table S2 Descriptive statistics for the prey mass distributions across 51 carnivorous terrestrial mammals.

Table S3 Descriptive statistics for the prey mass distributions across 57 carnivorous aquatic mammals.

Appendix S1 Database of predator body mass values and prey body mass values, including sources.

Acknowledgments

We would like to thank T. Ord for statistical advice. All silhouettes in Figure 2 were downloaded from http://phylopic.org and are available for reuse under the Public Domain Mark 1.0 license (Panthera) and the Creative Commons Attribution-ShareAlike 3.0 Unported license (Eubalaena).

Author Contributions

Conceived and designed the experiments: MAT TLR. Performed the experiments: MAT. Analyzed the data: MAT. Contributed reagents/ materials/analysis tools: MAT TLR. Contributed to the writing of the manuscript: MAT TLR.

References

1. Hayward MW, Kerley GIH (2005) Prey preferences of the lion (*Panthera leo*). Journal of Zoology 267: 309–322.
2. Hayward MW (2006) Prey preferences of the spotted hyaena (*Crocuta crocuta*) and degree of dietary overlap with the lion (*Panthera leo*). Journal of Zoology 270: 606–614.
3. Novaro AJ, Funes MnC, Susan Walker R (2000) Ecological extinction of native prey of a carnivore assemblage in Argentine Patagonia. Biological Conservation 92: 25–33.
4. Johnson CN, Isaac JL, Fisher DO (2007) Rarity of a top predator triggers continent-wide collapse of mammal prey: dingoes and marsupials in Australia. Proceedings of the Royal Society B: Biological Sciences 274: 341–346.
5. Daskalov GM, Grishin AN, Rodionov S, Mihneva V (2007) Trophic cascades triggered by overfishing reveal possible mechanisms of ecosystem regime shifts. Proceedings of the National Academy of Sciences 104: 10518–10523.
6. Fortin D, Beyer HL, Boyce MS, Smith DW, Duchesne T, et al. (2005) Wolves influence elk movements: behavior shapes a trophic cascade in Yellowstone National Park. Ecology 86: 1320–1330.
7. Tucker MA, Ord TJ, Rogers TL (2014) Evolutionary predictors of mammalian home range size: body mass, diet and the environment. Global Ecology and Biogeography: doi: 10.1111/geb.12194.
8. Ottaviani D, Cairns SC, Oliverio M, Boitani L (2006) Body mass as a predictive variable of home-range size among Italian mammals and birds. Journal of Zoology 269: 317–330.
9. Brose U (2010) Body-mass constraints on foraging behaviour determine population and food-web dynamics. Functional Ecology 24: 28–34.
10. Brandl R, Kristín A, Leisler B (1994) Dietary niche breadth in a local community of passerine birds: an analysis using phylogenetic contrasts. Oecologia 98: 109–116.
11. Costa GC, Vitt LJ, Pianka ER, Mesquita DO, Colli GR (2008) Optimal foraging constrains macroecological patterns: body size and dietary niche breadth in lizards. Global Ecology and Biogeography 17: 670–677.
12. Sinclair ARE, Mduma S, Brashares JS (2003) Patterns of predation in a diverse predator-prey system. Nature 425: 288–290.
13. Carbone C, Teacher A, Rowcliffe JM (2007) The costs of carnivory. PLoS Biology 5: 363–368.
14. Riede JO, Brose U, Ebenman B, Jacob U, Thompson R, et al. (2011) Stepping in Elton's footprints: a general scaling model for body masses and trophic levels across ecosystems. Ecology Letters 14: 169–178.
15. Scharf FS, Juanes F, Rountree RA (2000) Predator size-prey size relationships of marine fish predators: interspecific variation and effects of ontogeny and body size on trophic-niche breadth. Marine Ecology Progress Series 208: 229–248.
16. Costa GC (2009) Predator Size, Prey Size, and Dietary Niche Breadth Relationships in Marine Predators. Ecology 90: 2014–2019.
17. King RB (2002) Predicted and observed maximum prey size – snake size allometry. Functional Ecology 16: 766–772.
18. Jones KE, Bielby J, Cardillo M, Fritz SA, O'Dell J, et al. (2009) PanTHERIA: a species-level database of life history, ecology, and geography of extant and recently extinct mammals. Ecology 90: 2648–2648.
19. Fritz SA, Bininda-Emonds ORP, Purvis A (2009) Geographical variation in predictors of mammalian extinction risk: big is bad, but only in the tropics. Ecology Letters 12: 538–549.
20. Maddison WP, Maddison DR (2010) Mesquite: A modular system for evolutionary analysis. Version 2.74 <http://mesquiteproject.org/mesquite/mesquite.html>.
21. Caballero S, Jackson J, Mignucci-Giannoni AA, Barrios-Garrido H, Beltrán-Pedreros S, et al. (2008) Molecular systematics of South American dolphins Sotalia: Sister taxa determination and phylogenetic relationships, with insights into a multi-locus phylogeny of the Delphinidae. Molecular Phylogenetics and Evolution 46: 252–268.
22. Kuhn TS, Mooers AØ, Thomas GH (2011) A simple polytomy resolver for dated phylogenies. Methods in Ecology and Evolution 2: 427–436.
23. R Core Team (2013) R: A Language and Environment for Statistical Computing. R Foundation for Statistical Computing, Vienna, Austria: http://www.R-project.org/.
24. Drummond AJ, Suchard MA, Xie D, Rambaut A (2012) Bayesian phylogenetics with BEAUti and the BEAST 1.7. Molecular Biology and Evolution.
25. Johnson JB, Omland KS (2004) Model selection in ecology and evolution. Trends in Ecology and Evolution 19: 101–108.
26. Symonds MR, Moussalli A (2011) A brief guide to model selection, multimodel inference and model averaging in behavioural ecology using Akaike's information criterion. Behavioral Ecology and Sociobiology 65: 13–21.
27. Burnham KP, Anderson DR (2002) Model Selection and Multimodal Inference: A practical Information-Theoretical Approach. New York: Springer.
28. Burnham KP, Anderson DR, Huyvaert KP (2011) AIC model selection and multimodel inference in behavioral ecology: some background, observations, and comparisons. Behavioral Ecology and Sociobiology 65: 23–35.
29. Orme D, Freckleton R, Thomas G, Petzoldt T, Fritz S, et al. (2012) caper: Comparative Analyses of Phylogenetics and Evolution in R. R package version 0.5. http://CRAN.R-project.org/package = caper.
30. Freckleton RP, Harvey PH, Pagel M (2002) Phylogenetic Analysis and Comparative Data: A Test and Review of Evidence. The American Naturalist 160: 712–726.
31. Felsenstein J (1985) Phylogenies and the Comparative Method. The American Naturalist 125: 1–15.
32. Pinheiro JC, Bates DM (2000) Mixed-effects models in S and S-Plus. New York: Springer.
33. Bates D, Maechler M, Bolker B, Walker S (2013) lme4: Linear mixed-effects models using Eigen and S4. R package version 1.0-5. http://CRAN.R-project.org/package = lme4.
34. Pawar S, Dell AI, Savage VM (2012) Dimensionality of consumer search space drives trophic interaction strengths. Nature 486: 485–489.
35. Smith FA, Lyons SK (2011) How big should a mammal be? A macroecological look at mammalian body size over space and time. Philosophical Transactions of the Royal Society B: Biological Sciences 366: 2364–2378.
36. Brose U, Jonsson T, Berlow EL, Warren P, Banasek-Richter C, et al. (2006) Consumer–resource body-size relationships in natural food webs. Ecology 87: 2411–2417.
37. Hairston JNG, Hairston SNG (1993) Cause-effect relationships in energy flow, trophic structure, and interspecific interactions. The American Naturalist 142: 379–411.
38. Andersen KH, Beyer JE, Lundberg P (2009) Trophic and individual efficiencies of size-structured communities. Proceedings of the Royal Society B: Biological Sciences 276: 109–114.
39. Jonathan Davies T, Meiri S, Barraclough TG, Gittleman JL (2007) Species co-existence and character divergence across carnivores. Ecology Letters 10: 146–152.
40. Dell'Arte GL, Laaksonen T, Norrdahl K, Korpimäki E (2007) Variation in the diet composition of a generalist predator, the red fox, in relation to season and density of main prey. acta oecologica 31: 276–281.
41. Renaud S, Michaux J, Schmidt DN, Aguilar J-P, Mein P, et al. (2005) Morphological evolution, ecological diversification and climate change in rodents. Proceedings of the Royal Society B: Biological Sciences 272: 609–617.
42. Carbone C, Mace GM, Roberts SC, Macdonald DW (1999) Energetic constraints on the diet of terrestrial carnivores. Nature 402: 286–288.
43. Rasmussen GSA, Gusset M, Courchamp F, Macdonald DW (2008) Achilles' Heel of Sociality Revealed by Energetic Poverty Trap in Cursorial Hunters. The American Naturalist 172: 508–518.
44. Creel S (1997) Cooperative hunting and group size: assumptions and currencies. ANimal behaviour 54: 1319–1324.
45. Lührs M-L, Dammhahn M, Kappeler P (2013) Strength in numbers: males in a carnivore grow bigger when they associate and hunt cooperatively. Behavioral Ecology 24: 21–28.
46. Bromham L, Lanfear R, Cassey P, Gibb G, Cardillo M (2012) Reconstructing past species assemblages reveals the changing patterns and drivers of extinction through time. Proceedings of the Royal Society B: Biological Sciences 279: 4024–4032.
47. McCain CM, King SRB (2014) Body size and activity times mediate mammalian responses to climate change. Global Change Biology.
48. Moritz C, Patton JL, Conroy CJ, Parra JL, White GC, et al. (2008) Impact of a century of climate change on small-mammal communities in Yosemite National Park, USA. Science 322: 261–264.
49. McCain CM, Colwell RK (2011) Assessing the threat to montane biodiversity from discordant shifts in temperature and precipitation in a changing climate. Ecology Letters 14: 1236–1245.

9

Investigating Population Structure of Sea Lamprey (*Petromyzon marinus*, L.) in Western Iberian Peninsula Using Morphological Characters and Heart Fatty Acid Signature Analyses

Maria João Lança[1,2], **Maria Machado**[2], **Catarina S. Mateus**[3,4], **Marta Lourenço**[3], **Ana F. Ferreira**[3], **Bernardo R. Quintella**[3,5]*, **Pedro R. Almeida**[4,6]

1 Escola de Ciências e Tecnologia, Departamento de Zootecnia, Universidade de Évora, Évora, Portugal, **2** Instituto de Ciências Agrárias e Ambientais Mediterrânicas, Universidade de Évora, Évora, Portugal, **3** Centro de Oceanografia, Faculdade de Ciências, Universidade de Lisboa, Lisboa, Portugal, **4** Museu Nacional de História Natural e da Ciência & Centro de Biologia Ambiental, Universidade de Lisboa, Lisboa, Portugal, **5** Departamento de Biologia Animal, Faculdade de Ciências, Universidade de Lisboa, Lisboa, Portugal, **6** Escola de Ciências e Tecnologia, Departamento de Biologia, Universidade de Évora, Évora, Portugal

Abstract

This study hypothesizes the existence of three groups of sea lamprey *Petromyzon marinus* L. in Portugal (North/Central group, Tagus group, and Guadiana group), possibly promoted by seabed topography isolation during the oceanic phase of the life cycle. Within this context, our purpose was to analyze the existence of a stock structure on sea lamprey populations sampled in the major Portuguese river basins using both morphological characters and heart tissue fatty acid signature. In both cases, the multiple discriminant analysis revealed statistically significant differences among groups, and the overall corrected classification rate estimated from cross-validation procedure was particularly high for the cardiac muscle fatty acid profiles (i.e. 83.8%). Morphometric characters were much more useful than meristic ones to discriminate stocks, and the most important variables for group differentiation were eye length, second dorsal fin length and branchial length. Fatty acid analysis showed that all lampreys from the southern Guadiana group were correctly classified and not mixing with individuals from any other group, reflecting a typical heart fatty acid signature. Our results revealed that 89.5% and 72.2% of the individuals from the Tagus and North/Central groups, respectively, were also correctly classified, despite some degree of overlap between individuals from these groups. The fatty acids that contributed to the observed segregation were C16:0; C17:0; C18:1ω9; C20:3ω6 and C22:2ω6. Detected differences are probably related with environmental variables to which lampreys may have been exposed, which leaded to different patterns of gene expression. These results suggest the existence of three different sea lamprey stocks in Portugal, with implication in terms of management and conservation.

Editor: David William Pond, Scottish Association for Marine Science, United Kingdom

Funding: This work was financially supported by Foundation for Science and Technology (FCT) through project PTDC/BIA-BDE/71826/2006 and also by FEDER Funds through the Operational Programme for Competitiveness Factors - COMPETE and National Funds through FCT - Foundation for Science and Technology under the Strategic Project PEst-C/AGR/UI0115/2011. The funders had no role in study design, data collection and analysis, decision to publish, or preparation of the manuscript.

Competing Interests: The authors have declared that no competing interests exist.

* Email: bsquintella@fc.ul.pt

Introduction

European populations of sea lamprey (*Petromyzon marinus* L.) have declined over the last 30 years [1], [2], and several authors have pointed out a reduction in sea lamprey abundance in Portuguese rivers [3], [4]. Sea lampreys can be found in all major Portuguese river basins, being more abundant in the central and northern regions of the country [3]. Due to the reduction in population abundance and the anthropogenic pressures to which this species is subjected, in Portugal it is classified as "Vulnerable" in the Red List of Threatened Vertebrates [4].

Whereas the continental phase of lampreys' life cycle is well known, the oceanic phase remains a mystery, with available data resuming to a few accidental captures of host species with scars or, occasionally, lampreys still attached to the fish or cetaceans [5]. A limited record of 80 sea lampreys captured in the northwest Atlantic indicated that almost all individuals with less than 39 cm long where taken in bottom trawls on the continental shelf or in coastal trap nets, whereas most animals with more than 56 cm long were captured in mid-water trawls along the shelf edge or over the continental slope [6]. Evidence that sea lamprey might not show homing behaviour first emerged following a tagging study with a landlocked population of the Great Lakes [7], and was then corroborated using genetic analysis on anadromous populations captured along the east coast of North America [8], [9].

The anatomy and physiology of an individual is sensitive both to genetic and environmental factors, which are responsible for phenotypic variation reflecting morphological characteristics [10]. In meristic terms, the effect of abiotic factors during ontogeny may result in significant differences between individuals of the same population, [11]. Morphometric characters are exposed to the same abiotic factors for an even longer period of time, which may increase the susceptibility of having more differences [12]. If those differences are ecologically significant and constant in time, they may allow the identification of individuals of different populations or stocks [13]. Morphometric variables measured in the cephalic region of sea lamprey larvae were found to be more suitable for a morphological analysis of geographic variation between Portuguese river basins [14]. Meristic characters were also assessed but the discriminatory power between groups, i.e. river basins, was comparatively weaker.

The concept of stock is fundamental for both fisheries and endangered species management [15]. A stock can be defined as a population or portion of a population of which all members are characterized by similarities which are not heritable, but are induced by the environment, and which include members of several different subpopulations [16]. Unit stocks can also be defined as characteristic populations or sets of subpopulations within subareas of the geographic range of a species [17], or as "... an intraspecific group of randomly mating individuals with temporal and spatial integrity" [13].

Spawning areas are normally clearly distinguished among the different stocks, but since fish may undertake considerable migrations, catches may also consist of fish from several stocks. For this reason, much work has been carried out to find characters that can be used for stock identification [18]. Waldman et al.[19] suggested that stock identification could be based upon catch data, tag recoveries, meristics, morphometrics, scale morphology, parasites, and cytogenetic: protein electrophoresis, monogenetic, mitochondrial DNA and nuclear DNA.

One of the limitations when using fatty acid profiles of a tissue as biomarkers and/or to characterize species, subspecies, populations, or stocks, is that the fatty acid profile under analysis can be influenced by various environmental factors, including the diet [20]. However, when fatty acid profiles are used for identification, the assumption is that the composition of fatty acids in membrane phospholipids is genetically controlled and stable over time, and therefore the phospholipid fatty acids may be used as a natural marker over a longer timescale [21]. Several studies have indicated genetic control of the fatty acid composition in the heart lipids although the impact of environmental factors could not be excluded [18], [21]. The lipid composition of cardiac skeletal muscle has a high level of polar lipids incorporated in the membrane phospholipid pool, so its fatty acyl structure restricts the ability of the acyl chains to reflect diet [22], and because of the specialized functions of these lipids on membranes, this lipid class is relatively robust to dietary changes. For the reasons explained above, fatty acids of cardiac skeletal muscle may serve as natural markers for the identification of stocks [21], [23]. In the last decade, several reports have suggested that fatty acid composition of phospholipids in some body tissues (e.g. heart tissue, brain, eggs) have a stable genetics basis, making these tissues appropriate for stock identification [18], [20], [24], [25]. Many fish species such as herring (Clupea harengus L.), striped bass (Morone saxatilis Walbaum, 1792) and cod (Gadus morhua L.) had been studied with this approach looking for possible stock differences [18], [26], [27].

Within this context, we hypothesize the existence of three sea lamprey groups in Portugal, possibly promoted by the seabed topography isolation during the oceanic phase of the life cycle; three large abyssal plains, and adjacent continental slopes, occur off western Iberian Peninsula: the Iberia Abyssal Plain in the north, the Tagus Abyssal Plain in the centre and the Horseshoe Abyssal Plain in the south. The Iberia Abyssal Plain is separated from the Tagus Abyssal Plain by the Estremadura Spur and the Tore Seamount, and by the Nazaré Canyon (continental shelf). The Tagus Abyssal Plain is separated from the Horseshoe Abyssal Plain through the Gorringe Bank and the Setúbal Canyon (Fig. 1).The hypothesis presented in this study is associated with the assumption that the bulk of the parasitic attacks are directed towards benthic hosts, which are believed to have restricted dispersal capability when compared with the more mobile pelagic species, and thus exchanges between lamprey feeding areas (i.e. groups) are strongly reduced. This hypotheses is supported by two evidences: (i) recent discoveries indicate a shorter hematophagous feeding stage (~1 year) [28], first reported to last from 23 to 28 months [29]; and (ii) there are no capturing records of adult lampreys or fish with wounds, compatible with potential lamprey attachments, in the data collected from annual surveys performed in the Portuguese continental shelf by IPMA I. P., the Portuguese fisheries laboratory (Yorgos Stratoudakis, pers. comm.). The short adult feeding stage attributed to the sea lamprey reduces considerably their dispersion capability in the marine environment, and this fact, together with the absence of evidences of adult lampreys feeding in the continental shelf, support the present study hypotheses that postulates that the majority of the growth during the parasitic feeding stage are related with attachments to benthic hosts species that live in the continental slope and/or abyssal plains. To test this hypothesis we analysed (i) morphological differentiation, and (ii) heart tissue phospholipid fatty acid profile between sea lamprey adults from the main Portuguese river basins, divided in the three groups mentioned above. In parallel, we performed analysis of genetic differentiation among the exact same groups using 12 microsatellite loci (results not shown, unpublished data), but no differences were found among the three groups at this level. To end, we discuss on the possibility of the existence of three sea lamprey stocks off western Iberian Peninsula, distinct by segregation in the trophic phase of the life cycle, and make some considerations and recommendations for conservation.

Methods

Ethics statement

Sampled sea lampreys were transported alive to the laboratory in a 0.4 m^3 capacity tank equipped with proper life support system including aeration, external filter and temperature control. In the laboratory, the individuals were first immersed in cold water to minimize handling stress and pain sensibility and sacrificed through decapitation method. This study was carried out in strict accordance with the recommendations present in the Guide for the Care and Use of Laboratory Animals of the European Union – in Portugal represented by the Decree-Law n°129/92, Portaria n°1005/92. Approval by a named review board institution or ethics committee was not necessary as the final model for ethical experimentation using fish as biological models was not implemented in Portuguese research units at the time of experimentation. This work was conducted under an institutional license for animal experimentation and a personal license to first author Maria João Lança and the co-authors Pedro R. Almeida and Bernardo R. Quintella, issued by the Direcção-Geral de Veterinária (DGV), Portuguese Ministry of Agriculture, Rural Development and Fisheries.

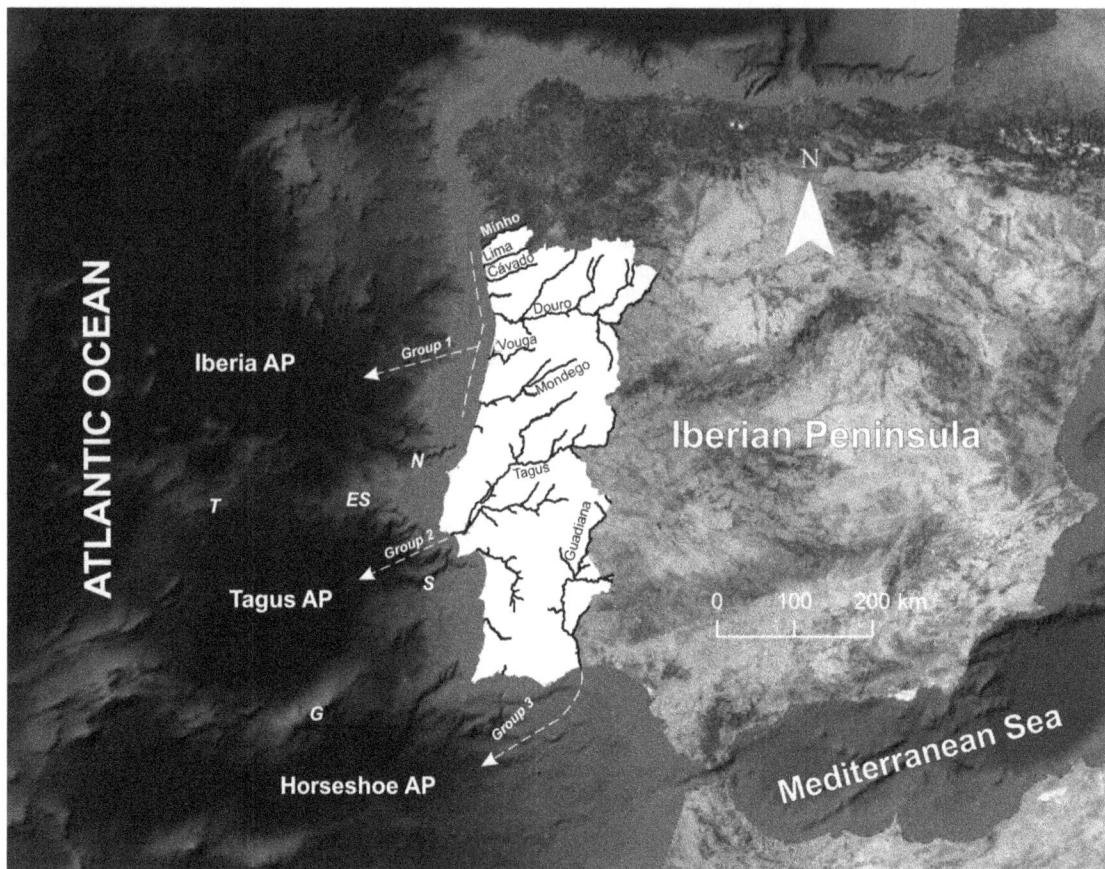

Figure 1. Location of the river basins from which sea lamprey individuals were collected. Formation of the three groups (testing hypothesis) based on the geographical location of the river mouth and the proximity to western Iberian oceanic areas with the representation of the seamounts and canyons that contour the three abyssal plains. Acronyms: Iberia AP - Iberia Abyssal Plain; Tagus AP – Tagus Abyssal Plain; Horsheshoe AP – Horsheshoe Abyssal Plain; T – Tore Seamount; ES – Estremadura Spur; G – Gorringe Bank.

Sampling

Sea lamprey spawners were captured in March 2008 during the peak of their spawning migration in eight river basins: Minho (41°52′N; 08°50′W), Lima (41°41′N; 08°49′W), Cávado (41°32′N; 08°47′W), Douro (41°08′N; 08°40′W), Vouga (40°39′N; 08°43′W), Mondego (40°8′N; 08°50′W), Tagus (39°03′N; 08°47′W) and Guadiana (37°38′N; 07°39′W). No specific permissions were required for sampling in these locations because the adult lampreys were captured by local fishermen in designated professional fishing areas. This study was conducted with a species considered "Vulnerable" by the Portuguese Red List of Threatened Vertebrates but general permits for field sampling fish (including *P. marinus*) were accredited by the Autoridade Florestal Nacional (AFN).

A total of 224 sea lampreys were collected, including about 30 individuals from each river basin. Specimens from each river basin were grouped *a priori* to test the hypothesis of stock fragmentation promoted by the seabed physiographic features during the oceanic parasitic phase. We defined three groups based on the geographical location of the river mouth, namely, the proximity to western Iberia oceanic areas (Fig. 1). Group 1 includes individuals captured in the Minho, Lima, Cávado, Douro, Vouga and Mondego river basins; Group 2 collects specimens captured in River Tagus; and Group 3 includes individuals from River Guadiana (Fig. 1).

Detailed temperature–salinity distribution in the Northeast Atlantic, the region encompassing the three large abyssal plains (i.e. Iberia, Tagus and Horseshoe), is available at the web-site of the Centre of Oceanography of the University of Lisbon (http://co.fc.ul.pt/en/data) [30].

Morphological characters

A total of 34 morphological characters were used: 24 morphometric and 10 meristic (Fig. 2), following [31]–[33].

All the 224 captured lampreys were used for the morphometric analysis. The morphometric characters were measured using graduated scales (±0.5 mm) and callipers (±0.5 mm; Fig. 2a). A sub sample of 201 lampreys was used for the meristic analysis. The oral disc of each individual was photographed (Kodak Z740) to count the meristic characters (Fig. 2b). To standardize the procedure, the picture was taken through an acrylic plate with the oral disc always opened to its maximum amplitude. A graphical scale was used to calibrate each image. The pictures were analyzed and processed using the Image J software [34] to count the number of teeth and rows in the anterior, laterals and posterior fields of the lamprey oral disc. The adopted teeth terminology (Fig. 2b) follows that proposed by Vladykov and Follett [31]. Trunk myomeres were counted between the anterior edge of the cloacal slit and the posterior edge of the last branchial opening, following [32]. All counts and measurements were made on the left side of the body.

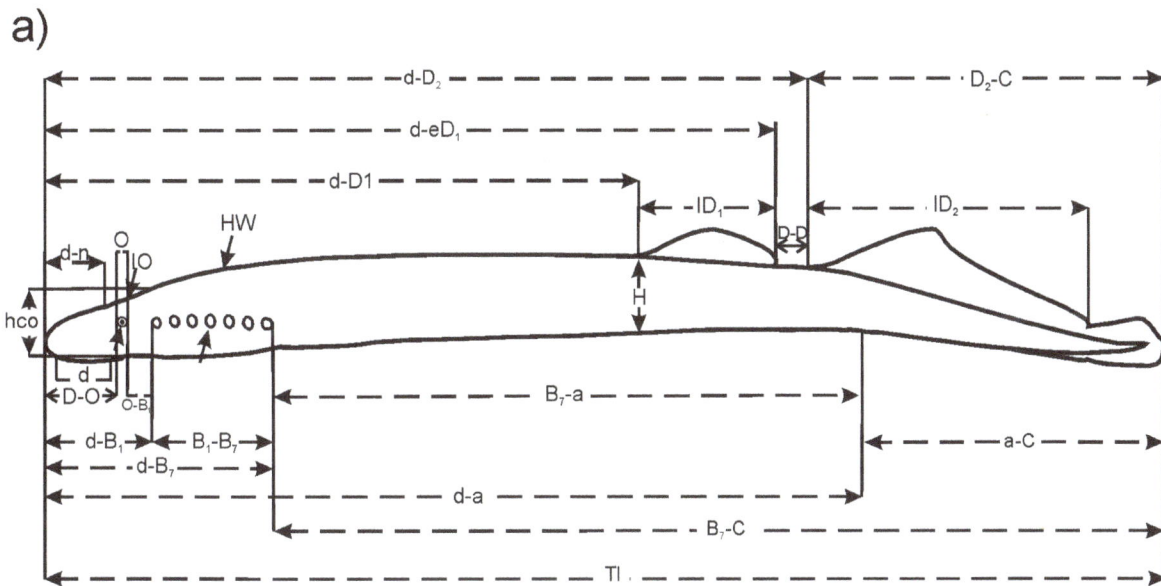

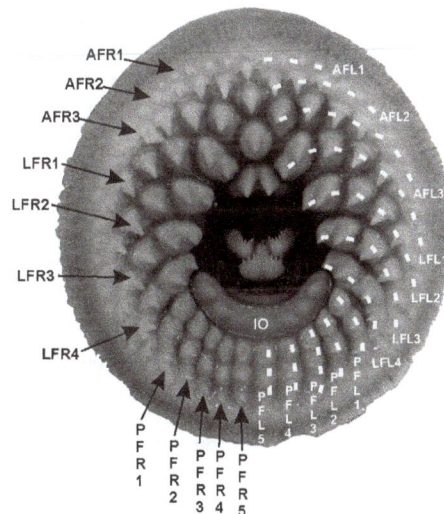

Figure 2. Schematic representation of the morphological features recorded for the analysis of geographic variation of sea lamprey in Portugal. (a) lateral view outline with the representation of the measured morphometric characters: TL, total length; d, disc length; d-a, distance between disc and anus; a-C, tail length; B_7-C, postbranchial length; B_7-a, trunk length; d-D_1, predorsal distance; d-eD_1, distance between disc and posterior end of first dorsal fin; d-D_2, distance between disc and base of second dorsal fin; D_2-C, dorsal part of caudal fin length; ID_1, first dorsal fin length; ID_2, second dorsal fin length; D-D, distance between dorsal fins; H, body depth; d-O, preocular distance; O, eye diameter; O-B_1, postocular length; Hco, head depth; d B_1, prebranchial length; B_1-B_7, branchial length; d-B_7, head length; d-n, prenostril length; IO, interocular distance; HW, head width; (b) photograph of the oral disc with the representation of the counted meristic characters: AF, anterior field; LF_R, lateral right field; LF_L, lateral left field; PF, posterior field; SO, supraoral lamina; L, lingual lamina; IO, infraoral lamina; TNteeth, total number of teeth; AFteeth, number of teeth in the anterior field; LFteeth, number of teeth in the lateral field; PFteeth, number of teeth in the posterior field; TNrows, total number of rows; AFrows, number of rows in the anterior field; LFrows, number of rows in the lateral field; PFrows, number of rows in the posterior field; IOcusps, number of cusps in the infra-oral lamina).

Total mass (± 0.01 g) of each individual was determined using a precision balance (Kern 440-36).

To test the hypothesis of sexual dimorphism in morphometric and meristic characters of *P. marinus*, the gender of all individuals was confirmed with histological slides, prepared with sections of reproductive organs according to the standard protocol of [35].

Histological slides were observed using a stereomicroscope (Leica DM 2000).

Tissue preparation and collection

Data on body total mass (TM, nearest g) and total length (TL, nearest mm) was registered for each sea lamprey.

Data on heart total mass (HTM, nearest g) and gender was reported for each sea lamprey (see previous section for gender determination). Sex ratio (male/female), heart total mass/body gutted mass ratio (HTM/BGM, expressed in percentage) and heart percentage of water loss (H_{water}) were also determined.

The heart was rapidly excised and rinsed in ice-cold 0.9% NaCl solution. Heart was then sliced and frozen between the tongues of an aluminum clamp that was cooled in liquid nitrogen. The frozen tissue heart samples were stored in aluminum canisters at $-80°C$ until laboratorial processing.

Heart tissue lipid extraction and fatty acid analysis

Pre-testing (random subsample of 8 individuals per river basin) determination of heart total lipids, neutral lipids and polar lipids revealed that more than 90 percent of the lipids present in heart tissue were polar lipids, resulting in all further analyses for fatty acid profile characterization were only done in polar lipid class. Lipid extraction was made according to the method described by Lança et al. [36]. Briefly, heart tissue total lipids were extracted using a Dionex 100 accelerated solvent extractor (ASE). To prepare for extraction, aliquots of 1 g portion of heart muscle samples were weighed on an analytical balance (Mettler AT201; Greifensee, Switzerland) and their masses were recorded to the nearest 0.01 mg. Tissue samples were then lyophilized until constant mass to determine the percentage of water loss and aliquots of heart tissue with 100 mg of dry weight were used. The total lipid sample was then extracted with a mixture of 60% chloroform and 40% methanol (Merck, Darmstadt, Germany) at $100°C$ and 13.8 MPa. The crude extract was then concentrated under a stream of nitrogen and vacuum and heart total lipid (HTLip, expressed in g per g of dry heart muscle) were determined. Each sample was reconstituted in 20–30 volumes of ice-cold acetone to separate neutral lipids from polar lipids Because the proportion of neutral lipids obtained were negligible, only the lower phase corresponding to polar lipids was saponified in methanolic NaOH 0.5 N at $70°C$ for 20 min. Fatty acids were then prepared with boron-trifluoride-methanol (14 g BF_3/L CH_3OH, Merk-Schuchardt, Germany) in order to give fatty methyl esters (FAME) according to the procedure of [37].

FAME were analysed by liquid-gas chromatography in a Hewlett Packard HP 6890 Series GC System according the chromatographic conditions described in [36].

The presence of C13:0 fatty acid on samples was confirmed using a GC-MS Bruker Scion 456 equipped with a BR-Swax $30×0,25$ µm column. Conditions were as follow: Inlet temperature $-250°$; Inlet mode - split 20 mL/min; He 1,2 mL/min column constant flow and oven temperature range from $120°C$ (for 5 min) to $240°C$ (for 10 min) with a ramping rate of $4°C$/min; Ionization source: 70 eV electron ionization; and the GC-MS operated in full scan mode from 40–450 Da. To detect the subject compound ion extraction m/z 74 the m/z 87 were made.

The unsaturation index (UI), a measure of the number of double bonds within a sample, was calculated as the sum of the percentage of each unsaturated fatty acid multiplied by the number of double bonds within that fatty acid [38].

Data analysis and interpretation: morphometric and meristic analysis

The statistical package SPSS for Windows (IBM, version 20.0) was used for data treatment and statistical analysis. Data transformations were used when appropriate.

The statistical analysis was applied following [14] which compared morphometric and meristic characters of P. marinus ammocoetes captured in several Portuguese river basins.

Briefly, analyses were performed separately for morphometric and meristic characters, and morphometric data were statistically adjusted to eliminate the influence of allometric growth as described in [14].

Outliers were detected by regression analysis of morphometric characters against total length, and by scatter plots of residual versus predicted values [39], resulting in the elimination of morphometric data for 19 lampreys (n = 18 from Group 1 and n = 1 from Group 3).

Of the 23 morphometric characters used in this analysis, 22 showed a linear relationship with total length ($P<0.05$). For character O, the only morphometric variable uncorrelated with total length ($P>0.05$), no size adjustment of the data was performed. For each of the 22 morphometric characters linearly related with total length, an analysis of covariance (ANCOVA) was employed to test for differences in allometric relationships among samples (i.e. geographical groups) and to estimate the common within-group regression slopes [39]. According to the ANCOVA analyses, within-group regression slopes were significantly different (df = 2, 201; $P<0.05$) for six of the morphometric characters ($d-D_1$; H; d; hco; B_1-B_7; HW); and thus size adjustment was based on the common within-group slopes and was performed following a modification of the allometric formula given by [40], as described in [14].

A multivariate analysis of variance (MANOVA) was used to assess the main and interaction effects of categorical variables (gender and geographical groups) on the 23 dependent morphometric variables. Highly significant differences ($P<0.001$) were found between gender and groups, but not for the interaction effect (gender × group; $P>0.05$) (Table 1). Consequently, 10 morphometric variables were removed from further analysis to eliminate the influence of sexual dimorphism among morphometric characters (Table 2).

No significant relationship ($P<0.05$) was found between the meristic characters and total length and thus, no size adjustment was performed. Outliers were detected by the SPSS Boxplot procedure following [14], which resulted in the elimination of 13 specimens (n = 7 Group 1, n = 3 Group 2 and n = 3 Group 3).

A permutational multivariate analysis of variance (PERMANOVA, two-way crossed design) was used to assess the main and interaction effects of two factors (gender and geographical groups) on the meristic variables. No sexual dimorphism effect ($P>0.05$) nor geographical group ($P>0.05$) was found among the meristic characters (Table 3), and consequently meristic data was not subsequently analysed with a multiple discriminant analysis (MDA).

Morphometric data from the three groups defined a priori were compared separately by means of a MDA.

Since the groups formed a priori varied markedly in size, a stratified (river/gender) random sample from the larger group (i.e. Group 1) was performed to reduce their size to a level comparable with the smaller groups (i.e. Groups 2 and 3), following recommendations for sample sizes in MDA analysis [41]. Consequently, the MDA was run with a subsample of 36 individuals from Group 1, all individuals from Group 2 (n = 28) and from Group 3 (n = 23). The computational method used to derive the discriminant function was the stepwise method with the selection rule to maximize Mahalanobis D^2 between groups [41], and remaining MDA related procedures followed [14].

Data analysis and interpretation: heart fatty acid profiles

The statistical package SPSS for Windows (IBM, version 20.0) was used for data treatment and statistical analysis. Data transformations were used when appropriate. Since the distribu-

Table 1. MANOVA multivariate test with sex and geographical groups as factors, and adjusted morphometric characters as dependent variables.

Effect	Pillai's Trace value	F	df	P	Partial ε^2	Obs. Power
Group	0.581	3.186	(46; 358)	<0.001	0.290	1.00
Sex	0.536	8.928	(23; 178)	<0.001	0.536	1.00
Group × Sex	0.245	1.086	(46; 358)	> 0.05	0.122	0.97

tion of fatty acid percentages is binomial, an arcsine transformation of fatty acid data was used prior to statistical analysis to meet assumptions of normality, independence and homocedasticity.

The integrated chromatogram values for each fatty acid were expressed as a percentage of the total sum of fatty acids identified in order to eliminate concentration effects.

Multivariate analysis of variance (MANOVA) was used to see the main interaction effects of categorical variable (gender and geographical groups) on multiple interval variables (fatty acids) and to test our hypothesis that distinct geographical groups of sea lampreys present distinct heart phospholipids fatty acid composition. Significance of the MANOVA was evaluated with Wilk's lambda. Multiple discriminant analysis (MDA) was used to identify which fatty acid contributed most to the differences in heart tissue composition among geographical groups.

Once again, for the MDA a stratified (river/gender) random sample from the larger group (i.e. Group 1) was performed to reduce their size to a level comparable with the smaller groups (i.e. Groups 2 and 3). Consequently, the MDA was run with a subsample of 36 individuals from Group 1, all individuals from Group 2 (n = 19) and from Group 3 (n = 19). Also, the number of independent variables must not exceed the smallest group size [41], and consequently, the MDA was run with a subsample of 19 fatty acids instead of the 30 fatty acids previously identified in each sample group. So, one must select few fatty acids for use in a particular analysis, generally choosing fatty acids that are expected to vary based on biological functions or, if no a priori hypotheses exist, simply have the greatest abundance in the sample set. Because, in phosphoglycerides, the most common of the phospholipids that constitute animal cell membranes seldom contain significant amounts of saturated fatty acids other than

Table 2. Mean of adjusted morphometric characters used for the morphological analysis of P. marinus.

Morphometric	Male	Female	MANOVA (F statistic; df = 1)
d	3.978	3.945	9.246***
d-a	6.449	6.455	3.667^{NS}
a-C	5.463	5.448	3.603^{NS}
B_7-C	6.541	6.545	12.366**
B_7-a	6.133	6.145	12.195**
d-D_1	6.099	6.104	0.057^{NS}
d-eD_1	6.313	6.316	0.000^{NS}
d-D_2	6.379	6.384	0.260^{NS}
D_2-C	5.643	5.630	1.970^{NS}
ID_1	4.671	4.670	0.096^{NS}
ID_2	5.385	5.369	0.484^{NS}
D-D	3.682	3.710	2.452^{NS}
H	3.995	4.068	50.131***
d-O	4.130	4.093	31.630***
O	2.185	2.176	0.107^{NS}
O-B_1	3.023	3.006	0.524^{NS}
Hco	3.790	3.773	7.432**
d-B_1	4.486	4.457	25.592***
B_1-B_7	4.470	4.466	0.631^{NS}
d- B_7	5.165	5.149	12.228**
d-n	4.010	3.972	18.027***
IO	3.857	3.832	15.328***
HW	4.000	3.990	1.912^{NS}

Tests of Between-Subjects Effects from the MANOVA for the factor sex (presented in Table 1), are also presented.
Acronyms of variables as defined in Figure 2; NS P>0.05;* P<0.05;** P<0.01; *** P<0.001.

Table 3. PERMANOVA results for the two-way crossed design, with geographical group and sex as factors, and meristic characters as variables.

Source	Df	SS	MS	Pseudo-F	P(perm)	Unique perms
Group	2	0.821	0.410	1.498	0.193	9942
Sex	1	0.569	0.569	2.078	0.124	9951
Group × Sex	2	0.517	0.258	0.943	0.437	9955
Residual	181	49.592	0.274			
Total	186	51.493				

16:0, 18:0, and to a lesser extent 20:0 [42] so all fatty acids either with chain length smaller than 10 carbons, or with chain length greater than 22 carbons were excluded; in the pool of monounsaturated fatty acids the C14:1 and the only odd chain fatty acid were excluded and in the pool of polyunsaturated fatty acids the C18:3ω3 and C18:3ω6 were excluded since they were not detected in any of the groups.

The remaining procedures regarding the MDA with the fatty acids were similar to the analysis performed with the morphometric characters.

The Pearson correlation test was used to analyse the relationship between heart total mass (HTM) and heart total lipid content (HTLip) for lampreys of each geographical group.

Results

Sexual dimorphism

From the 224 adult lampreys captured, 109 were males and 115 were females. The total length (TL) and body total mass (TM) of the sampled lampreys ranged from 63.9 cm to 97.9 cm (mean TL = 86.4 cm) and from 770 g to 1806 g (mean $M_T = 1188$ g), respectively. Significant differences were found (ANCOVA; $F_{(1, 221)} = 8.153$, $P < 0.01$) when comparing male ($y = 0.0898x^{2.1209}$; $r^2 = 0.61$; $d.f. = 107$; $P < 0.001$) and female length-weight relationship ($y = 0.0568x^{2.2342}$; $r^2 = 0.72$; $d.f. = 113$; $P < 0.001$). Generally, males tend to be longer while females are heavier, and differences tend to increase with length.

Gender related differences were found in 10 of the 23 morphometric characters analysed (Table 1; MANOVA, $P < 0.001$). In general, males have larger cephalic regions including longer d, d-O, hco, d-B_1, d-n and IO; while females have a tendency to longer and larger trunks (B_7-a and H, respectively) (Table 2). No significant differences were found between genders for the analysed meristic characters (Table 3; PERMANOVA, $P > 0.05$).

Morphometric analysis

The regressions for Z scores from discriminant functions 1 and 2 of discriminant analysis against total length were not significant ($r^2 = 0.011$, df = 86, $P = 0.340$; and $r^2 = 0.021$, df = 86, $P = 0.176$), indicating that size effects had been removed from the adjusted morphometric variables. Discriminant functions are statistically significant (Wilk's lambda, $P < 0.001$), and all pairs of groups showed statistically significant differences (df = 3, 82; $P < 0.05$). The stepwise analysis revealed that three morphometric characters contributed significantly to the MDA (O, lD_2, B_1-B_7). The Z scores and centroids from discriminant functions 1 and 2 were plotted against each other to develop a graphic representation of the relationship among groups (Fig. 3a). The two discriminant functions account for 59.3% and 40.7% of total variation. The

total classification rate estimated from cross-validation procedure was 54%, ranging from 38.9 to 73.9% (Table 4). Press's Q test revealed that the classification accuracy is statistically significant better than chance (Press's Q = 16.759, df = 1, $P < 0.001$).

The interpretation of the plot in figure 3a indicates that Function 1 is the primary source of separation between Group 2 (river Tagus' lampreys) and Group 3 (river Guadiana's lampreys); whereas Function 2 discriminates Group 1 (Northern river basins' lampreys) from the remaining. Discriminant loadings and potency index were used to assess the contributions of the three discriminant morphometric variables (Table 5). High correlations between the discriminant loadings of the variable O with the first function, and lD_2 and B_1-B_7 with the second function identified the variables with the best discriminatory power for each axis (Table 5). The character O was the variable with the highest potency index value and can be considered the most important morphometric character to distinguish adult *P. marinus* entering Tagus (mean O = 2.15) and Guadiana (mean O = 2.22) river basins (Table 5). Group 1 lampreys have intermediate size eyes (mean O = 2.18), longer second dorsal fins (mean $1D_2 = 5.39$) and shorter branchial lengths (mean B_1-$B_7 = 4.46$) when compared with lampreys entering rivers Tagus ($1D_2 = 5.37$; B_1-$B_7 = 4.48$) and Guadiana ($1D_2 = 5.36$; B_1-$B_7 = 4.47$).

Meristic analysis

No significant differences were found between meristic characters of lampreys belonging to the *a priori* defined geographical groups (Table 3; PERMANOVA, $P > 0.05$) and no significant interaction between gender and group was detected (Table 3; PERMANOVA, $P > 0.05$). The AFteeth, LFteeth and PFteeth showed high variation between specimens, whereas the AFrows and LFrows did not show variation among individuals (Table 6).

Tissue fatty acid profile

Individuals of Group 2 showed significantly higher values of heart total mass (HTM) ($P < 0.001$) than the other two groups. However, no significant differences were detected for HTM/BGM ratio among individuals of the three groups ($P > 0.05$; Table 7). The values of the HTM/BGM ratio were 0.25% for groups 1 and 2 and 0.24% for Group 3 (Table 7).

The heart total lipid content (HTLip) revealed significant differences ($P < 0.001$) among the individuals of the three groups with sea lampreys of Group 2 showing the lowest value (15.3%), and the individuals of Group 3 presenting the highest value (30.8%) (Table 7). For each of the three groups, no significant correlation was found between HTM and HTLip.

The fatty acid profile of the heart tissue phospholipids varied among individuals of the three groups (Table 8). Gender had no significant effect (MANOVA, $P > 0.05$) in fatty acid relative

a)

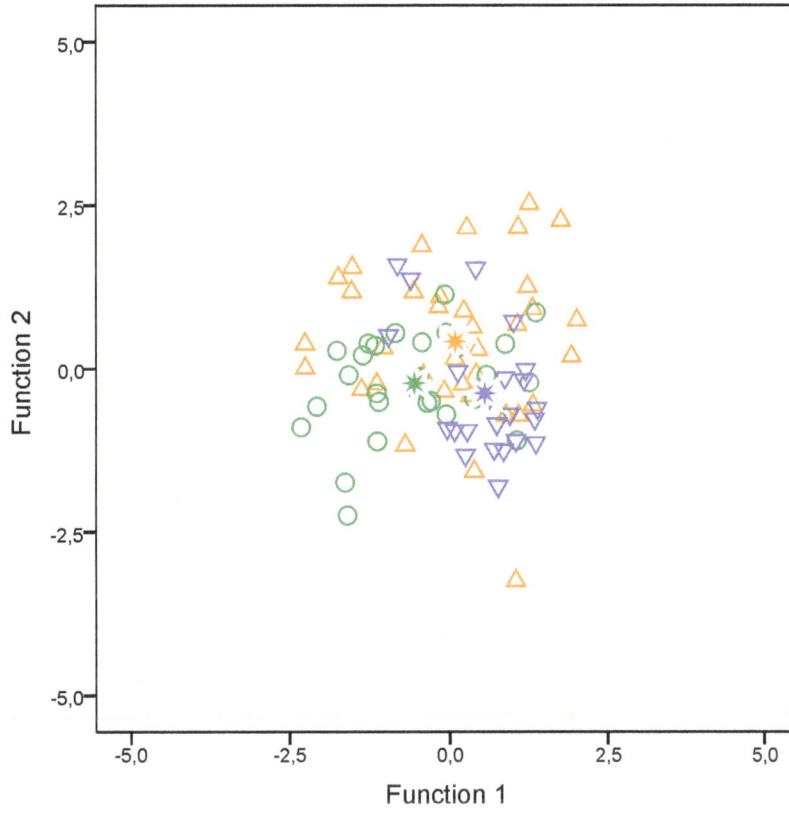

b)

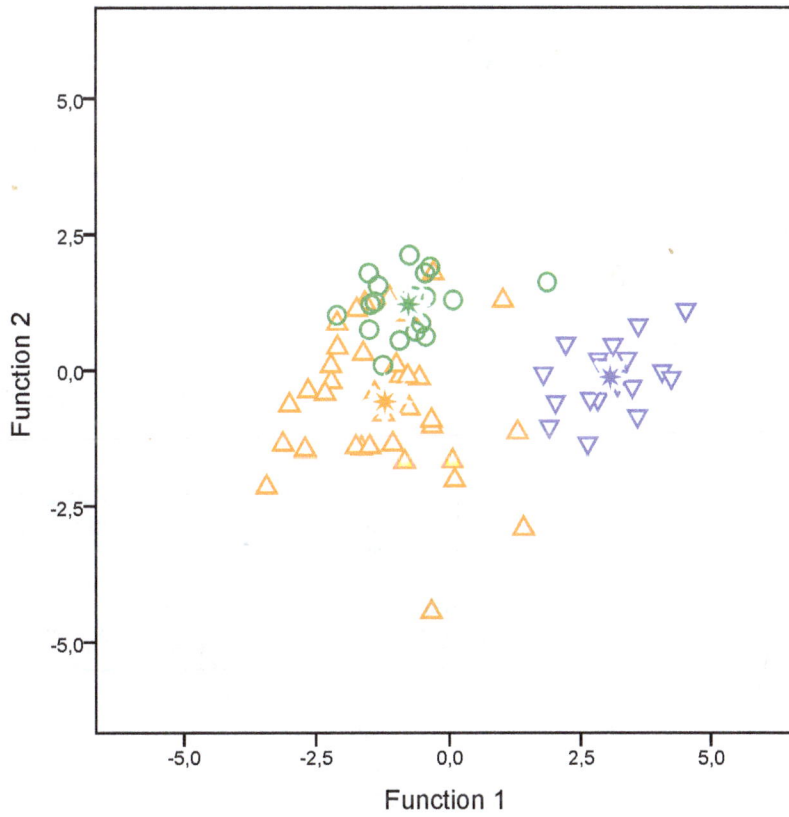

Figure 3. **Plot of the discriminant *Z* scores and group centroids of discriminate functions 1 and 2 for the three groups of adult sea lampreys based on (a) morphometric characters and (b) fatty acid composition of heart total lipids.** △ - Group 1: Minho, Lima, Cávado, Douro, Vouga and Mondego basins; ○ - Group 2: Tagus; ▽ - Group 3: Guadiana.

composition of phospholipids of heart tissue, whereas geographical groups exhibited a significant effect (MANOVA, $P \leq 0.001$). The interaction gender*geographical group had no significant effect in this variable (MANOVA, $P > 0.05$).

The predominant class of fatty acids in heart tissue was the saturated fatty acids (SFA), with percentages that ranged from 37.92% in individuals from Group 3, to 39.52% in individuals of Group 1 (Tab. 8). The major exception was recorded in sea lampreys from Group 2, where the percentage of SFA was similar to that of monounsaturated fatty acids (MUFA). In what concerns MUFA, individuals of Group 3 were characterized by the lowest values (19.67%), and individuals of Group 2 were characterized by the highest values (38.33%). Polyunsaturated fatty acid (PUFA) relative percentages varied between 17.72% (Group 2) and 21.38% (Group 3) (Table 8).

The predominant fatty acids detected were C16:0, C18:0 and C13:0 for SFA; C18:1ω9 and C16:1 for MUFA; and C22:6ω3 (DHA), C22:5ω3 (DPA) and C20:5ω3 (EPA) for PUFA. The percentage content of EPA and DHA demonstrates the dominance of DHA over EPA (DHA/EPA ratio) Although occurring at relative low amounts, several odd chain fatty acids like C13:0; C15:0; C17:0 and C17:1, were also present in heart tissue fatty acid profiles (Table 8). The unsaturated-to-saturated ratio (UFA/SFA) was used as an indirect indicator of the membrane fluidity (Table 8), since it has been previously reported that membranes with high UFA/SFA ratio show a high fluidity [43]. Individuals from Group 2 presented the highest value (1.45), and individuals of Group 3 exhibited the lowest value (1.24), but no significant differences were observed ($P > 0.05$). The unsaturation index (UI) was higher in Group 3 (132), followed by Group 1 (125) and by Group 2 (116), but no significant differences were observed ($P > 0.05$).

The MDA for the 19 fatty acids proved to be statistically significant and the overall corrected classification rate estimated from cross-validation procedure was 83.8% (Fig. 3b). All lampreys from Group 3 were correctly classified (100%) and not mixing with the sea lampreys from any of the other groups, reflecting a typical heart fatty acid signature for individuals of this group (Fig. 3b, Table 9). It is also interesting to note that, 89.5% and 72.2% of the individuals from groups 2 and 1, respectively, were also correctly classified (Table 9). The fatty acids that contributed to the segregation of groups were the SFA C17:0 (35.9%) and C16:0 (25%); the MUFA C18:1ω9 (14.1%); and the PUFA C20:3ω6 (4.7%) and C22:2ω6 (3.3%). The fatty acid with the highest

potency index that contributed for the separation of the sea lamprey groups was C17:0. Press's Q test revealed that the classification accuracy was significantly better than chance (Press's $Q = 754.735$ $gdl = 1$; $P \leq 0.001$).

Discussion

Morphological variation and potential adaptation in the proposed groups

Sexual dimorphism among mature lampreys is well known and appears to be similar in all lamprey species [44]. This study data showed that male adult sea lampreys captured in the beginning of the spawning migration weight less than females, and although some additional subtle morphometric differences between genders were found, no obvious secondary sex characters were detected. Males show an increased prebranchial length and oral disc size, while females have a longer and wider trunk, similar to the findings of [45] for the Arctic lamprey *Lethenteron camtschaticum* (Tilesius, 1811). A larger head in males is most likely related with distinctive behaviours during the spawning period, in particular nest construction and agonistic behaviour when competing for females. A stronger suctorial capacity provided by larger oral discs may be an important characteristic to maximize reproductive success among males. Similar results were described for Southern Hemisphere species of the genera *Geotria* and *Mordacia*, of which an increase in size of the male oral disc, associated with an extension in the length of the preorbital region, was observed [46], [47]. The larger trunk observed in females is most likely related with the maximization of space for the development of the gonads, thus increasing fecundity. Elongated trunk in female lampreys was also described by [48] for the European river lamprey *Lampetra fluviatilis* L.

Morphologically, the classification rate estimated for animals from River Tagus (Group 2) was 57%, which means that most of the animals that entered this river to spawn presented similar characteristics, even though a considerable number of lampreys showed a morphological profile compatible with the other two groups. A poor discrimination rate (39%) was found for the northern lampreys pooled in Group 1 (i.e., rivers Minho, Lima, Cávado, Douro, Vouga and Mondego). This result indicates that specimens from this geographical region are morphologically more diverse when compared with sea lampreys entering Tagus and Guadiana basins. The wider range of abiotic scenarios like depth (see Fig. 1 for sea floor topography), observed in the oceanic area

Table 4. Classification results obtained with the stepwise discriminant analysis cross-validation for morphometric characters to determine the predictive accuracy level of the discriminant functions.

Groups	*N*	Percent correct	Predicted Group Membership (count)		
			Group 1	Group 2	Group 3
Group 1	36	38.9	14	11	11
Group 2	28	57.1	7	16	5
Group 3	23	73.9	5	1	17
Total	87	54.0			

N, number of individuals.

Table 5. Summary of discriminant loadings and potency index for morphometric and meristic variables.

Variables	Discriminant loadings		Potency index
	Function 1	Function 2	
O	0.74*	−0.13	0.33
ID_2	−0.03	0.81*	0.27
B_1-B_7	−0.32	−0.60*	0.21

Acronyms of variables as defined in Figure 2; * largest absolute correlation between each variable and any discriminant function.

where northern lampreys occur, when compared with the other oceanic feeding areas discussed here, may lead to a higher morphological variability during the parasitic phase and, consequently, a lower classification rate of lampreys from Group 1. Also, this northern group may have lampreys originally from other oceanic regions, namely the Galicia Interior Basin or even the Biscay Abyssal Plain region, thus contributing to the low predicted group membership.

Three morphometric variables were considered significant in discriminating the three groups: eye length, second dorsal fin length and branchial length. The larger eye and longer branchial lengths of the Guadiana lampreys can be indicative of deeper feeding grounds, and the need for a more efficient mechanism of blood oxygenation. Comparatively, the relatively smaller eye and branchial length of the Northern lampreys may reveal shallower feeding grounds, and a less stressful demand for oxygen. Interestingly, the Northern lampreys with the lower classification rate, for both morphometric and heart fatty acid characters, was composed by individuals with longer second dorsal fins and thus, in theory, more capable for longer dispersions. The Guadiana lampreys had the highest classification rates among the three groups and the shortest second dorsal fin.

Heart total mass/body gutted mass ratio in the proposed groups

The values of HTM/BGM ratio obtained for all individuals of the three geographical groups were similar (~0.25%). This value is consistent with previous studies also with sea lamprey (~0.3%; [49]), and similar to the characteristic values obtained for other poikilothermic vertebrates (0.08%–0.30% for fish, and 0.19% for

amphibians and reptiles), but distant from the values usually determined in mammals (0.64%) [50].

In vertebrates this ratio reflects a direct relationship between the size of heart and oxygen consumption [51]. One would expect that lampreys from Group 3, with longer branchial lengths that may maximize oxygen uptake in less oxygenated habitats, would have a higher HTM/BGM ratio, but this was not the case.

Our results showed that individuals of Group 3 were characterized by the highest HTLip content. This result can be an indicator of increased rate of oxidative metabolism of fatty acids, since in cardiac muscle over 70% of energy consumed for electro-mechanical activity is covered by mitochondrial oxidation of fatty acids [52]. A study among 16 species of teleost fish revealed that glucose metabolism and fatty acid utilization increase with the increased energy demand [53]. According to this, our result suggests a higher demand for fatty acid oxidative metabolism of animals sampled in Group 3, comparatively to the individuals belonging to groups 1 and 2.

A lack of correlation between HTM and HTLip was expected because under physiological conditions, myocardial triacylglycerol stored in lipid droplets in the cytoplasm of the cardiac muscle cells is in a steady state condition, where no major alterations in the absolute amount of triacylglycerol fatty acids occur [52].

Heart tissue fatty acid profile in the proposed groups

For all three groups, heart tissue fatty acid profile showed that SFA were the most representative. In groups 1 and 2 these were followed by MUFA and then PUFA whereas in Group 3, the MUFA and PUFA order was inverted. Moreover, since cardiac skeletal muscle contains a significant content of phospholipids, C16:0, C18:0, C16:1 and C18:1ω9 high relative amounts are in

Table 6. Summary statistics for the meristic characters analysed in the sub sample of 201 sea lamprey individuals included.

Meristic	Mean	SD	Min.	Max
Myo	73.1	1.6	68	78
TNteeth	148.6	8.0	128	170
AFteeth	39.5	3.8	29	49
LFteeth	62.4	2.9	54	69
PFteeth	37.2	2.7	31	46
TNrows	25.2	0.4	24	27
AFrows	7.0	0.0	7	7
LFrows	8.0	0.0	8	8
PFrows	9.2	0.4	8	11
IOcusps	7.5	0.7	6	10

Acronyms of variables as defined in Figure 2; *SD*, standard deviation; *Min*, minimum; *Max*, maximum.

Table 7. Mean (± standard deviation) heart total mass (HTM, g), heart total mass/body gutted mass ratio (HTM/BGM expressed in percentage), mean heart total lipids (HTLip, expressed in g *per* g of dry tissue), heart water loss (H$_{Water}$, expressed as percentage) and sex ratio of sea lamprey individuals analysed.

Variables	Group 1	Group 2	Group 3
HTM	2.89±0.5^{*b}	3.34±0.4^{*a,c}	2.76±0.3^{*b}
HTM/BGM	0.25	0.25	0.24
HTLip	0.24^{*b,c}	0.15^{*a,c}	0.31^{*a,b}
H$_{Water}$	73.07±2.7	71.23±1.9	75.32±1.2
Sex Ratio (male/female)	0.48	0.46	0.68

Cases in which the relative amounts of a fatty acid are significantly different ($P < 0.001$) among the groups are marked with signs: *a: significantly different from Group 1; *b: significantly different from Group 2; *c significantly different from Group 3.

accordance with the fact that those fatty acids are the most common in the *sn*-1 position of phospholipids [38], [54]. Omega-3 fatty acids were also present In fact, the stable genetic basis of the fatty acid composition of heart tissue phospholipids [18], [20], [24], [25] in addition to the clear tendency for certain types of fatty acids to be incorporated into the *sn*-1 and *sn*-2 positions of the structural phospholipids, restrict the ability of the acyl chains to reflect diet.

Based on that, the fatty acid pool of polar lipids is considered stable over time and studies done with two stocks of reared Atlantic cod had already demonstrated it [18].The C13:0 showed higher values than expected and the explanation for these results is not known. However, this fatty acid could be associated to the presence of microbial sources considering some authors that C13:0 could be a useful of microbial presence on detritus [55–58].

Although ectothermic animals appear to increase the membrane content of unsaturated fatty acids in response to colder temperatures, a clear and direct relationship between specific unsaturated fatty acids and quantitative measurements of membrane fluidity has not been demonstrated [54]. In fact, a given overall fluidity level can be met by various fatty acids compositions, and often the fluidity of cellular membranes can be adjusted by converting SFA to MUFA, while the PUFA levels remain unaltered [59]. Because our results revealed that the unsaturated/saturated fatty acid ratio and unsaturation index were not significantly different among groups, this could mean that different fatty acid signatures were not caused by a direct effect of temperature on the phospholipids of heart tissue, but it is reasonable to believe that fatty acid profiles are phenotypic characters that must be correlated with differences in the abiotic factors that characterized different habitats and results from adaptation processes.

Physiological adaptation to environmental conditions

Fish species inhabiting areas where environmental conditions are relatively stable and constant may develop a specialization of their membranes phospholipids that allow them to adapt to the environment where they live [60]. Habitats are different in many ways (e.g. temperature, salinity, pressure), so it is reasonable to hypothesize that environmental differences may lead to adjustments of the expression of several genes, resulting, for instance, in distinct heart tissue fatty acid signatures. Then, considering the heart tissue fatty acid profile as a phenotypic variation, the presence of different fatty acid signatures likely indicates limited mixing among groups and may offer a practical measure for stock discrimination [18], [23]. The differences in the heart tissue fatty acid profile of individuals of the geographical groups seem to result

from the influence of environmental factors during the oceanic trophic phase of the lampreys' life cycle and to the geographical isolation promoted by seabed topography.

If this hypothesis is correct, it is possible that the oceanic phase of the sea lamprey life cycle, following the dispersion period during the juvenile trophic migration, is represented by a much less mobile adult stage restricting the mixture of lampreys from different geographical groups. This limited dispersal in marine environment was also highlighted in a recent work by Spice et al. [61] with Pacific lamprey (*Entosphenus tridentatus* Gairdner in Richardson 1836) and supported by Silva *et al.* [28] that suggest that at least a fraction of the sea lamprey population can reach adult size in approximately 14 months of hematophagous feeding. Moreover, part of this parasitic period can be spent feeding in rivers and estuaries before the trophic migration to the sea [62], so the marine stage might be even shorter than one year.

Marine trophic phase and target host species

Dispersal is often density-dependent in a wide variety of *taxa* [63]. Due to population density, dispersal may relieve pressure for resources in an ecosystem, and competition for these resources may be a selection factor for dispersal mechanisms [64].

It is likely that the absence of large pelagic fish species (inexistence of salmonids and drastic reduction of shads *Alosa* sp.) in the southwest and south coasts of Portugal, induces sea lamprey juveniles undergoing their marine trophic phase to target benthic fish species [65]. On the other hand, the northern Portuguese river basins, which have a higher proportion of individuals in relation to Tagus and Guadiana groups [14], [66], can prey upon pelagic fish and thus experience a wider dispersion throughout the neighbouring marine areas and, consequently, river basins. Sea lampreys from the west coast of Portugal apparently present some clues that indicate the existence of two different trophic pathways, one typical of a top predator of a marine food web with a planktonic base, and the other including both planktonic and benthonic species [65]. Since benthonic fish are usually less vagrant in the adult stage when compared with the more mobile pelagic species [67], the migrations of adult sea lampreys between feeding areas (i.e. stocks) would be attenuated, and differences in the fatty acid profile of heart most likely arise under those circumstances.

Juvenile dispersal, feeding areas and spawning migration

The three sea lamprey groups here identified are probably associated to the three isolated abyssal plains (and/or nearby continental slopes) off western Iberian Peninsula, and it is likely that they constitute three different stocks. Throughout the

Table 8. Relative amounts, as percentage of sum (mean ± sd), of fatty acids in heart tissue total lipids of sea lamprey individuals analysed.

	Fatty acid	Group 1	Group 2	Group 3
SFA	C10:0	2.84±2.09[*b]	0.74±1.10[*a;c]	3.36±1.54[*b]
	C12:0	0.08±0.16[*c]	0.06±0.05	ND
	C13:0	7.51±5.76[*b;c]	2.20±4.22[*a;c]	13.47±5.99[*a;b]
	C14:0	1.50±0.99[*c]	1.75±0.46[*c]	0.70±0.32[*a;b]
	C15:0	ND	ND	0.15±0.14[*a;b]
	C16:0	15.92±4.86[*b;c]	20.39±2.26[*a;c]	10.97±1.90[*a;b]
	C17:0	ND	ND	0.90±0.23[*a;b]
	C18:0	7.06±1.48[*b;c]	9.13±1.49[*a;c]	5.55±0.95[*a;b]
	C20:0	0.20±1.63[*b;c]	0.88±2.10[*a;c]	ND
	C22:0	0.22±0.62	0.28±0.30	ND
	ΣSFA	39.52	39.32	37.92
MUFA	C14:1	ND	0.06±0.06	ND
	C16:1	9.92±4.2[*b;c]	13.69±2.59[*a;c]	6.09±1.69[*a;b]
	C18:1ω9	16.64±6.23[*b]	22.95±2.67[*a;c]	12.86±2.55[*b]
	C20:1ω9	0.49±0.47[*b;c]	1.0±0.46[*a;c]	0.28±0.09[*a;b]
	C22:1ω9	ND	0.14±0.19	ND
	ΣMUFA	28.40	38.33	19.67
PUFA	C18:2ω6	0.29±0.23	0.35±0.14	0.30±0.21
	C20:2	0.14±1.20[*b;c]	0.32±3.11[*a;c]	ND
	C20:3ω6	ND	0.16±0.33	ND
	C20:3ω3	2. 69±1.48	3.10±1.07	3. 15±0.95
	C20:4ω6	ND	0.19±0.18	ND
	C20:5ω3	2.60±0.99	2.34±0.61[*c]	3.11±0.54[*b]
	C22:2ω6	0.12±0.52	ND	ND
	C22:5ω3	3.46±1.73	3.45±0.92	4.25±0.78
	C22:6ω3	8.21±4.62	6.32±1.74[*;c]	10.45±3.14[*b]
	ΣPUFA	18.83	17.72	21.38
	Σ(PUFA+MUFA)	47.23	56.05	41.05
	ΣUFA/ΣSFA	1.29	1.45	1.24
	Σω3	16.96	15.21	20.96
	Σω6	0.29	0.70	0.30
	C22:6ω3/C20:5ω3	3.16	2.70	3.36
	ΣEPA+DPA+DHA	14.27	12.11	17.81

SFA, saturated fatty acids; MUFA, monounsaturated fatty acids; PUFA, polyunsaturated fatty acids. Cases in which the relative amounts of a fatty acid are significantly different ($P< 0.001$) among the groups are marked with signs: *a: significantly different from group 1; *b: significantly different from group 2; *c significantly different from group 3. Fatty acids C6:0 and C8:0 are not presented because were not detected in each one of the three groups. ND, not detected.

Table 9. Classification results obtained with the stepwise MDA cross-validation for heart tissue fatty acids to determine the predictive accuracy level of the discriminant functions.

Number of individuals classified into group

Groups	N	Percent Correct	North/Central	Tagus	Guadiana
North/Central	36	72.2	26	8	2
Tagus	19	89.5	1	17	1
Guadiana	19	100	0	0	19
Total	74	83.4	–	–	–

juveniles' trophic migration it is likely that some mixture between groups occurs, particularly between the northern (Group 1) and the Tagus (Group 2) stocks, which would be in agreement with the dispersal phase of some marine fish species during the juvenile stage in the same geographical area [68]. The lack of genetic differentiation between groups (results not shown, unpublished data) corroborates this scenario: since adults present significant levels of differentiation at the morphological and physiological levels, and there is genetic mixing between groups, the juvenile migration is most likely accompanied by dispersal among basins. During the spawning migration, lampreys seem to preferentially move north, probably attracted by the exceptional freshwater flow originated in northern river basins, particularly those north of river Douro, inclusive. In fact, eight animals sampled in the northern group presented characteristics of the Tagus' group.

The bulk of the juvenile lampreys from the isolated Guadiana river basin (Group 3) probably migrate to the feeding areas located at the Horshoe Abyssal Plain or nearby areas, which is located on the southern Iberian margin off western the Mediterranean Sea, and return to spawn in their river of origin. The impact of the Mediterranean Outflow Water (MOW) in the potential feeding area of animals entering the River Guadiana is particularly evident between 500 to 1400 m and shows higher temperatures and salinities than the North Atlantic Central Water (NACW) [69]. The unique conditions caused by the MOW influence may be responsible for the high distinct heart tissue fatty acid profile found in lampreys from group 3, as revealed by the 100% predictive accuracy level. In the oceanic zone over the continental slope, from December to February the dominant current (depth up to 1200 m) is oriented northward (Ana Teles-Machado and Álvaro Peliz, unpublished data). This may impel adult sea lampreys approaching the continent in the beginning of the spawning migration to the north with the prevailing current. Moreover, near the continental shelf, the dominant current is southward, and migrating sea lampreys may be once again oriented northward attracted by the odours transported from the northern rivers basins, which present higher ammocoete densities and river discharges than the Tagus or Guadiana river basins. This might explain why the North-Central group showed the occurrence of eight lampreys from the Tagus group and two from the Guadiana group.

In conclusion, the significant morphological and physiological differences found between groups are most likely the result from the influence of environmental factors to which lampreys may have been exposed during the oceanic trophic phase of the life cycle, rather than derived from a genetic basis. This would imply that the oceanic phase of the sea lamprey life cycle is composed by a dispersion period during the juvenile migration, followed by a much less mobile adult stage, which will restrict the mixture of adult lampreys from different geographical groups, segregated by seabed topography.

Implications for conservation

The population structure put in evidence in this work have important implications in terms of management and conservation of *P. marinus* in Portugal, where it is considered threatened. Three stocks of this species are apparently present in Atlantic waters off country: the northern, the Tagus and the Guadiana stocks. The first includes individuals from Minho, Lima, Cávado, Douro, Vouga and Mondego river basins and, possibly, from North-western Spain (Galician rivers; not included in this study). A considerable number of lampreys still use the above referred basins for reproduction [66], except in River Douro, where apparently there are no suitable conditions for nest building in the available 20 km of river stretch downstream of the first obstacle [3]. The probable existence of a common stock in north-western Iberian waters reinforces the need for international joint efforts to manage this halieutic resource, commercially exploited both in Portuguese and Spanish watersheds.

Tagus and Guadiana stocks are, however, priority in conservation terms. The number of lampreys entering these basins, particularly in the southern Guadiana river basin, is very scarce. The existence of a lamprey stock composed mainly by sea lampreys originally from the Guadiana basin raises some concerns about the future of the species in its southern limit of distribution, mainly due to the hydric stress known to occur in this basin, and exacerbated by the potential effects of climate change.

Acknowledgments

This work was financially supported by Foundation for Science and Technology (FCT) through project PTDC/BIA-BDE/71826/2006 and also by FEDER Funds through the Operational Programme for Competitiveness Factors - COMPETE and National Funds through FCT - Foundation for Science and Technology under the Strategic Project PEst-C/AGR/UI0115/2011.

We thank our colleagues Carlos Alexandre, Sílvia Pedro and Filipe Romão for helping with tissue sampling and preparation.

Author Contributions

Conceived and designed the experiments: MJL BQ PA. Performed the experiments: MJL MM CM ML AF BQ PA. Analyzed the data: MJL BQ PA. Contributed reagents/materials/analysis tools: MJL MM CM ML AF. Wrote the paper: MJL CM BQ PA.

References

1. Lelek A (1987) The Freshwater Fishes of Europe, Volume 9: Threatened fishes of Europe. Wiesbaden: Aula-Verlag. 343 p.
2. Renaud CB (1997) Conservation status of Northern Hemisphere lampreys (Petromyzontidae). J Appl Icthyol 13: 143–148.
3. Mateus CS, Rodríguez-Muñoz R, Quintella BR, Alves MJ, Almeida PR (2012) Lampreys of the Iberian Peninsula: distribution, population status and conservation. Endanger Species Res 16: 183–198.
4. Rogado L, Alexandrino P, Almeida PR, Alves J, Bochechas J, et al. (2005) Peixes. In: Cabral MJ, et al., editors. Livro vermelho dos vertebrados de Portugal. Lisboa: Instituto de Conservação da Natureza. pp. 63–114.
5. Farmer GJ (1980) Biology and physiology of feeding in adult lamprey. Can J Fish Aquat Sci 37: 1751–1761.
6. Halliday RC (1991) Marine Distribution of the Sea Lamprey (Petromyzon marinus) in the Northwest Atlantic. Can J Fish Aquat Sci 48: 832–842.
7. Bergstedt RA, Steelye JG (1995) Evidence for lack of homing by sea lampreys. Trans Am Fish Soc 124: 235–239.
8. Genner MJ, Hillman R, McHugh M, Hawkins SJ, Lucas MC (2012) Contrasting demographic histories of European and North American sea lamprey (Petromyzon marinus) populations inferred from mitochondrial DNA sequence variation. Mar Freshw Res 63: 827–833.
9. Waldman J, Grundwald C, Wirgin I (2008) Sea lamprey Petromyzon marinus: an exception to the rule of homing in anadromous fishes. Biol Lett 4: 659–662.
10. Barlow GW (1961) Causes and significance of morphological variation in fishes. Syst Zool 10: 105–117.
11. Tåning AV (1952) Experimental study of meristic characters in fishes. Biol Rev Camb Philos Soc 27: 169–193.
12. Melvin GD, Dadswell MJ, McKenzie A (1992) Usefulness of meristic and morphometric characters in discriminating populations of American shad (Alosa sapidissima) (Ostreichthyes: Cluoeidae) inhabiting a marine environment. Can J Fish Aquat Sci 49: 266–280.
13. Ihssen PE, Booke HE, Casselman JM, McGlade JM, Payne NR, et al. (1981) Stock identification: materials and methods. Can J Fish Aquat Sci 38: 1838–1855.
14. Almeida PR, Tomaz G, Andrade NO, Quintella BR (2008) Morphological analysis of geographic variation of sea lamprey ammocoetes in Portuguese river basins. Hydrobiologia 602: 47–59.

15. Begg GA, Waldman JR (1999) An holistic approach to fish stock identification. Fish Res 43: 35–44.

16. Marr JC (1957) The problem of defining and recognizing subpopulations of fishes. In: Marr JC. Contributions to the studies of subpopulations of fishes, Special Scientific Report n° 208. Washington D.C.: United States Department of Interior, Fish and Wild Life Services. pp. 1–6.

17. Saila S, Jones C (1983) Fishery Science and the Stock Concept - Final Report P.O. NA83-B-A-0078 (MS). US National Marine Fisheries Service, Northeast Fisheries Center, Woods Hole, M.A. 02543.

18. Joensen H, Steingrund P, Fjallstein I, Grahl-Nielsen O (2000) Discrimination between two reared stocks of cod (Gadus morhua) from the Faroe Islands by chemometry of the fatty acid composition in the heart tissue. Mar Biol 136 (3): 573–80.

19. Waldman JR, Grossfield J, Wirgin I (1988) Review of stock discrimination techniques for striped bass. N Am J Fish Manag 8: 410–425.

20. Joensen H, Grahl-Nielsen O (2000) Discrimination of Sebastes viviparus, Sebastes marinus and Sebastes mentella from Faroe Islands by chemometry of the fatty acid profile in the heart and gill tissues and in the skull oil. Comp Biochem Physiol Biochem Mol Biol 126: 69–79.

21. Grahl-Nielsen O (2005) Fatty acid profiles as natural marks for stock identification. In: Cadrin SX, Friedland KD, Waldman J, editors. Stock identification methods: applications in fishery science. Amsterdam: Elsevier Academic Press. pp.247–271.

22. Hishikawa D, Shindou H, Kobayashi S, Nakanishi H, Taguchi R, et al. (2008) Discovery of a lysophospholipid acyltransferase family essential for membrane asymmetry and diversity. Proc Natl Acad Sci U.S.A. 105: 2830–2835.

23. Joensen H, Grahl-Nielsen O (2004) Stock structure of Sebastes mentella in the North Atlantic revealed by chemometry of the fatty acid profile in heart tissue. ICES J Mar Sci 61: 113–126.

24. Grahl-Nielsen O, Averina E, Pronin N, Radnaeva L, Käkelä R (2011) Fatty acid profile in different fish species in Lake Baikal. Aquat Biol 13: 1–10.

25. Joensen H, Grahl-Nielsen O (2001) The redfish species Sebastes viviparus, Sebastes marinus and Sebastes mentella have different composition of their tissue fatty acids. Comp Biochem Physiol Biochem Mol Biol 129, 73–85.

26. Grahl-Nielsen O, Mjaavatten O (1992) Discrimination of striped bass stocks: a new method based on chemometry of the fatty acid profile in heart tissue. Trans Am Fish Soc 121: 307–314.

27. Grahl-Nielsen O, Ulvund KA (1990) Distinguishing populations of herring by chemometry of fatty acids. Am Fish Soc Sympos 7: 566–571.

28. Silva S, Servia MJ, Vieira-Lanero R, Barca S, Cobo F (2013) Life cycle of the sea lamprey Petromyzon marinus: duration of and growth in the marine life stage. Aq Biol 18: 59–62.

29. Beamish FWH (1980) Biology of the North American anadromous sea lamprey, Petromyzon marinus. Can J Fish Aquat Sci 37: 1924–1943.

30. Bashmachnikov I, Neves F, Nascimento A, Medeiros J, Ambar I, et al. (2014) Detailed temperature–salinity distribution in the Northeast Atlantic from ship and Argo vertical casts. Ocean Sci Discuss 11: 1473–1517, doi:10.5194/osd-11-1473-2014, 2014.

31. Vladykov VD, Follett WI (1967) The teeth of lampreys (Petromyzonidae): their terminology and use in a key to the holarctic genera. J Fish Res Board Can 24: 1067–1075.

32. Holčík J (1986) Determination criteria. In: Holčík J, editor. The freshwater fishes of Europe Vol. 1, Part I – Petromyzontiformes. Wiesbaden: Aula-Verlag. pp.24–32.

33. Gill HS, Renaud CB, Chapleau F, Mayden RL, Potter IC (2003) Phylogeny of living parasitic lampreys (Petromyzontiformes) based on morphological data. Copeia 4 (2003): 687–703.

34. Collins TJ (2007) Image J for microscopy. BioTech 43: 25–30.

35. Bancroft J, Gamble M (2002) Theory and Practice of Histological Techniques 5th Edition. London: Churchill Livingstone. 725 p.

36. Lança MJ, Rosado C, Machado M, Ferreira R, Alves-Pereira I, et al. (2011) Can muscle fatty acid signature be used to distinguish diets during the marine trophic phase of sea lamprey (Petromyzon marinus, L.)? Comparative Biochemistry and Physiology, Part B 159: 26–39.

37. Morrison WR, Smith LM (1964) Preparation of fatty acid methyl esters and dimethylacetals from lipids with boron fluoride-methanol. J Lipid Res 5: 600–608.

38. Logue JA, De Vries AL, Fodor E, Cossins AR (2000) Lipid compositional correlates of temperature-adaptative interspecific differences in membrane physical structure. J Exp Biol 203: 2105–2115.

39. Schaefer KM (1991) Geographic variation in morphometric characters and gill-raker counts of Yellowfin Tuna, Thunnus albacares, from the Pacific Ocean. Fish Bull 89: 289–297.

40. Claytor RR, MacCrimmon HR (1987) Partitioning size from morphometric data: a comparison of five statistical procedures used in fisheries stock identification research. Canada: Canadian Technical Report of Fisheries and Aquatic Services n°1531, Minister of Supply and Services.

41. Hair JF, Anderson RE, Tatham RL, Black WC (1998) Multivariate Data Analysis, 5th edition. USA: Prentice Hall. 768 p.

42. Tocher DR (2003) Metabolism and functions of lipids and fatty acids in fish. Rev Fish Sci 11: 107–184.

43. Casadei MA, Mañas P, Niven G, Needs E, Mackey BM (2002) Role of membrane fluidity in pressure resistence of Escherichia coli NCTC 8164. Appl Environ Microbiol 68: 5965–5972.

44. Hardisty MW, Potter IC (1971) The general biology of adult lampreys. In: Hardisty MW, Potter IC, editors. The biology of lampreys Vol. 1. London: Academic Press. pp.127–206.

45. Kucheryavyi AV, Savvaitova KA, Gruzdeva MA, Pavlov DS (2007) Sexual dimorphism and some special traits of spawning behavior of the Arctic Lamprey Lethenteron camtschaticum. J Ichthyol 47: 481–485.

46. Potter IC, Strahan R (1968) The taxonomy of the lampreys Geotria and Mordacia and their distribution in Australia. Proc Linn Soc Lond 179: 229–240.

47. Potter IC, Lanzing WJR, Strahan R (1968) Morphometric and meristic studies on populations of Australian lampreys of the genus Mordacia. Zool J Linn Soc 47: 533–546.

48. Hardisty MW (1986) Lampetra fluviatilis (Linnaeus, 1758). In: Holčík J, editor. The freshwater fishes of Europe Vol 1, Part I – Petromyzontiformes. Wiesbaden: Aula-Verlag. pp. 249–278.

49. Claridge PN, Potter IC (1974) Heart Ratios at Different Stages in the Life Cycle of Lampreys. Acta Zool 55: 61–69.

50. Poupa A, Ostadal B (1969) Experimental cardiomegalies and cardiomegalies in free living animals. Ann N Y Acad Sci 156: 445–468.

51. Hardisty MW, Potter IC (1972) The circulatory system. In: Hardisty MW, Potter IC editors. The biology of lampreys, Vol. 2. London: Academic Press. pp.241–259.

52. van der Vusse GJ, Van Bilsen M, Glatz JFC (2000) Cardiac fatty acid uptake and transport in health and disease. Cardiovasc Res 45: 279–293.

53. Sidell BD, Driedzic WR, Stowe DB, Johnston IA (1987) Biochemical correlations of power development and metabolic fuel preferenda in fish hearts. Physiol Zool 60: 221–232.

54. Arts MT, Kohler CC (2009) Health and condition in fish: the influence of lipids on membrane competency and immune response. In: Arts MT, Brett MT, Kainz MJ, editors. Lipids in aquatic ecosystems. New York: Springer. pp. 237–255.

55. Kaneda T (1991) Iso- and Anteiso-Fatty Acids in Bacteria: Biosynthesis, Function, and Taxonomic Significance. Microbiol Rev 55: 288–302.

56. Mayzaud P, Chanut JP, Ackman RG (1989) Seasonal changes of the biochemical composition of marine particulate matter with special reference to fatty acids and sterols. Mar Ecol Prog Ser 56: 189–204.

57. Lechevalier MP (1982) Lipids in bacterial taxonomy. In: Laskin AI, Lechevalier HA, editors. CRC Handbook of Microbiology. Boca Raton, Florida: CRC Press, Inc. pp.435–541.

58. Rajendran N, Suwa Y, Urushigawa Y (1993) Distribution of phospholipid ester-linked fatty acids biomarkers for bacteria in the sediment of Ise Bay. Mar Chem 42: 39–56.

59. Brooks S, Clark GT, Wright SM, Trueman RJ, Postle AD, et al. (2002) Electrospray ionization mass spectrometric analysis of lipid restructuring in the carp (Cyprinus carpio L.) during cold acclimation. J Exp Biol 205: 3989–3997.

60. Cossins AR, Prosser CL (1978) Evolutionary adaptation of membranes to temperature. Proc Natl Acad Sci U.S.A. 75: 2040–2043.

61. Spice EK, Goodman DH, Reid SB, Docker MF (2012) Neither philopatric nor panmictic: microsatellite and mtDNA evidence suggests lack of natal homing but limits to dispersal in Pacific lamprey. Mol Ecol 21: 2916–2930.

62. Silva S, Servia MJ, Vieira-Lanero R, Cobo F (2013) Downstream migration and hematophagous feeding of newly metamorphosed sea lampreys (Petromyzon marinus Linnaeus, 1758). Hydrobiologia 700: 277–286.

63. Amarasekare P (2004) The role of density-dependent dispersal in source–sink dynamics. J Theor Biol 226: 159–168.

64. Irwin AJ, Taylor PD (2000) Evolution of dispersal in a stepping-stone population with overlapping generations. Theor Popul Biol 58: 321–328.

65. Lança MJ, Machado M, Ferreira R, Alves-Pereira I, Quintella BR, et al. (2013) Feeding strategy assessment through fatty acid profiles in muscles of adult sea lampreys from the western Iberian coast. Sci Mar 77: 281–291.

66. Quintella BR (2006) Biology and conservation of the sea lamprey (Petromyzon marinus L.). Portugal: PhD Thesis, University of Lisbon.

67. Helfman GS, Collette BB, Facey DE (1997) The diversity of fishes. Massachusetts: Blackwell Science. 528 p.

68. Tanner SE, Reis-Santos P, Vasconcelos RP, Thorrold SR, Cabral HR (2013) Population connectivity of Solea solea and Solea senegalensis over time. J Sea Res 76: 82–88.

69. Ambar I, Serra N, Brogueira MJ, Cabeçadas G, Abrantes F, et al. (2002) Physical, chemical and sedimentological aspects of the Mediterranean outflow off Iberia. Deep Sea Res Part II Top Stud Oceanogr 49: 4163–4177.

10

Multi Year Observations Reveal Variability in Residence of a Tropical Demersal Fish, *Lethrinus nebulosus*: Implications for Spatial Management

Richard D. Pillans[1]*, **Douglas Bearham**[3], **Andrew Boomer**[4], **Ryan Downie**[2], **Toby A. Patterson**[2], **Damian P. Thomson**[3], **Russel C. Babcock**[1]

1 CSIRO Marine and Atmospheric Research, Brisbane, Queensland, Australia, **2** CSIRO Marine and Atmospheric Research, Hobart, Tasmania, Australia, **3** CSIRO Marine and Atmospheric Research, Floreat, Perth, Western Australia, Australia, **4** SIMS/IMOS Animal Tagging and Monitoring, Mosman, New South Wales, Australia

Abstract

Off the Ningaloo coast of North West Western Australia, Spangled Emperor *Lethrinus nebulosus* are among the most highly targeted recreational fish species. The Ningaloo Reef Marine Park comprises an area of 4,566 km^2 of which 34% is protected from fishing by 18 no-take sanctuary zones ranging in size from 0.08–44.8 km^2. To better understand Spangled Emperor movements and the adequacy of sanctuary zones within the Ningaloo Reef Marine Park for this species, 84 Spangled Emperor of a broad spectrum of maturity and sex were tagged using internal acoustic tags in a range of lagoon and reef slope habitats both inside and adjacent to the Mangrove Bay Sanctuary zone. Kernel Utilisation Distribution (KUD) was calculated for 39 resident individuals that were detected for more than 30 days. There was no relationship with fish size and movement or site fidelity. Average home range (95% KUD) for residents was 8.5±0.5 km^2 compared to average sanctuary zone size of 30 km^2. Calculated home range was stable over time resulting in resident animals tagged inside the sanctuary zone spending ~80% of time within the sanctuary boundaries. The number of fish remaining within the array of receivers declined steadily over time and after one year more than 60% of tagged fish had moved outside the sanctuary zone and also beyond the 28 km^2 array of receivers. Long term monitoring identified the importance of shifting home range and was essential for understanding overall residency within protected areas and also for identifying spawning related movements. This study indicates that despite exhibiting stable and small home ranges over periods of one to two years, more than half the population of spangled emperor move at scales greater than average sanctuary size within the Ningaloo Reef Marine Park.

Editor: Sebastian C. A. Ferse, Leibniz Center for Tropical Marine Ecology, Germany

Funding: This study was funded by the Wealth from Ocean Flagship of CSIRO and the Western Australia Marine Science Institute (WAMSI). The Australian Institute of Marine Science provided in kind vessel support. The Ningaloo Reef Ecosystem Tracking Array is a fully co-invested installation co-funded by CSIRO and the Australian Animal Tracking and Monitoring System (AATAMS), a facility of Australia's Integrated Marine Observing System (IMOS) whose support made this project possible. The funders had no role in study design, data collection and analysis, decision to publish, or preparation of the manuscript.

Competing Interests: The authors have declared that no competing interests exist.

* Email: Richard.Pillans@csiro.au

Introduction

The cumulative behaviour of individuals in determining net population movement patterns has significant implications for sustainable management of harvested species. Individual animal movements are particularly important for both spatial management and Ecosystem Based Fisheries Management (EBFM) which are increasingly used as means of preserving biodiversity, maintaining habitat structure and ecosystem function, and for preservation of genetic diversity [1,2,3]. While ecologically fundamental, observing, understanding and predicting an animal's habitat requirements has been difficult. The difficulties in observing the movements of marine animals have been addressed by technological developments in the field of animal telemetry [4]. For reef fish the use of implanted acoustic tags has been particularly significant [5,6]. This study demonstrates the importance of long term monitoring and the need to consider movement at multiple spatial and temporal scales in determining the spatial usage in a reef fish population.

Marine spatial management, whether for conservation or EBFM, is complex because the effectiveness of any spatial management measure will be a function of the size of spatial management units and the scale of movement of the species in question. The proportion of populations that will be protected, by no-take MPAs for example, is likely to be positively correlated with reserve size and vary inversely with the species mobility [7,8,9,10]. Consequently while there is a clear expectation that some effect of protection on exploited species will be offered by no-take zones, it is highly uncertain which species will respond, or to what extent, resulting in uncertainty around the effectiveness of no-take areas in protecting multiple species. Variability both in the effectiveness of MPA protection and any benefits to fisheries via spillover are likely to be a function of variability in individual movement patterns and habitat requirements combined with the size, shape and habitat encompassed by the MPA [9,11,12,13,14,15,16].

Marine Protected Areas (MPAs) and spatial management are both important tools for Ecosystem Based Fishery Management (EBFM) and fisheries management [1,2,17,18]. There is mounting evidence for long term positive conservation outcomes within no-take MPAs with increases recorded in the biomass, density, individual size and species richness in the majority of studies irrespective of study latitude and reserve size (see [2,19,20]). In addition to the benefits for single species which gain protection from MPAs there is also growing evidence for MPAs restoring ecosystem function through cascading trophic interactions resulting from protection of target species (see [3]). Although the evidence for spillover is less well documented, due to the complexity associated with its measurement, advances in telemetry and stable isotope technology have allowed some studies to document spillover of adults and larvae into surrounding non-reserve areas [20,21,22](Goñi et al. 2006, Lester et al. 2009, Harrison et al. 2012).

In order to deliver any of their potential benefits MPAs need to afford long-term protection of species in order to maintain populations of large, highly fecund individuals. If MPAs are too small, resident fish that roam into adjacent areas or that have multiple home ranges will be more susceptible to capture [5,8,13,23]. The majority of reef fish species studied to date have shown limited movement [5,24,25,26,27]but also large variability in home range. Parrotfish, surgeonfish and goatfish tagged in Hawaii demonstrated movements from 0.1–0.6 km with a very few individuals moving up to 2.0 km from the tagging site across sand channels between reef habitats [5].In New Caledonia, parrotfish and coral groupers moved up to 5.0 km from the tagging site and crossed areas of sand channel between reefs [25]. Within Australia, few studies have been conducted in coral reef environments. Coral trout *Plectropomus leopardus* demonstrated limited movement up to 250 m outside the spawning season [24] and larger movements up to 5.2 km from established home range during spawning periods [28]. More recent work on short term movement of steephead parrotfish *Chlorurus microrhinus* on the Great Barrier Reef has demonstrated movements of less than 400 m from the point of capture [27].

While individual variations in habitat use or movement and mobility may be of intrinsic ecological interest, management of fish species is pursued at the level of populations. It is therefore the cumulative behaviour of individual fish and how they scale to overall population level movement patterns that has significant implications for sustainable management and/or conservation of fished species. Despite the importance of movement, very few studies have addressed population level movement of a number of animals from different age classes tagged in a variety of habitats [16], and none in a coral reef environment.

The Spangled Emperor *Lethrinus nebulosus* (Försskal, 1775) is widespread throughout the Indo-West Pacific, including the Red Sea, Persian Gulf, East Africa to Southern Japan and Samoa [29]. It occurs in nearshore and offshore areas and in coral and rocky reefs, coralline lagoons, seagrass beds, mangrove swamps, and coastal sand and rock areas to depths of at least 75 m [29]. Adults are found alone or in small schools, whereas juveniles form large schools. It feeds mostly on echinoderms, molluscs, crustaceans, and to a lesser extent on polychaetes and fishes [29]. This species has a non functional protogynous hermaphroditic life history strategy [30] that prevents size-based sex determination.

Within the Ningaloo Reef Marine Park (NRMP), *L. nebulosus* is an important predator of grazing invertebrates along the Ningaloo Reef, and is likely to play an important role in maintaining ecosystem function [31]. A creel survey of recreational fishers in the NMP showed that members of the family Lethrinidae were the most targeted species [32] and this, combined with their longevity and slow growth along the west Australian coastline [30,33], makes them susceptible to over-exploitation [34,35]. Indeed,

despite only being captured by recreational fishers within the NMP, there is evidence of population decline as well as a reduction in the modal age and percentage of old fish [33] along the Ningaloo coastline. During the period in which this decline occurred, the area of no take MPA's in the Ningaloo Marine Park was increased from approximately 10% to 34% [36].

Conventional mark-recapture tagging of lethrinids within the Ningaloo Marine Park supports the theory of limited movement. Moran et al. [37] tagged 1781 *L. nebulosus* and *L. atkinsoni* with 60 fish recaptured over three years. Of these recaptures, 66% were made within 5.5 km of the tagging site, and 75% within 10 km, even in the third year after tagging. A few fish had moved 110 km within three months of the release and no fish were recaptured more than 148 km away. Almost all recaptures occurred within the lagoon (Ross Marriott, Pers. Comm.) with the recapture data supporting the theory that most individuals within the lagoon have a relatively small area of occupancy, even over long periods at liberty. However, the spatial resolution of these data is coarse (11.1 km), especially in relation to the average size of no-take zones at NMP (mean linear extent 9.0 km).

Using an array of acoustic receivers and surgically implanting acoustic tags into 84 *L. nebulosus*, the primary objective of this study were to 1) monitor movement patterns of a wide size range of *L. nebulosus* captured in different habitats within the Ningaloo reef and determine the site fidelity and size of activity centres ("home range") of individuals, 2) using estimates of home range investigate variability in habitat usage and residency relevant to population level movement patterns and 3) using movement characteristics, assess the adequacy of MPA zoning within the NRMP.

Methods

Study site

The Ningaloo Reef Marine Park (NRMP) encompasses Australia's largest fringing reef which is one of the world's largest fringing reefs [36], covering a total area of 4,566 km^2. It runs along 300 km of Western Australia's coastline from Bundegi in the Exmouth Gulf to Red Bluff in the south (Figure 1). In 2006 the NMP was substantially extended and 34% of total Park area was incorporated into limited or no-take Sanctuary Zones under the revised Management Plan (2005–2015). Areas where no fishing is allowed are referred to as sanctuary zones. The NRMP is zoned for multiple uses and although no commercial fishing is allowed fishing is permitted in recreation zones with recreational fishing catch controlled by possession and size limits. Although areas of protection were chosen based on the best available knowledge and with a view to including the full spectrum of representative marine habitats, many of these decisions were made without in-depth knowledge of the biological communities that reside there and without data on the movement patterns of any fish species.

Acoustic monitoring system

An array of acoustic receivers was located within and adjacent to the Mangrove Bay sanctuary zone (695 ha) and extended from <1 m of water near the shoreline to the reef slope in ~50 m of water (Figure 1). Receivers were spaced 200–800 m apart and detection ranges generally did not overlap. The array encompassed habitats including mangrove-lined shores, limestone pavement, coral reefs interspersed with expanses of sand as well as areas of flat hard substratum dominated by macroalgae (predominantly *Sargassum* and other fucalean algae) within the lagoon. A continuous fringing reef creates a barrier to movement out of the lagoon at low tide and during times of high swell

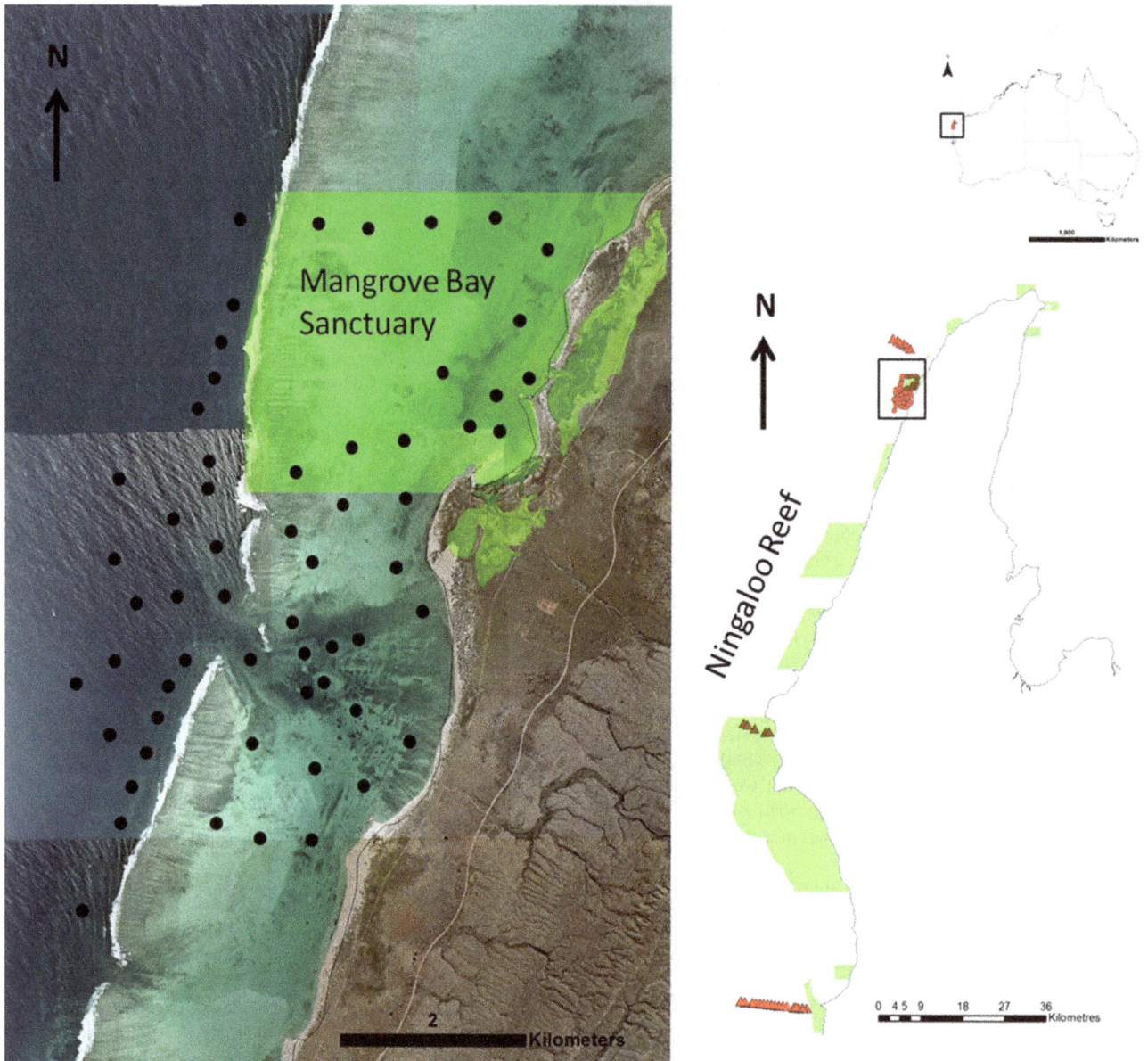

Figure 1. Map of study area. Map of study area with the red triangles showing the lines of cross shelf receivers. From north to south: Tantabiddi, Coral Bay and Winderbandi Point. The Coral Bay receivers covered both lagoon and reef slope habitat. Sanctuary zones in the Ningaloo Marine Park are shown as light green shading. The black circles show the acoustic receivers within and adjacent to the Mangrove Bay Sanctuary.

however a reef pass provides direct access to deeper reef slope waters. Several large *Porites*-dominated patch reefs are present in the reef pass. The reef slope consists of spur and groove habitat as well as areas of limestone reef interspersed with sand patches. Beyond 35 m, the substratum is predominantly sandy sediment with occasional flat limestone reef.

An array of Vemco VR2 and VR2W acoustic receivers was used to monitor the movements of tagged fish. The Mangrove Bay array consisted of 50 acoustic receivers from December 2007–May 2008 and 60 acoustic receivers from May 2008–May 2010. In addition to the Mangrove Bay array, there were three cross shelf lines of receivers extending from the reef slope (~12 m) to the 200 m isobath located along the Ningaloo Reef (Figure 1).

Individual *L. nebulosus* were internally tagged with Vemco coded transmitters (tags). Depending on fish size, individuals were tagged with a V9-2L, V9-2H, V13-1H or V16-4H transmitter. These transmitters range in size from 9×29 mm to 16×68 mm and weighed between 2.9–26 g in water. The pulse rate of transmitters varied from 40–320 s and battery life varied from 185–2020 d depending on the frequency of each ping and the power output of the tag. Each successfully decoded pulse train was recorded as a single detection in the memory of the individual VR2 as the transmitter's identification number, date and time. Receivers were downloaded every three to four months throughout the study, and the batteries were changed at least every six months.

Range tests were done by placing one of each of the four transmitter types used at 50, 100, 150, 200, 300, 400 and 500 m away from individual receivers in a straight line. This was also done in at least two directions more than $90°$ apart on 20 receivers

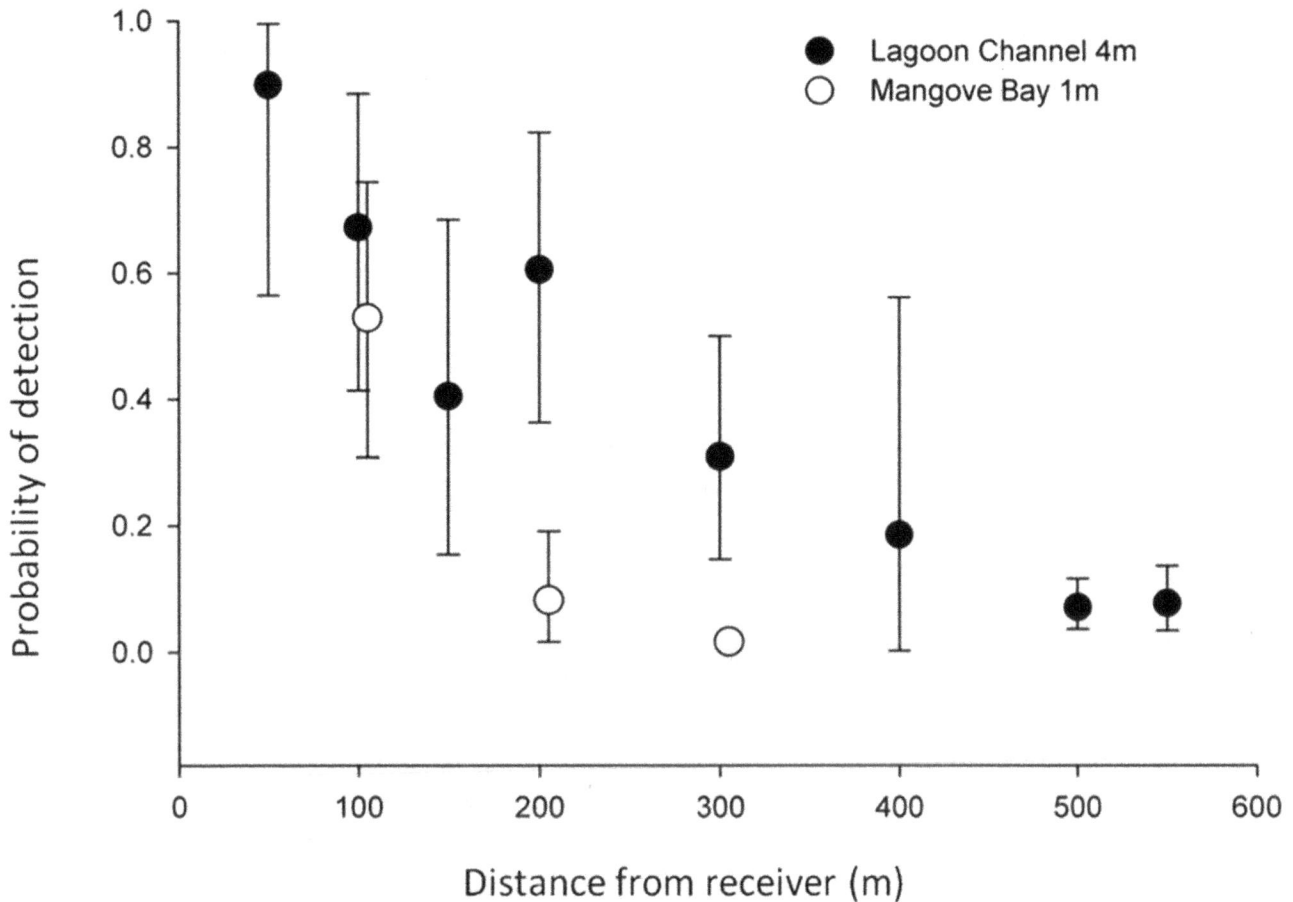

Figure 2. Mangrove Bay array range testing. Proportion of detections received by VR2 receivers at increasing distances from the test transmitter in the lagoon channel (filled circles, 4.0 m water depth) and at Mangrove Bay (white circles, 1.0 m water depth).

within the array. Range testing was conducted in a variety of wind conditions ranging 0–30 knots. These data were used to determine a probability of detection within the array (Figure 2).

Capture and tagging

Capture and tagging of fish was conducted under CSIRO Brisbane Animal Ethics Permit (permit A2/07). Fish were captured with rod and line between November 2007–November 2009. Fish chosen for tagging were placed in a 120 l tub containing 30 mg.l^{-1} of AQUI-S in seawater. Fish remained in the tub until they reached stage III anaesthesia at which time they were placed on their dorsal surface into a V-shaped piece of foam lined with plastic. After removing a few scales, a small incision was made slightly off the mid-line between the pelvic fins and anus. Transmitters that had been soaking in an antiseptic bath (povidone iodine and distilled water 5:100) for at least 30 min were then inserted into the peritoneal cavity. Depending on the transmitter size, either two or three dissolving sutures were used to close the wound. Following surgery, fish were measured and externally tagged with a plastic dart tag (Hallprint, South Australia). A mass dependent dose of ENGEMYCIN, (100 mg Oxytetracycline.ml^{-1}) was administered intramuscularly in the dorsal surface. Fish were allowed to recover in a 120 l tub filled with continuously replenished seawater. The average time from capture to completion of surgery was six to seven minutes while recovery times

varied ranged from 10–30 min. Once fully recovered, fish were released at the site of capture.

Detection and spatial analysis

The detection span of each individual was calculated as the date from first detection to last detection whereas days detected was the total number of days on which each individual was detected. Kernel distribution was calculated for those animals that were detected for more than 30 days and on at least one receiver. Area utilisation was estimated using the utilisation distribution [38] and its estimates with kernel techniques [39]. Utilisation distribution is a probability density function that quantifies an individual's relative use of space [40]. It depicts the probability of an animal occurring at a location within its home range as a function of relocation points (data obtained from receiver detections) [41].

Kernel distribution (50 and 95%) was calculated using the Hawth's tools extension for ArcMap. Kernel area was calculated using the animal movement extension for Arcview (ANME). The data from acoustic receivers are not explicitly spatial in the sense that the location of the receiver which detects an individual is obviously not the true location of that individual. However, the frequency of detection at a receiver is directly related to the location of the individual. As a result estimation of a smoothing factor (Hlscv) from treating acoustic detection data identically to continuous tracking data (e.g. from a GPS) will greatly underestimate home range size and result in an improbable series of discrete kernels. In our case smoothing factors calculated using ANME in Arcview, were between 100–500 and resulted in discreet kernels around individual receivers. Determining how to address these problems is an open research question for acoustic data. Centre of activity methods (e.g. [42]) rely on overlapping receivers and similarly ad hoc treatments of time (choice of time step). Other approaches applied in this area – such as Brownian bridge models (see [43]) also have limitations and bring strong assumptions about the nature of movements between receivers which are difficult to test using the available data.

Given that our primary goal was to examine movement in relation to usage of an MPA, there was a need to apply a suitably precautionary approach. In this instance, and given the likely detection radius established from range testing, this entails the choice of a smoothing factor which will be robust given the detection radius of an animal. A range of smoothing factors from 100–1200 were tested, however, a value of 1000 provided the most realistic estimates of KUD given the detection range and spacing of receivers combined with where animals had been detected. Using this larger smoothing factor allowed for variability in detectability within the array and gaps in the detection radii of individual receivers to be accounted for given that the probability of detection varied considerably (from 0–100%) depending on the distance of a transmitter to a receiver/s within the array. Although higher smoothing parameters will overestimate kernel distributions and also resulted in less variation between animals, it provided a more robust and precautionary assessment of spatial utilisation at a population level as opposed to what may be misleading representations of multiple kernels centred on individual receivers. To evaluate overlap of home range with the Mangrove Bay Sanctuary, the area of 95% kernel inside the sanctuary boundary was calculated for individual fish tagged inside and outside the MPA.

Behaviour at the individual level as characterised by the 50 and 95% kernel density and distance from kernel centre to tagging location were summarised at the population level in order to evaluate movement characteristics, illustrate the scale of movement and habitat use, and the range of variation in behaviour within the population. To obtain an estimate of the distance that fish moved, the maximum distance moved between receivers within the array was measured as the maximum straight line distance between receivers that had detected individual fish. Maximum distance moved was only calculated for fish that were detected on more than one receiver. Although this measure could potentially be biased by the array design, given the length of time fish spent in the array and the number of receivers detecting individuals (Table 1), it provides a measure of scale of detectable movement within the array.

Temporal analysis

A multiple stepwise backwards elimination regression was used to evaluate the effect of fish size, battery life, transmission interval, tag range, habitat tagged and the shortest distance from tagging location to array edge on detection and movement parameters. Habitats in which fish were tagged included; reef slope, reef flat, coral outcrops within the lagoon, Mangrove Bay and shoreline pavement. Movement parameters included total number of detections, number of receivers detecting each individual, detection span, number of days an individual was detected, proportion of days detected, days detected/battery life, detection span/battery life, distance from tag location to 50% kernel centre, 50% kernel area (km^2) and 95% kernel area (km^2). These analyses allowed us to assess whether measurements of fish behaviour were biased or affected by non-biological technical characteristics of the tracking

Table 1. Tag detection relationships with tag technical specifications and fish biological characteristics.

		Individual Parameters						Overall Regression		
		Battery Life	Transmission interval (min)	Tag range (avg)	Habitat tagged in	Fork Length (cm)	Distance to array edge	R-square	F Value	p
Total Detections	F (p)	3.55 (0.0630)	-	-	-	-	-	0.0415	3.55	0.0630
Span of detections (days)	F (p)	11.31 (0.0012)	-	-	-	-	-	0.1212	11.31	0.0012
Days detected (No.)	F (p)	-	-	-	-	-	-	-	-	-
Proportion days detected (days/span)	F (p)	-	-	-	6.06 (0.0159)	3.21 (0.0767)	-	0.1390	6.54	0.0023
Tag location to Kernel centre*	F (p)	5.47 (0.0242)	5.55 (0.0232)	7.43 (0.0093)	-	-	-	0.1678	2.82	0.0502
Receivers detecting (No.)	F (p)	-	-	-	-	13.45 (0.0004)	-	0.1409	13.45	0.0004
Kernel area (50%)*	F (p)	-	-	-	28.33 (<.0001)	4.90 (0.0326)	-	0.4223	14.62	<.0001
Kernel area (95%)*	F (p)	-	-	-	76.85 (<.0001)	-	-	0.6709	40.78	<.0001

Multiple backward elimination stepwise regression of animal, tag and habitat parameters against detection and movement measurements for 84 *L. nebulosus* tagged and monitored with the Mangrove Bay array. Overall regression R-squared, F and P values represent the overall significance of the combined influence of all factors. Significant interactions between parameters are illustrated by F and P values, - = Non-significant interaction.
* = subset of data above for which kernels could be calculated, distances to kernel centre of >10 entered as 10.

systems as well as their relationships to biological and ecological parameters.

Long term residency

The proportion of animals remaining within the array after tagging was calculated using the detection span of each individual. When an animal was not detected within the array for more than one week, it was classified as having left the array. Animals that left the array and then returned were incorporated into the calculation at each time period. Two animals that were known to have been captured by recreational fishers and another two that were continuously recorded by one receiver were excluded from calculations at the date of capture or when the tag was only detected on one receiver. For comparisons of residency inside and outside the sanctuary zone, only data from animals tagged inside the lagoon were utilised as the Mangrove Bay Sanctuary does not extend past the reef crest.

Results

General trends in detection and kernel distribution

A total of 84 *L. nebulosus* ranging in size from 26–67 cm FL were tagged with acoustic tags within a large array of acoustic receivers (Appendix S1) and animals monitored for up to 864 days. Nine animals were not detected at all following release and the fate of these animals is uncertain. These nine animals were not included in subsequent analysis. The size range of animals tagged encompassed all sex and maturity stages, however the sex of tagged animals could not be determined. Animals were tagged at 34 locations on 17 occasions between 30/11/2007–6/11/2009 either in the lagoon or along the reef slope. This period spanned different intra-annual seasons and spawning periods. A range of habitats including, reef flats, lagoon coral outcrops and mangroves are present in the lagoon where fish were captured and tagged.

The results of our tagging program have revealed significant information relating to site fidelity and residency of individuals and also the variability in movement within the tagged population. Tagged *L. nebulosus* that were recorded at least once following release were detected up to 73,980 times on as many as 27 receivers over a maximum period of 891 days (Appendix S1). Throughout the detection period, over 70% of tagged fish were detected on less than 10 receivers (median = 5.0). Thirty nine of the 84 L. *nebulosus* that were tagged were detected over a period of 30 days, allowing a kernel distribution to be calculated (Appendix S1). Of these, two were thought to have died or been eaten close to a receiver resulting in the tag falling out of the animal and being continuously detected by one receiver only. Eight of the *L. nebulosus* for which kernels were calculated remained undetected for periods of several months following tagging. An additional 36 (48%) were detected following release but were detected for too short a time period and/or on too few receivers to warrant calculating kernel distributions. Although the kernel area for these 36 animals was not calculated, we have assumed that the distance from tag location to their centre of distribution was greater than 10 km. This assumption was based on the maximum distance between receivers on the edge of the array. Therefore animals that have moved greater than 10 km from any position within the array cannot be detected by receivers within the array.

Regression analyses demonstrated that the total number of detections (R-square = 0.044) and the span of the detection period (R-square = 0.124) were both significantly related to battery life which was expected. There was no significant relationship between the total number of days on which detections occurred and the

range of parameters tested, however the proportion of days detected was significantly related to fish size and the habitat in which the fish was tagged. Although overall R-square was low (R-square = 0.14), larger fish tended to be detected on fewer days than smaller fish. Fish tagged in the lagoon were detected for longer periods than fish tagged on the reef slope (Table 1).

The number of receivers on which a fish was detected was significantly related to fish size (R-square = 0.14), with the largest fish tending to be detected on more receivers. The 50% and 95% kernel area were significantly related to habitat in which fish were tagged and explained far higher proportions of the total variability in the data (R-square of 0.42 and 0.67 respectively). Overall, fish tagged in the lagoon had smaller kernel area than fish tagged on the reef slope. Only a small amount of the variability (14%) in the 50% kernel area was explained by fish size (FL cm). Tag technical characteristics such as battery life, transmission interval and average detection range were not significant in regressions with kernel area.

Kernel area and residence

A detection history sufficient to allow calculation of kernel density distributions was obtained for 39 individuals (Figure 3) whose kernel distributions were calculated (Appendix S1). These 39 fish were detected regularly within the array with the median percentage of days detected being 78 (Appendix S1). The calculated 50 and 95% kernel areas ranged from 1.4–5.8 km^2 (average 2.3±0.1 SE) and 4.9–21.1 km^2 (average 8.6±0.6 SE) respectively. Given the large smoothing factor (1000) used to calculate KUD, these values are likely to be an overestimation of home range, but provide greater confidence in estimates of MPA effectiveness. Had we been attempting to estimate fine scale habitat use, these values would not be appropriate. There was no difference in the size of kernel area for fish tagged inside or outside the sanctuary zone boundary (Figure 4). For fish tagged inside the Mangrove Bay sanctuary zone (n = 16), the average proportion of the 95% kernel distribution that occurred within the sanctuary zone boundary was 81% (± 8 SE) with only a small degree of overlap (19%) with fished areas. Of those fish tagged outside the sanctuary (n = 23), the average proportion of the 95% kernel distribution that occurred inside the sanctuary zone boundary was 17% (± 6 SE) with the majority of kernel area within the fished area. Roughly half of all resident fish were detected on five receivers or less (Appendix S1). The remainder were detected on between 6–15 receivers with only a few individuals (10%) ranging more widely. This suggests that given the maximum of approximately 800 m separation between receivers, the vast majority of these fish use less than 5.8 km^2.

Indeed, the majority of animals remained within the array during their detection period; however seven animals were detected frequently after tagging and were then not detected for periods varying from 47 to 209 days before again being detected. After the period of not being detected, all of these animals were detected on receivers where they had previously been detected, indicating a return to near the original location. Although fish 8046 was detected on the same receivers, it was also detected on a different suite of receivers approximately 3.0 km north of its previous activity centre.

Five of the fish that returned to the array were part of a group of 12 fish that ceased being detected between 5–22 February 2008. This period coincided with a Category three tropical cyclone "Nicholas" which crossed the Ningaloo coast on 19 February 2008. These five fish returned to the same area between the 13 April–27 June 2008. Seven fish were not detected again and presumably moved outside the array. The 12 fish that moved

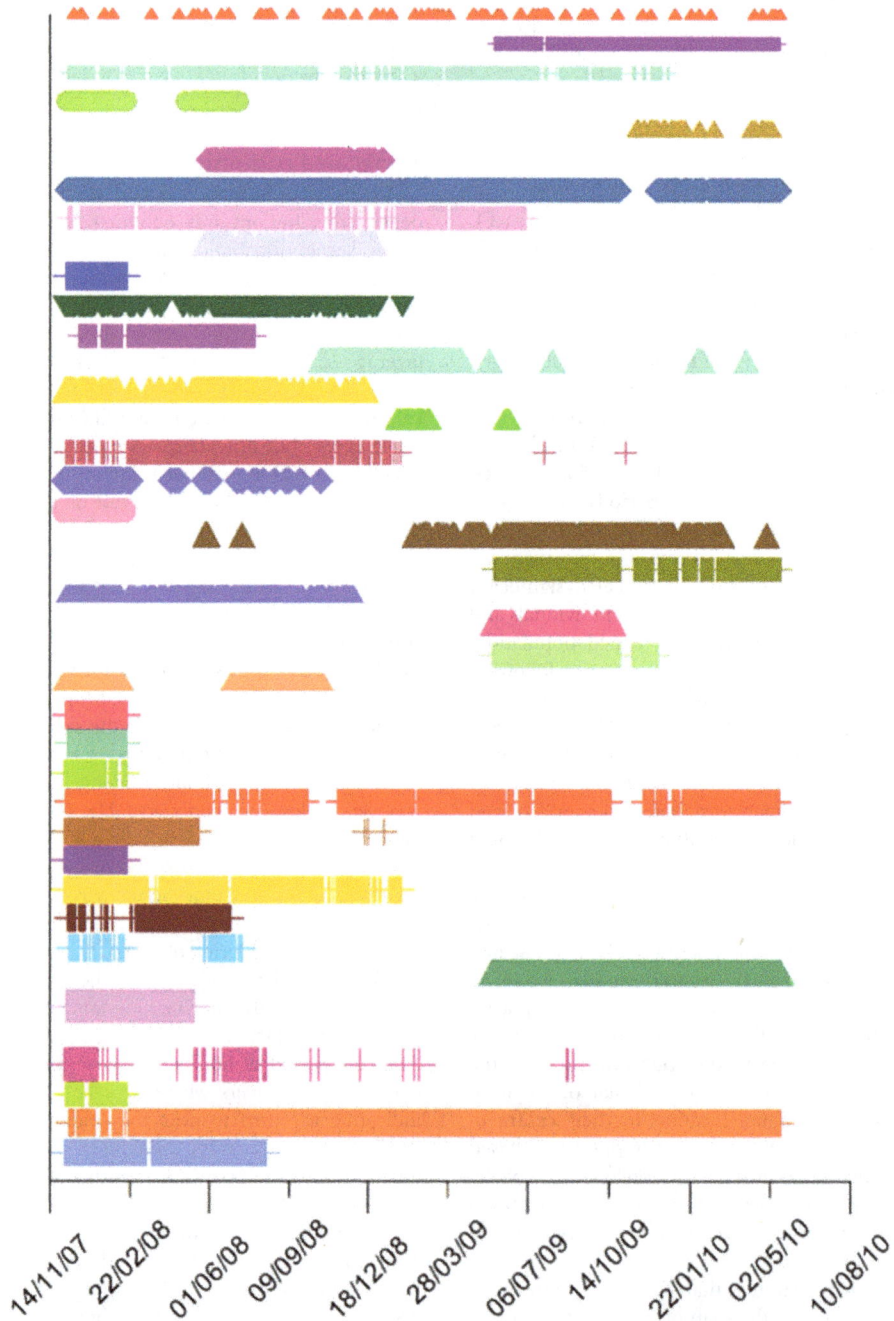

Figure 3. Detection plot of *Lethrinus nebulosus.* Detection plot of 39 *L. nebulosus* from date of tagging to last detection. Legend order from top left down equates to order of plot from top to bottom and matches order of Table 1.

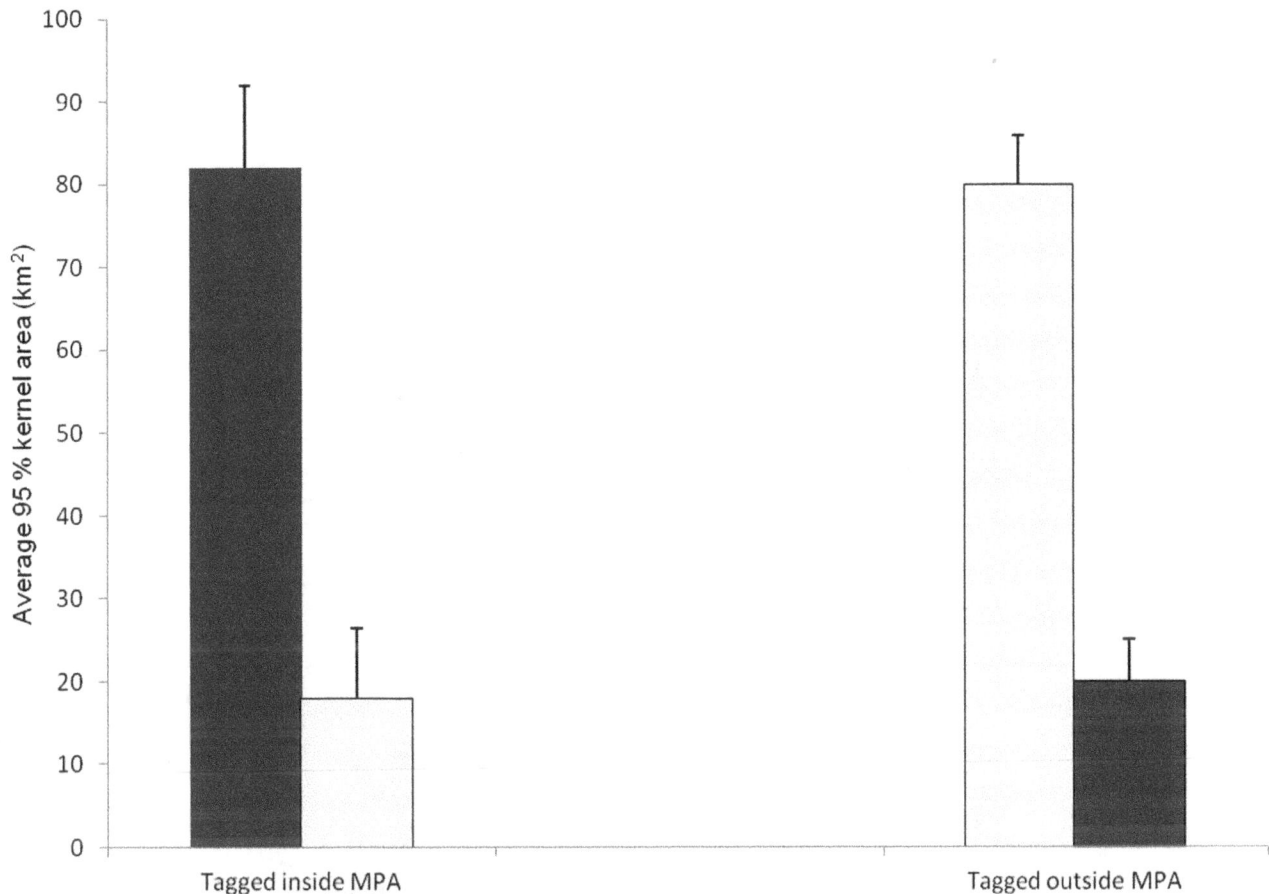

Figure 4. Kernel overlap with sanctuary zone. The average (+ SE) area of the 95% kernel distribution that occurred inside and outside the Mangrove Bay Sanctuary Zone for fish tagged within (n = 16) and outside (n = 23) the Sanctuary Zone boundary. Open bars: average 95% Kernel area inside MPA; dark bars: average 95% kernel outside MPA.

outside the array in February ranged in size from 26.5–49.5 cm FL with the returning fish 32.5–49.5 cm FL.

Seventeen fish were captured and tagged during spawning aggregations on the reef slope in December 2007, in the centre of the array adjacent to the reef pass. Following the end of the spawning season, the majority of these fish disappeared from the array, except for 8171 which returned to the lagoon where it remained for the next nine months before again moving offshore to the reef slope between October–December in 2008 and 2009. Of the fish that were not detected after 10 December 2007, several fish were detected by receivers on the reef slope in October, November and December in 2008 but not in 2009 whereas others were detected in 2009 but not 2008 suggesting that these fish returned to the same spawning site in some but not all years. Four fish tagged during spawning aggregations were detected on the line of cross shelf receivers off Tantabiddi (8074, 8111, 8165, 8171) with one fish (8139) also detected on receivers near Coral Bay. These individuals were recorded by the cross shelf lines immediately prior to and after the spawning season and were the only fish detected by receivers other than those within the Mangrove Bay array.

A range of behaviours and patterns of habitat use were displayed by fish resident in the array. A 56 cm FL adult fish tagged during a November spawning aggregation returned to the lagoon in December 2007 and spent the rest of 2008 inside the lagoon until the following spawning season when it returned to the

reef slope between October and December (Figure 5A). In contrast a 38.5 cm FL fish also tagged on the reef slope in May 2008 remained in the vicinity of its release over a period of 222 days (Figure 5B). This fish was one of a few fish on the reef slope that remained in the area where it was tagged for a long period of time. A 27 cm FL tagged in the lagoon in December 2007 was detected every day for 77 days before disappearing (Figure 5C). This fish was one of a number of fish that departed the array in February 2008 (Figure 3) around the time of tropical cyclone "Nicholas". During the same time period, a 34 cm FL fish tagged in December 2007 was detected over a period of 420 days (Figure 5D) until the tag battery presumably ran out as prior to 2010, VEMCO coded acoustic tags did not have a defined life span.

Variability in behaviour and long term residency

The proportion of fish tagged in the lagoon and on the reef slope remaining within the array over time was determined from detection data (Figure 6). For all tagged fish (n = 84), there was a gradual decline in the number of animals that remained within the array. Fish tagged in the lagoon were more than twice as likely to remain within the array for long periods of time than fish tagged on the reef slope. For fish tagged in the lagoon, after 101–150 days and 301–350 days, 46% and 33%, respectively, were still being detected by the array. For fish tagged on the reef slope, 18% and 30% were still being detected after 101–150 and 301–350 days,

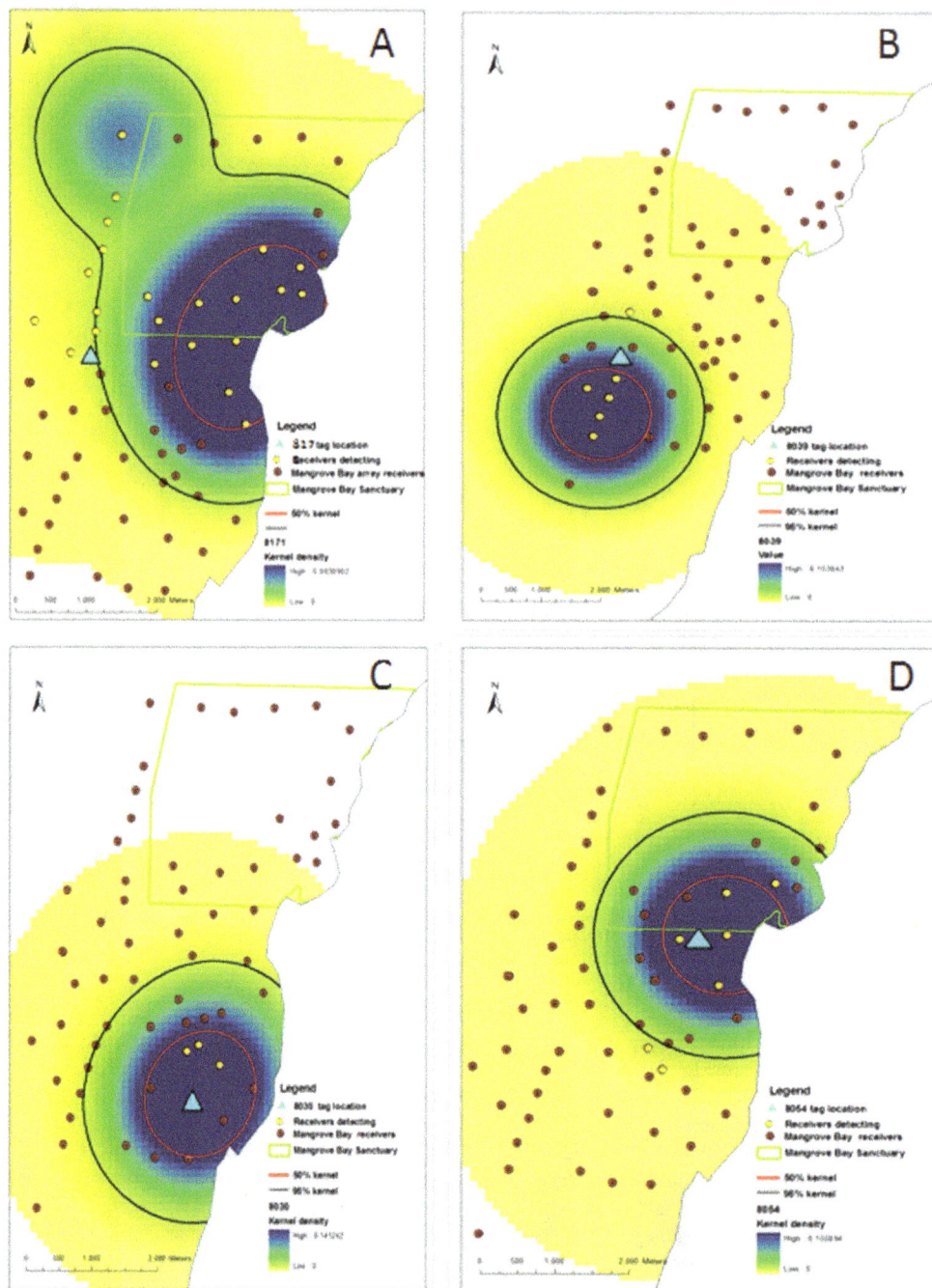

Figure 5. Kernel density of four individual *L. nebulosus.* Fixed kernel density of two *L. nebulosus* tagged on the reef slope (tag number 8171 (A), 8039 (B) and two tagged near coral outcrops in the lagoon 8030(C) and 8054 (D). The tagging location, receivers detecting the fish and all receivers within the array as well as the 50 and 95% kernel densities and fixed kernel density are shown. The boundary of the Mangrove Bay sanctuary is also shown.

respectively. The increase in the proportion of fish being detected on the reef slope after 301–400 days and again after 601–800 days was due to fish returning during the spawning season (October–December). When fish tagged on the reef slope were further divided into fish tagged during the spawning season and fish tagged outside the spawning season, fish tagged outside the spawning season were up to four times more likely to remain in the array for longer periods.

Long term residency was similar for fish tagged inside and outside the sanctuary zone. For fish tagged inside the sanctuary zone, after 101–150 days with 46% of those tagged inside the sanctuary and 36% of those tagged outside the sanctuary were still being detected. After 301–350 days, 27% of fish tagged inside and outside the sanctuary were still being detected.

Maximum distance moved between receivers was calculated for fish tagged inside the lagoon and on the reef slope. For animals tagged inside the lagoon, average (± SE) maximum distance

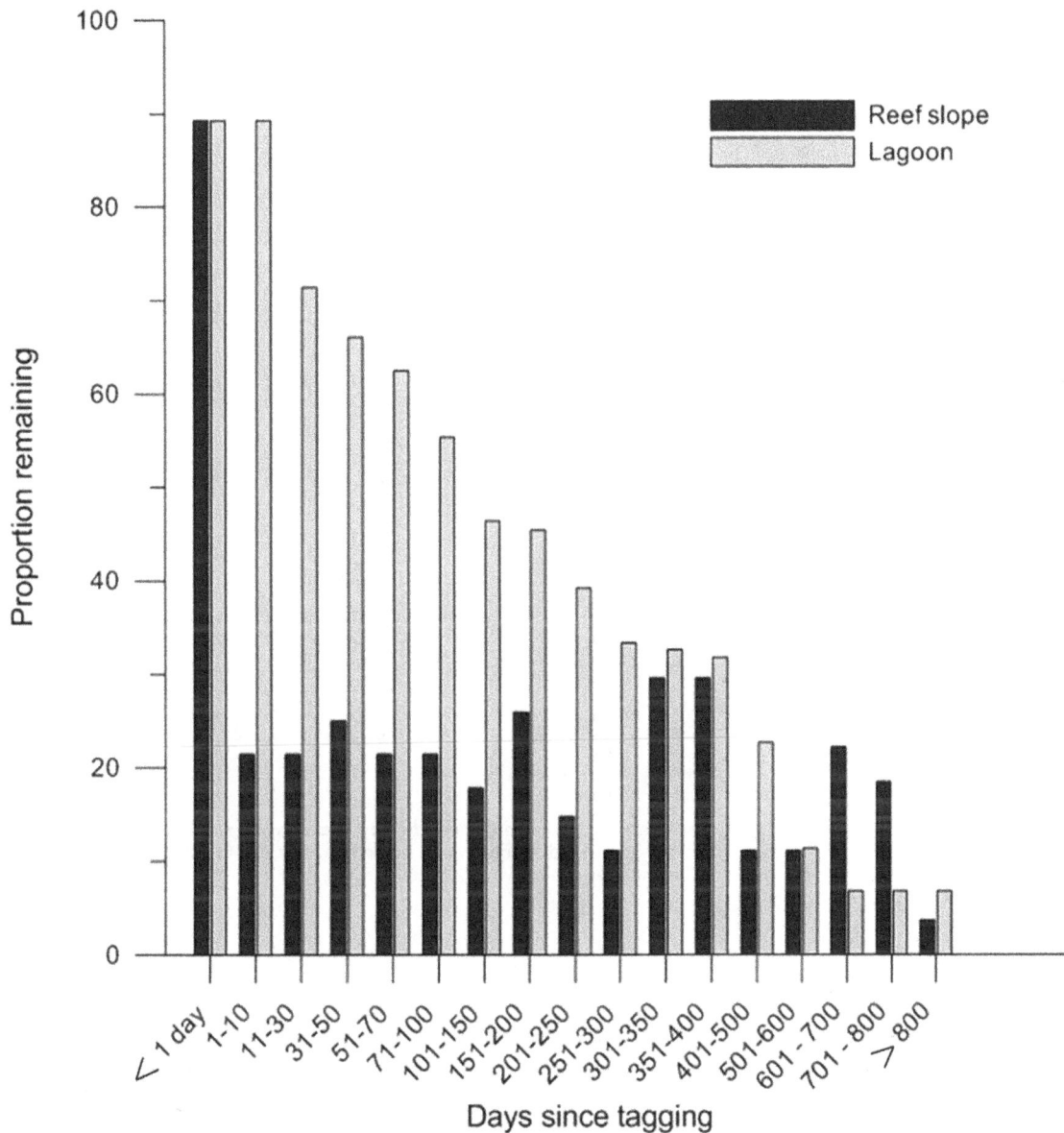

Figure 6. Long term decline in the number of individual fish remaining in the Mangrove Bay array. The percentage of *L. nebulosus* tagged in the lagoon and on the reef slope that were detected within the array at 1–800 days after tagging. The spawning period coincides with the time periods 301–350, 351–400 and 601–700 days since tagging.

moved was 2.92 (0.28) km which was significantly less (p = 0.009) than animals tagged on the reef slope 4.21 (0.46) suggesting that fish on the reef slope cover greater linear distances. For fish tagged inside the lagoon and on the reef slope, the average maximum distance moved did not differ significantly (p>0.6) between animals that were in the array for less than 30 days and animals that were detected for long periods suggesting that fish that moved out of the array were not pre-disposed to moving larger linear distances. Similarly, distance from position tagged to kernel centre was also smaller for animals tagged in the lagoon than those tagged on the reef slope (Figure 7A) as was average kernel area. When habitat-related trends of lagoon fish were examined in more detail it was clear that individuals associated with lagoon coral outcrops and mangroves had particularly high levels of site attachment, explaining much of the trend between lagoon and reef slope, since distances moved by reef flat and shoreline reef tagged

individuals were more similar to those from fish tagged on the reef slope (Figure 7B).

Discussion

Using acoustic telemetry we were able to examine the fine scale movements and residence behaviours of individual *L. nebulosus* within the NRMP over three years. Our results have implications both for the specifics of marine spatial management at the NRMP and also in how ecologists characterise home range, residency and spatial usage of reef dwelling marine fishes.

Residency

The high resolution data gathered using acoustic telemetry revealed that *L. nebulosus* exhibit significant variability in their movement patterns and home range. Despite our methods

A

B

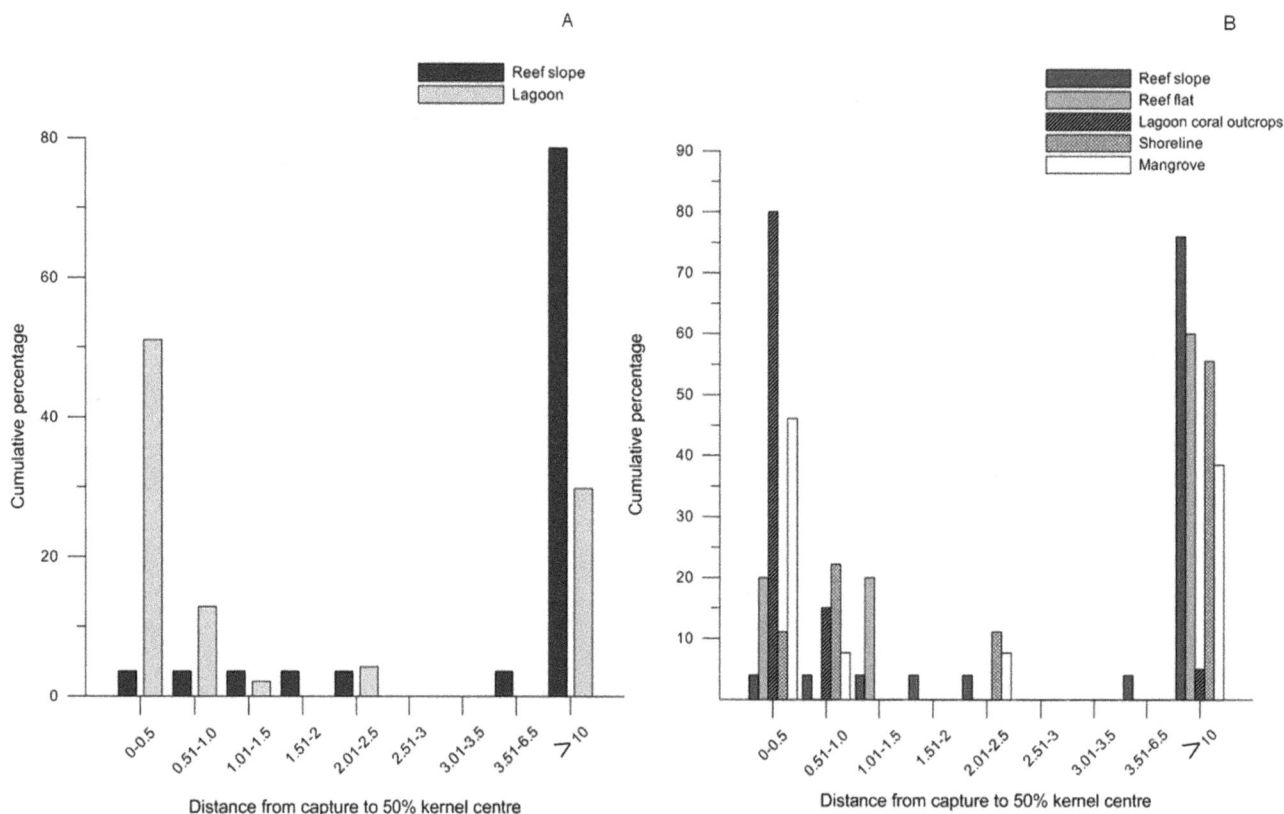

Figure 7. Distance from tag location to 50% kernel centre of 75 animals detected by the array of receivers. Cumulative percentage of distance from tag location to 50% kernel centre of 75 animals detected at some stage within the array. 7A) fish tagged on the reef slope and all lagoon habitats, 7B) reef slope and four different lagoon habitats. Sample size was 22, 7, 20, 9 and 16, respectively for reef slope, reef flat, lagoon coral outcrops, shoreline and mangrove.

resulting in home range being overestimated, a large proportion of the animals we tagged had a relatively small home range (average 95% kernel of 8.6 ± 0.5 km^2) and residency that could persist for more than 2.5 years. Long term stability in home range has been demonstrated in parrotfish *C. microrhinus* with home range persisting over 1.9 years [44] and *Scarus iserti* and *Sparismosa viride* with home range persisting for up to 3.5 years [45,46]. However, despite the fact that some *L. nebulosus* remained in an area for several months to years, there is evidence of fish moving to areas outside the array with some returning to where they were tagged. However, several fish left the array entirely and did not return. These fish appear to be either nomadic without a defined home range or, more likely, have established home range/s outside the array. Chateau & Wantiez [47] demonstrated that *L. nebulosus* moved up to 1.7 km between reefs in New Caledonia and multiple activity centres have been described in *Pagrus auratus* [15].

To give these figures some context, our estimates of home range are 2–350 times greater than other coral reef species where data are available. Herbivorous unicorn fish (*Naso unicornis* and *N. litturatus*) have been shown to have a home range of 3.2–3.8 km^2 in Guam [6] whereas steephead parrotfish (*C. microrhinus*) on the Great Barrier Reef had an average home range less than 1.0 km^2 [27] which is up to 41 times greater than home range estimates of the surf parrotfish *Sparismosa viride* in the Caribbean [46]. Coral trout (*Plectropomus leopardus*) on the Great Barrier Reef had home ranges of 1.0–1.8 km^2 [28] and a single Maori Wrasse had a home range of 5.0 km^2 in New Caledonia [48]. In contrast more

mobile green jobfish *Aprion viriscens* frequently moved more than 9.0 km between receivers [49] indicative of a home range several times larger than spangled emperor in the current study.

As noted in the methods, the characterisation of spatial utilisation from acoustic detection data is problematic. But despite the general shortcomings in the KUD method for characterising spatial distribution, our use of this approach should be robust to negatively biased estimates of home range size due to our choice of a large smoothing factor which should go some way to accounting for the detections arising from somewhere within a large radius around each receiver. Additionally our use of data only from individuals detected for more than 30 days means that the KUD estimates should be less prone to biases which might occur when individuals are observed for only very short or sporadic periods. The obvious downside of our method (and many others in current usage in similar studies) is that it does not permit formal statistical estimation of parameters describing the spatial distribution of individuals. Further investigation using recently developed state-space methods (e.g. [50]) may be useful in refining these estimates and will be investigated in future studies.

With respect to adequacy of the sanctuary zone size (areas where fishing is not permitted), the average size of sanctuary zones within the Ningaloo Marine Park is approximately 49 km^2 however, over 70% are less than 30 km^2 due to the large size of two of the sanctuary zones within the NRMP. The Mangrove Bay sanctuary is 11 km^2 which is only slightly greater than the average 95% KUD for spangled emperor (8.5 km^2). In the absence of long term monitoring and given that the dimensions of our array were

28 km^2, this suggests that even some of the smallest sanctuary zones within the NRMP will provide some degree of protection for this species. However, after one year of monitoring, more than 60% of the fish tagged in the lagoon had moved beyond the detection range of the array of receivers. This indicates that sanctuary zones larger than 30 km^2 are required to protect a significant proportion (>50%) of the population within the lagoon, as these fish tend to be less mobile than fish tagged on the reef slope.

Variability in behaviour and long term residency

Our data indicate that while *L. nebulosus* populations include significant proportions of resident fish, making them a good candidate for MPA protection. The fact that many fish may be nomadic or utilise more than one core area suggests that networks of MPA's are likely to be necessary to protect a larger proportion of the population. Variability in movement behaviour within populations has also been described in other reef fish, such as *Pagrus auratus* and *Sebastes melanops* where between a third to more than half the population has been shown to exhibit high levels of movement [51,52] or to have multiple activity centres [15]. Meyer et al. [5] demonstrated permanent emigration and excursions from Kealakekua Bay MPA by surgeonfish, parrotfish and goatfish providing evidence that relatively small reef fishes are less resident and more mobile than previously thought, although only 3% of all fish were detected moving two km or more from the point of capture. For *L. nebulosus*, it is not yet clear whether the contrast between residence at a single core area, as opposed to nomadism or occupation of multiple core areas represents two modes of behaviour, or a continuum of increasing home range size. A combination of additional telemetry data from an array of receivers significantly larger than that employed in this study combined with modelling approaches are required to better evaluate the network of MPA's within the NRMP.

Meyer et al. [5] showed that tagged reef fish relocated their home range within the array with individuals displaying shifts in space use and changes in the detection frequency at single receivers. However if multiple core areas exist for the *L. nebulosus* in our study they occur at scales either smaller than the range of a single receiver or larger than the Mangrove Bay array as there was no evidence of this in our study at the scales over which we were able to examine it. Nevertheless it is possible that individuals that left and then returned to the Mangrove Bay array had shifted their core areas displaying multiple activity centres. One returning fish (8046) appeared to have shifted core areas.

The departure of 12 fish in February 2008 coincided with the approach of a Category three tropical cylone "Nicholas". It is possible that the weather conditions and falling barometric pressure associated with this cyclone caused some individuals to move to deeper waters. A similar response was demonstrated by young of the year Blacktip Sharks (*Carcharhinus limbatus*) in Florida where all 13 animals being monitored by an array of receivers departed the shallow waters of Terra Ceia Bay a few hours before the arrival of a tropical storm and then returned several hours after the storm had passed [53]. The departure and return of these Blacktip sharks was very closely aligned to the arrival and departure of the storm, in contrast to the behaviour of juvenile and adult *L. nebulosus* that departed over a longer time period. Similarly, fish that returned did so over a period of several weeks to four months.

Fisheries based estimates of mortality for *L. nebulosus* in the region range between $Z = 0.285$–0.384 [30] which would mean that between 57 and 45% of tagged individuals would be expected to survive for two years after tagging. While these are probably minimum estimates (being derived from populations outside protected areas), the number of tagged animals remaining in the array after two years was much lower than these values (Figure 6). Emigration from the study site is therefore considered likely to be the largest component of declining number of animals detected over time, with declining numbers being a net product of both departures and returns of tagged individuals.

Previous studies have demonstrated declining numbers of detections in tagged cohorts beginning shortly after tagging and continuing to decline throughout the monitoring period [5,25,54,55,56]. These trends have been evident within time spans significantly less than expected battery life. The interpretation of such declines is problematic and they have been attributed to a combination of the following: 1) premature transmitter failure, 2) tagging effects (elevated predation, increased mortality or transmitter expulsion, 3) mortality unrelated to tagging (natural of fishing mortality) and 4) emigration from the study site [5]. Viewed in aggregate the residency patterns of *L. nebulosus* at Mangrove Bay appear to produce a picture that resembles the result of dispersive processes (Figure 6) with the number of tagged individuals in the array gradually declining, and we consider this the likely explanation for the majority of the decline. There was some evidence of fishing mortality at Mangrove Bay with two animals recaptured by recreational fishers (as demonstrated also by [25,54,55,56]). There was also evidence of natural mortality; characteristic patterns of tag detections are recorded where animals either died or were consumed close to a receiver resulting in the tag falling out and being continually detected by one receiver. There was also one instance where an animal displayed limited movement over a period of months, followed by weeks of large scale roaming within the array until the tag was detected continuously on one receiver. This was consistent with movement patters of black tip reef sharks (*Carcharhinus melanopterus*) tagged as part of this study and we suspect this individual was consumed by a black tip reef shark. Eventual eversion of the stomach is likely to have caused the tag to fall to the bottom near a receiver as it is common for carcharhinid sharks to periodically evert their stomachs completely into the oral cavity, possibly to expel undigested food or gastric parasites [57].

Effects of zoning and implications for reserve design

It is clear that approximately half the *L. nebulosus* population in the Mangrove Bay sanctuary zone at any time receives any lasting protection from it. But despite the high residency shown by *L. nebulosus* in Mangrove Bay over short periods (weeks to months), our long-term movement data illustrate that even these resident fish will have a greater than 20% chance of crossing the reserve boundary (Figure 6). For non-resident fish the likelihood of spending time outside the sanctuary zones is of course even greater. While home range size of tracked fish is significantly smaller than the average MPA size within the Ningaloo Reef Marine Park, both maximum linear distance moved by fish within the array as well as evidence of a large proportion of fish moving beyond the limits of the acoustic array suggest large sanctuary zones will be needed to protect a larger proportion of the population within the NRMP. Surveys of the NRMP have shown higher biomass of *L. nebulosus* in no-take zones [31,58] but it has not been possible to date to demonstrate a relationship between reserve size and *L. nebulosus* abundance. For sanctuary zones established in 1986 neither reserve size (area, perimeter, sea perimeter (perimeter excluding the terrestrial and shoreline boundary), area/perimeter and area/sea perimeter) were correlated with the relative abundance of *L. nebulosus* [31].

Results of the current study showed that fish tagged on the reef slope displayed greater levels of movement than fish tagged in the lagoon, with fish on the reef slope more likely to move outside the monitored area. These movement data are supported by surveys of *L. nebulosus* biomass within the NRMP. Biomass was higher inside the lagoon than on the reef slope with significantly higher biomass inside protected areas within the lagoon than adjacent fished areas [31]. In contrast, biomass of *L. nebulosus* on the reef slope was higher in fished areas than protected areas. Although the array could detect long-shore movement of fish on the reef slope, it was not primarily designed to detect offshore movements with the deepest receivers at approximately 50 m depth and two km from the reef crest. Movement of fish on the reef slope to deep water habitat more than 10 km offshore therefore requires additional research.

There are several lines of evidence consistent with the conclusion that the sizes of many sanctuary zones at Ningaloo are too small to achieve conservation outcomes (i.e. maintain or restore impacted populations). The response to no-take status in terms of effect size (no-take:fished biomass) is relatively small (1.5 times greater biomass), even for sanctuary zones established in 1986 [31] and at the one site where such comparisons are possible, the abundance of *L. nebulosus* has declined several fold both inside and outside the Osprey Sanctuary Zone [31]. Using an age-based demographic analysis of fish captured along the Ningaloo Reef in 1989–1991 and 2007–2008, Marriott et al. [33] showed that under 2007–2008 fishing levels, *L. nebulosus* were at risk of overfishing. Furthermore, levels of fishing pressure have had a dramatic effect on fish stocks in this region with the modal age of fish declining from six to five years and the percentage of fish older than six years declined from ~50% in 1989–1991 to ~28% in 2007–2008.

The Mangrove Bay Sanctuary zone is one of the smaller no-take zones in the Ningaloo Marine Park and, for resident *L. nebulosus*, our results show that such a no-take zone can offer reasonably high levels of protection. However, at least 48% of fish moved distances great enough to take them not only out of the reserve but out of the array altogether.

Differences in residence times between the lagoon and reef slope are likely to be due to the fact that animals tagged on the reef slope during spawning periods had moved to this area as part of a spawning migration. There was evidence of this with one fish returning to the lagoon after the spawning season and remaining within the lagoon for 10 months before moving offshore in October–December. Similarly, several adult fish tagged during the spawning season were only recorded by the array between October–December and then only on receivers on the reef slope. Five sexually mature fish tagged during spawning aggregations were detected on lines of cross shelf receivers during spawning periods (October–December). Four fish were detected off Tantabiddi (~10 km north) and one off Coral Bay (~80 km south). These detections suggest fish undergo long distance movements associated with spawning that are significantly greater than home range estimates. Long distance spawning movements are common even in highly site attached species such as coral groupers [59]. The scale and frequency of these movements requires additional research. Unfortunately these lines of cross-shelf receivers did not extend into the lagoon preventing us detecting large scale movements within the lagoon.

Fish tagged on the reef slope had a greater average kernel area, moved greater linear distances between receivers than fish tagged inside the lagoon. These differences are most likely due to the habitat on the reef slope being confined to a narrow band from the reef crest to ~30 m. This may result in animals either having a longer home range or shifting home range more frequently. For coral trout in the GBR, reef width accounted for 65% of the variability in home range area for animals on fringing reefs, whereas individuals tracked around patch reefs had a larger home range due primarily to the width of estimated home range [24].

Higher rates of movement and larger activity centres of individual *L. nebulosus* in reef slope habitats, and the apparent targeted use of this habitat for spawning by at least some of the population, indicates the need for increased size of no-take zones in these habitats is even greater. But the reef slope at Mangrove Bay is not protected by a no-take zone, a situation that is typical of most sanctuary zones in the northern section of the NMP where deeper reef habitats are under-represented.

Influence of tag parameters

The battery life, transmission interval and measured range of tags in this study varied from 185–2020 days, 40–320 s and 50->500 m, respectively. That there was no relationship between battery life, tag transmission interval and average tag range and ecological measurements such as kernel size suggests that biological and ecological factors are by far the most important factors in determining the dimensions of fish activity centres in our tracking data. Furthermore, the configuration of the array (as indexed by distance to array edge) was not a significant parameter in any of the regressions, suggesting that our conclusions are not biased by the size of the array or the location of tagging with respect to the array location. These findings have important implications for the design of acoustic arrays and suggest that the results of tagging studies are extremely robust and that the behaviour of individual fish is far more important than the behaviour or characteristics of tags used to monitor individuals. We did not find any indication that the non-overlapping nature of the receivers in the array detracted from its effectiveness in informing our understanding of the movements of *L. nebulosus*.

There is an increasing body of literature on the movement and home range of reef fish that attempts to relate tracking data to MPA effectiveness; this has resulted in considerable variation in the species studied, spatial and temporal scale, study duration and sampling methods used to generate movement data. While there is mounting evidence that the majority of coral reef fish remain within a small core area, studies are commonly limited to few individual animals of the same size and/or, that are monitored by few receivers (normally fewer than 10) that have limited coverage of available habitat (usually less than 5 km²) and have been conducted for a short time period (3–6 months) [5,25,27,60,61,62]. For example, a study of *C. microrhinus* tracked animals for less than five days and described home ranges as having stabilised after 3–4 days with no evidence of home range increasing over time [27]. If such short time frames had been applied to our studies of *L. nebulosus* we would have come to quite different conclusions to those we have drawn over a period of years.

While we acknowledge the usefulness of these studies and the fact that battery life, tag size and cost have increased the ability to conduct long term studies on larger numbers of fish, it is important to be aware that important variability exists within populations at a range of spatial, temporal and ontogenetic levels. Management decisions made on the basis of relatively limited and short term observations may be compromised if they do not include the full range of information required for application at a population level. This is particularly true in the case of designing reserves for properties such as "spillover" which is by definition a complex product of movements into and out of no-take areas, growth, recruitment, reproduction and fishing behaviours which needs to be modelled [16] to assess the likelihood of effective net results on either side of the conservation/fisheries ledger.

Conclusions

The behaviour of *L. nebulosus* within the Ningaloo Reef Marine Park was variable and complex. The majority of individuals had home ranges that were smaller than current MPA's, however after one year of monitoring more than half the fish had moved beyond the monitored area and outside the MPA boundary. Within NRMP, MPA's may need to be expanded to offer greater protection to the species, particularly with respect to fish on the reef slope that were more mobile. Furthermore, annual spawning aggregations adjacent to reef passes have important implications for MPA design and planning.

Acknowledgments

Logistical assistance in Exmouth was provided by WA Department of Conservation (DEC).

Author Contributions

Conceived and designed the experiments: RP RB. Performed the experiments: RP DB AB RD TP DT RB. Analyzed the data: RP TP RB. Contributed reagents/materials/analysis tools: RP TP RB. Wrote the paper: RP TP RB.

References

1. Worm B, Lotze HK, Myers RA (2003) Predator diversity hotspots in the blue ocean. Proc Natl Acad Sci USA 100(17): 9884–9888.
2. Halpern BS (2003) The impact of marine reserves: do reserves work and does reserve size matter? Ecol Appl 13(1):S117–S137.
3. Babcock RC, Shears NT, Alcala AC, Barrett NS, Edgar GJ, et al. (2010) Decadal trends in marine reserves reveal differential rates of change in direct and indirect effects. Proc Natl Acad Sci USA 107:18256–18261.
4. Cooke SJ, Hinch SG, Wikelski M, Andrews RD, Kuchel LJ, et al. (2004) Biotelemetry: a mechanistic approach to ecology. Trends Ecol Evol 19(6): 334–343.
5. Meyer CG, Papastamatiou YP, Clark TB (2010) Differential movement patterns and site fidelity among trophic groups of reef fishes in a Hawaiian marine protected area. Mar Biol 157:1499–1511.
6. Marshall A, Mills JS, Rhodes KL, McIlwain J (2011) Passive acoustic telemetry reveals highly variable home range and movement patterns among unicornfish within a marine reserve. Coral Reefs 30(3):631–642.
7. Polacheck T (1990) Year around closed areas as a management tool. Nat Resour Model 4:327–353.
8. DeMartini EE (1993) Modeling the potential of fishery reserves for managing Pacific coral-reef fishes. Fish Bull 91(3):414–427.
9. Kramer DL, Chapman MR (1999) Implications of fish home range size and relocation for marine reserve function. Env Biol Fish 55(1–2):65–79.
10. Gerber LR, Kareiva PM, Bascompte J (2002) The interplay of life history attributes and fishing pressure in evaluating efficacy of marine reserves. Biol Conserv 106:11–18.
11. Attwood CG, Bennett BA (1995) Modelling the effect of marine reserves on the recreational shore-fishery of the South-Western Cape. S Afr J Marine Sci 16(1): 227–240.
12. Starr RM, O'Connell V, Ralston S (2004) Movements of lingcod (*Ophiodon elongatus*) in southeast Alaska: potential for increased conservation and yield from marine reserves. Can J Fish Aquat Sci 61:1083–1094.
13. Rakitin A, Kramer DL (1996) Effect of a marine reserve on the distribution of coral reef fishes in Barbados. Mar Ecol Prog Ser 131: 97–113.
14. Claudet J, Osenberg CW, Domenici P, Badalamenti F, Milazzo M, et al. (2010) Marine reserves: fish life history and ecological traits matter. Ecol Appl 20(3):830–839.
15. Parsons DM, Morrison MA, Slater MJ (2010) Responses to marine reserves: decreased dispersion of the sparid *Pagrus auratus* (snapper). Biol Cons 143:2039–2048.
16. Babcock RC, Egli DP, Attwood CG (2012) Individual-Based model of reef fish populations in temperate marine reserves based on acoustic telemetry and census data. Environ Conserv 39(3):282–294.
17. White C, Kendall BE, Gaines S, Siegel DA, Costello C (2008) Marine reserve effects on fishery profit. Ecology Letters 11: 370–379.
18. Hobday AJ, Flint N, Stone T, Gunn JS (2009) Electronic tagging data supporting flexible spatial management in an Australian longline fishery. In: Nielsen JL, Arrizabalaga H, Fragoso N, Hobday A, Lutcavage M, Sibert J editors. Tagging and Tracking of Marine Animals with Electronic Devices. Reviews: Methods and Technologies in Fish Biology and Fisheries 9. New York. pp. 381–403.
19. Micheli F, Halpern BS, Botsford LW, Warner RR (2004) Trajectories and correlates of community change in no-take marine reserves. Ecol Appl 14:1709–1723.
20. Lester SE, Halpern BS, Grorud-Colvert K, Lubchenco J, Ruttenberg BI, et al. (2009) Biological effects within no-take marine reserves: A global synthesis. Mar Ecol Prog Ser 384:33–46.
21. Goñi R, Quetglas A, Reñones O (2006) Spillover of lobster *Palinurus elephas* (Fabricius 1787) from a Western Mediterranean marine reserve. Mar Ecol Prog Ser 308:207–219.
22. Harrison HB, Williamson DH, Evans RD, Almany GR, Thorrold SR, et al. (2012) Larval Export from Marine Reserves and the Recruitment Benefit for Fish and Fisheries. Curr Biol 22:1023–1028.
23. Nowlis JS, Roberts CM (1999) Fisheries benefits and optimal design of marine reserves. Fish Bull 97:604–616.
24. Zeller DC (1997) Home range and activity patterns of the coral trout *Plectropomus leopardus* (Serranidae). Mar Ecol Prog Ser 154:65–77.
25. Chateau O, Wantiez L (2009) Movement of four coral reef fish species in a fragmented habitat in New Caledonia: Implications for the design of marine protected area networks. ICES J Mar Sci 66:50–55.
26. Meyer CG, Holland KN (2005) Movement patterns, home range size and habitat utilization of the bluespine unicornfish, *Naso unicornis* (Acanthuridae) in a Hawaiian marine reserve. Environ Biol Fish 73:201–210.
27. Welsh JQ, Bellwood DR (2012) Spatial ecology of the steephead parrotfish (*Chlorurus microrhinos*): an evaluation using acoustic telemetry. Coral Reefs 31:55–65.
28. Zeller DC (1998) Spawning aggregations: patterns of movement of the coral trout *Plectropomus leopardus* (Serranidae) as determined by ultrasonic telemetry. Mar Ecol Prog Ser 162:253–263.
29. Carpenter KE (2001) Family Lethrinidae Emperors In: Carpenter KE, Niem VH editors. FAO species identification guide for fishery purposes. The living marine resources of the Western Central PacificVolume 5, Bony fishes part 3 Menidae to Pomacentridae. Rome. pp 3037.
30. Marriott RJ, Jarvis NDC, Adams DJ, Gallash AE, Norriss J, et al. (2010) Maturation and sexual ontogeny in the spangled emperor, *Lethrinus nebulosus*. J Fish Biol 76:1396–1414.
31. Babcock R, Haywood M, Vanderklift M, Clapin G, Kleczkowski M, et al. (2008) Ecosystem Impacts of Human Usage and the Effectiveness of Zoning for Biodiversity conservation: Broad-scale Fish Census. Final Analysis and Recommendations 2007. Hobart: CSIRO Marine and Atmospheric Research. 99 p.
32. Sumner NR, Williamson PC, Malseed BE (2002) A 12-Month Survey of Recreational Fishing in the Gascoyne Bioregion of Western Australia During 1998–99. Western Australia: Fisheries Research Report No. 139, Department of Fisheries WA.
33. Marriott RJ, Adams DJ, Jarvis NDC, Moran MJ, Newman SJ, et al. (2011) Age-based demographic assessment of fished stocks of *Lethrinus nebulosus* in the Gascoyne Bioregion of Western Austalia. Fisheries Manag Ecol 18:89–103.
34. Adams PB (1980) Life history patterns in marine fishes and their consequences for fisheries management. Fish Bull 78:1–12.
35. Parent S, Schriml LM (1995) A model for the determination of fish species at risk based upon life-history traits and ecological data. Can J Fish Aquat Sci 52:1768–1781.
36. CALM MPRA (2005) Management plan for the Ningaloo Marine Park and Muiron Islands Marine Management Area: 2005–2015. Management Plan Number 52. Perth, Western Australia Department of Conservation and Land Management. 111 p.
37. Moran M, Edmonds J, Jenke J, Cassels G, Burton C (1993) Fisheries biology of emperors (Lethrinidae) in North-West Australian coastal waters. FRDC Project 89/20 Final Report, Fisheries Department of Western Australia, Perth. 58 p.
38. van Winkle W (1975) Comparison of several probabilistic home-range models. J Wildlife Manage 39:118–123.
39. Worton BJ (1989) Kernel methods for estimating the utilization distribution in home-range studies. Ecology 70(1):164–168.
40. Kernohan BJ, Gitzen RA, Millspaugh JJ (2001) Analysis of animal space use and movements. In: Millspaugh JJ, Marzluff JM (eds) Radio Tracking and Animal Populations. Academic Press, San Diego, California. pp. 126–166.
41. White GC, Garrott GA (1990) Analysis of wildlife radio-tracking data. Academic Press, Inc., San Diego, California.
42. Simpfendorfer CA, Heupel MR, Hueter RE (2002) Estimation of short-term centres of activity from an array of omnidirectional hydrophones and its use in studying animal movements. Can J Fish Aquat Sci 59: 23–32.
43. Horne JS, Garton EO, Krone SM, Lewis JS (2007) Analysing animal movements using Brownian bridges. Ecology 88(9): 2354–2363.

44. Bellwood DR (1985) The functional morphology, systematics and behavioural ecology of parrotfishes (family Scaridae). PhD dissertation, James Cook University, Townsville, Australia.

45. Ogden JC, Buckman NS (1973) Movements, foraging groups, and diurnal migrations of striped parrotfish Scarus croicensis Bloch (Scaridae). Ecology 54(3):589–596.

46. van Rooij MJ, Kroon FJ, Videler JJ (1996) The social and mating system of the herbivorous reef fish Sparisoma viride: One-male versus multi-male groups. Environ Biol Fish 47:353–378.

47. Chateau O, Wantiez L (2008) Human impacts on residency behaviour of spangled emperor, Lethrinus nebulosus, in a marine protected area, as determined by acoustic telemetry. J Mar Biol Assoc UK 88(4): 825:829.

48. Chateau O, Wantiez L (2007) Site fidelity and activity patterns of a humphead wrasse, Cheilinus undulatus (Labridae), as determined by acoustic telemetry. Environ Biol Fish 80:503–508.

49. Meyer CG, Papastamtiou YP, Holland KN (2007) Seasonal, diel and tidal movements of green jobfish (Aprion virescens) at remote Hawaiian atolls: implications for marine protected area design. Mar Biol 151:2133–2143.

50. Pederson MW, Weng KC (2013) Estimating individual animal movement from observation networks. Methods Ecol Evol 4: 920–929.

51. Egli DP, Babcock RC (2004) Ultrasonic tracking reveals multiple behavioural modes of snapper (Pagrus auratus) in a temperate no-take marine reserve. ICES J Mar Sci 61:1137–1143.

52. Green KM, Starr RM (2011) Movement of small adult black rockfish: implications for the design of MPAs. Mar Ecol Prog Ser 14:219–230.

53. Heupel MR, Simpfendorfer CA, Hueter RE (2003) Running before the storm: blacktip sharks respond to falling barometric pressure associated with Tropical Storm Gabrielle. J Fish Biol 63:1357–1363.

54. Lindholm J, Kauffman L, Miller S, Wagschal A, Newville M (2005) Movement of yellowtail snapper (Ocyurus chrysurus Bloch 1790) and black grouper (Mycteroperca bonaci Poey 1860) in the northern Florida Keys National Marine Sanctuary as determined by acoustic telemetry. Marine sanctuaries conservation series MSD-05-4. U.S. Department of Commerce, National Oceanic and Atmospheric Administration, Marine Sanctuaries Division, Silver Spring, MD. 17 p.

55. Lindholm J, Knight A, Kauffman L, Miller S (2006a) Site fidelity and movement of parrotfishes Scarus coeruleus and Scarus taeniopterus at Conch Reef (northern Florida Keys). Carrib J Sci 42:138–133.

56. Lindholm J, Knight A, Kaufman L, Miller S (2006b) A pilot study of hogfish (Lachnolaimus maximus Walbaum 1792) movement at the conch reef research only area (northern Florida keys). National marine sanctuary program NMSP-06-06. US Departmentof Commerce, National Oceanic and Atmospheric Administration, National Marine Sanctuary Program, Silver Spring MD. 14 p.

57. Andrews PLR, Young JZ (1993) Gastric motility patterns for digestion and vomiting evoked by sympathetic nerve stimulation and 5-dehydroxytryptamine in the dogfish Scyliorhinus canicula. Philos T R Soc B 342:363–380

58. Westera M, Lavery P, Hyndes G (2003) Differences in recreationally targeted fishes between protected and fished areas of a coral reef marine park. J Exp Mar Biol Ecol 294: 145–168.

59. Rhodes KL, McIlwain J, Joseph E, Nemeth RS (2012) Reproductive movement, residency and fisheries vulnerability of brown-marbled grouper, Epinephelus fuscoguttatus (Forsskål, 1775). Coral Reefs 31: 443–453.

60. Lowe CG, Topping DT, Catamil DP, Papastamtiou YP (2003) Movement patterns, home range, and habitat utilisation of adult kelp bass Paralabrax clathratus in a temperate no take marine reserve. Mar Ecol Prog Ser 256:205–216.

61. Nanami A, Yamada H (2009) Site fidelity, and spatial arrangement of daytime home range of thumbprint emperor Lethrinus harak (Lethrinidae). Fish Sci 75:1109–1116.

62. March D, Alós J, Grau A, Palmer M (2011) Short-term residence and movement patterns of the annular seabream Diplodus annularis in a temperate marine reserve. Estuar Coast Shelf S 92:581–587.

11

Long-Distance Dispersal via Ocean Currents Connects Omani Clownfish Populations throughout Entire Species Range

Stephen D. Simpson[1*⑨], **Hugo B. Harrison**[2⑨], **Michel R. Claereboudt**[3], **Serge Planes**[4,5]

1 Biosciences, College of Life and Environmental Sciences, University of Exeter, Exeter, United Kingdom, **2** Australian Research Council Centre of Excellence for Coral Reef Studies, James Cook University, Townsville, Queensland, Australia, **3** Department of Marine Science and Fisheries, Sultan Qaboos University, Al-Khod, Oman, **4** Le Centre de Biologie et d'Ecologie Tropicale et Méditerranéenne, l'Université de Perpignan, Perpignan, Pyrénées-Orientales, France, **5** Laboratoire d'Excellence "CORAIL", Centre de Recherches Insulaires et Observatoire de l'Environnement, Moorea, French Polynesia

Abstract

Dispersal is a crucial ecological process, driving population dynamics and defining the structure and persistence of populations. Measuring demographic connectivity between discreet populations remains a long-standing challenge for most marine organisms because it involves tracking the movement of pelagic larvae. Recent studies demonstrate local connectivity of reef fish populations via the dispersal of planktonic larvae, while biogeography indicates some larvae must disperse 100–1000 s kilometres. To date, empirical measures of long-distance dispersal are lacking and the full scale of dispersal is unknown. Here we provide the first measure of long-distance dispersal in a coral reef fish, the Omani clownfish *Amphiprion omanensis*, throughout its entire species range. Using genetic assignment tests we demonstrate bidirectional exchange of first generation migrants, with subsequent social and reproductive integration, between two populations separated by over 400 km. Immigration was 5.4% and 0.7% in each region, suggesting a biased southward exchange, and matched predictions from a physically-coupled dispersal model. This rare opportunity to measure long-distance dispersal demonstrates connectivity of isolated marine populations over distances of 100 s of kilometres and provides a unique insight into the processes of biogeography, speciation and adaptation.

Editor: John A. Craft, Glasgow Caledonian University, United Kingdom

Funding: SDS was supported by a NERC Postdoctorate Fellowship (NE/B501720/1), a NERC Knowledge Exchange Fellowship (NE/J500616/2), a Royal Society Exchange grant and an EPHE Fellowship. The Project NEMO Expedition Team was funded by the Davis Trust, University of Edinburgh Development Trust, Carnegie Trust, BS-AC Jubilee Trust, Weir Trust, Genetics Society, British Association and Shell Oman. The funders had no role in study design, data collection and analysis, decision to publish, or preparation of the manuscript.

Competing Interests: The authors have declared that no competing interests exist

* Email: S.Simpson@exeter.ac.uk

⑨ These authors contributed equally to this work.

Introduction

Dispersal drives population dynamics, allows the replenishment of harvested marine species and defines the structure and persistence of marine populations across fragmented and often ephemeral habitat landscapes [1–3]. Since most coastal marine organisms are site-attached as adults, connectivity between discreet populations depends on the successful dispersal of planktonic larvae [4]. However, larvae spend days to months developing in the open ocean before settling to new habitats, thus tracking the dispersal trajectories of individuals during the larval stage is not feasible, and the full scale of dispersal remains largely unknown [5,6]. In recent years, substantial effort has been made to understand patterns of connectivity in marine species between discreet patches of coastal habitat in order to inform conservation efforts [7] and better manage natural resources [8]. At small spatial scales (10 s km), dispersal can facilitate the replenishment of local fished areas by neighbouring protected populations [9,10]. Over large spatial scales (1000 s km) dispersal can drive the spread of invasive species [11–13] and facilitate species range shifts in response to climate change [14]. At intermediate scales (100 s km) dispersal is predicted to allow the recolonisation of disturbed and depleted populations [7], however direct measurements of the successful movement and colonisation of individual larvae at this scale have been thus far elusive.

Many recent studies have focussed on coral reef fish, which are generally highly site-specific to naturally fragmented habitats, thus providing a valuable model system for studying dispersal and connectivity in marine ecosystems. Findings show that recruitment on coral reefs is largely driven by the retention of larvae within their populations of origin [15,16] combined with immigration from neighbouring populations [9,17–21]. However, the behavioural and ontogenetic characteristics of coral reef fish larvae [4,22], broad-scale genetic homogeneity seen in some species [23–25] and predictions of coupled-biophysical models [26–30] all suggest larvae also have the potential to undertake long migrations during their pelagic phase. Indeed, the vast species ranges seen in many reef fish, some spanning entire ocean basins, suggest that occasional long-distance dispersal or background gene flow along regional stepping-stones must occur to maintain genetic coherence of species and prevent speciation. To date, evidence for long-distance dispersal (100–1000 s km) comes from studies of evolutionary processes that measure historical gene flow [23–25], while attempts to infer large-scale patterns of dispersal using oceanographic models [26,29–31] are, as yet, unvalidated estimates of demographic connectivity.

Combining direct measurements of successful colonisation following long-distance dispersal with validated estimates of demographic connectivity, we investigate the potential for coral reef fishes to disperse over long distances during only a short pelagic phase. Our study focuses on the Omani clownfish *Amphiprion omanensis* (Fig. 1), which is endemic to the Arabian Sea and found only on shallow coral reef dominated habitat located in two regions of the southern coast of Oman, separated by over 400 km of high exposure sandy shores [32]. This provides a unique opportunity to investigate dispersal between distant populations spanning an entire species range. Over large spatial scales, where gene flow is restricted, genetic assignment tests can be used to identify the origin of individuals provided that discrete populations are genetically distinct and that all populations have been sampled [33–35]. Using population-specific assignment thresholds, we determined whether individuals were local-type, first-generation migrants or local-migrant hybrids within each region. We compared our empirical findings to predictions from a simple physically-coupled individual-based dispersal model to determine whether oceanographic current flows could predict long distance dispersal events.

Materials and Methods

a) Field sampling of *Amphiprion omanensis* (including Ethics Statement)

During an expedition in December 2006 and January 2007, we sampled 136 clownfish in the northern province of Ash Sharqiyah and 260 in the southern province of Dhofar (Table S1). Fish were collected and released without causing lasting harm and with specific approval for this study from both the University of Edinburgh Ethical Review Committee and the University of Edinburgh Expedition Committee, and with permission from the Ministry of Regional Municipalities, Environment and Water Resources, Sultanate of Oman and Department of Nature Conservation and Wildlife, Governate of Dhofar. A small section of the pectoral fin was removed from each individual and preserved in 95% ethanol. Based on available habitat in each region we estimate that we sampled 5–20% of all individuals in each region.

b) DNA extraction and microsatellite genotyping

Genomic DNA was extracted from fin tissue samples using standard proteinase K digestion (Gentra Puregene, QIAGEN). Samples were genotyped with a panel of six microsatellite loci previously described for two congeneric species [36,37] (Table S2) and optimised for *Amphiprion omanensis* in two multiplex reactions. Each locus was sequenced to confirm the presence of simple sequence repeats and integrated to multiplex PCRs on the basis of fragment sizes, heterodimer duplexing and complementarity of melting temperatures. Multiplex PCRs consisted of 1 µl genomic DNA, 0.2 µl $MgCl_2$, 1.5 µl dNTPs, 1 µl buffer, 0.5 µl *Taq* polymerase, and 1 µl H_2O. Multiplex A contained 0.8 µl forward and reverse primers for loci *Ao120*, *Ao84* and *AoCF3*. Multiplex B contained 0.6 µl forward and reverse primers for loci *Ao55* and *Ao22*, and 1.2 µl of each primer for locus *AoCF11*. All

PCR reactions were performed on a Eppendorf Mastercycler ep with the following cycling conditions: an initial denaturation of 2 min at 94°C followed by 10 cycles of 45 s at 94°C, 45 s at 60°C (multiplex A) or 66°C (multiplex B), and 45 s at 72°C, followed by 20 cycles of 45 s at 94°C, 45 s at 55°C (multiplex A) or 61°C (multiplex B), and 45 s at 72°C, with a final elongation period of 45 min at 72°C. Amplified PCR products were analysed on a CEQ 8000 Genetic Analysis System (Beckman-Coulter, Fullerton CA) and the resulting electrophoregrams were scored using CEQ 8000 software. Alleles were scored manually (twice, blind) to eliminate scoring errors.

c) Population genetics statistical analysis

Samples were grouped within northern and southern regions. Nei's unbiased expected and observed heterozygosity [38], number of alleles and allelic richness over all samples, and Weir & Cockerham's estimator of inbreeding F_{IS} [39] were computed in FSTAT v.2.9.3.2 [40]. Observed genotypes were tested for linkage disequilibrium and departures from Hardy-Weinberg equilibrium (HWE) due to heterozygote deficiency at each locus with genotypes randomised among samples as implemented in ARLEQUIN v3.5 [41]. Significance of multiple tests was assessed with sequential Bonferroni corrections applied for multiple comparisons [42]. The occurrence of null alleles and large allele drop-outs were assessed at each locus using MICROCHECKER v.2.2.3 [43]. Analysis of molecular variance (AMOVA) and pairwise genetic distances within and between provinces (F_{ST}) were computed in ARLEQUIN. Missing data accounted for 0.9% and 1.5% of the data in northern and southern populations, respectively.

d) Assignment tests and study-specific assignment thresholds

We applied a model-based Bayesian clustering method implemented in STRUCTURE 2.3.3 [35], using a Markov Chain Monte Carlo (MCMC) resampling procedure, to estimate the most parsimonious allocation of samples to distinct genetic clusters and distinguish between local-type, first-generation migrants and local-migrant hybrids. With the optimum solution (K = 2), we performed 10 independent runs using 100,000 MCMC iterations with a burn-in period of 50,000 steps and computed the arithmetic mean of posterior probabilities of assignment amongst runs.

To determine population-specific thresholds of assignment we first used a conservative threshold of 0.9 [44] to identify northern and southern-type individuals and determine unbiased allelic frequencies for each region. We then randomly selected 100 individuals from each region and simulated 5,000 local-type individuals for each population and 10,000 north-south hybrid individuals using HYBRIDLAB v1.0 [45]. By examining the distribution of posterior probabilities of all 20,000 simulated individuals (5 runs with K = 2, 50,000 iterations and burn-in of 50,000) we identified population-specific thresholds of assignment that give a 95% probability of correct allocation of individuals to local-type, first-generation migrant and north-south hybrid classes in our samples (Fig. S1).

e) Physically-coupled individual-based Lagrangian stochastic dispersal model

We constructed and parameterised an *a priori* dispersal model to simulate dispersal of the larval stage of *A. omanensis*. The model was forced using daily surface current data obtained from the Navy Coastal Ocean Model (US Naval Research Laboratory) for the period 2005–2008, and simulations of larvae released from

Figure 1. The Omani clownfish, *Amphiprion omanensis,* **is endemic to the southern coast of Oman.** Adults provide high levels of parental care to their young that hatch with well-developed swimming and sensory capabilities before embarking on a <3 week pelagic larval phase, during which time they may disperse over long distances (>400 km).

both northern and southern locations were made for the *A. omanensis* spawning period during three successive seasons (see Information S1). The proportions of larvae that were retained locally, were dispersed to distant reefs or did not settle were retained from each simulation for comparison with our empirical measurements of long-distance dispersal.

Results

A total of 136 *A. omanensis* individuals were collected in the northern province of Ash Sharqiyah and 260 individuals from the southern province of Dhofar. These two populations are separated by over 400 km of high exposure sandy shore with no suitable habitat and representing the limits of the species range. We found significant genetic differentiation between the two regions ($F_{ST} = 0.042$; Information S1; Fig. S2), with higher genetic diversity in the southern population (Table S3). Significant departures from HWE due to heterozygote deficiency were observed for three of six loci in both northern and southern regions after Bonferroni correction (Table S3). Locus *Ao84* showed significant departure from HWE in the northern populations only, whereas *AoCF11* showed evidence of heterozygote deficiency in the southern populations. Only *Ao120* showed no significant deviation from HWE expectations, although expected heterozygosity was marginally higher than expected in

the southern population. The underlying causes of heterozygote deficiencies were likely due the presence of genetic sub-structure within southern and northern populations that were not captured in our sample. Allelic richness was higher in the southern population, however the mean observed heterozygosity was comparable between regions. Tests for genotypic disequilibrium identified 2 of 30 pairwise comparisons as showing significant linkage after Bonferroni correction ($p < 0.0017$).

Using assignment tests, we detected an asymmetrical dispersal pattern between the two regions with a higher occurrence of southward dispersal than vice versa (Fig. 2a). Our analysis revealed a total of 14 migrants (5.4% of locally sampled individuals) in the southern province of Dhofar that had originated in the northern province of Ash Sharqiyah, but only 1 individual (0.7% of locally sampled individuals) identified as having made the opposite journey (Fig. 2b). Study-specific thresholds were 0.741 and 0.765 for the northern and southern regions respectively, and minimum and maximum thresholds to distinguish hybrid individuals from local-type were 0.413 and 0.711 respectively. Study-specific thresholds were 0.741 and 0.765 for the northern and southern regions respectively, and minimum and maximum thresholds to distinguish hybrid individuals from local-type were 0.413 and 0.711 respectively. Individuals identified as migrants were collected in multiple locations and included both adults and juveniles, demonstrating that some migrating larvae had subse-

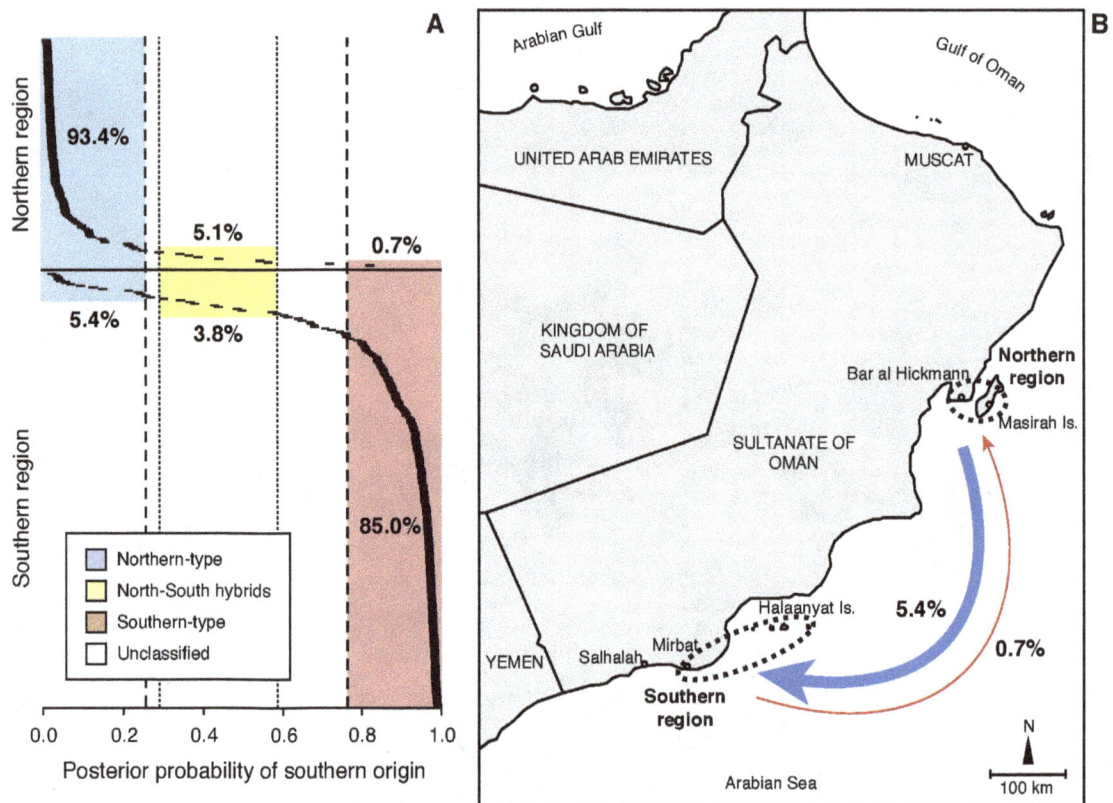

Figure 2. Long-distance dispersal of the Omani clownfish. (a) Bayesian clustering analysis identified individuals that had migrated to distant populations over 400 km from their natal origins. Dashed lines indicate the thresholds of assignment of individuals to northern and southern types. Dotted lines indicate the thresholds of assignment of individuals as north-south hybrids. Percentage values indicate proportion of assigned individuals to different types in each region. (b) The southern Omani coastline has two regions of coral reefs separated by over high exposure sandy shores. The Indian summer monsoon is the main driver of both atmospheric and oceanic regimes causing strong southeasterly winds during summer months (the Khareef monsoon), which then reverse to weaker northeasterly winds during the winter monsoon. These current regimes favour southward dispersal of the Omani clownfish.

quently become successfully integrated into local populations since maturation in clownfish is socially mediated [36].

In addition to first-generation migrants, we identified 17 north-south hybrids, 10 (3.8% of locally sampled individuals) in the south and 7 (5.1% of locally sampled individuals) in the north. The majority of individuals sampled in each population (221 or 85.0% in the south and 127 or 93.4% in the north) were identified as local-type (Fig. 2a). An additional 15 individuals in the south and 1 in the north could not be confidently distinguished from local and local-migrant hybrid types based on population-specific modelled assignment thresholds. The greater number of unclassified individuals in the southern population is most likely due to the limited proportion of individuals sampled (we estimate 5–20%), greater genetic diversity and higher immigration rates in this population pertaining to a less distinct genetic signature than the northern population and a more conservative minimum threshold of assignment. While migrants and local-migrant hybrids are minority components of the overall population, this study highlights the dispersal potential of pelagic larvae and demonstrates that exchange of larvae between distant populations is likely to be a regular background process rather than one restricted to rare and stochastic events.

The measured southward bias of dispersal matched *a priori* predictions of a physically-coupled individual-based dispersal model parameterised for the larval stage of *A. omanensis* and forced using seasonal current data. Movement of larvae released

into the model between December to February, when *A. omanensis* usually reproduces, was predominantly southward as a result of the winter monsoon (Table S4). Conversely, realisations of the model outside of the clownfish spawning season (March/April) predicted northward movement of larvae due to the onset of the summer Khareef monsoon. Depending on the date (between 1 December to 30 January) and year of release, a small percentage (ranging from 0.00 to 5.10%) of larvae were exchanged between the northern and southern populations within a single month and 'settled' to appropriate habitat within the 40 days competency period. Averaged over the three winter spawning seasons, the transport of larvae from the northern region to southern reefs was 0.06% to Mirbat and 0.32% to the Halaaniyat Islands, while northward exchange was ≤0.01% (Table S4). Although quantitatively sensitive to initial population size and associated output of larvae, which is not known, qualitatively the model consistently predicted higher migration rates from the northern to the southern populations as seen in our direct measures of long-distance dispersal. Our combined empirical and modelling results for *A. omanensis* suggest that regional current regimes directly influence the dispersal of larvae between southern and northern extremes of its species range.

Discussion

Geography, oceanography, ecology and behaviour combined favourably in this study system to provide a unique opportunity to empirically measure connectivity between distant populations of a coral reef fish. We found demographically relevant bidirectional exchange of *A. omnanensis* individuals between two regions of suitable coral reef habitat separated by >400 km of high exposure sandy shore, representing the longest direct measure of larval dispersal for any marine fish species to date. Local immigration rates were 5.4% in the south and 0.7% in the north, indicating a biased southward exchange, and these rates matched *a priori* predictions from a physically-coupled individual-based dispersal model (maximum connectivity in a single spawning event 5.1% and 0.1% respectively), which simulated passive larval transport in realistic current fields during the reproductive season. Local migrants were found at all life stages suggesting that these long-distance dispersal events occurred over multiple events and are a regular phenomenon rather than a process limited to chance [46]. Additionally, we identified individuals with mixed north-south genotypes in both regions demonstrating social and reproductive integration of migrants into local breeding populations.

The combination of rarity and conspicuousness meant that our team of 22 divers over 92 dives sampled an estimated 5–20% of all *A. omnanensis* individuals, providing subsequent analyses with sufficient discriminatory power. With more markers and a higher proportion of individuals sampled we would expect intra-regional patterns in dispersal and retention to also become evident as in previous local-scale studies of congeneric species [15,16,19–21,36], and would allow the 16 individuals that could not be distinguished between local and local-migrant hybrid types to be confidently assigned. Nonetheless our approach, based on qualifying the unique genetic signature of regional populations, provides a powerful tool to identify local migrants and measure connectivity in marine organisms that is applicable to a wide range of species.

Harvested and threatened populations under spatially variable levels of exploitation, or exposed to environmental stressors including fluctuations in habitat quality and climate change, are offered substantial resilience if replenishment from far beyond the range of local impact is possible. While there are obvious benefits to staying close to home [16], the spatially heterogeneous landscape of coral reef habitats may further confer selective advantage to long-distance dispersers that carry novel mutations across vast stretches of open water. Theory suggests that low rates of migration can rescue individual populations from local extinction and ensure long-term prevalence of species [47–48], although until now the nature of long-distance larval exchange in coral reef fishes has remained elusive. Our finding of regular exchange between distant populations explains how seemingly isolated populations maintain their genetic coherence, preventing local adaptation-driven ecological separation that leads to reproductive incompatibility and ultimately speciation. Additional studies using the same geographic case study but comparing a range of contrasting species would enable the influences of life history and oceanography on long-distance dispersal to be determined.

The persistence of spatially structured populations is a factor of both local growth rates and connectivity between populations [49,50]. Coral reef associated organisms are naturally fragmented and dispersal amongst patches is a fundamental aspect of population dynamics. For the Omani clownfish, the persistence of populations in the northern and southern provinces of the Arabian Sea appears heavily dependent on larval connectivity and larval retention within each province. However, even the low

levels of immigration observed between provinces have importance consequences for demographic processes. The greater immigration from north to south indicate a shortfall in local retention in the southern population [49,50] and a dependence on immigration from the northern population, which itself is comparatively more isolated. In the context of marine spatial planning, these asymmetries in connectivity patterns would warrant a need for greater protection of the northern population. If such high levels of demographic connectivity can be expected between two populations separated by over 400 km, populations in continuous reef habitats or complex barrier reef systems are likely to be much more homogeneous than previously assumed.

Our study also demonstrates that simple physically-forced models can give valuable predictions for realised patterns of connectivity. It is likely that if ocean currents are important for driving the long-distance dispersal events we observed, for fishes with prolonged pelagic larval durations (weeks to months), including many commercially important coastal fishes, long-distance dispersal may be even more prevalent [27,51,52]. Further development of the model would allow an ensemble of parameter sets, capturing temporal variability in circulation and the full range of phenological and larval biological traits to be used to generate distributions of self-recruitment and long-distance dispersal for other species in the region. The ability to predict the degree of connectivity between fragmented populations will allow fisheries and conservation managers and marine spatial planners charged with developing national and regional networks of marine protected areas to better manage these populations [53,54]. Furthermore, better characterisation of the tails of dispersal kernels would provide valuable insight into the ecological and evolutionary processes of biogeography, speciation and potential for species to adapt to current and future climate change.

Supporting Information

Figure S1 Frequency distributions of simulated genotypes for the derivation of population-specific assignment thresholds. A total of 5,000 northern-type, 5,000 southern-type and 10,000 north-south hybrid genotypes were simulated using location specific allelic frequencies. Bayesian clustering analysis determined the posterior probability of assignment of each simulated individual to either northern or southern populations.

Figure S2 Scatterplot of the two main components of the discriminant analysis of principal components in four populations of the Omani clownfish *Amphiprion omanensis*. Sampled populations are shown using different colours and 95% inertia ellipses and dots represent each individual in the sample. The x-axis represents 82.9% and the y-axis represents 9.6% of genetic information retained in each discriminant function (inset).

Table S1 Sampling sites and number of samples collected.

Table S2 Details of six polymorphic dinucleotide microsatellite loci developed for *Amphiprion omanensis*.

Table S3 Details of genetic analysis of each marker. Number of alleles (*Na*), observed heterozygosity (*Ho*), expected heterozygosity (*He*), the inbreeding coefficient (*F*), and departure from Hardy-Weinberg's equilibrium (HWE) were calculated for each locus.

Table S4 Connectivity matrices (% of larvae successfully reaching a reef) for simulated dispersal events.

The upper three panes are yearly averages (10 realisations); the lower panes are overall means, maximum values, standard deviation, coefficient of variation of connectivity and number of non-zero larval transport events over all 30 realisations. Matrices read from row to-column; grey cells indicate long-distance dispersal. Mirbat and Halaaniyat combine to form the southern region; Masirah and Bar Al Hickmann (MasBAH) combine to form the northern region (see Fig. 2B).

Information S1 Physically-coupled individual-based Lagrangian stochastic dispersal model; Graphical representation of genetic variation in the Omani clownfish; Supplementary References.

References

1. Clobert J, Danchin E, Dhondt AA, Nichols JD (2001) Dispersal. Oxford: Oxford University Press. 480 p.
2. Hanski I (1999) Metapopulation ecology. Oxford: Oxford University Press. 324 p.
3. Davies NB, Krebs JR, West SA (2012) An introduction to behavioural ecology, 4th edition. Chichester: Wiley-Blackwell. 506 p.
4. Shanks AL (2009) Pelagic larval duration and dispersal distance revisited. Biol Bull 216: 373–385.
5. Jones GP, Almany GR, Russ GR, Sale PF, Steneck RS, et al. (2009) Larval retention and connectivity among populations of corals and reef fishes: history, advances and challenges. Coral Reefs 28: 307–325. (DOI 10.1007/s00338-009-0469-9).
6. Levin LA (2006) Recent progress in understanding larval dispersal: new directions and digressions. Integr Comp Biol 46: 282–297. (DOI 10.1093/icb/icj024).
7. Trakhtenbrot A, Nathan R, Perry G, Richardson DM (2005) The importance of long-distance dispersal in biodiversity conservation. Divers Distrib 11: 173–181. (DOI 10.1111/j.1366-9516.2005.00156.x).
8. McCook LJ, Almany GR, Berumen ML, Day JC, Green AL, et al. (2009) Management under uncertainty: guide-lines for incorporating connectivity into the protection of coral reefs. Coral Reefs 28: 353–366. (DOI 10.1007/s00338-008-0463-7).
9. Harrison HB, Williamson DH, Evans RD, Almany GR, Thorrold SR, et al. (2012) Larval export from marine reserves and the recruitment benefit for fish and fisheries. Curr Biol 22: 1023–1028. (DOI 10.1016/j.cub.2012.04.008).
10. Almany GR, Hamilton RJ, Bode M, Matawai M, Potuku T, et al. (2013) Dispersal of grouper larvae drives local resource sharing in a coral reef fishery. Curr Biol 23: 626–630. (DOI 10.1016/j.cub.2013.03.006).
11. Gaither MR, Bowen BW, Toonen RJ, Planes S, Messmer V, et al. (2010) Genetic consequences of introducing allopatric lineages of Bluestriped Snapper (Lutjanus kasmira) to Hawaii. Mol Ecol 19: 1107–1121. (DOI 10.1111/j.1365-294X.2010.04535.x).
12. Gaither MR, Jones SA, Kelley C, Newman SJ, Sorenson L, et al. (2011) High connectivity in the deepwater snapper Pristipomoides filamentosus (Lutjanidae) across the Indo-Pacific with isolation of the Hawaiian archipelago. PLOS ONE 6: e28913. (DOI 10.1371/journal.pone.0028913).
13. Mooney HA, Hobbs RJ (2000) Invasive species in a changing world. Washington DC: Island Press. 384 p.
14. Simpson SD, Jennings S, Johnson MP, Blanchard JL, Schön P-J, et al. (2011) Continental shelf-wide response of a fish assemblage to rapid warming of the sea. Curr Biol 21: 1565–1570. (DOI 10.1016/j.cub.2011.08.016).
15. Almany GR, Berumen ML, Thorrold SR, Planes S, Jones GP (2007) Local replenishment of coral reef fish populations in a marine reserve. Science 316: 742–744. (DOI 10.1126/science.1140597).
16. Jones GP, Planes S, Thorrold SR (2005) Coral reef fish larvae settle close to home. Curr Biol 15: 1314–1318. (DOI 10.1016/j.cub.2005.06.061).
17. Berumen ML, Almany GR, Planes S, Jones GP, Saenz-Agudelo P, et al. (2012) Persistence of self-recruitment and patterns of larval connectivity in a marine protected area network. Ecol Evol 2: 444–452. (DOI 10.1002/ece3.208).
18. Christie MR, Tissot BN, Albins MA, Beets JP, Jia Y, et al. (2010) Larval connectivity in an effective network of marine protected areas. PLOS ONE 5: e15715. (10.1371/journal.pone.0015715).

19. Planes S, Jones GP, Thorrold SR (2009) Larval dispersal connects fish populations in a network of marine protected areas. Proc Natl Acad Sci USA 106: 5693–5697. (DOI 10.1073/pnas.0808007106).
20. Saenz-Agudelo P, Jones GP, Thorrold SR, Planes S (2011) Connectivity dominates larval replenishment in a coastal reef fish metapopulation. Proc Roy Soc B Biol Sci 278: 2954–2961. (DOI 10.1098/rspb.2010.2780).
21. Saenz-Agudelo P, Jones GP, Thorrold SR, Planes S (2012) Patterns and persistence of larval retention and connectivity in a marine fish metapopulation. Mol Ecol 21: 4695–4705. (DOI 10.1111/j.1365-294X.2012.05726.x).
22. Leis JM, Hay AC, Howarth GJ (2009) Ontogeny of in situ behaviours relevant to dispersal and population connectivity in larvae of coral-reef fishes. Mar Ecol Prog Ser 379: 163–179. (DOI 10.3354/meps07904).
23. Mora C, Treml EA, Roberts J, Crosby K, Roy D, et al. (2011) High connectivity among habitats precludes the relationship between dispersal and range size in tropical reef fishes. Ecography 35: 89–96. (DOI 10.1111/j.1600-0587.2011.06874.x).
24. Purcell JFH, Cowen RK, Hughes CR, Williams DA (2006) Weak genetic structure indicates strong dispersal limits: a tale of two coral reef fish. Proc Roy Soc B Biol Sci 273: 1483–1490. (DOI 10.1098/rspb.2006.3470).
25. Underwood JN, Travers MJ, Gilmour JP (2012) Subtle genetic structure reveals restricted connectivity among populations of a coral reef fish inhabiting remote atolls. Ecol Evol 2: 666–679. (DOI 10.1002/ece3.80).
26. Cowen RK, Paris CB, Srinivasan A (2006) Scaling of connectivity in marine populations. Science 311: 522–527. (DOI 10.1126/science.1122039).
27. Roberts CM (1997) Connectivity and management of Caribbean coral reefs. Science 278: 1454–1457. (DOI 10.1126/science.278.5342.1454).
28. Treml EA, Halpin PN, Urban DL, Pratson LF (2008) Modeling population connectivity by ocean currents, a graph-theoretic approach for marine conservation. Landscape Ecol 23: 19–36. (DOI 10.1007/s10980-007-9138-y).
29. Foster NL, Paris CB, Kool JT, Baums IB, Stevens JR, et al. (2012) Connectivity of Caribbean coral populations: complementary insights from empirical and modelled gene flow. Mol Ecol 21: 1143–1157. (DOI 10.1111/j.1365-294X.2012.05455.x).
30. Wood S, Paris CB, Ridgwell A, Hendy EJ (2014) Modelling dispersal and connectivity of broadcast spawning corals at the global scale. Global Ecol Biogeogr 23: 1–11. (DOI 10.1111/geb.12101).
31. Kool JT, Paris CB, Barber PH, Cowen RK (2011) Connectivity and the development of population genetic structure in Indo-West Pacific coral reef communities. Global Ecol Biogeogr 20: 695–706. (DOI 10.1111/j.1466-8238.2010.00637.x).
32. Schils T, Coppejans E (2003) Phytogeography of upwelling areas in the Arabian Sea. J Biogeogr 30: 1339–1356. (DOI 10.1046/j.1365-2699.2003.00933.x).
33. Cornuet J-M, Piry S, Luikart G, Estoup A, Solignac M (1999) New methods employing multilocus genotypes to select or exclude populations as origins of individuals. Genetics 153: 1989–2000.
34. Guillot G (2005) A spatial statistical model for landscape genetics. Genetics 170: 1261–1280. (DOI 10.1534/genetics.104.033803).
35. Pritchard JK, Stephens M, Donnelly P (2000) Inference of population structure using multilocus genotype data. Genetics 155: 945–959.
36. Buston PM, Bogdanowicz SM, Wong A, Harrison RG (2007) Are clownfish groups composed of close relatives? An analysis of microsatellite DNA variation in Amphiprion percula. Mol Ecol 16: 3671–3678. (DOI 10.1111/j.1365-294X.2007.03421.x).

Acknowledgments

We thank Jennifer McIlwain and Andrew Halford for first introducing SDS to the Southern Coast of Oman, and for logistical support on early scoping trips. We thank the other 22 members of the Project NEMO Expedition Team and Jonathan Smith for assistance in the field, Elisabeth Rochel and Cecile Fauvelot for assistance in the lab, Jonathan Kool for assistance with the statistical analyses and Phil Munday, Mark Meekan, Mark Beaumont, Martin Genner and Andy Radford for comments on the manuscript. We also thank Henry Carson and an anonymous reviewer for constructive comments. We are grateful to Ali Al-Kiyumi (Director General of Nature Conservation, Ministry of Regional Municipalities, Environment and Water Resources), Ali Salem Bait Said (Director of Nature Conservation and Wildlife, Governate of Dhofar) and Anton McLachlan (Dean of Science, Sultan Qaboos University) for authorising the work.

Author Contributions

Conceived and designed the experiments: SDS. Performed the experiments: SDS HBH MC SP. Analyzed the data: SDS HBH MC SP. Contributed reagents/materials/analysis tools: SDS HBH MC SP. Contributed to the writing of the manuscript: SDS HBH MC SP.

37. Quenouille B, Bouchenak-Khelladi Y, Hervet C, Planes S (2004) Eleven microsatellite loci for the saddleback clownfish *Amphiprion polymnus*. Mol Ecol Notes 4: 291–293. (DOI 10.1111/j.1471-8286.2004.00646.x).

38. Nei M (1987) Molecular evolutionary genetics. New York: Columbia University Press. 512 p.

39. Weir BS, Cockerham CC (1984) Estimating F-statistics for the analysis of population structure. Evolution 38: 1358–1370. (DOI 10.2307/2408641).

40. Goudet J (1995) FSTAT (Version 1.2): A computer program to calculate F-statistics. J. Heredity 86: 485–486.

41. Excoffier L, Estoup A, Cornuet JM (2005) Bayesian analysis of an admixture model with mutations and arbitrarily linked markers. Genetics 169: 1727–1738. (DOI 10.1534/genetics.104.036236).

42. Rice WR (1989) Analyzing tables of statistical tests. Evolution 43: 223–225. (DOI 0.2307/2409177).

43. van Oosterhout C, Hutchinson WF, Wills DPM, Shipley P (2004) MICRO-CHECKER: software for identifying and correcting genotyping errors in microsatellite data. Mol Ecol Notes 4: 535–538. (DOI 10.1111/j.1471-8286.2004.00684.x).

44. Vähä JP, Primmer CR (2006) Efficiency of model-based Bayesian methods for detecting hybrid individuals under different hybridization scenarios and with different numbers of loci. Mol Ecol 15: 63–72. (DOI 10.1111/j.1365-294X.2005.02773.x).

45. Nielsen EEG, Bach LA, Kotlicki P (2006) HYBRIDLAB (version 1.0): a program for generating simulated hybrids from population samples. Mol Ecol Notes 6: 971–973. (DOI 10.1111/j.1471-8286.2006.01433.x).

46. Siegel DA, Mitarai S, Costello CJ, Gaines SD, Kendall BE, et al. (2008) The stochastic nature of larval connectivity among nearshore marine populations. Proc Natl Acad Sci USA 105: 8974–8979. (DOI 10.1073/pnas.0802544105).

47. Gonzalez A, Lawton JH, Gilbert FS, Blackburn TM, Evans-Freke I (1998) Metapopulation dynamics, abundance, and distribution in a microecosystem. Science 281: 2045–2047. (DOI 10.1126/science.281.5385.2045).

48. Hill MF, Hastings A, Botsford LW (2002) The effects of small dispersal rates on extinction times in structured metapopulation models. Am Nat 160: 389–402. (DOI 10.1086/341526).

49. Hastings A, Botsford LW (2006) Persistence of spatial populations depends on returning home. Proc Natl Acad Sci USA 103: 6067–6072. (DOI 10.1073/pnas.0506651103).

50. Carson HS, Cook GS, López-Duarte PC, Levin LA (2011) Evaluating the importance of demographic connectivity in a marine metapopulation. Ecology 92: 1972–1984. (DOI 10.1890/11-0488.1).

51. Galarza JA, Carreras-Carbonell J, Macpherson E, Pascual M, Roques S, et al. (2009) The influence of oceanographic fronts and early-life-history traits on connectivity among littoral fish species. Proc Natl Acad Sci USA 106: 1473–1478. (DOI 10.1073/pnas.0806804106).

52. White C, Selkoe KA, Watson J, Siegel DA, Zacherl DC, et al. (2010) Ocean currents help explain population genetic structure. Proc Roy Soc B Biol Sci. 277: 1685–1694. (DOI 10.1098/rspb.2009.2214).

53. Sale PF, Cowen RK, Danilowicz BS, Jones GP, Kritzer JP, et al. (2005) Critical science gaps impede use of no-take fishery reserves. Trends Ecol Evol 20: 74–80. (DOI 10.1016/j.tree.2004.11.007).

54. Green AL, Fernandes L, Almany G, Abesamis R, McLeod E, et al. (2014) Designing marine reserves for fisheries management, biodiversity conservation, and climate change adaptation. Coast Manage 42: 143–159. (DOI 10.1080/08920753.2014.877763).

12

Rearing in Seawater Mesocosms Improves the Spawning Performance of Growth Hormone Transgenic and Wild-Type Coho Salmon

Rosalind A. Leggatt, Tanya Hollo, Wendy E. Vandersteen, Kassandra McFarlane, Benjamin Goh, Joelle Prevost, Robert H. Devlin*

Fisheries and Oceans Canada, West Vancouver Laboratories, West Vancouver, BC, Canada

Abstract

Growth hormone (GH) transgenes can significantly accelerate growth rates in fish and cause associated alterations to their physiology and behaviour. Concern exists regarding potential environmental risks of GH transgenic fish, should they enter natural ecosystems. In particular, whether they can reproduce and generate viable offspring under natural conditions is poorly understood. In previous studies, GH transgenic salmon grown under contained culture conditions had lower spawning behaviour and reproductive success relative to wild-type fish reared in nature. However, wild-type salmon cultured in equal conditions also had limited reproductive success. As such, whether decreased reproductive success of GH transgenic salmon is due to the action of the transgene or to secondary effects of culture (or a combination) has not been fully ascertained. Hence, salmon were reared in large (350,000 L), semi-natural, seawater tanks (termed mesocosms) designed to minimize effects of standard laboratory culture conditions, and the reproductive success of wild-type and GH transgenic coho salmon from mesocosms were compared with that of wild-type fish from nature. Mesocosm rearing partially restored spawning behaviour and success of wild-type fish relative to culture rearing, but remained lower overall than those reared in nature. GH transgenic salmon reared in the mesocosm had similar spawning behaviour and success as wild-type fish reared in the mesocosm when in full competition and without competition, but had lower success in male-only competition experiments. There was evidence of genotype×environmental interactions on spawning success, so that spawning success of transgenic fish, should they escape to natural systems in early life, cannot be predicted with low uncertainty. Under the present conditions, we found no evidence to support enhanced mating capabilities of GH transgenic coho salmon compared to wild-type salmon. However, it is clear that GH transgenic salmon are capable of successful spawning, and can reproduce with wild-type fish from natural systems.

Editor: Paul E. Witten, Ghent University, Belgium

Funding: Funding for this project was through a Canadian Regulatory System for Biotechnology grant to RHD. The funders had no role in study design, data collection and analysis, decision to publish, or preparation of the manuscript.

Competing Interests: The authors have declared that no competing interests exist.

* Email: Robert.devlin@dfo-mpo.gc.ca

Introduction

Increasing growth rates of fish is one of the primary goals for advancement of aquaculture production. Selective breeding can increase growth rates over many generations and its use has been well established in aquaculture. In recent decades there has also been interest in using transgenic technologies to increase production. In particular, insertion of growth hormone (GH) transgenes has been demonstrated to dramatically increase growth rates in a number of fish species [1–7]. Atlantic salmon containing a chinook salmon growth hormone gene fused to an ocean pout antifreeze promoter is currently under consideration by the United States of America's Food and Drug Administration for potential approval for human consumption [8]. If approved, it would become the first commercial transgenic animal used for human consumption. As transgenic technologies are relatively new, and phenotypic effects can be large, concern has been expressed regarding the potential environmental risks transgenic fish may

pose to natural ecosystems. In particular, whether transgenic fish could breed with wild fish, thereby introducing the transgene to wild populations, or establish themselves in natural environments and potentially alter ecosystem food chains, is of concern [9]. The frequency of a transgene in populations will depend on both its rate of introduction and its effects on survival and reproduction (fitness) under different environmental conditions. GH transgenic fish can grow very fast, and in some cases can possess an adult body size greater than wild-type, which has been hypothesized to have the potential to provide a mating advantage [6,10,11]. Some phenotypes caused by GH transgenesis can also be advantageous under specific conditions (e.g. competitive foraging success [12]), whereas others cause negative fitness effects (e.g. reduced disease resistance and predator avoidance [13–15]). Previous modelling has found that a GH transgene conferring large effects on one of these fitness components could result in elimination or expansion of the transgene in populations [16–20]. Further, combinations of positive and negative pleiotropic effects (e.g. a mating advantage coupled with reduced viability) could theoretically cause population extinctions [11,21], although genetic background of wild-type fish may provide counter-selection to restore population-level fitness [20]. Hence, understanding the ability of GH transgenic strains to reproduce is one critical component of estimating overall net effects on their fitness [22].

The reproductive ability of cultured GH transgenic coho and Atlantic salmon has been previously examined. Data thus far demonstrate that cultured GH transgenic salmon have the ability to show appropriate spawning behaviour and successfully spawn, but their reproductive success and level of behaviour is greatly decreased compared to wild-type salmon reared in nature [23–25]. In particular, when in competition with wild-type fish reared in nature, male and female GH transgenic coho salmon only contributed 3.9% to total F_1 offspring generated in mixed-genotype spawning trials [24], and GH Atlantic and coho salmon males participated in 10% and 0% respectively of spawning events with nature-reared wild-type females [23,25]. In paired trials, GH transgenic coho had 14–60% the success of nature-reared wild-type fish depending on the pairing [23], and male GH transgenic Atlantic salmon had 50% the spawning success of wild Atlantic salmon males [25]. In addition, female GH transgenic coho salmon performed fewer diggings and coverings in competition or in pairs [23,24], and male GH transgenic coho and Atlantic salmon had greatly decreased aggressive behaviour and inconsistently decreased courtship behaviour compared to nature-reared wild-type fish [23–25].

The above information taken by itself suggests GH transgenic salmon have greatly decreased spawning abilities. However, where examined, there was also a very large effect of rearing conditions on wild-type salmon grown in the same laboratory culture conditions (as necessary for GH transgenic coho salmon) causing greatly reduced spawning success. In competitive trials with hatchery males reared in nature, cultured wild-type males contributed only 12.6% of offspring [24], and in paired spawning trials only 46–55% of cultured wild-type coho successfully spawned relative to nature-reared fish [23]. Further, cultured wild-type males and females showed decreased aggressive and courtship behaviours in competition with nature-reared fish [24], although they had equal male quivers and greater female digs in paired trials [23]. In addition, wild-type fish grown in equal culture conditions as GH transgenic fish had lower body weight and length compared to those raised in natural conditions from smolt, often had delayed maturation, and neither wild-type nor transgenic fish raised in culture had the mature red colouration or male kype and visible teeth of nature-reared fish [23]. This concurs with other studies that found juvenile [26–28] and lifetime [29–31] rearing in culture decreased spawning success in salmon. As such, whether the poor reproductive success of GH transgenic fish observed in previous studies is due to culture effects or to effects of the transgene (and their interactions) is unknown. Direct comparisons of transgenic and wild-type fish raised in equal conditions are limited and conflicting. Bessey et al. [23] found wild-type and GH transgenic coho males had equally poor success in competition for a single nature-raised female, while Moreau et al. [25] found GH transgenic Atlantic salmon mature male parr had lower spawning success than equally-raised wild-type mature male parr siblings in competition for a nature-reared female. Success of GH transgenic fish seems to be highly dependent on species and/or experimental conditions. For example, GH transgenic catfish and common carp were found to have approximately equal spawning success as wild-type fish [32,33], GH transgenic medaka had increased [11], equal [22], or decreased [34] mating advantage over wild-type medaka depend-

ing on the study, conditions, and/or strain, and GH transgenic zebrafish had lower reproductive success than wild-type zebrafish [35]. As well, Pennington and Kapuscincki [36] found the reproductive success of male GH transgenic medaka relative to wild-type males was influenced by earlier rearing environments (i.e. level of food availability and presence of predators), suggesting genotype×environmental interactions influence the spawning success of GH transgenic fish.

Experiments in GH transgenic medaka found fitness data could accurately predicted the potential for an invading transgene to persist in a population, assuming fitness would not be influenced by the invaded ecosystem [34]. However, whether the relative capabilities for reproductive success between transgenic and wild-type salmon remain the same or differ when animals are reared in different environmental conditions (e.g. are there significant genotype-by-environment interactions affecting this phenotype?) is not known. Intentional release of GH transgenic fish into natural ecosystems to determine their spawning ability when raised in nature is not an acceptable experimental approach given the near impossibility of removing experimental animals to mitigate negative effects should they occur. In GH transgenic coho salmon, rearing in semi-natural contained stream conditions was found to dramatically restore wild-type phenotype and behaviour of juveniles [37]. In the present study, in an attempt to minimize the effects of culture on spawning success, we reared wild-type and GH transgenic coho salmon (*Oncorhynchus kisutch*) in large (350,000 L), semi-natural seawater tanks (hereafter termed mesocosms) from smolt to maturity. The mesocosms were designed to minimize culture effects by more closely mimicking nature than typical rearing in smaller tanks: natural water supply and lighting, low rearing densities, and minimum daily human-to-fish interactions. The purpose of our study was threefold: 1) to determine if and to what extent wild-type spawning success could be restored by mesocosm rearing, 2) to determine how GH transgenic spawning success and behaviour compares to wild-type salmon after seawater rearing in the mesocosm, and 3) to determine if the spawning success of GH transgenic salmon in nature can be extrapolated from this and previous data (i.e. are there genotype-by-environmental interactions in spawning success of GH transgenic and wild-type fish). For this we compared the spawning success and behaviour of mesocosm-reared transgenic and wild-type fish, as well as nature-reared wild-type fish, in three experimental conditions: as spawning male-female pairs (No Competition), with mixed male and female fish together competing for spawning sites and mates (Full Competition), and with two male types competing for wild-type nature-reared female mates (High Male Competition).

Materials and Methods

Experimental Animals

Spawning experiments took place at Fisheries and Oceans Canada's Centre for Aquaculture and Environmental Research (CAER), West Vancouver, BC, Canada (49°20′N, 123°14′W), under requirements established by the Canadian Council for Animal Care. Approval and permits for these experiments were granted by the Pacific Region Animal Care Committee (Permit numbers: 09-009, 10-016, 11-016). All fish were from, or derived from, the sea-ranched Chehalis River hatchery population located in Southwestern British Columbia. Three main groups of fish were examined: GH transgenic fish raised in the mesocosm from smolt (termed T Mesocosm fish), wild-type fish raised in the mesocosm from smolt (NT Mesocosm fish), and wild-type fish raised in nature from smolt (NT Nature fish). NT Nature fish were hatchery-raised as freshwater juveniles at the Chehalis River Enhancement Facility, Agassiz, BC, Canada (49°16′N, 121°15′W, operated by and under authority from Fisheries and Oceans Canada), released to a natural river as smolts, and re-caught as mature salmon returning to the hatchery area to spawn at age 3 yrs. Transfers of fish from the Chehalis River Enhancement Facility to CAER were conducted under permits from Pacific Region Introductions and Transfers Committee (ITC), which is composed of authorities from Fisheries and Oceans Canada and the British Columbia Ministry of the Environment. NT Mesocosm fish were fish raised to smolt in hatchery conditions at CAER (2012) or a mix of those with fish raised in freshwater under hatchery conditions at the Chehalis River Hatchery (2009, 2010, 2013), followed by rearing in the seawater mesocosms until maturation at age 3 yrs. T Mesocosm fish were coho salmon hemizygous for the OnMTGH1 transgene (see [3,38] for details), produced by crosses of wild-type Chehalis coho with strain M77 coho hemizygous or homozygous for the OnMTGH1 transgene. T Mesocosm fish were raised in hatchery conditions at CAER under one of two juvenile feeding level regimes: 1) For the Full Competition experiments, T fish were fed to satiation as juveniles to achieve accelerated growth and maturity at two-years old (these fish acquired smolt status in their first year and were transitioned to seawater in late summer along with their NT counterparts that were one year older), and 2) For the High Male Competition and No Competition experiments, T fish were pair fed a ration restricted to that of the NT fish during the juvenile phase (freshwater period prior to smolt) in order to prevent accelerated growth during this stage. These latter transgenic fish reach maturity at the normal three years of age rather than two years for satiated transgenic salmon. These T and their NT counterparts were transitioned to seawater in late spring before transfer to a mesocosm, closer to the normal smolt time for coho salmon. In one year (2010 spawning year, see Table 1 for relevant arenas), both NT and T Mesocosm fish were temporarily held in 12,800 L or 4000 L seawater tanks for 6 months prior to mesocosm entry (due to facility and operational issues). In Full Competition experiments, a fourth fish group was included: NT Culture fish were 3- to 4-year old fish raised in hatchery conditions at CAER to the smolt stage, then reared in standard culture seawater-fed tanks (4000 L) from smolt until maturity. Growth and survival during seawater mesocosm rearing are to be reported elsewhere.

Mesocosm Rearing

At the smolt stage, all fish to be raised in the mesocosms were implanted with a Passive Integrated Transponder (PIT) tag, fin-clipped for genotype confirmation (as per [24]), weighed and measured, transitioned to seawater over 8–10 days, and then transferred to the mesocosm for seawater rearing until maturity. Mesocosm tanks were semi-natural, circular, 12.2 m in diameter and 3 m deep, for a total volume of 350,000 L each. Mesocosm conditions were designed to minimize culture effects associated with typical rearing in smaller tanks. This was attempted by maintaining abiotic factors closer to natural than under standard culture conditions. Mesocosms were fed with ambient-tempera-ture, sand-filtered seawater from Burrard Inlet, BC. Unidirectional water inflow maintained a constant current within the tank to stimulate continual swimming by the fish. Lighting was natural, filtered through a translucent white tent cover. The mesocosms were fitted with a 1 m high screen, which had a primary purpose of minimizing visual perception of humans by the fish within the mesocosm building. Antibiotics and vaccinations were not administered with the exception of a single dorsal sinus injection of 20 mg/kg oxytetracycline administered to one half the

mesocosm-reared fish from the 2012 spawning experiment 10 months after mesocosm entry (as part of an examination of survival in the mesocosm, to be reported elsewhere). Fish in the mesocosms were reared at very low densities atypical of normal culture conditions, with a maximum density of 0.58–1.79 kg/m^3 (2009: 0.58 kg/m^3; 2010: 0.78 kg/m^3, 2013: 1.72 kg/m^3, 2012: 1.79 kg/m^3), with much lower densities throughout most of the mesocosm culture. In contrast, density of NT Culture fish reared in the 4000 L tanks was 4.3 kg/m^3 at maturity. Fish were hand fed 2 times per day to satiation with commercial salmonid feed (Skretting Canada) using size adjusted feeds appropriate for specific developmental stages. In addition, fish from spawning years 2009, 2010, and 2012 were supplemented with feed from an automatic feeder 5x/day.

Prior to all spawning experiments, fish were seine-netted out of the mesocosms, lightly anaesthetised with tricaine methanesulfo-nate (100 mg/L, buffered with 200 mg sodium bicarbonate/L), weight and length recorded, and near-mature fish sorted into one of the following containers fed with well water at 10°C: 1) an artificial spawning channel, 2) a 9000 L holding tank (Full and High Male Competition experiments), or 3) a 12,500 L holding tank (No Competition experiments). Fish were held in well water for a minimum of one week to acclimate. Mature and near-mature NT Nature fish were collected as needed from the Chehalis River Hatchery in late December through January and held in freshwater at the laboratory until the start of the spawning trials.

Spawning Experiments

Three main experiments were performed with varying levels of competition (see Table 1 and descriptions below for details): I) a No Competition experiment with single-paired male and female fish to determine if NT and T Mesocosm fish were capable of spawning, could display appropriate spawning behaviour, and whether they could spawn with NT Nature fish, II) a Full Competition experiment (both sexes of two types of fish) to determine the spawning success and behaviour of NT and T Mesocosm fish in competition, and III) a High Male Competition experiment (two types of males paired with one type of female) to examine the spawning behaviour and success of male NT and T Mesocosm fish in competition for NT Nature females. Male success was examined in greater depth, as male spawning success has been previously shown to be affected to a greater degree by external and internal factors than for female fish [26,27,39–41].

I. No Competition Experiments. No Competition experi-ments (i.e. single male and female pairs) took place from January to February, 2012 (see Table 1 for details). Experiments were conducted in eight 4.3×0.9 m spawning channels with an approximate water depth of 30 cm (excluding gravel), contained within a tent building with a translucent white cover. One submersible pump per channel created a unidirectional flow that was maintained at approximately 14 cm/sec in the middle of the channel during spawning experiments. Gravel size of the spawning channels was a mix of 2.5 and 3.8 cm average diameter, and was set at an average depth of 13 cm. Video equipment linked to a computer with Milestone XProtect Video Recording Software was mounted above every two channels for continuous behaviour recording. Channels were shielded with blue tarps at the sides and ceiling to minimize glare off the water during video recording. One 60-watt bulb was placed on the ceiling at the front of every two channels for continual dim lighting, in addition to natural diurnal light filtered through the tent and tarps. Water supply to the channels was well water at 10°C.

Experimental fish were removed from the holding tank and assessed for maturity. Only those fish with soft abdomens that

Table 1. Experimental design of spawning experiments.

Experiment	Trial	n[1]
I. No Competition (conducted in 2012)		
	NT Mesocosm ♀×NT Mesocosm ♂	n = 7
	T Mesocosm ♀×T Mesocosm ♂	n = 7
	NT Nature ♀×NT Nature ♂	n = 7
	NT Nature ♀×NT Mesocosm ♂	n = 7
	NT Mesocosm ♀×NT Nature ♂	n = 7
	NT Nature ♀×T Mesocosm ♂	n = 8
	T Mesocosm ♀×NT Nature ♂	n = 8
II. Full Competition (conducted in 2009, 2010, 2013)		
	i. NT Mesocosm + T Mesocosm	n = 4 (1 in 2009[2], 1 in 2010, 2 in 2013)
	ii. NT Mesocosm + NT Nature	n = 4 (2 in 2009, 2 in 2010)
	iii. T Mesocosm + NT Nature	n = 2 (2 in 2013)
	iv. NT Mesocosm + NT Culture	n = 1 (1 in 2009)
III. High Male Competition (conducted in 2012)		
	i. NT Mesocosm ♂ vs. T Mesocosm ♂	n = 3
	ii. NT Mesocosm ♂ vs. NT Nature ♂	n = 3
	iii. T Mesocosm ♂ vs. NT Nature ♂	n = 2

Fish groups are wild-type (NT) or growth hormone transgenic (T) coho salmon reared from smolt in a 350,000 L seawater Mesocosm, in Nature, or in standard Culture (4000 L tank). **I.** No Competition experiments consisted of single male×female pairs in spawning channels for 48 h. **II.** Full Competition experiments consisted of mixed male and female fish of two fish groups, 4 fish per fish group and sex, for a total of 16 fish in each spawning arena. **III.** High Male Competition experiments consisted of 4 male fish each of two different fish groups, competing for 4 NT Nature females, for a total of 12 fish in each spawning arena. For experiments **I.** and **III.** transgenic fish were ration restricted as juveniles (age at maturity = 3 years), and all NT fish were reared as juveniles at CAER. For experiment **II.** transgenic fish were fully fed as juveniles (age at maturity = 2 years), and NT fish were a mix of fish reared as juvenile at CAER, and those reared as juveniles at the Chehalis Enhancement Facility. Age of NT fish at maturity was 3 years (Mesocosm and Nature-reared) or 3–4 years (Culture-reared).
[1]Experimental unit for I = one spawning pair, experimental units for II and III = one arena.
[2]Due to limited fish numbers, this arena consisted of 3 fish per genotype and sex for a total of 12 fish.

expelled eggs (females), or free-flowing milt (males) when lightly pressed were used. Each spawning trial was initiated by placing one male and one female fish in each channel. The pair remained in the channels undisturbed for 48 h, during which time behaviour was continually recorded by the digital video camera system. At the end of 48 h, fish were removed, lightly anaesthetized, and measured for weight and length.

Behaviour of each spawning pair recorded on video was assessed for 5 min, every two hours for 48 h or until one of the pair had died. Video was then screened to identify all spawning events over the entire 48 hr period. Time to first spawn was recorded for each successful pair. For each spawning event behaviour was assessed for 5 min each at 30 min, 20 min, and 10 min prior to spawn, at spawn (0 min), and 10 min and 20 min post-spawn. As well, whether egg and/or milt release were visible at spawn was recorded for each spawning event if visibility allowed. Behaviours recorded were as follows: total time male attended or pursued female per 5 min, total time female maintained position over a nest per 5 min, number of quivers or gapes by male and female, and number of digs and covers by female.

Artificial Spawning Channel for Full and High Male Competition Experiments. Full and High Male Competition experiments (see below) were conducted in an artificial spawning channel. Total spawning channel dimensions were 2.1×30.5 m, with water depth 25–50 cm deep (excluding gravel). Two external, variable speed pumps created unidirectional flow that was maintained at approximately 12 cm/sec during spawning trials. Gravel size of the spawning channel averaged 3.8 cm in diameter,

and was set at an average depth of 15 cm. The spawning channel was located within a tent with a white translucent cover that allowed natural lighting to filter in. The channel was supplied with flow-through fresh well water at 10°C, with the exception of the Full Competition experiments in 2009 where the channel was supplied with ambient temperature creek water. The channel was divided into 4 spawning arenas of 7.3×2.1 m in size (Full Competition experiments), or 8 arenas of 3.7×2.1 m in size (High Male Competition experiments), resulting in approximately 1.9 m² per spawning female. This density is within the range observed for coho salmon in nature [42], although was less than the average redd size for coho (2.8 m², [43]), resulting in medium to high density of spawning females as defined by Fleming and Gross [26]. Removable screens of 2.4 cm² square wire mesh divided the arenas during spawning, which were covered with fine mesh (1.5 mm²) when eggs were estimated to have reached the eyed stage, in order to retain emerging fry within their respective arenas. One edge of the channel was shielded by a dark blue tarp with several slits cut per arena to allow for behavioural observations while minimizing observer effects.

II. Full Competition Experiments. Full Competition experiments consisted of equal numbers of male and female fish from two different fish groups per arena. Four males and four females each of two groups (defined by genotype (T or NT) and rearing condition (Mesocosm, Nature, or Culture) of fish were place in arenas (for a total of 16 fish per arena), resulting in four potential types of matings (1st Group ♀×1st Group ♂, 1st Group ♀×2nd Group ♂, 2nd Group ♀×1st Group ♂, and 2nd Group ♀×2nd

Group ♂). Full Competition experiments were divided into for different trials as outlined in Table 1.

Experimental animals were lightly anaesthetized, and maturity assessed as in No Competition experiments above. In 2013, NT Mesocosm fish had low overall growth rates, and thus in this year larger mature fish from the NT population were chosen. Otherwise fish were chosen at random from available mature fish. Weight and length of mature fish was recorded, a section of fin was removed and placed in 95% ethanol for pedigree analysis, and fish were tagged with a Petersen tag colour and number coded for fish type, sex, and individual. The day prior to the start of the trials the fish were sorted into their appropriate arenas, with a screen separating male and female fish. The trials were initiated by removing the separating screen. Behavioural observations were made for 5 min per arena conducted four times per day starting at 9 am, 11 am, 1 pm and 3 pm. Behaviours recorded were aggressive behaviours (chases, bites), courtship behaviours (males attending females, quivers and gapes by males and females, digs and covers by females, see [23] for details), and spawning occurrences (milt and egg releases). Coho salmon are a semelparous species (i.e. they die shortly after spawning), and fish were allowed to undergo natural spawning mortality in the stream. All deceased fish were removed once daily, identification taken, egg weight recorded and remaining egg number estimated for females, and testes condition recorded for males. Behavioural observations continued in each arena until 1 day after all of one sex of fish had died within the arena

Fecundity (total egg weight, and estimate of total egg number) was taken on any surplus NT Mesocosm, T Mesocosm, and NT Nature female fish to estimate expected egg mass of female fish used in the spawning trial. In the 2013 year, these data were also used to estimate the expected number of offspring produced per female, as well as an estimation of % offspring survival. This calculation was not included for 2009 and 2010 year classes due to incomplete data on spawning females used in these trials.

When fry first emerged, water velocity was decreased to approximately 4.7 cm/sec. Fry were fed 2–4 times a day commercial crumb or mash diet (Skretting Canada). Approximately 1 month after emergence, offspring were removed from arenas and euthanized by an overdose of anaesthetic (200 mg/L tricaine methanesulfonate, 400 mg/L sodium bicarbonate). Total offspring numbers per arena were assessed. In 2009 and 2010, a random subset of euthanized fry were bled (caudal sever) into microtitre plate wells containing 100 μL of 0.01 N NaOH, and the remaining fish stored in 95% ethanol. In 2013, all euthanized fry were placed in ethanol, then a random subset removed and tail fins placed in 0.01 N NaOH. Tissues in NaOH were heated to 99°C for 5 min to liberate DNA and denature nucleases in preparation for pedigree analysis.

III. High Male Competition Experiments for NT Nature Females. High Male Competition experiments (i.e. equal numbers of two types of male fish competing for limited number of NT Nature female fish) took place from January - February, 2012. Experimental animals were tagged as above in Full Competition experiments. To initiate the trials, four NT Nature females were place in arenas, followed by four males each of two different groups (4x ♂ 1st Fish Group+4x ♂ 2nd Fish Group+4x NT Wild ♀ = 12 fish total with a male:female ration of 2:1). High Male Competition experiments were divided into three different trials as outlined in Table 1.

Behavioural observations and final fish processing were as described above for the Full Competition experiments. Offspring were sampled for pedigree analysis as in 2009 and 2010 Full Competition experiments above.

Pedigree Analysis

Parentage of a random subset of offspring from each arena in Full Competition and High Male Competition experiments were determined by microsatellite analysis. Parent tissue DNA was extracted into distilled water using DNeasy Kits (Qiagen Inc., Germantown, MD), or by placing fin clip in 0.01 N NaOH and heating to 99°C for 5 min. Parents were screened to identify informative loci and unique microsatellite alleles. All primers and reagents were obtained from Life Technologies (Austin, TX)/ Applied Biosystems (Foster City, CA). Three to five of the following microsatellites primers were used per arena: one111 [44]; ots101 [45]; omm1008 [46]; omm1128, omm1135 [47]; omm1231, omm1270 [48]; omm1322 [49]; omm1399 [50]; omm5007, omm5008, omm5030, omm5090, omm5092 [51]; omm5132 [52]; and ssa407 [53]. Forward primers were tagged with one of four fluorescent probes (6FAM, VIC, NED, PET), and reverse primers contained a 7 bp tail (GTGTCTT). Amplification of microsatellites was conducted via PCR reactions in 96-well plates using a GeneAmp PCR system 2720 thermal cycler (Applied Biosystems). PCR reactions contained either 0.3–0.5 μL extracted tissue in 0.01 N NaOH or 1 μL extracted DNA in water, and 10 μL reaction mix containing 1x PCR buffer, 0.2 mM dNTPs, 2 mM MgCl$_2$, 1.2–1.5 μM each forward and reverse primers, and 0.05 U Taq. PCR reactions were as follows: 1 cycle of 95°C for 10 min (denaturing); 30 cycles of 94°C for 30 sec (denaturing), 48–62°C for 30 sec (annealing), 72°C for 1 min (extending); 1 cycle of 72°C for 7 min (extending). Annealing temperatures for primers were as follows: 48°C (omm5090), 54°C (ots101), 56°C (one111), 60°C (omm5132), 61°C (ssa407), 62°C (omm5008), and 58°C (all remaining primers). After amplification, 0.75 μL of PCR reactions were combined with 10 μL HiDi Formamide containing 0.35–0.5 μL GS-LIZ500 size standard in a 96-well plate. Samples were heated to 99°C for 3 min then immediately cooled on ice. The PCR products were then detected and sized using a 3130x Genetic Analyzer (Applied Biosystems). Parents of offspring were determined using WhichParent software (available at http://bml.ucdavis.edu/research/research-programs/ conservation/salmon-research/salmon-genetics-software/).

Statistical Analyses

Statistical comparisons among fish groups for most spawning behaviour and success variables were through 1-way ANOVA, followed by Student-Neuman-Keuls (SNK) post-hoc test, where spawning pairs (No Competition experiments) or arenas (Full and Male Competition experiments) were considered experimental units. For proportional data (e.g. proportion of a fish group that successfully spawned), data were arcsine transformed prior to analysis. If Normality or Equal Variance tests failed, data were ln, reciprocal, or square root transformed and reanalyzed. If transformation failed to bring about Normality and Equal Variance, data were analyzed by Kruskal-Wallis ANOVA on Ranks, followed by Dunn's post-hoc test.

For variables where more than one factor was of interest, 2-way ANOVA's were performed, with transformations as above where appropriate. Analyses by 2-way ANOVA included variables in the Full Competition experiments where effects of both sex and fish group were of interest (e.g. % of offspring that fish contributed to, aggressive behaviour, etc.), and behaviour measurements during spawning events in No Competition experiments where fish group and time were factors. Weight and length were analyzed with fish group and year as factors. In addition, whether there were significant Arena×Fish Group interactions on raw data in Full and Male Competition experiments was examined (i.e. did the relative behaviour and success of fish groups differ among arenas). In the results, Arena effect is only addressed if there were significant Arena×Fish Group interactions.

In No Competition experiments, proportion analysis was done by Chi-squared analysis, or where values were small by Fisher-Exact test. In Full Competition experiments, whether there was significant assortative mating for each arena was analyzed by 2×2 contingency analysis with Yates correction [28]. In Full and High Male Competition experiments, whether there was significant influence of fish length on spawning success (number of offspring or number of mates per fish) was determined by linear regression.

In T Mesocosm + NT Nature Full and High Male Competition trials, n = 2 arenas resulted in poor statistical power for between-arena comparisons. In these trials, there were no significant Arena×Fish Group interactions (p-values for behaviour and spawning success ranged from 0.101 to 0.882). Therefore, data from the two arenas were combined and analyzed with individual fish as experimental units. A statement of difference for each comparison is made in the text only if p<0.05. Statistical analyses were performed using SigmaStat (San Jose, CA). Data are presented as mean ± standard error of the mean.

Results

Fish Size and Morphology

Fish weight, length, and condition factor (CF) used in the spawning trials are given in Figure 1A–C. In all years transgenic (T) Mesocosm fish were greater in weight and length than wild-type (NT) Mesocosm fish (p<0.001). NT Mesocosm fish were similar in weight to NT Nature fish in 2009 and 2010, but smaller in weight in 2012 and 2013 (p<0.001). This was primarily due to increased weight of NT Nature fish in later years (p<0.001), as the weight of NT Mesocosm fish was not significantly different between years (p = 0.406). In all years NT Mesocosm fish were shorter in length than NT Nature fish (p<0.001). T Mesocosm fish were significantly larger in weight than NT Nature fish in 2009, 2010, and 2012 (p<0.001) but similar in weight in 2013. T Mesocosm fish were larger in length in 2009 and 2010, and smaller in length than NT Nature fish in 2012 and 2013 (p< 0.001). The changing relative sizes of T Mesocosm and NT Nature fish were due to both an increase in NT Nature fish size and a decrease in T Mesocosm size in later years. In general, CF was ranked in order of T Mesocosm > NT Mesocosm > NT Nature, although differences were not significant in 2013 between NT Mesocosm and NT Nature fish. In 2009, NT Culture fish were considerably smaller than all other groups in both weight and length (p<0.001), but only differed in CF from NT Nature fish.

Spawning morphology of representative fish used in the experiments are given in Figure 2. NT Nature fish displayed typical spawning morphology of coho salmon: red colouration of sides in males and females, and elongated and hooked jaw and humped back in males. Mesocosm-raised fish developed some of the red colouration observed in NT Nature fish, although colouration tended to be dark brown rather than red in most fish. Male NT Mesocosm and T Mesocosm fish developed the hooked jaw (kype) of NT Nature males, although without the elongated jaw observed in nature-reared fish. Minor hump development associated with sexual maturation was observed in only some NT Mesocosm males, and not observed in T Mesocosm males. Mesocosm-raised fish tended to have less developed or more eroded tails than NT Nature fish, particularly T Mesocosm fish. While NT Mesocosm fish approached the fusiform shape of NT Nature fish, T Mesocosm fish tended to be deeper bodied than both groups of fish. In 2010, the deeper body shape of T Mesocosm fish was extremely exacerbated in a few fish (Figure 2,

extreme). In addition, excessive cranial growth was observed in some T Mesocosm fish.

I. No Competition Experiments

To determine the influence of GH transgenesis and mesocosm culture on the ability of salmon to spawn with a member of their same group and with NT Nature fish, and to display appropriate spawning behaviour in the absence of competition, we examined spawning success and behaviour of T Mesocosm, NT Mesocosm, and NT Nature fish in No Competition (paired) experiments over 48 h.

Ia. Spawning Success. There were no significant differences in the proportion of pairs that spawned in any of the crosses examined (p = 0.419, Table 2), although these results should be interpreted with caution as the power of the test for this comparison was low (0.396). When the percent of fish that spawned was summed over male partner, there were no significant differences in spawning success of females (p = 0.830, power = 0.077, Figure 3). When the percent of fish that spawned was summed over female partner, NT Nature males had 2.2-fold greater spawning success than T Mesocosm males and 1.7-fold greater spawning success than NT Mesocosm males, although these differences were not significant (p = 0.119, Figure 3). However, these results also should be interpreted with caution due to low power of the test for this comparison (0.425). The time to first spawn did not differ between different types of spawning pairs (p = 0.764, see Table 2), female type summed over male (p = 0.286), or male type summed over female (p = 0.684). The percent of spawning events where egg release was visible did not differ among spawning pairs (p = 0.533, Table 2) or among female type summed over male partner (p = 0.757, average 46.7% with visible eggs). However, the percent of spawning events where milt release was visible did differ among spawning pairs (p = 0.046, Table 2), where NT Nature female×NT Nature male and NT Mesocosm female×NT Nature male pairs had greater percent of spawning events with visible milt than NT Nature female×NT Mesocosm male. As well, when male type summed over female partner was examined, NT Nature males had approximately 2-fold more events with visible milt release than NT Mesocosm males (88% and 44.4% respectively, p = 0.017). T Mesocosm males had similar percent of events with visible milt release as NT Mesocosm males (50%), but did not differ significantly from NT Nature males (p = 0.127).

Ib. Behaviour at Perispawn. During spawning events, there were no differences between fish types in any observed behaviour when summed over 30 min (Table 3). When individual times surrounding the spawning events were examined, T Mesocosm females performed approximately 30% less digs than NT Nature or NT Mesocosm fish overall prior to the spawning event when summed over male fish (p<0.001, Figure 4A), although did not differ at individual time points. As well, immediately after spawn all three groups differed in number of covers by females where NT Mesocosm > NT Nature > T Mesocosm (p = 0.001). There were no significant differences between fish groups in the number of quivers performed by males around the spawning event (p = 0.829, Figure 4B). However, both NT Nature and NT Mesocosm males had typical pattern of quivers reported for other salmonids (e.g. [54]), where number of quivers increased with time up to the spawning event, and then greatly decreased after the spawning event. This pattern of quivers over time was not observed in T Mesocosm males, who decreased quivers over time on average up to the spawning event, and had a high number of quivers 10 min post-spawn (see Figure 4B).

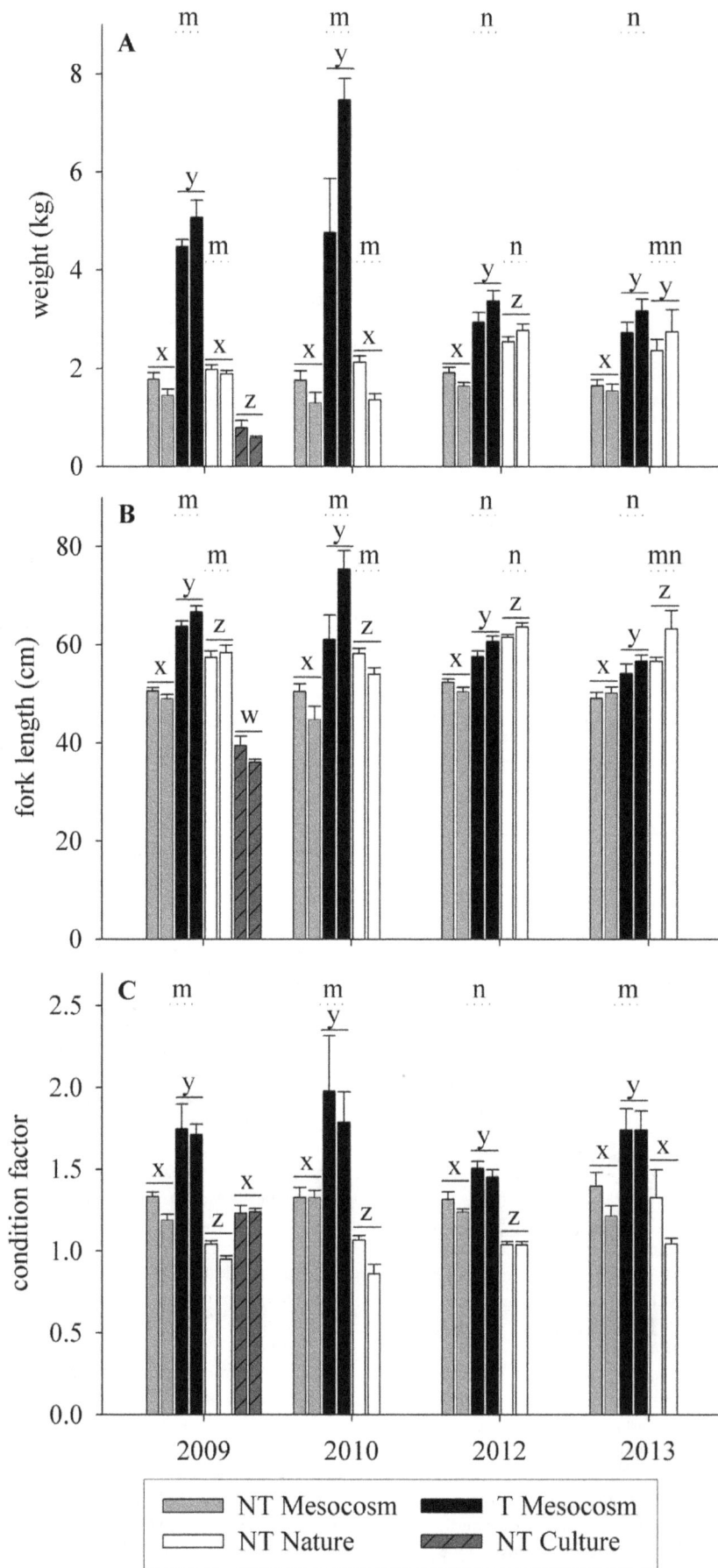

Figure 1. Size of fish used in spawning trials. A) Weight (kg), **B)** fork length (cm), and **C)** condition factor. Fish groups are wild-type fish raised in the mesocosm (NT Mesocosm), in natural conditions (NT Nature), or standard culture conditions (NT Culture) from smolt, and GH transgenic fish raised in the mesocosm from smolt (T Mesocosm). Data are given as female then male for each fish type. Fish in years 2009, 2010, and 2013 were used in the Full Competition experiments, measurements taken before entry in the spawning channel, and transgenic fish were fully fed as juveniles. Fish in year 2012 were used in the High Male Competition and No Competition experiments, measurements taken at time of mortality or post-trial respectively, and transgenic fish were fed a wild-type ration as juveniles. w,x,y,z over a solid line indicates differences among fish groups within years summed over sex, and m,n over a dotted line indicates differences within fish groups among years, summed over sex.

Ic. Courtship Behaviour over 48 h. Different crosses did not differ in the overall time females maintained position over their nests (p = 0.641), the number of locations that females maintained nests (p = 0.457), or number of male quivers performed (p = 0.129, Table 4) per 5 min interval averaged every 2 h for 48 h. When fish were paired with a member of their same group, NT Nature females performed 4.6-fold and 3.4-fold more digs and covers than NT Mesocosm or T Mesocosm females respectively (p = 0.013), and NT Nature males spent 2.4-fold more time attending females than NT Mesocosm or T Mesocosm males (p = 0.018, see Table 4).

II. Full Competition Experiments

To determine whether GH transgenesis and/or mesocosm rearing influences the spawning success of salmon during competition, we compared the spawning success of T Mesocosm, NT Mesocosm, and NT Nature fish in mixed male and female competitions (Full Competition).

IIa. Spawning Success. *Effect of Transgene (NT Mesocosm + T Mesocosm):* Refer to section i in Figure panels for relevant data. All groups of fish showed the ability to mate, however the pattern of percent offspring from matings in NT Mesocosm + T Mesocosm spawning trials differed greatly between years and arenas (Figure 5 Insert). In two arenas, NT Mesocosm female×NT Mesocosm male matings produced the most offspring,

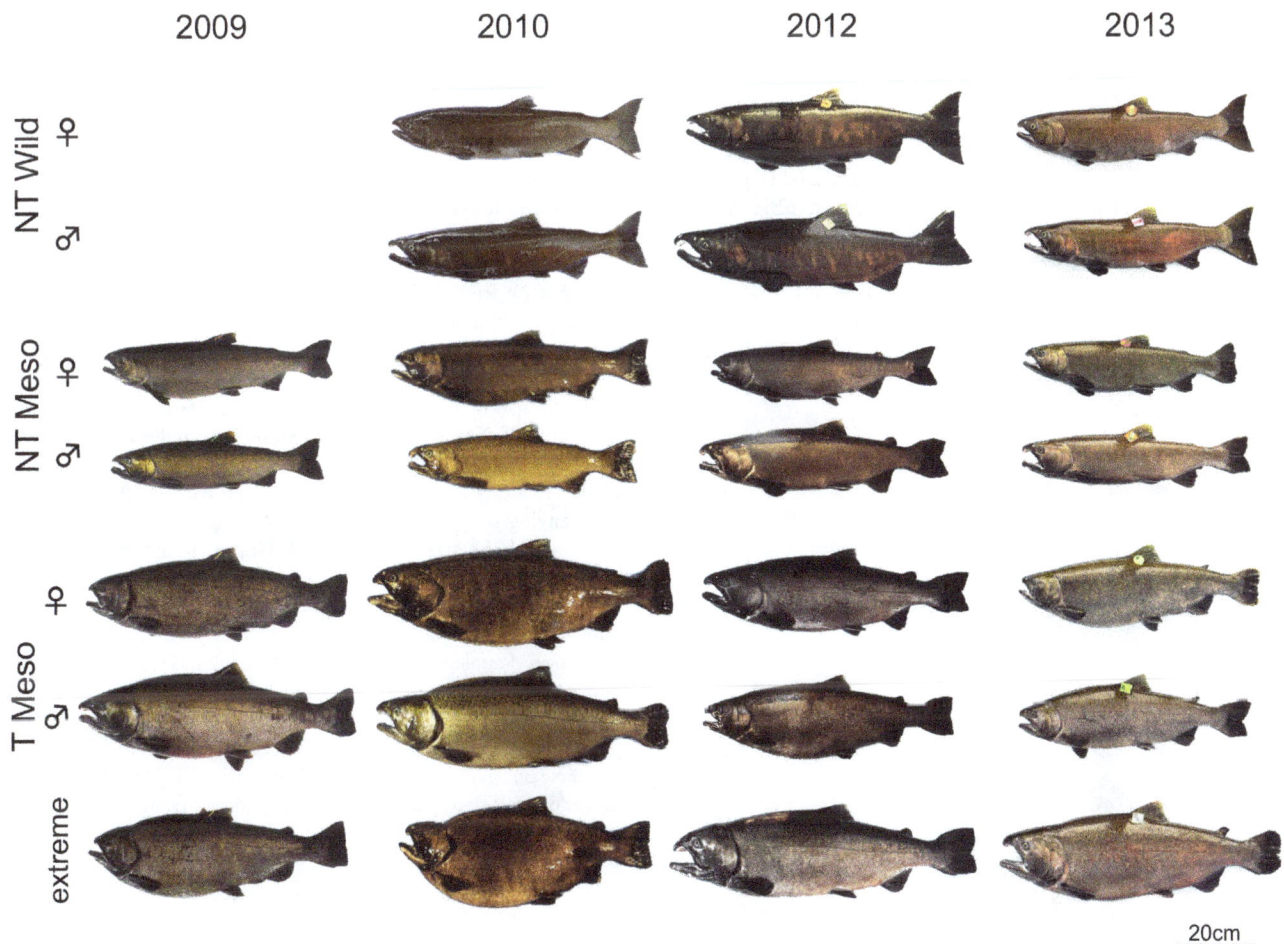

Figure 2. Representative morphology of fish used in spawning trials. Fish groups are wild-type fish raised in natural conditions from smolt (NT Nature), or in the mesocosm from smolt (NT Meso), and GH transgenic fish raised in the mesocosm from smolt (T Meso). Representative fish expressing the most extreme morphology of T Meso fish are also given in each year. In years 2009, 2010, and 2013, transgenic fish were fully fed as juveniles, and in 2012 were fed a ration restricted to that of NT fish as juveniles. Morphology of NT Nature fish for 2009 spawning year was not available.

Table 2. Number of crosses used, spawning success (% of pairs that spawned), and % of spawning events where egg or milt release is visible in No Competition experiments of NT Mesocosm, T Mesocosm, and NT Nature coho salmon.

Cross	% of pairs that spawned[1]	Time (h) to first spawn	% with visible egg release	% with visible milt release
NT Mesocosm ♀×NT Mesocosm ♂	42.9 (41.7)	23.3±7.5	33.3	66.7[ab]
T Mesocosm ♀×T Mesocosm ♂	28.6 (20)	26.6±17.3	50	50[ab]
NT Nature ♀×NT Nature ♂	71.4 (90.9)	11.8±4.0	31.3	87.5[a]
NT Nature ♀×NT Mesocosm ♂	28.6 (50)	13.4±5.9	66.7	0[b]
NT Mesocosm ♀×NT Nature ♂	57.1 (n/a)	21.2±6.4	50	100[a]
NT Nature ♀×T Mesocosm ♂	25.0 (55.6)	17.9±7.5	100	50[ab]
T Mesocosm ♀×NT Nature ♂	50 (12.5)	18.0±5.8	60	80[ab]

NT = wild-type coho salmon, T = growth hormone transgenic coho salmon, Mesocosm = fish were raised from smolt in a mesocosm, Nature = fish were raised from smolt in natural conditions.
[1]number in parentheses is equivalent % of pairs that spawned in cultured fish from [23].
[a,b]indicates significant differences between pairs, p<0.05.

while in the other two arenas, either T Mesocosm female×NT Mesocosm male or NT Mesocosm female×T Mesocosm male matings produced the most offspring. In all years T Mesocosm female×T Mesocosm male matings produced the fewest offspring, but not by a significant margin. When data from all arenas were averaged, there were no significant differences in percent offspring produced by the different matings (p = 0.218, Figure 5i). In addition, there were no differences between T Mesocosm and NT Mesocosm fish in the % of individuals that spawned (p = 0.382, Figure 6Ai) or the percent of total available partners that individual fish spawned with (p = 0.801, Figure 6Bi). Neither offspring number nor percent of total available partners individual fish spawned with significantly correlated with fish length in any group (p = 0.400 to 0.947). Contingency tables to analyze for assortative mating were not significant in any arena for this or any other Full Competition trial (ii, iii, and iv below, p = 0.492 to

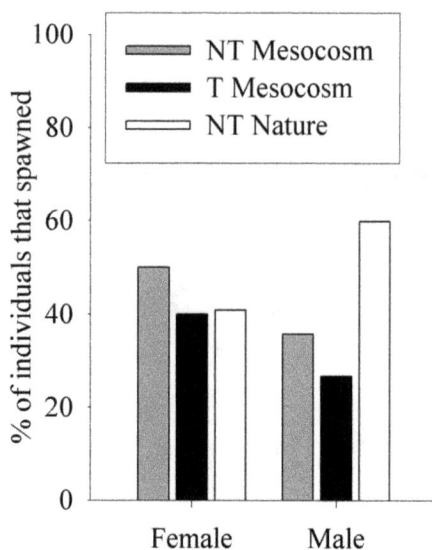

Figure 3. Spawning success (% that spawned) of fish groups in No Competition experiments. Fish groups are wild-type (NT) and growth hormone transgenic (T) coho salmon raised from smolt in a Mesocosm or in natural (Nature) conditions, summed over male or female fish.

0.876). The percent expected egg mass remaining in NT Mesocosm and T Mesocosm females post spawning mortality did not differ (p = 0.383, Figure 6Ci). Of the fifteen NT Mesocosm females used in total, all but one contributed offspring to the spawning trials, and this one female had an egg mass at mortality that was consistent with that of an unspawned NT Mesocosm female (predicted from fecundity-body size relationships, data not shown). Of the fifteen T Mesocosm females used, four did not contribute offspring to the spawning trials. Of these, three females had smaller egg mass than expected for an unspawned T Mesocosm female (0–67% of expected mass, average 24.2±21.6%).

Effect of Mesocosm Relative to Nature Rearing (NT Mesocosm + NT Nature): Refer to section ii in Figure panels for relevant data. In three of four arenas, offspring from NT Nature female×NT Nature male matings far surpassed all other matings in number (82–90% total offspring), while in the remaining arena there were similar offspring numbers from NT Nature female×NT Nature male and NT Nature female×NT Mesocosm males matings (40% and 37% total offspring respectively). It should be noted that in this latter arena all NT Nature males died early in the spawning trial, leaving only NT Mesocosm males available to mate in the later half of the trial. Overall, NT Nature female×NT Nature male matings accounted for the majority of offspring (74.8%, p< 0.001, Figure 5ii), with other matings accounting for approximately 11% (NT Nature female×NT Mesocosm male, and NT Mesocosm female×NT Nature male) or 2.3% (NT Mesocosm female×NT Mesocosm male) of offspring. In addition, twice as many NT Nature fish spawned than NT Mesocosm fish (p = 0.003, see Figure 6Aii), and NT Nature fish had 35% more partners on average than NT Mesocosm fish (p = 0.009, see Figure 6Bii). There was a weak but significant positive correlation between fish length and number of offspring produced by NT Nature female fish (p = 0.010, $R^2 = 0.39$), as well as between fish length and number of mates in NT Nature male and female fish (p = 0.024 $R^2 = 0.31$, and p = 0.043 $R^2 = 0.26$ respectively), but not for NT Mesocosm fish. NT Mesocosm females had 8.6-fold greater percent of expected egg mass remaining post spawning mortality (p = 0.015), and all but one NT Nature female had no or very few eggs remaining post spawning mortality (Figure 6Cii). Of the sixteen NT Mesocosm females used in total, nine did not contribute offspring to the spawning trial. Of these, six NT Mesocosm females had smaller egg mass at mortality than

Table 3. Spawning behaviour during No Competition experiments summed over 30 min measured in 5 min intervals at −30, −20, −10, 0, +10, +20 min from a spawning event.

Cross	Time that ♀ Maintains nest (min)	Time that ♂ attends ♀ (min)	Number of digs by ♀	Number of quivers by ♂
NT Nature ♀×NT Nature ♂	25.4±1.7	26.7±0.8	25.2±1.6	4.6±0.9
NT Mesocosm ♀×NT Mesocosm ♂	24.9±1.5	24.4±2.2	23.7±1.2	8.8±4.3
T Mesocosm ♀×T Mesocosm ♂	25.7±3.1	25.9±0.9	18.0±1.0	7.0±3.0
NT Nature ♀×NT Mesocosm ♂	24.3±0.8	28.5±1.5	21.5±5.5	1.5±0.5
NT Mesocosm ♀×NT Nature ♂	29.3±0.7	25.6±0.7	27.1±1.7	5.1±1.3
NT Nature ♀×T Mesocosm ♂	24.0±0.7	27.7±1.6	16.5±0.5	4.5±2.5
T Mesocosm ♀×NT Nature ♂	26.8±1.8	23.8±2.9	14.9±4.4	5.4±1.3

NT = wild-type coho salmon, T = growth hormone transgenic coho salmon, Mesocosm = fish were raised from smolt in a mesocosm, Nature = fish were raised from smolt in natural conditions.

expected for unspawned NT Mesocosm females (2–79% of expected mass, average 51.1±10.7%). Of the sixteen NT Nature females used, all but one female contributed offspring to the spawning trial, and this one female had an egg mass at mortality consistent with what was expected for an unspawned NT Nature female fish.

Effect of Transgene and Mesocosm Rearing (T Mesocosm + NT Nature): Refer to section iii in Figure panels for relevant data. In the two arenas examined, offspring from NT Nature female×NT Nature male matings accounted for the majority of offspring (p = 0.038, 72.5%), while other matings accounted for 19.1% (T Mesocosm female×NT Nature male), 14.2% (NT Nature female×T Mesocosm male), or 1.3% (T Mesocosm female×T Mesocosm male) of total offspring (Figure 5iii). In addition, 3 times as many NT Nature fish spawned than T Mesocosm fish (p = 0.007, Figure 6Aiii), and NT Nature fish had 5 times as many partners on average than T Mesocosm fish (p = 0.001, Figure 6-

Biii). There was a significant positive correlation between fish length and number of offspring for T Mesocosm female fish (p = 0.015, $R^2 = 0.66$), but no other significant correlations between length and spawning success were noted. T Mesocosm females had more that 100-fold greater percent of expected egg mass remaining post spawning mortality than NT Nature females (p = 0.003), and all NT Nature females had no or very few eggs remaining post spawning mortality (Figure 6Ciii). Of the eight T Mesocosm female fish used, five did not contribute offspring to the spawning trial, and all of these five fish had lower egg mass at mortality than expected for an unspawned T Mesocosm female (28–72% of expected, average 55.1±8.2%). Of the eight NT Nature females used, all contributed offspring to the spawning trial.

Effect of Mesocosm Rearing Relative to Standard Culture (NT Mesocosm + NT Culture): In the one arena examined, matings with NT Mesocosm paternal parents had 1.9-fold greater offspring

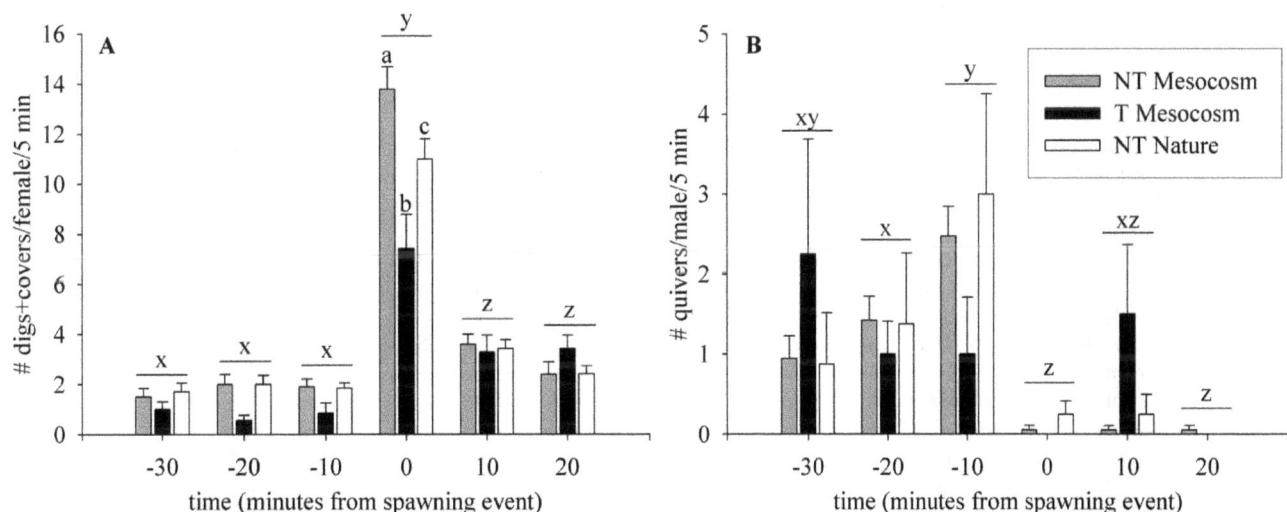

Figure 4. Behaviour of fish groups during spawning events in No Competition experiments. Fish groups are wild-type (NT) and growth hormone transgenic (T) coho salmon raised from smolt in a Mesocosm or in natural (Nature) conditions. **A)** Number of digs by females during 5 min intervals and **B)** number of quivers by males during 5 min intervals measured −30, −20, −10, 0, +10, and +20 min from spawning events. Data are given as means over females or males ± standard error of the mean. Significant differences between groups within time period are indicated by (a,b,c), and significant differences between time points summed over groups are indicated by (x,y,z), p<0.05.

Table 4. Courtship and spawning behaviour during No Competition experiments over 5 min, average every 2 h over 48 h or until one fish died.

Cross	Time (sec) that ♀ maintains nest	Time (sec) that ♂ attends ♀	Number of digs by ♀	Number of quivers by ♂	Number of nests ♀ maintains over 48 h
NT Nature ♀×NT Nature ♂	105.8±32.8	181.2±26.6[a]	1.39±0.37[a]	0.06±0.06	2.0±0.5
NT Mesocosm ♀×NT Mesocosm ♂	56.1±20.0	79.0±21.1[b]	0.31±0.10[b]	0.07±0.03	1.3±0.2
T Mesocosm ♀×T Mesocosm ♂	76.5±20.7	72.4±30.2[b]	0.41±0.13[b]	0.01±0.01	1.4±0.3
NT Nature ♀×NT Mesocosm ♂	81.9±24.4	131.9±37.6	0.67±0.20	0.07±0.04	1.3±0.2
NT Mesocosm ♀×NT Nature ♂	82.4±26.6	131.2±35.8	0.83±0.26	0	1.1±0.3
NT Nature ♀×T Mesocosm ♂	68.1±5.0	108.6±23.6	0.31±0.05	0	1.8±0.2
T Mesocosm ♀×NT Nature ♂	48.6±22.0	121.0±26.7	0.35±0.23	0.06±0.03	1.1±0.4

NT = wild-type coho salmon, T = growth hormone transgenic coho salmon, Mesocosm = fish were raised from smolt in a mesocosm, Nature = fish were raised from smolt in natural conditions.
[a,b]indicates significant differences between pairs when fish are paired with a member of their same group, p<0.05.

numbers than those with NT Culture paternal parents (68.3% and 31.7% total offspring respectively, p<0.001), while matings with NT Culture maternal parents had 1.2x greater offspring numbers than those with NT Mesocosm maternal parents (54.3% and 45.7% total offspring respectively, p = 0.029). There was no difference between the two groups of fish in % of individuals that spawned (75% and 100% for NT Mesocosm and NT Culture respectively, p = 1.00) or % of total available partners individual fish spawned with (31.3% and 37.5% for NT Mesocosm and NT Culture respectively, p = 0.623). NT Mesocosm and NT Culture females had similar percent egg mass remaining at spawning mortality (29.2±19.6% and 29.0±18.0% respectively, p = 0.769). Of the four NT Mesocosm and four NT Culture females used, only one NT Mesocosm female did not contribute offspring to the trial. This one fish had lower egg mass at mortality than expected for an unspawned NT Mesocosm female (13.0% of expected).

IIb. Behaviour. *Effect of Transgene (NT Mesocosm + T Mesocosm):* NT Mesocosm and T Mesocosm fish did not differ in the number of aggressive actions given (p = 0.762) or received (p = 0.132), or in the number of attending behaviours given by males (p = 0.311) or received by females (p = 0.209, see Figures 7Ai–Di). However, the average time at which females received attending behaviour within the spawning trial differed, with T Mesocosm females receiving attention before NT Mesocosm females (48% and 62% through the spawning trial respectively, p = 0.022). NT Mesocosm and T Mesocosm fish did not differ significantly in number of quivers given (0.38±0.28 and 0.03±0.02 quivers/fish/5 min interval respectively, p = 0.343) or received (0.32±0.29 and 0.05±0.04 quivers received/fish/5 min interval respectively, p = 0.402), or in the number of digs and covers observed (p = 0.686, Figure 7Ei). There was a significant interaction between fish group and arena on aggressive actions given (p = 0.030), but no individual differences.

Effect of Mesocosm Rearing Relative to Nature (NT Mesocosm + NT Nature): NT Mesocosm fish performed 13.1% of the number of aggressive behaviours performed by NT Nature fish (p = 0.015, Figure 7Aii), and received 1.7-fold more aggressive action than NT Nature fish (p = 0.041, Figure 7Bii). NT Mesocosm fish performed 36.7% of attending behaviour (p = 0.032, Figure 7Cii), and received 25.9% of the attending behaviour of NT Nature fish (p = 0.032, Figure 7Dii). NT Mesocosm fish performed 11.5% of the digs and covers that NT Nature fish performed (p = 0.029, Figure 7Eii), although the two groups of fish did not differ in

number of quivers/gapes given (0.22±0.15 and 0.16±0.10 quivers+gapes/fish/5 min interval respectively, p = 0.751) or received (0.31±0.18 and 0.08±0.05 quivers+gapes received/ fish/5 min interval respectively, p = 0.261).

Effect of Transgene and Mesocosm Rearing (T Mesocosm + NT Nature): T Mesocosm fish performed only 31.6% of the aggressive behaviours of NT Nature fish (p<0.001, Figure 7Aiii), and received 2.3-fold more aggressive action than NT Nature fish (p = 0.023, Figure 7Biii). T Mesocosm males performed 32.7% of the attending behaviours of NT Nature males (p = 0.005, Figure 7Ciii). There was no difference between the two groups in the number of attending behaviours received by females (p = 0.333, Figure 7Diii), although NT Nature females received attending behaviours on average earlier than T Mesocosm females (at 37% and 69% of the way through the spawning trial respectively, p<0.001). NT Nature and T Mesocosm fish did not differ in number of quivers given (0.11±0.11 and 0.11±0.05 quivers/fish/5 min interval respectively, p = 0.663) or received (0.06±0.03 and 0.16±0.13 quivers received/fish/5 min interval respectively, p = 0.618), nor in numbers of digs and covers performed by females (p = 0.295, Figure 7Eiii). It should be noted that in one of the two T Mesocosm×NT Nature arenas, all NT Nature females had died by 72% of the way through the trial, leaving only T Mesocosm females to spawn and interact with. However, the only difference in spawning success and behaviour between the two arenas was that in the arena where all NT Nature females died early, no T Mesocosm females were observed digging, while in the other arena T Mesocosm fish had 2.4-fold the observed digging as NT Nature fish (0.295 and 0.125 digs/fish/ 5 min interval respectively), although they performed the digs later in the trial than NT Nature females (at 49% and 22% of the way through the spawning trial respectively).

Effect of Mesocosm Rearing Relative to Standard Culture (NT Mesocosm + NT Culture): In the one arena examined, there were no differences between fish groups in number of aggressive actions given (p = 0.861) or received (p = 0.856), although male fish received more aggressive actions than female fish (p = 0.021, see Table 5 for all behaviour). NT Mesocosm males performed 2.2-fold more attending behaviours than NT Culture fish (p = 0.006). There was no difference between fish groups in the number of attending behaviours received (p = 0.729), although NT Mesocosm females received attention on average earlier than NT Culture females (p<0.001). NT Culture females performed more digs and

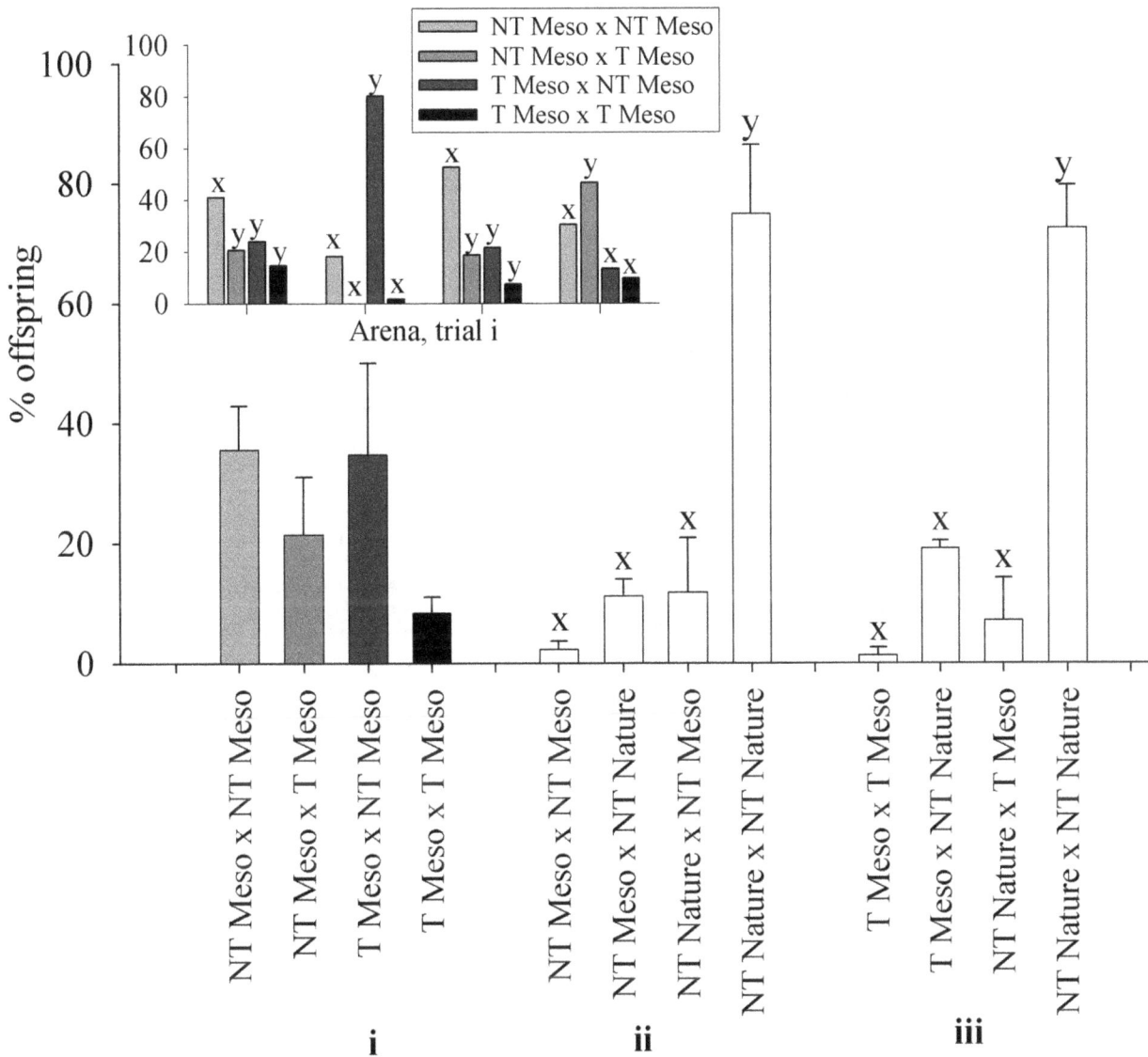

Figure 5. Percent of offspring produced by different matings during Full Competition experiments. Fish groups are wild-type (NT) and growth hormone transgenic (T) coho salmon raised in a mesocosm (Meso) or natural (Nature) conditions from smolt. Each spawning arena contained two groups of fish, four fish of each sex/group, competing for spawning sites and mates. Matings (bars) are given as Female×Male parent. Trials were: **i)** *Effect of transgene:* NT Mesocosm + T Mesocosm (n = 4 arenas), **ii)** *Effect of mesocosm rearing:* NT Mesocosm + NT Nature (n = 4 arenas), and **iii)** *Effect of transgene and mesocosm rearing:* T Mesocosm + NT Nature (n = 2 arenas). For trial i, the **Insert** provides % offspring for individual arenas. Significant differences among matings within trials or arenas are indicated by letters (x,y), p<0.05.

covers than NT Mesocosm females (p = 0.003). There was no difference between fish groups in number of quivers given (p = 0.650), or received (p = 0.057), although NT Mesocosm females received quivers on average before NT Culture females (p<0.001).

IIc. Offspring Survival. Offspring survival (total offspring recorded as percent of expected offspring calculated from spawned egg mass) was estimated for the 2013-year class only, due to incomplete fecundity data for spawning females in other years. In this year, offspring from NT Mesocosm females had 5.1-fold greater survival than offspring from T Mesocosm females (50.4±12.0% and 9.9±2.9% respectively, p = 0.002). Offspring survival from NT Nature females 26.7±5.2%) did not differ significantly from either Mesocosm group in this year.

III. High Male Competition Experiments for NT Nature Females

To specifically examine whether GH transgenesis and/or mesocosm rearing influences the spawning success of male salmon during competition, we compared the spawning success of T Mesocosm, NT Mesocosm, and NT Nature males (two types per trial) in competition for NT Nature females.

IIIa. Spawning Success. *Effect of Transgene (NT Mesocosm ♂ vs T Mesocosm ♂):* NT Mesocosm males had 11-fold more offspring than T Mesocosm males in competition for NT Nature females (p<0.001, Figure 8Ai). As well, NT Mesocosm males had 4 times more partners on average than T Mesocosm males (p = 0.015, Figure 8Bi), but the two groups did not differ significantly in the % of males that spawned (p = 0.055,

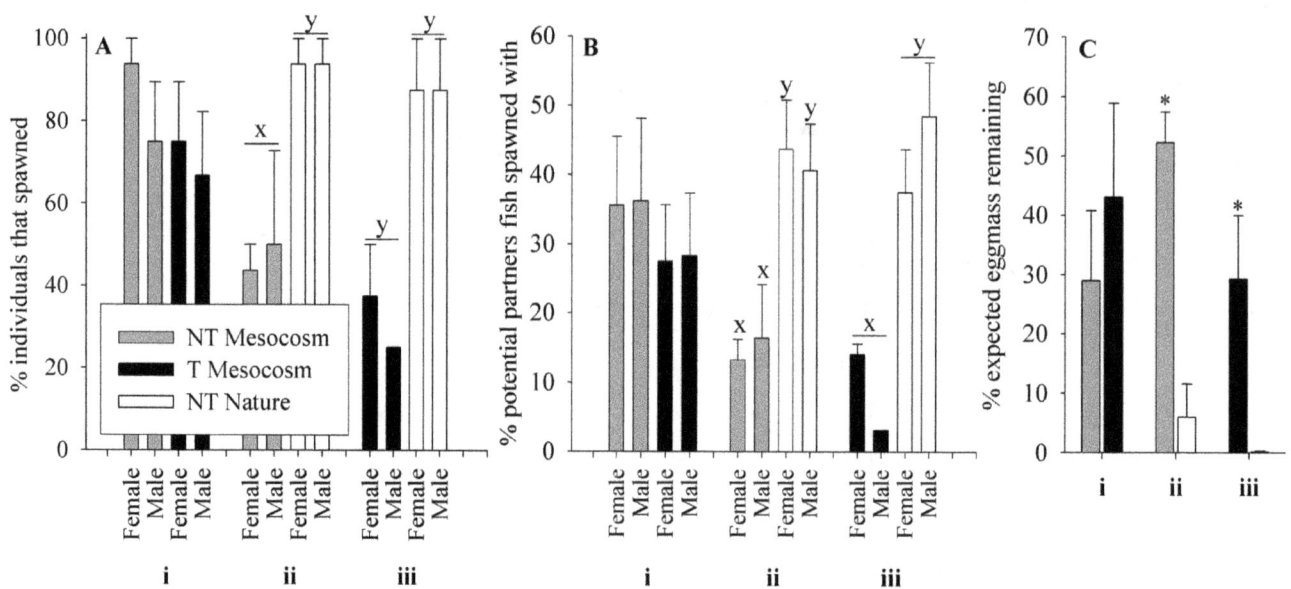

Figure 6. Mating success of fish groups in Full Competition experiments. Fish groups are wild-type (NT) and growth hormone transgenic (T) coho salmon raised in a Mesocosm or natural (Nature) conditions from smolt. Spawning arenas contained two groups of fish, four fish of each sex/ group, competing for spawning sites and mates. Trials were: **i)** *Effect of transgene*: NT Mesocosm + T Mesocosm (n = 4 arenas), **ii)** *Effect of mesocosm rearing*: NT Mesocosm + NT Nature (n = 4 arenas), and **iii)** *Effect of transgene and mesocosm rearing*: T Mesocosm + NT Nature (n = 2 arenas). **A)** % of individuals that spawned, **B)** % of available partners individual fish spawned with, **C)** % of expected egg mass remaining at time of spawning mortality. Data are given as means over arenas ± standard error of the mean. Significant differences within trial type between groups/sex are indicated by letters (x,y), p<0.05.

Figure 8Ci). There was a weak but significant correlation between fish length and number of offspring produced or proportion of available partners individual fish spawned with in T Mesocosm males only (p = 0.032 R^2 = 0.38, and p = 0.043 R^2 = 0.35 respectively).

Effect of Mesocosm Rearing Relative to Nature (NT Mesocosm ♂ vs NT Nature ♂): NT Mesocosm males had 5.5% of the offspring that NT Nature males had in competition for NT Nature females (p<0.001, Figure 8Aii). As well, NT Mesocosm males spawned with 25.8% of the available females that NT Nature males spawned with (p = 0.027, Figure 8Bii), but did not differ significantly in the % of males that spawned (p = 0.200, Figure 8Cii). Neither offspring number nor percent of total available partners individual fish spawned with significantly correlated with fish length in any group (p = 0.363 to 0.972).

Effect of Transgene and Mesocosm Rearing (T Mesocosm ♂ vs NT Nature ♂): When NT Nature males were in competition with T Mesocosm males for NT Nature females, only NT Nature males successfully spawned (Figure 8Aiii–Ciii). Neither offspring number nor percent of total available partners individual fish spawned with significantly correlated with fish length (p = 0.338 and p = 0.274 respectively for NT Nature males).

IIIb. Behaviour. *Effect of Transgene (NT Mesocosm ♂ vs T Mesocosm ♂)*: NT Mesocosm and T Mesocosm males did not differ in overall number of aggressive actions given (p = 0.594) or received (p = 0.071), or in the number of attending behaviours (p = 0.087) or quivers (p = 0.317) given when in competition for NT Nature females (Figure 9Ai–Di). However, there was a significant interaction between fish group and arena in aggressive actions received (p = 0.046), where NT Mesocosm males received fewer aggressive actions than T Mesocosm males in one arena, but not in the other two arenas.

Effect of Mesocosm Rearing Relative to Nature (NT Mesocosm ♂ vs NT Nature ♂): NT Mesocosm males performed only 17.5%

of the aggressive actions that NT Nature males did (p<0.001, Figure 9Aii), while NT Mesocosm males received overall 4.6-fold more aggressive actions than NT Nature males (p<0.001, Figure 9Bii). However, there was a strong fish group×arena interaction for aggressive actions received (p = 0.002), where the greater aggressive actions received by NT Mesocosm males were only significant for one arena. NT Nature males performed 3 times the average number of the attending behaviours of NT Mesocosm males (p = 0.006, Figure 9Cii), although the two groups did not differ significantly in number of quivers given (p = 0.169, Figure 9Dii).

Effect of Transgenic and Mesocosm Rearing (T Mesocosm ♂ vs NT Nature ♂): T Mesocosm males performed only 5.1% of the aggressive actions that NT Nature males did (p<0.001, Figure 9Aiii), but did not differ significantly in number of aggressive actions received (p = 0.097, Figure 9Biii). T Mesocosm males also performed only 20.1% of the attending behaviours of NT Nature males (p = 0.001 Figure 9Ciii), and unlike NT Nature males they did not perform any observed quivers to NT Nature females (p< 0.001, Figure 9Diii).

Discussion

Influence of Mesocosm Rearing on Spawning Success of Wild-type Fish

The spawning success and spawning behaviour (aggressive, courtship, and digging behaviours) of wild-type (NT) coho salmon raised in large (350,000 L), semi-natural seawater mesocosm approached that of nature-reared coho salmon under conditions without competition, but was much lower than nature-reared fish under competitive conditions (see Table 6 for an overall comparison of NT Mesocosm and NT Nature fish). NT Mesocosm fish displayed some of the spawning morphology of NT Nature fish, but were shorter in length, indicating mesocosm rearing only

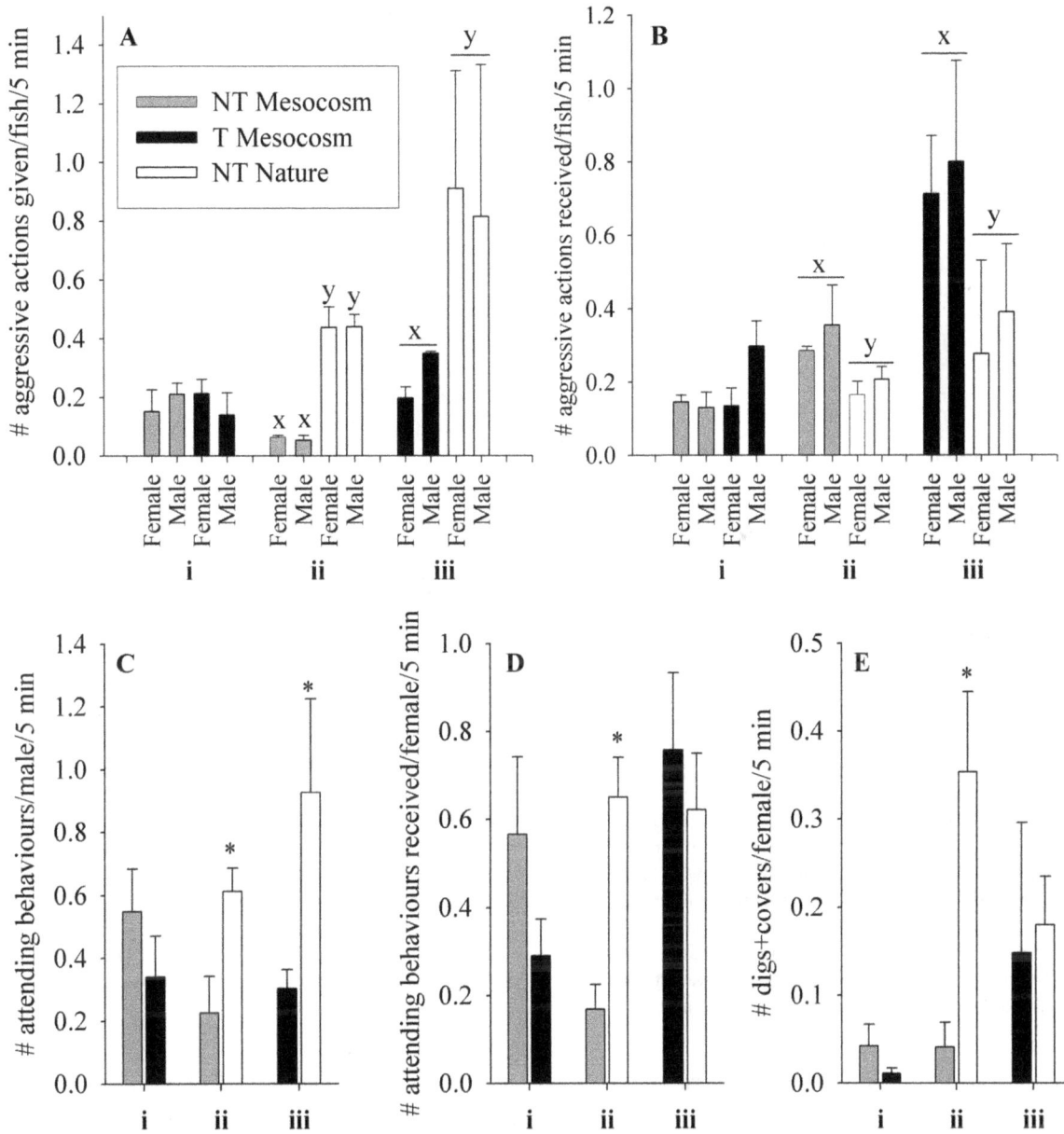

Figure 7. Spawning behaviour of fish groups during Full Competition experiments. Fish groups are wild-type (NT) and growth hormone transgenic (T) coho salmon raised in a Mesocosm or natural (Nature) conditions from smolt. Each spawning arena contained two groups of fish, four fish of each sex/group, competing for spawning sites and mates. Trials were: **i)** *Effect of transgene:* NT Mesocosm + T Mesocosm (n = 4 arenas), **ii)** *Effect of mesocosm rearing:* NT Mesocosm + NT Nature (n = 4 arenas), and **iii)** *Effect of transgene and mesocosm rearing:* T Mesocosm + NT Nature (n = 2 arenas). Behaviour is averaged over four daily 5 min daily intervals measured starting at 9 am, 11 am, 1 pm, and 3 pm. **A)** number of aggressive actions (chases and bites) given, **B)** number of aggressive actions received, **C)** numbers of times males attend females, **D)** number of times females are attended by males, **E)** number of times females dig or cover. Data are given as means over arenas ± standard error of the mean. Significant differences within trial type between groups/sex are indicated by letter (x,y) or by * on largest value, p<0.05.

partially restored typical morphology of nature-reared spawning fish compared to rearing in smaller (4000 L) culture containers [23]. In competition trials (both Full Competition and High Male) comparing NT Mesocosm and NT Nature fish, NT Mesocosm fish had much lower spawning success and behaviour. This concurs with previous studies that found other types of culture-rearing reduced spawning behaviour and success in Pacific salmon [29–31]. In contrast, in experiments without inter-male competition (single male and female pairs), NT Mesocosm fish tended to have similar spawning success and behaviour as NT Nature fish.

While rearing in a seawater mesocosm did not fully restore spawning success of wild-type fish, comparisons with previous experiments indicate that mesocosm rearing improved some aspects of spawning success and behaviour of wild-type fish over rearing in small culture tanks. In the one trial examined in the present study, NT Mesocosm males had greater spawning success and behaviour than NT Culture males, although NT Mesocosm females had slightly lower spawning success (and inconsistent relative behaviour) compared to NT Culture females. When results of the current Full Competition experiments were compared with that of culture-reared wild-type fish in Fitzpatrick

Table 5. Courtship and spawning behaviour of NT Mesocosm and NT Culture fish in a Full Competition Spawning experiment.

Behaviour	NT Mesocosm ♀	NT Mesocosm ♂	NT Culture ♀	NT Culture ♂
Number of aggressive behaviours given/fish/5 min	0.08±0.04	0.10±0.04	0.13±0.04	0.07±0.05
Number of aggressive behaviours received/fish/5 min	0.04±0.02	0.14±0.04	0.05±0.03	0.14±0.05
Number of attending behaviours given/male/5 min	n/a	0.54±0.08[a]	n/a	0.25±0.09[b]
Number of attending behaviours received/female/5 min	0.32±0.08	n/a	0.54±0.19	n/a
Average time of attending behaviours received as % of total time	5.5±0.7%[a]	n/a	14.0±0.7[b]	n/a
Number of digs/female/5 min	0[a]	n/a	0.26±0.11[b]	n/a
Number of quivers given/male/5 min	n/a	0.21±0.09	n/a	0.31±0.18
Number of quivers received /female/5 min	0.08±0.05	n/a	0.43±0.19	n/a
Average time quivers received as % of total time	4.5±0.9%[a]	n/a	13.3±0.6[b]	n/a

NT = wild-type coho salmon, Mesocosm = fish were raised from smolt in a mesocosm, Culture = fish were raised from smolt in standard culture conditions. Four fish of each sex and fish group were represented in one arena. Behaviour is averaged over four 5 min daily intervals starting at 9 am, 11 am, 1 pm, and 3 pm during the spawning trial.
[a,b]indicates significant differences between fish groups, p<0.05.

et al. [24], mesocosm-reared fish had better spawning success than culture-reared fish when in competition with NT Nature fish (25% and 13% of nature-reared fish respectively), although had similar patterns of suppressed spawning and courtship behaviour. The present study showed no significant differences in paired (No Competition) spawning success between wild-type mesocosm-reared fish and nature-reared fish, while Bessey et al. [23] found culture-reared wild-type females had lower spawning success than full nature-reared pairs, although the patterns of spawning success were similar between the two studies. During spawning, patterns of quivers surrounding a spawning event differed between the current study and Bessey et al. [23], where only mesocosm-raised wild-

type fish shared the pattern displayed by nature-reared fish of increasing quivers up to a spawning event, then decreasing quivers post-spawning [54]. Overall, seawater mesocosm rearing appeared to increase spawning success of wild-type coho over standard culture conditions while in competition with nature-reared fish, but improvements in spawning success in No Competition experiments are less clear. The factors associated with mesocosm culture that prevent full restoration of spawning success and behaviour are not known, but likely include lack of habitat complexity and limited spatial scope in the mesocosm environment. As well, diet differences between mesocosm- and nature-reared fish (commercial versus natural) may have influenced

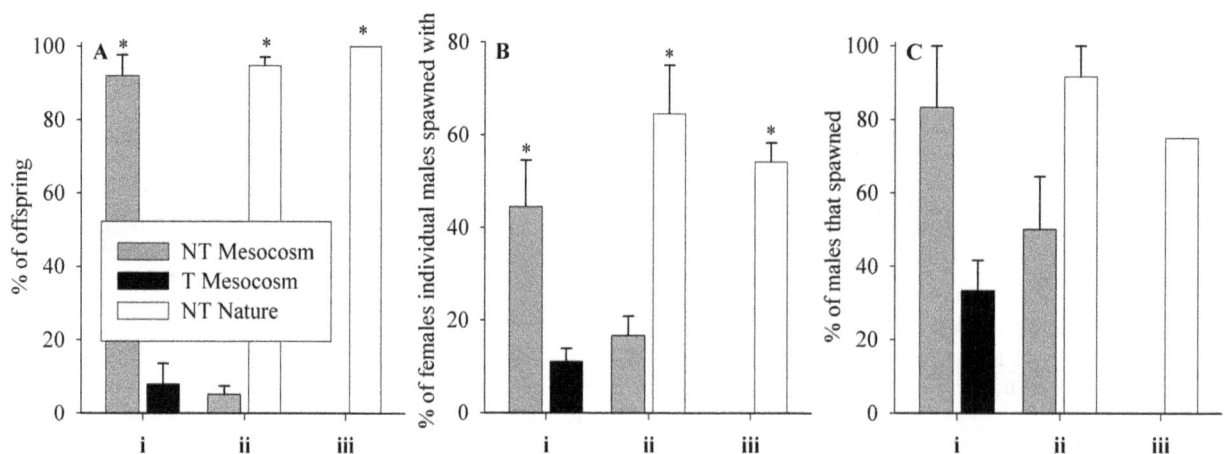

Figure 8. Spawning success of male fish groups during High Male Competition experiments for NT Nature females. Fish groups are wild-type (NT) and growth hormone transgenic (T) coho salmon raised in a Mesocosm or natural (Nature) conditions from smolt. Each spawning arena contained two groups of four male fish, in spawning competition for four NT Nature females (2:1 Male:Female). Trials were **i)** *Effect of transgene:* NT Mesocosm ♂ vs T Mesocosm ♂ (n = 3 arenas), **ii)** *Effect of mesocosm rearing:* NT Mesocosm ♂ vs NT Nature ♂ (n = 3 arenas), and **iii)** *Effect of transgene and mesocosm rearing:* T Mesocosm ♂ vs NT Nature ♂ (n = 2 arenas). **A)** % of offspring male fish contributed to, **B)** % of available NT Nature female partners individual male fish spawned with, **C)** % of male individuals that spawned. Data are given as means over arenas ± standard error of the mean. Significant differences within trial type between groups/sex are indicated by * on the largest value, p<0.05.

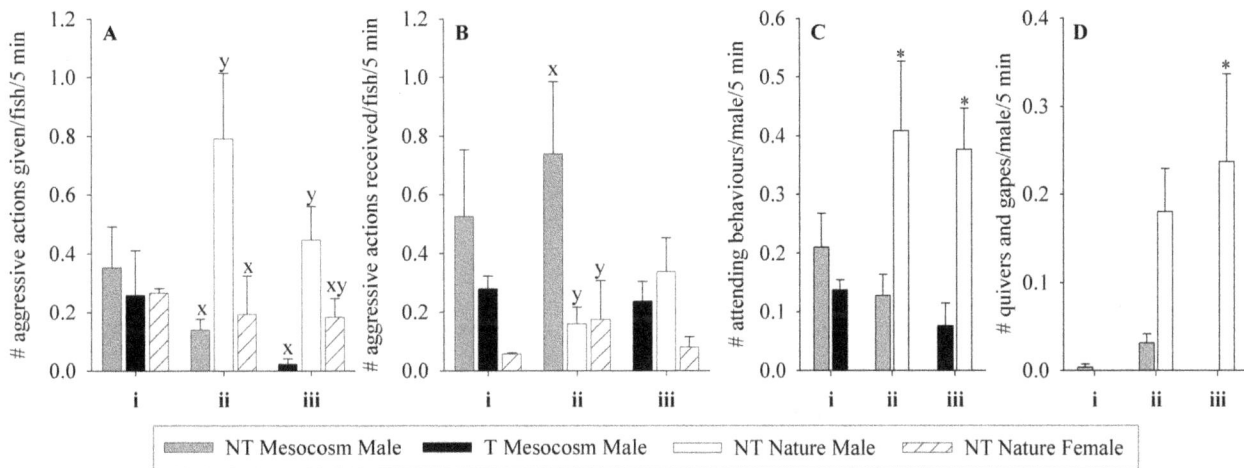

Figure 9. Spawning behaviour of male fish groups during High Male Competition experiments for NT Nature females. Fish groups are wild-type (NT) and growth hormone transgenic (T) coho salmon males raised in a Mesocosm or natural (Nature) conditions from smolt. Each arena contained two groups of four male fish in spawning competition for four NT Nature females in an artificial spawning arena. Trials were **i)** *Effect of transgene:* NT Mesocosm ♂ vs T Mesocosm ♂ (n = 3 arenas), **ii)** *Effect of mesocosm rearing:* NT Mesocosm ♂ vs NT Nature ♂ (n = 3 arenas), and **iii)** *Effect of transgene and mesocosm rearing:* T Mesocosm ♂ vs NT Nature ♂ (n = 2 arenas). Behaviour is averaged over four 5 min daily intervals measured starting at 9 am, 11 am, 1 pm, and 3 pm during the experiment. **A)** number of aggressive actions (chases and bites) given, **B)** number of aggressive actions received, **C)** numbers of times male attends NT Nature females, **D)** number of quivers given by male. Data are given as means over arenas ± standard error of the mean. Significant differences within trial type between groups/sex are indicated by letter (x,y) or by * over largest value, p<0.05.

spawning colouration, and consequent attractiveness to potential mates. In Full Competition experiments with NT Mesocosm fish, spawning success was significantly correlated to body length in NT Nature fish. While this correlation was not observed in NT Mesocosm fish, the smaller length of NT Mesocosm fish may have decreased their mating advantage with NT Nature fish. This would concur with other studies in Pacific salmon where size has been reported to have significant impact on mating advantage in male and, to a lesser extent, female fish [26,28].

Spawning Success of Mesocosm-reared GH Transgenic Salmon

As observed in other GH transgenic salmon studies [23–25], T Mesocosm fish had very low reproductive success relative to wild-type salmon reared in nature. However, when compared to wild-type fish grown in the mesocosm, T Mesocosm fish had equal spawning success in Full Competition and No Competition experiments, and displayed similar overall spawning behaviour (see Table 6 for an overall comparison of T Mesocosm and NT fish). In Full Competition, this held true whether T and NT Mesocosm fish were in direct competition, or in competition with

Table 6. Summary of relative spawning success and behaviour of NT and T coho salmon reared in a semi-natural mesocosm or natural systems, in three different competition experiments (No, Full, and High Male Competition).

	Trials		
	NT Meso vs T Meso	NT Meso vs NT Nature	T Meso vs NT Nature
OVERALL	≥	<	<
I. No Competition			
a. Spawning success	=	≈	≈
b. Spawning behaviour	≈	≤	≤
II. Full Competition			
a. Spawning success	=	<	<
b. Spawning behaviour	=	<	<
III. Male Competition			
a. Spawning success	>	<	<
b. Spawning behaviour	=	<	<

NT = wild-type coho salmon. T = growth hormone transgenic coho salmon. Meso = reared in a semi-natural 350,000 L seawater tank from smolt. Nature = reared in natural systems from smolt. Spawning success = proportion of offspring contribute to, of individuals that spawned, and of mates spawned with. Spawning behaviour = aggressive, courtship (attending, quivers, gapes), and digging behaviours.

NT Nature fish. This concurs with studies in other fish models, where GH transgenic catfish, medaka, and carp had similar reproductive success as wild-type fish grown in equal conditions [22,32,33]. In contrast, other studies found GH transgenic Atlantic salmon parr, zebrafish, and medaka had lower reproductive success compared to wild-type fish grown in equal conditions [25,34–36], although GH transgenic medaka males have inconsistently higher reported mating success compared to wild-type [11]. There was no evidence for assortative mating of T Mesocosm and NT fish in any trial examined. In the present study, T Mesocosm fish differed from NT Mesocosm fish in a few ways that could indicate this strain of GH transgenic fish, particularly males, have decreased spawning success in some circumstances. NT Nature females chose NT Mesocosm males over T Mesocosm males in High Male Competition, indicating differences between wild-type and transgenic fish were present that were of significance to mate choice of NT Nature fish or to spawning success of males. As well, during High Male Competition with NT Nature males, T Mesocosm males failed to spawn, were not observed to quiver, and participated in much fewer aggressive behaviours than NT Mesocosm males in competition with NT Nature males. While behaviour of T Mesocosm and NT Mesocosm males in the High Male Competition experiments did not significantly differ, T Mesocosm fish in general did deviate more from typical spawning morphology than NT Mesocosm fish, which could have played a role in mate choice by NT Nature females, and/or competitive ability of T fish. The excessive cranial growth in some transgenic strains under standard culture conditions [23,55,56] is not, prior to maturation, pronounced in the strain under study (M77 strain), however these abnormalities became apparent in some T mesocosm fish at maturation. In addition, the extreme deep body of some T Mesocosm fish in the 2010 year has not been previously observed (and curiously was seen only in one year in the present studies), but such changes in overall body proportion would likely influence reproductive success compared to a wild phenotype. Despite their larger size, T Mesocosm fish did not have significantly greater spawning success than NT Mesocosm fish indicating size alone does not influence spawning success in coho, although larger T fish tended to have greater spawning success in competition than smaller T fish. As well, transgenic coho salmon reared in standard culture have lower ejaculate density and sperm mobility [24], potentially exacerbating the lower spawning ability of male transgenic fish. The greater impact of GH transgenesis on spawning success of male versus female salmon concurs with other salmonid models where spawning success of male fish was more suppressed by hatchery rearing, or domestication combined with culture-rearing, than in female fish [26,27,39–41], likely due to a more intense breeding competition in male relative to female salmonids (see [57]). As well, transgenic fish used in the High Male Competition experiments were ration restricted during juvenile growth. In GH transgenic medaka, males had lower relative spawning success after rearing on low versus high food availability [36]. As such, poor reproductive performance of T Mesocosm males in the High Male Competition experiments may have been exacerbated by restricted juvenile growth in these fish.

When NT and T Mesocosm fish were in full competition, there were a number of transgenic females that appeared to spawn some of their egg mass, but did not contribute to the final offspring population. Offspring survival as percent of expected eggs spawned was much lower in T Mesocosm females than NT Mesocosm females. As such, T Mesocosm females appear to have increased incidence of unfertilized eggs and/or egg mortality. This could be due to unsynchronized dropping of eggs and milt in a spawning pair, or nest destruction due to poorly covered eggs.

During a spawning event in No Competition experiments, T Mesocosm males did not have a typical pattern of quivers up to and after a spawning event. This may have interfered with synchronization of spawning events, although T Mesocosm fish did not differ from NT Mesocosm fish in the percent of spawning events with visible eggs and/or milt in No Competition experiments. As well, T Mesocosm females covered less than NT fish immediately after a spawning event. In Full Competition experiments, spawning areas provided were 1.92 m^2/spawning female, which is 30% less than the reported average nest size for coho salmon in nature (2.8 m^2/female, [43]). As well, we found most nests were located in the downstream section of the arenas with several nests superimposed and overlapping. The poor covering by T Mesocosm fish (as expressed by low number of digs immediately after spawning) may indicate their nests were more susceptible to damage in such a highly competitive arena, resulting in decreased survival of offspring. In addition, differential mortality of T offspring has been noted in different strains of GH transgenic coho [58] and well as in GH transgenic carp [33], which could result in overall lower survival of transgenic female offspring. Transgenic fish tend to have higher fecundity in coho salmon and medaka [22,23], but potential for poor nest covering, unsynchronized spawning, and differential mortality of transgenic offspring may offset this advantage. Both NT and T Mesocosm females retained more eggs than NT Nature females (assessed at death post-spawning), and several mesocosm-raised individuals did not spawn at all. This concurs with Berejikian et al. [31] who first noted such effects, finding captive-reared wild-strain chinook salmon deposited 50% fewer eggs than nature-reared females, and many captive-reared fish did not participate in spawning events.

Genotype-by-Environmental Interactions on Spawning Success

The spawning success of GH transgenic coho salmon relative to equally reared wild-type fish appears to be influenced by rearing background, as well as type of spawning trial. In full competition, GH transgenic fish raised in standard culture had only 30% of the spawning success of equally-raised wild-type fish as measured by percent of offspring [24], but T Mesocosm and NT Mesocosm fish had approximately equal spawning success in competition with NT Nature fish in the current study (see Figure 10 for reaction norms of spawning success in competition). This suggests the relative spawning success of GH transgenic and wild-type fish can be influenced by the environmental conditions of seawater rearing and or experimental assessment. One difference between the current study and Fitzpatrick et al. [24] is spawning competition was examined under lower spawning female density (1.92 m^2/female versus 1.13 m^2/female, respectively). As such, the greater success of T Mesocosm fish relative to culture-raised transgenic fish may be due in part to transgenic fish performing poorly in the more competitive environment of Fitzpatrick et al. [24], rather than entirely from benefits due to mesocosm rearing.

In male competition, Bessey et al. [23] found wild-type and transgenic culture-reared males had equally poor spawning success when in direct competition for a single nature-reared female, while the current study showed NT Mesocosm males greatly outcompeted T Mesocosm males in competition for NT Nature females. This indicates that wild-type males benefited more from mesocosm rearing over standard culture than did transgenic males, in contrast to the results of the Full Competition experiments. These effects also may be due to differences in spawning trial (2 males to 1 female in [23], versus 8 males to 4 females in the current study). As well, the transgenic males used in the current High Male Competition experiments were ration restricted during freshwater

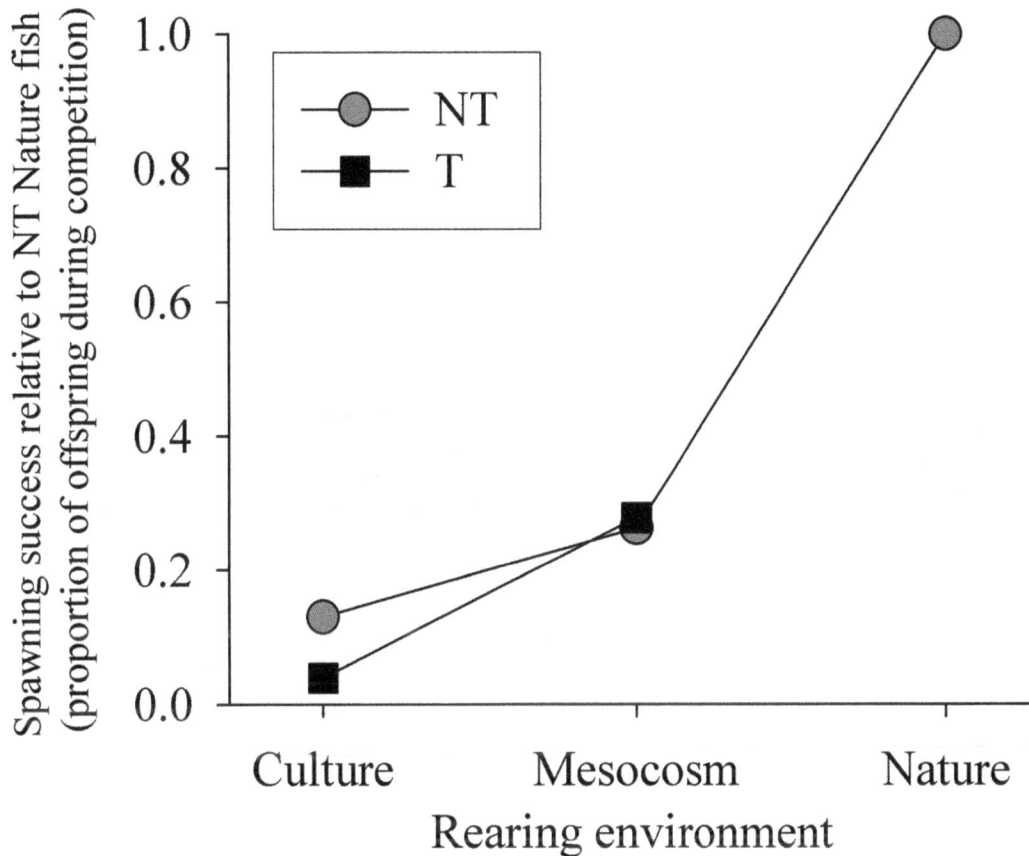

Figure 10. Genotype × Environmental reaction norms of spawning success of coho salmon in competition. Genotypes are wild-type (NT) and growth hormone transgenic (T) coho salmon reared in environments of standard Culture (4000 L tank, data taken from [24]), a semi-natural Mesocosm (350,000 L), or natural environments (Nature, NT only) from smolt. Spawning success is given as proportion of offspring fish group contributed to when in full competition (mixed male and female fish) with NT Nature fish, relative to the success of NT Nature fish. The lack of parallel reaction norms of NT and T fish between culture and mesocosm environments, and the unknown slope of the reaction norm for T fish between Mesocosm and Nature complicates predictions of spawning success of T fish in nature.

growth, while previous studies and the Full Competition experiments in the current study used transgenic fish that were fully-fed during juvenile rearing, which may have influenced their spawning ability (see above).

In No Competition experiments, while there were no significant differences in spawning success of NT Nature, NT Mesocosm or T Mesocosm fish, the general trends of spawning success followed that reported in Bessey et al. [23]. One exception to this is that transgenic females raised in culture had low spawning success with nature-reared males (12.5% spawned; [23]), while in the current study, mesocosm-reared transgenic females had medium spawning success with nature-reared males (50%), suggesting GH transgenic females may have greater spawning success without competition if raised in a mesocosm relative to standard culture conditions. The presence of genotype-by-environmental interactions on spawning success of GH transgenic fish concurs with Pennington and Kapuscinski [36] who found interacting influences of food availability and predator presence on spawning success of GH transgenic medaka. As such, extrapolating existing spawning data of GH transgenic fish to new environments (e.g. nature) is complicated by the non-parallel reaction norms seen for reproductive success among different environmental conditions.

Conclusions

Overall, mesocosm rearing from smolt to adulthood only partially restored spawning success to wild-type coho salmon, but critically did increase their reproductive capabilities to a level where relative comparisons to GH transgenic salmon can be made. Among all experiments conducted to date with strain M77 GH transgenic and reference wild-type coho salmon reared and studied under a range of environmental conditions, the data does not support this strain of GH transgenic fish possessing any mating advantage over wild-type. Under some conditions GH transgenic salmon have reduced reproductive success, and overall it is possible, but not certain, their fitness would be lower than that of wild-type. From an evolutionary perspective, wild-type salmon, which have significant capacity to select for different growth rates, have adjusted growth to maximize fitness under natural conditions, and it is unlikely that an anthropogenically-induced adjustment to such a phenotype would result in an enhancement of fitness for the animal in its present niche. However, it is also possible that quantum shift in growth and behaviour caused by GH transgenesis could provide access to a phenotype not accessible to wild strains (due to strong balancing selection), or could cause effects on other traits (e.g. feeding competition) that provide compensating benefits to net fitness [22]. While theoretical considerations are critical, determination of fitness effects of GH transgenesis and environmental influences on them need to be empirically determined for assessment of actual risk. Indeed, previous and current studies indicate significant genotype-by-environmental interactions exist, where the type of seawater rearing appears to exert differing influences on the spawning success of wild-type and GH transgenic coho salmon (see Figure 10). The relative spawning success of wild-type and transgenic fish was also greatly influenced by the type of spawning study, indicating care must be taken when extrapolating spawning success from a single study type. Together, existing data indicate escaped GH transgenic salmon would be capable of reproducing in the wild and successfully spawning with wild populations, although the extent to which they may do this is not fully known. If mesocosm rearing had restored the reproductive success of wild-type coho salmon to that seen for wild salmon from nature, we may have felt more confident that the reproductive success of transgenic salmon reared in the same conditions would more closely approximate that for the same strain should it be derived from nature. Thus, until we have improved ability to predict the phenotype of transgenic salmon in nature, insufficient data exists

to extrapolate with low uncertainty what the true spawning success, and hence risk, of these fish would be should they escape early in life and spend smolt to adulthood in natural conditions. Given the scale and scope of the present experiments, overcoming these limitations in empirical assessment will be challenging.

Acknowledgments

Thanks to Carlo Biagi, Krista Woodward, Dionne Sakhrani, Hamid Akbarishandiz, Peter Raven, Linda Poon, Robert Dominelli, Lisa Skinner, Lucas Halliday, Neil MacKinnon, Kate McGivney, and Jin-Hyoung Kim for assistance in set-up of spawning trials, Samantha White for assistance in set-up and editing the manuscript, and Larry Kahl and Don Johnson from the Chehalis River Enhancement Facility.

Author Contributions

Conceived and designed the experiments: RHD RAL. Performed the experiments: RAL TH WEV KM JP. Analyzed the data: RAL BG. Contributed reagents/materials/analysis tools: RHD. Contributed to the writing of the manuscript: RAL RHD.

References

1. Zhu Z, Li G, He L, Chen S (1985) Novel gene transfer into the fertilized eggs of gold fish (Carassius auratus L. 1758). J Appl Ichthyol 1: 31–34.
2. Du SJ, Gong Z, Fletcher GL, Shears MA, King MJ, et al. (1992) Growth enhancement in transgenic Atlantic salmon by the use of an "all fish" chimeric growth hormone gene construct. Bio/technology 10: 176–181.
3. Devlin RH, Yesaki TY, Biagi CA, Donaldson EM, Swanson P, et al. (1994) Extraordinary salmon growth. Nature 371: 209–210.
4. Rahman MA, Mak R, Ayad H, Smith A, McLean N (1998) Expression of a novel piscine growth hormone gene results in growth enhancement in transgenic tilapia (Oreochromis niloticus). Transgenic Res 7: 357–369.
5. Devlin RH, Yesaki TY, Donaldson EM, Du SJ, Hew C-L (1995) Production of germline transgenic Pacific salmonids with dramatically increased growth performance. Can J Fish Aquat Sci 52: 1376–1384.
6. Nam YK, Noh JK, Cho YS, Cho HJ, Cho KN, et al. (2001) Dramatically accelerated growth and extraordinary gigantism of transgenic mud loach Misgurnus mizolepis. Transgenic Res 10: 353–362.
7. Kobayashi S, Alimuddin, Morita T, Miwa M, Lu J, et al. (2007) Transgenic Nile tilapia (Oreochromis niloticus) over-expressing growth hormone show reduced ammonia excretion. Aquaculture 270: 427–435.
8. DFO (2013) Summary of the environmental and indirect human health risk assessment of AquAdvantage salmon. DFO Can Sci Advis Sec Sci Rep 2013/023. 26 p.
9. Kapuscinski AR, Hayes KR, Li S, Dana G, editors (2007) Environmental Risk Assessment of Genetically Modified Organisms: Methodologies for Transgenic Fish. Oxfordshire: CABI International. 304 p.
10. Devlin RH, Biagi CA, Yesaki TY, Smailus DE, Byatt JC (2001) Growth of domesticated transgenic fish. Nature 409: 781–782.
11. Howard RD, DeWoody JA, Muir WM (2004) Transgenic male mating advantage provides opportunity for Trojan gene effect in a fish. Proc Natl Acad Sci U S A 101: 2934–2938.
12. Devlin RH, Johnsson JI, Smailus DE, Biagi CA, Jönsson E, et al. (1999) Increased ability to compete for food by growth hormone-transgenic coho salmon Oncorhynchus kisutch (Walbaum). Aquaculture Res 30: 479–482.
13. Jhingan E, Devlin RH, Iwama GK (2003) Disease resistance, stress response and effects of triploidy in growth hormone transgenic coho salmon. J Fish Biol 63: 806–823.
14. Sundström LF, Löhmus M, Johnsson JI, Devlin RH (2004) Growth hormone transgenic salmon pay for growth potential with increased predation mortality. Proc R Soc Lond B 271: S350–S352.
15. Kim J-H, Balfry S, Devlin RH (2013) Disease resistance and health parameters of growth-hormone transgenic and wild-type coho salmon, Oncorhynchus kisutch. Fish Shellfish Immunol 34: 1553–1559.
16. Maclean N, Laight RJ (2000) Transgenic fish: an evaluation of benefits and risks. Fish Fish 1: 146–172.
17. Hedrick PW (2001) Invasion of transgenes from salmon or other genetically modified organisms into natural populations. Can J Fish Aquat Sci 58: 841–844.
18. Aikio S, Valosaari K-R, Kaitala V (2008) Mating preference in the invasion of growth enhanced fish. Oikos 117: 406–414.
19. Valosaari K-R, Aikio S, Kaitala V (2008) Male mating strategy and the introgression of a growth hormone transgene. Evol Appl 1: 608–619.
20. Ahrens RNM, Devlin RH (2011) Standing genetic variation and compensatory evolution in transgenic organisms: a growth-enhanced salmon simulation. Transgenic Res 20: 583–597.
21. Muir WM, Howard RD (1999) Possible ecological risks of transgenic organism release when transgenes affect mating success: sexual selection and Trojan gene hypothesis. Proc Natl Acad Sci U S A 96: 13853–13856.
22. Muir WM, Howard RD (2001) Fitness components and ecological risk of transgenic release: a model using Japanese medaka (Oryzias latipes). Am Nat 158: 1–16.
23. Bessey C, Devlin RH, Liley NR, Biagi CA (2004) Reproductive performance of growth-enhanced transgenic coho salmon. Trans Am Fish Soc 133: 1205–1220.
24. Fitzpatrick JL, Akbarashandiz H, Sakhrani D, Biagi CA, Pitcher TE, et al. (2011) Cultured growth hormone transgenic salmon are reproductively outcompeted by wild-reared salmon in semi-natural mating arenas. Aquaculture 312: 185–191.
25. Moreau DTR, Conway C, Fleming IA (2011) Reproductive performance of alternative male phenotypes of growth hormone transgenic Atlantic salmon (Salmo salar). Evol Appl 4: 736–748.
26. Fleming IA, Gross MR (1993) Breeding success of hatchery and wild coho salmon (Oncorhynchus kisutch) in competition. Ecol Appl 3: 230–245.
27. Fleming IA, Lamberg A, Jonsson B (1997) Effects of early experience on the reproductive performance of Atlantic salmon. Behav Ecol 8: 470–480.
28. Berejikian BA, Van Doornik DM, Scheurer JA, Bush R (2009) Reproductive behavior and relative reproductive success of natural- and hatchery-origin Hood Canal summer chum salmon (Oncorhynchus keta). Can J Fish Aquat Sci 66: 781–789.
29. Berejikian BA, Tezak EP, Schroder SL, Knudsen CM, Hard JJ (1997) Reproductive behavioral interactions between wild and captively reared coho salmon (Oncorhynchus kisutch). ICES J Mar Sci 54: 1040–1050.
30. Berejikian BA, Tezak EP, Park L, LaHood E, Schroder SL, et al. (2001) Male competition and breeding success in captively reared and wild coho salmon (Oncorhynchus kisutch). Can J Fish Aquat Sci 58: 804–810.
31. Berejikian BA, Tezak EP, Schroder SL (2001) Reproductive behavior and breeding success of captively reared chinook salmon. N Am J Fish Manag 21: 255–260.
32. Dunham RA, Chen TT, Powers DA, Nichols A, Argue B, et al. (1995) Predator avoidance, spawning, and foraging ability of transgenic channel catfish with rainbow trout growth hormone gene. In: Levin M, Grim C, Angle JS, editors. Biotechnology Risk Assessment: Proceedings of the Biotechnology Risk Assessment Symposium, June 6–8, 1995. Reston, VA: TechniGraphix. 127–139.
33. Lian H, Hu W, Huang R, Du F, Liao L, et al. (2013) Transgenic common carp do not have the ability to expand populations. PloS ONE 8: e65506.
34. Pennington KM, Kapuscinski AR, Morton MS, Cooper AM, Miller LM (2010) Full life-cycle assessment of gene flow consistent with fitness differences in transgenic and wild-type Japanese medaka fish (Oryzias latipes). Environ Biosafety Res 9: 41–57.
35. Figueiredo MA, Fernandes RV, Studzinski AL, Rosa CE, Corcini CD, et al. (2013) GH overexpression decreases spermatic parameters and reproductive success in two-years-old transgenic zebrafish males. Anim Reprod Sci 139: 162–167.
36. Pennington KM, Kapuscinski AR (2011) Predation and food limitation influence fitness traits of growth-enhanced transgenic and wild-type fish. Trans Am Fish Soc 140: 221–234.
37. Sundström LF, Löhmus M, Tymchuk WE, Devlin RH (2007) Gene-environment interactions influence ecological consequences of transgenic animals. Proc Natl Acad Sci U S A 104: 3889–3894.

38. Devlin RH, Biagi CA, Yesaki TY (2004) Growth, viability and genetic characteristics of GH transgenic coho salmon strains. Aquaculture 236: 607–632.

39. Fleming IA, Jonsson B, Gross MR, Lamberg A (1996) An experimental study of the reproductive behaviour and success of farmed and wild Atlantic salmon (*Salmo salar*). J Appl Ecol 33: 893–905.

40. Fleming IA, Hindar K, Mjølnerød IB, Jonsson B, Balstad T, et al. (2000) Lifetime success and interactions of farm salmon invading a native population. Proc R Soc Lond B 267: 1517–1523.

41. Anderson JH, Faulds PL, Atlas WI, Quinn TP (2013) Reproductive success of captively bred and naturally spawned Chinook salmon colonizing newly accessible habitat. Evol Appl 6: 165–179.

42. Fleming IA, Gross MR (1989) Evolution of adult female life history and morphology in a Pacific salmon (coho: *Oncorhynchus kisutch*). Evolution 43: 141–157.

43. Burner CJ (1951) Characteristics of spawning nests of Columbia River salmon. Fish Bull 52: 95–110.

44. Olsen JB, Wilson SL, Kretschmer EJ, Jones KC, Seeb JE (2000) Characterization of 14 tetranucleotide microsatellite loci derived from sockeye salmon. Mol Ecol 9: 2185–2187.

45. Small MP, Beacham TD, Withler RE, Nelson RJ (1998) Discriminating coho salmon (*Oncorhynchus kisutch*) populations within the Fraser River, British Columbia, using microsatellite DNA markers. Mol Ecol 7: 141–155.

46. Rexroad CE III, Coleman RL, Gustafson AL, Hershberger WK, Killefer J (2002) Development of rainbow trout microsatellite markers from repeat enriched libraries. Mar Biotechnol 3: 12–16.

47. Rexroad CE III, Coleman RL, Martin AM, Hershberger WK, Killefer J (2001) Thirty-five polymorphic microsatellite markers for rainbow trout (*Oncorhynchus mykiss*). Anim Genet 32: 317–319.

48. Rexroad CE III, Palti Y (2003) Development of ninety-seven polymorphic microsatellite markers for rainbow trout. Trans Am Fish Soc 132: 1214–1221.

49. Palti Y, Fincham MR, Rexroad CE III (2002) Characterization of 38 polymorphic microsatellite markers for rainbow trout (*Oncorhynchus mykiss*). Mol Ecol Notes 2: 449–452.

50. Rodriguez F, Rexroad CE III, Palti Y (2003) Characterization of twenty-four microsatellite markers for rainbow trout (*Oncorhynchus mykiss*). Mol Ecol Notes 3: 619–622.

51. Rexroad CE III, Rodriguez MF, Coulibaly I, Gharbi K, Danzmann RG, et al. (2005) Comparative mapping of expressed sequence tags containing microsatellites in rainbow trout (*Oncorhynchus mykiss*). BMC Genomics 6: 54.

52. Coulibaly I, Gharbi K, Danzmann RG, Yao J, Rexroad CE III (2005) Characterization and comparison of microsatellites derived from repeat-enriched libraries and expressed sequence tags. Anim Genet 36: 309–315.

53. Cairney M, Taggart JB, Høyheim B (2000) Characterization of microsatellite and minisatellite loci in Atlantic salmon (*Salmo salar* L.) and cross-species amplification in other salmonids. Mol Ecol 9: 2175–2178.

54. Berejikian BA, Tezak EP, LaRae AL (2000) Female mate choice and spawning behaviour of chinook salmon under experimental conditions. J Fish Biol 57: 647–661.

55. Devlin RH, Yesaki TY, Donaldson EM, Hew C-L (1995) Transmission and phenotypic effects of an antifreeze/GH gene construct in coho salmon (*Oncorhynchus kisutch*). Aquaculture 137: 161–170.

56. Ostenfeld TH, McLean E, Devlin RH (1998) Transgenesis changes body and head shape in Pacific salmon. J Fish Biol 52: 850–854.

57. Fleming IA, Petersson E (2001) The ability of released, hatchery salmonids to breed and contribute to the natural productivity of wild populations. Nord J Freshw Res 75: 71–98.

58. Leggatt RA, Biagi CA, Smith JL, Devlin RH (2012) Growth of growth hormone transgenic coho salmon *Oncorhynchus kisutch* is influenced by construct promoter type and family line. Aquaculture 356–357: 193–199.

13

DNA Barcoding for Species Assignment: The Case of Mediterranean Marine Fishes

Monica Landi[1]*, **Mark Dimech**[2¤], **Marco Arculeo**[3], **Girolama Biondo**[3], **Rogelia Martins**[4], **Miguel Carneiro**[4], **Gary Robert Carvalho**[5], **Sabrina Lo Brutto**[3], **Filipe O. Costa**[1]

1 Centre of Molecular and Environmental Biology (CBMA), Department of Biology, University of Minho, Braga, Portugal, 2 Malta Centre for Fisheries Science (MCFS), Fort San Lucjan Marsaxlokk, Malta, 3 Dipartimento di Scienze e Tecnologie Biologiche, Chimiche e Farmaceutiche (STEBICEF), University of Palermo, Palermo, Italy, 4 Modelling and Management Fishery Resources Division (DIV-RP), Instituto Português do Mar e da Atmosfera, Lisboa, Portugal, 5 Molecular Ecology and Fisheries Genetics Laboratory, School of Biological Sciences, Bangor University, Bangor, United Kingdom

Abstract

Background: DNA barcoding enhances the prospects for species-level identifications globally using a standardized and authenticated DNA-based approach. Reference libraries comprising validated DNA barcodes (COI) constitute robust datasets for testing query sequences, providing considerable utility to identify marine fish and other organisms. Here we test the feasibility of using DNA barcoding to assign species to tissue samples from fish collected in the central Mediterranean Sea, a major contributor to the European marine ichthyofaunal diversity.

Methodology/Principal Findings: A dataset of 1278 DNA barcodes, representing 218 marine fish species, was used to test the utility of DNA barcodes to assign species from query sequences. We tested query sequences against 1) a reference library of ranked DNA barcodes from the neighbouring North East Atlantic, and 2) the public databases BOLD and GenBank. In the first case, a reference library comprising DNA barcodes with reliability grades for 146 fish species was used as diagnostic dataset to screen 486 query DNA sequences from fish specimens collected in the central basin of the Mediterranean Sea. Of all query sequences suitable for comparisons 98% were unambiguously confirmed through complete match with reference DNA barcodes. In the second case, it was possible to assign species to 83% (BOLD-IDS) and 72% (GenBank) of the sequences from the Mediterranean. Relatively high intraspecific genetic distances were found in 7 species (2.2%–18.74%), most of them of high commercial relevance, suggesting possible cryptic species.

Conclusion/Significance: We emphasize the discriminatory power of COI barcodes and their application to cases requiring species level resolution starting from query sequences. Results highlight the value of public reference libraries of reliability grade-annotated DNA barcodes, to identify species from different geographical origins. The ability to assign species with high precision from DNA samples of disparate quality and origin has major utility in several fields, from fisheries and conservation programs to control of fish products authenticity.

Editor: Sean Rogers, University of Calgary, Canada

Funding: Research carried out at University of Palermo was supported by "Fondi di Ateneo ex 60% Università di Palermo". This work was supported by FEDER through POFCCOMPETE and by national funds from "Fundaçãopara a Ciência e a Tecnologia (FCT)" in the scope of the grants, FCOMP-01-0124-FEDER-010596 and PEst-OE/BIA/UI4050/2014. ML's work was supported by the fellowship Ref: SFRH/BPD/45246/2008 from Fundaçãopara a Ciência e a Tecnologia. The funders had no role in study design, data collection and analysis, decision to publish, or preparation of the manuscript.

Competing Interests: The authors have declared that no competing interests exist.

* Email: mlandi@bio.uminho.pt

¤ Current address: Fisheries and Aquaculture Department, Food and Agriculture Organisation of the United Nations, Athens, Greece

Introduction

The Mediterranean Sea is a semi-enclosed basin that embraces the marine area from the North East Atlantic Ocean, at West, to the Aegean Sea, at East. The confluences of marine ichthyofauna migrating from the Atlantic Ocean through the Strait of Gibraltar, from the Red Sea and the Indian Ocean through the Suez Channel, and from the Sea of Marmara and Black Sea through the Dardanelles, depicts a picture of the Mediterranean marine biodiversity characterized by a high species richness and peculiarities, including tropical species as well as endemisms [1], [2], [3].

Hosting 7% of the global marine ichthyofauna [4] the Mediterranean Sea is a fascinating prosperous biodiversity hotspot [5], [6] that captured the interest of numerous marine scientists since ancient times (e.g. Aristoteles) [1].

Holding such richness, the Mediterranean Sea can be elected as a very important scientific cradle in marine sciences. The considerable natural variation driven by distinctive regional evolutionary histories and dynamic anthropogenic pressures [7] presents major challenges in local biodiversity monitoring programmes.

Over the past decade, DNA barcoding has played a facilitatory role for accurate identification of marine ichthyofauna, thanks to the integration of molecular and traditional taxonomic methods

[8]. Such DNA-based method provides a robust and standardized approach for marine species identification, as witnessed by the remarkable boost of species identified [9], as well as its use for various applications [10], as for example fisheries and conservation programs [11]. DNA barcoding has been adopted in numerous studies illustrating its speed, reliability and accessibility [12], [13].

The possibility of compiling taxonomic and molecular data into a globally accessible public database (Barcode of Life Data System, BOLD, http://www.barcodinglife.org) [14], [15], comprising taxonomically diverse reference libraries (e.g. Costa et al. [16]), allows usage by a diverse community of scientists and end-users [12]. Such wide-scale adoption enables global comparisons of putative cosmopolitan species [17] facilitating opportunities for comparisons of different marine environments, as well as tackling issues relating to molecular evolution [18]. The availability of a detailed reference library comprising validated DNA barcodes [16] constitutes a robust platform against which to test query sequences, and it represents a valid tool for attributing species to unknown sequences [19].

The DNA barcoding methodology has been applied recently to identify 98 marine species inhabiting the eastern basin of the Mediterranean Sea [20]. Such studies are yet to be extended to marine ichthyofauna populating one of the highest biodiversity richness spots of the Mediterranean, namely the sea around Sicily and Malta.

Here, we present an extensive account of DNA barcodes for Mediterranean fishes based on the mitochondrial *cytochrome c oxidase subunit I* (COI). We used a query dataset composed of 486 specimens identified morphologically from the central basin of the Mediterranean Sea. DNA barcodes generated from these specimens were then screened against the reference dataset of fish from Portugal, as well as against public databases. There were several reasons to choose a reference library from a different location to our target area. First, the marine ichthyofauna of Portugal and of the extension of the Portuguese Continental Shelf is taxonomically well documented [21], and widely characterised using molecular genetic approaches [16], [22]. Specifically, a published reference library for 102 fish species from Portugal, built on COI data was evaluated for taxonomic reliability and attributed to reliability grades [16]. Despite comparing taxa from two differently highly dynamic areas shaped by the confluence of different seas [1], [16], our approach derives from the considerable overlap, especially of exploited species, in the ichthyofauna from Portugal and the central Mediterranean (www.fishbase.org) [23]. The connection between the Mediterranean Sea and the Atlantic Ocean through the Strait of Gibraltar underpins considerable taxonomic similarity, with more than 50 percent of the Mediterranean taxa being of Atlantic origin [24], together with ongoing gene flow in some species [25]. Concomitantly, we examine intra-species population divergence, since similar comparisons have revealed considerable lineage divergence or suggested the occurrence of cryptic species [26]. The universality of the DNA barcodes is, in part, based on the typical low within-species divergence regardless of geographic separation (see Kochzius et al. [27], Ward [28]). The detection of significant divergence among populations is particularly relevant in the present study, given the clear genetic separation previously reported for several fish species across the Atlantic-Mediterranean transition [25]. It also provides empirical scientific support for conservation measures to tackle biodiversity loss and for sustainable exploitation of shared marine fishery resources among southern European countries.

Materials and Methods

In this study we used a reference library of DNA barcodes of fishes from the temperate North East Atlantic, as a core "reference dataset", to assign species names to a set of fish collected in the central Mediterranean basin, hereafter referred to as "query dataset". The reference dataset was built using a collection of DNA barcodes of fish from Portugal (Fig. 1), to which taxonomic reliability grades were attributed [16]. The query dataset was composed of DNA barcodes obtained from fish specimens collected in the waters of Malta and Sicily (Italy). In all cases, fish were collected using trawling fishing methods, either on board of research vessels of governmental fisheries research agencies, or directly from legal fisheries landings. No endangered or protected species were sampled. Specific details on specimen collection and DNA barcode generation methods are provided below, separately for the reference and query datasets. Collection and sequencing details for all specimens examined in this study are available in the public project South European Marine Fish: MP (SEFMP), project codes CSFOM, FCFMT, MLFP, lodged in the Barcode of Life Data System (BOLD) [14].

Preparation of the reference dataset

The reference dataset comprises 792 DNA barcodes from 146 marine fish species collected along the Portuguese continental coast (Fig. 1). 659 DNA barcodes distributed among 102 species of this collection have been previously described, analyzed and verified for their taxonomic reliability through attribution of reliability grades [16]. Five DNA barcodes of the species *Zenion hololepis* have been previously described also [29]. DNA barcodes of the remaining 43 species, and additional 3 genera and 1 family not identified to species level ($n = 128$), collected off Portugal during 2009–2011, were here obtained for the first time, as described in Costa et al. [16]. All DNA sequences generated were characterized by the absence of stop codons, insertions or deletions. Table 1 provides details of the partitioning of the number of species, sequences and GenBank accession numbers among the BOLD projects above mentioned.

Preparation of the Mediterranean query dataset

Specimen collection. 486 specimens, representing 141 marine fish species, were collected from two Mediterranean sub-areas (Fig. 1). 219 specimens were collected from fisheries landings in Sicily, in 2006–2008. 267 samples, analyzed from the Malta area, were collected through the Mediterranean international Bottom trawl Survey (MEDITS), and the Annual national Fisheries data collection program (EC 199/08), in 2006–2007. Specimens were first identified to species level immediately after collection and later verified in the laboratory with the support of taxonomic keys [30], [31], [32], [33]. For each specimen, *c.* 0.5 g of skeletal muscle was dissected with a sterile blade and stored in 96% ethanol.

Molecular analyses. Fish tissue samples collected by Malta Centre for Fisheries Science were extracted and amplified as described in Costa et al. [16]. Specimens collected in Sicily were processed at University of Palermo. Total DNA was extracted from the muscle tissue of each specimen, using the DNeasy extraction kit (Qiagen). Sequences of the query dataset were obtained by amplification and sequencing of a 652 bp fragment of 5′ end of the mitochondrial gene *cytochrome c oxidase I* (COI-5P), using the primer pairs FishF1 and FishR1 selected from the primer cocktails described by Ward et al. [17]. Standard PCR reactions were carried out in 12.5 uL total volume, containing about 20 ng of DNA template, 6.25 uL of 10% trehalose, 2 uL of ultrapure

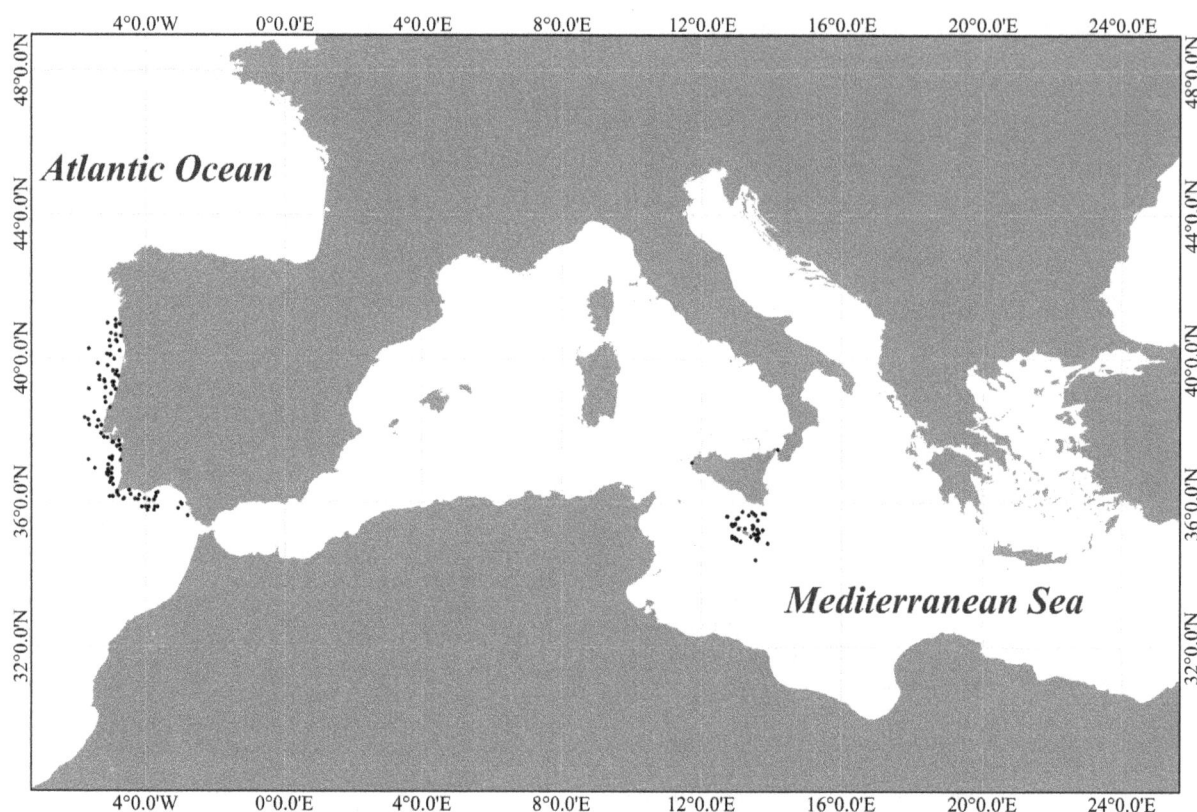

Figure 1. Location of sampling sites. Dots point areas of collection of marine fish specimens along the Portuguese coasts and in the central Mediterranean basin.

water, 1.25 uL of 10X PCR buffer (200 mMTris-HCl pH 8.4, 500 mMKCl), 0.625 uL MgCl2 (50 mM), 0.125 uL of each primer (0.01 mM), 0.0625 uL of each dNTP (10 mM), 0.060 uL of Platinum Taq Polymerase (Invitrogen). The following PCR cycling conditions were employed: 2 min at 95°C; 35 cycles of 0.5 min at 94°C, 0.5 min at 52°C, and 1 min at 72°C; 10 min at 72°C.

PCR products of the query dataset were visualized in a 1% agarose gel and subsequently purified using 10 U of Exonuclease I and 1 U Shrimp Alkaline Phosphatase at 37°C for 15 min, followed by 15 min at 80°C. Both forward and reverse DNA strands were sequenced by using sequencing primers FishF1 or FishR1 [17] and the BigDye Terminator v.3.1 Cycle Sequencing

Kit (Applied Biosystems, Inc.) on an ABI 3730 capillary sequencer following manufacturer's instructions.

COI-5P sequences were edited and aligned using MEGA version 5.05 [34], and characterized by the absence of stop codons, insertions or deletions. Sequence data were submitted to BOLD [14], and then deposited in GenBank (http://www.ncbi.nlm.nih.gov), corresponding to accession numbers: KJ709687-KJ709952; KJ709462-KJ709680; KJ768197-KJ768324.

Data Analyses

Reference dataset. The taxonomic reliability of the 128 barcodes, newly added to the reference library of DNA barcodes from the North East Atlantic, was empirically evaluated according

Table 1. Detailed partitioning of the number of species, COI-5P sequences, and GenBank accession numbers among the BOLD projects examined in this study.

Country	BOLD Project Code	Number of species	Number of COI-5P sequences	Dataset type	Source	GenBank Accessions
Portugal	FCFOP	102	659	Reference	Costa et al. 2012	JQ774505-JQ775163
Portugal	MLFPZ	1	5	Reference	Martins et al. 2012	JF718831-JF718835
Portugal	SEFMP-MLFP	73	128	Reference	Current study	KJ768197-KJ768324
Italy	SEFMP-CSFOM	109	219	Query	Current study	KJ709462-KJ709680
Malta	SEFMP-FCFMT	78	267	Query	Current study	KJ709687-KJ709952

to the ranking system described in Costa et al. [16]. The cases of congruence and lack of ambiguities of our barcode data, when compared with specimens retrieved from other public data or projects available in BOLD identification system (BOLD-IDS) [14], conferred a Grade A to that entry. Internal concordance observed only within our dataset, among a minimum of 3 specimens, was equivalent to a Grade B, where the taxonomic reliability decreases due to lack of matching sequences available in BOLD-IDS. Intraspecific sequences with a maximum of 2% (patristic) sequence divergence were attributed to Grade A or Grade B. Intraspecific distances in the COI barcode region have been comprehensively examined in multiple studies involving thousands of marine fish species [35]. These studies consistently revealed that, for the vast majority of the fish species examined (e.g. >98% according to Ward [26] and Knebelsberger et al. [36]), intraspecific distances were <2%. On the other hand, intraspecific distance >2% observed among at least 3 specimens, indicates sub-optimal concordance, and the specimen was then assigned to a Grade C. Lower grade of reliability is expressed by Grades D and E, attributed to sequences represented by a low number of individuals analyzed (1 or 2) and lacking of matching sequence available in BOLD-IDS (Grade D), or sequences with discordant species assignment, from matches with a different species or displaying paraphyly or polyphyly [16].

Pair-wise distances at different taxonomic levels (conspecific, congeneric, and confamilial) were estimated for the reference dataset by using the Kimura 2-parameter (K2P) distance model [37], implemented in BOLD (Distance Summary tool). The analyses were initially run by using the whole dataset of validated sequences, and then repeated by considering only sequences assigned to Grades A and B (689 entries), in order to exclude less reliable entries [16].

Molecular identification of species from the query dataset. To assure the blind use of query sequences in molecular assignments, taxonomic identifications initially attributed to the specimens of the query dataset were temporarily removed. First, we used only the reference dataset to build a Neighbour-Joining tree (NJ) based on the K2P distance model and generated using MEGA version 5.05 [34]. Bootstrap values for each branch node were estimated by 1000 replications. Subsequently, we added the query sequences and inspected their position in the tree. Whenever a query sequence occurred within a monophyletic cluster with <2% intra-cluster divergence, it would be assigned to the reference species forming that cluster. Sequences lacking similarities and not matching within existing clusters were representative sequences for a new cluster.

In the case of entries from the query dataset that would not match with reference sequences ranked to grade A [16], additional external confirmation was required. Species assignments were therefore verified by submitting the query sequence to the search engines BOLD Identification System (BOLD-IDS) [14] and GenBank's BLAST [38]. To avoid cross internal verification with our reference dataset, sequences already publicly available from Costa et al. [16], were not considered for matches. Specimen assignment to species was based on a minimum of 98% pair-wise sequence identity over the whole length of the barcode. If more than one matching species was found within the 98–100% identity range, the assignment was made to the one showing the highest identity. Sequences lacking matching clusters or outside the minimum similarity threshold were assumed as first-time COI-5P sequenced species and, therefore, constitute original DNA barcode additions to the global fish barcode library. In those few cases the original species assignment by means of morphological characters was retained, and later confirmed from matching sequences.

Finally, we screened the dataset in order to flag species showing intraspecific distances >2% K2P.

Results

Reference dataset

The reference dataset comprised COI-5P sequences for 146 species of marine fish (792 specimens), distributed across 115 genera, 70 families, and 25 orders. Eighty-seven species were represented by 3 to 20 individuals per species. Twelve species of the reference dataset constituted new additions to the global library of published COI-5P barcodes for marine fish (Table S1). 122 species, corresponding to 689 DNA barcodes, were assigned to grades A and B.

For the reference dataset, within-species K2P mean distance was 22x lower than average congeneric distance (0.39% and 8.91%, respectively). Average confamilial distance was 15.89% (Table 2A). The maximum intraspecific distance (18.74%) was observed for the species *Scorpaena notata*. Minimum congeneric distance (1.09%) was observed for the genus *Trachurus*. When estimating genetic distances at different taxonomic levels by using the subset of 689 sequences graded as A and B (Table 2B), mean intra-specific distance (0.28%) was 29x lower than mean inter-specific distance (8.38%), while within-family distance was 15.61%. The genus *Microchirus* had the highest within-genus distance (23.06%), while the genus *Trachurus* had the lowest (1.09%).

Query dataset

The query dataset was formed by 141 putative species (based on morphology) belonging to 110 genera, 67 families, and 26 orders, with 1 to 16 specimens per species. Sixty-five species were represented by 3–16 specimens. Seven of the species analysed from the Mediterranean Sea, namely *Aphanius fasciatus* (11 specimens), *Leucoraja melitensis* (1 specimen), *Pomatoschistus tortonesei* (2 specimens), *Raja radula* (1 specimen), *Solea aegyptiaca* (1 specimen), *Squatina aculeata* (1 specimen), and *Tetrapturus belone* (1 specimen), occur exclusively in this area [23]. An additional species *Scorpaenodes arenai* (1 specimen) has a narrow distribution limited to the Azores, in the North East Atlantic, and to the Strait of Messina, in the Mediterranean Sea [23]. Twelve species of the query dataset constitute first-time additions to the public marine fish DNA barcode library.

Reference and Query datasets – joint analyses

Based on morphological identification, the merged dataset resulted in 218 species distributed over 160 genera and 91 families. Overall, 34.9% of the species (76 species) were collected in the North East Atlantic only, 33.1% (72 species) only from the Mediterranean Sea, and 32.0% (70) were common to both areas (Table S1). To assess the rate of identification success of the molecular assignments here tested, species names were temporarily removed from all query sequences, and subsequently tested against the DNA barcodes of the reference dataset.

Species identification through molecular assignment

The global NJ tree yielded 233 monophyletic clusters with less than 2% within-cluster divergence (Fig. S1). Sixty-four single-entry clusters were generated, 29 belonging to the reference dataset and 35 to the query dataset. 65% of the query barcodes ($n = 314$) matched with a reference barcode, among which 98% (308 specimens) were unambiguously confirmed through complete match (100% similarity) (Fig. S1). Three clusters ($n = 6$) showed species mismatches (*Diplodus vulgaris/D. sargus*, *Epinephelus*

Table 2. Pair-wise COI-5P barcode distances (expressed in %; K2P model) of marine fish species from the reference barcode library, at different taxonomic levels.

A

Comparison (Intra-)	Comparison (N)	Minimum Distance	Mean Distance± SE	Maximum Distance
Species	3452	0	0.39±0	18.74
Genus	2019	1.09	8.91±0	23.06
Family	4884	3.12	15.89±0	31.28

B

Comparison (Intra-)	Comparison (N)	Minimum Distance	Mean Distance± SE	Maximum Distance
Species	3321	0	0.28±0	3
Genus	1905	1.09	8.38±0	23.06
Family	4199	3.12	15.61±0	31.28

A. Values calculated using all 792 COI-5P barcodes representing 146 species from the reference library. B. Values calculated using the subset of 689 COI-5P barcodes (122 species) assigned to Grades A and B.

costae/*Mycteroperca rubra*, and *Diaphus holti/Lobianchia gemellarii*) when the reference cluster assignments were compared with the original morphology-based identifications of the query sequences. Query specimens could not be assigned to species in 35.4% (172 sequences) of the cases, for the lack of corresponding species in the reference dataset. When query sequences were checked for their similarity with homologous sequences in public databases, BOLD-IDS returned 83% (143 entries) of the query sequences as unequivocally identified, by a minimum of 98% similarity with a matching species. 2% (3 entries distributed across 3 species) of the query sequences showed species mismatch, while 15% (26 entries/8 species) had no matching species (Table 3). BLAST searches allowed confirmation of the identity of 72% of the query sequences (n = 123), while 8% (14 entries/6 species), matched with a different species, and 20%, equivalent to 35 sequences representing 9 species, had no matching sequences (Table 3).

Cases of relatively high intraspecific genetic distances

Nine species showed intraspecific distances greater than 2% K2P and also displayed two well supported sub-specific clusters in the NJ tree (Fig. S1), namely: *Sarda sarda* (2.2%), *Raja montagui* (3.0%), *Coris julis* (3.3%), *Nezumia sclerorhynchus* (4.3%), *Diplodus annularis* (6.0%), *Scorpaena scrofa* (6.3%), *Diplodus sargus* (8.0%), *Spicara maena* (10.9%), and *Scorpaena notata* (18.74%). Moreover, in five of these species (*C. julis, D. annularis, N. sclerorhyncus, S. maena,* and *R. montagui*) the specimens sorted among the two clusters according to their geographic origin, i.e. specimens from NE Atlantic all grouped in one cluster while Mediterranean specimens all grouped in the other.

Discussion

DNA barcode-based assignments

We present a test case to the DNA barcode-based identification of 141 putative marine fish species collected in one of the richest marine biodiversity spots of the Mediterranean basin, the Sea around Sicily and Malta. We used a public reference library of DNA barcodes representing 146 marine species from the temperate NE Atlantic to assign species to 486 query sequences from specimens collected in the central Mediterranean. 65% of query sequences matched with the reference barcodes. Successful unambiguous species assignments were obtained in the majority of cases (98%) where matching sequences (and matching species)

were available in the reference dataset. Because the reference dataset is composed of a high percentage of species-barcodes annotated with grade A or B (87%), these assignments can be considered very robust.

The successful assignments typically displayed query sequences embedded within the reference haplotypes' cluster, thus showing little or no divergence between North East Atlantic and Mediterranean populations (66 species). These results support our initial premise on the feasibility of using an annotated reference library from the temperate North East Atlantic for species assignments across a neighbouring oceanic basin (i.e. the Mediterranean Sea), and strengthens the robustness of DNA barcode-based approaches for fish species identifications regardless of geographic distance, as observed elsewhere [28], [36], [39].

However, because the overlap of the species analyzed in the two datasets (reference *versus* query) was only partial, a number of sequences could not be assigned using reference library from the North East Atlantic. Nevertheless, through BOLD-IDS and GenBank searches it remained possible to assign species to 83% and 72%, respectively, of these sequences. For both BOLD-IDS and GenBank, however, the reliability of the assignments could not be confirmed or verified due to the presence of unpublished sequences in these databases.

Mismatches and ambiguities in species assignments

The small percentage of mismatches and ambiguous assignments detected (1.91%) are not necessarily failure of the DNA barcodes to discriminate among species. On the contrary, as discussed further below, most mismatches probably result from species complexes with non-stabilized taxonomic classifications, pending taxonomic revisions and clarifications, morphology-based misidentifications and differential interpretations of the validity of synonyms. A sub-set of the mismatches may have resulted from potential cryptic species. Several studies examining DNA barcodes of the ichthyofauna from other oceanic regions have found similar mismatches and ambiguities between DNA barcode data and current taxonomic knowledge in a low percentage of the species examined [36], [40], [41]. In fact, we did not find any case in our marine fish species dataset with an apparent inability of COI to distinguish species. For example, despite the set of three congeneric species of *Trachurus* in our dataset (*Trachurus picturatus, Trachurus mediterraneus* and *Trachurus trachurus*) exhibiting atypically low average congeneric distances (*Trachurus* spp., mean distance 2.28%, minimum distance 1.09%), each

Table 3. Nearest matches (BOLD, GenBank) in the identification of the specimens from the central Mediterranean query dataset.

Species	BOLD	GenBank
Aphanius fasciatus	*Aphanius fasciatus* (100%)	*Aphanius anatoliae* (87%)
Carapus acus	*Carapus bermudensis* (94%)	*Carapus bermudensis* (94%)
Diaphus holti	*Diaphus rafinesquii* (99%)	*Diaphus rafinesquii* (99%)
Epigonus telescopus	*Epigonus denticulatus* (100%)/	*Epigonus denticulatus* (99%)
	Howellas herborni (100%)	
Epinephelus caninus	*Mycteroperca rosacea* (93%)	*Epinephelus poecilonotus* (93%)
Hymenocephalus italicus	*Hymenocephalus longiceps* (83%)	*Pseudonezumia flagellicauda* (82%)
Hyporthodus haifensis	*Epinephelus chabaudi* (98%)	*Epinephelus chabaudi* (98%)
Labrus merula	*Labrus merula* (100%)	*Labrus merula* (100%)
Leucoraja melitensis	*Leucoraja melitensis* (100%)	Rajiidae (99%)
Mustelus asterias	*Mustelus asterias* (100%)/	*Mustelus palumbes* (99%)/
	M. palumbes (100%)/*M. sp.* (100%)	*M. lenticulatus* (99%)
Mycteroperca rubra	*Epinephelus costae* (100%)	*Epinephelus costae* (99%)
Pegusa impar	*Pegusa lascaris* (95%)	*Pegusa lascaris* (98%)
Scorpaenodes arenai	*Scorpaenodes sp.* SGP-2010 (88%)	*Scorpaenodes sp.* SGP-2010 (88%)
Syngnathus acus	*Syngnathus typhle* (96%)	*Symphodus tinca* (99%)
Tetrapturus belone	*Tetrapturus belone* (100%)/	*Tetrapturus pfluegeri* (100%)
	T. angustirostris (100%)/	
	T. pfluegeri (100%)	
Trachinus radiates	*Trachinus radiatus* (100%)	*Trachinus draco* (90%)
Trisopterus capelanus	*Trisopterus luscus* (96%)	*Trisopterus luscus* (96%)

Only species that did not match any sequence of the reference dataset are reported.
Values within parenthesis express the percentage of similarity (BOLD) and of identity (GenBank).

species formed an independent branch in the NJ tree, with a unique set of COI-5P haplotypes. Apparently, the low sequence differentiation among species of *Trachurus* is not an exclusive feature of the COI-5P region, and equally low distances have been observed at other locus, namely cytochrome b [42].

Through the analyses of BOLD tools (namely BOLD-IDS generated trees, data not shown), as well as published sequences and literature, we were able to infer the most likely explanation for the observed mismatches and ambiguities. For example, the mismatches involving the pair *Diplodus sargus*/*Diplodus vulgaris* apparently resulted from morphology based-misidentification of the former. DNA barcodes separate clearly these two species, as displayed in the BOLD-IDS tree (data not shown), where each species forms a separate branch with specimens from multiple locations. Hence, in this case, it appears likely that the two specimens from Portugal previously published as *D. sargus* [16] are indeed *D. vulgaris* that match with conspecifics of the query dataset from Sicily.

A number of other mismatches appear to be associated with unstable taxonomic status of some species complexes and their synonyms, as for example with the pairs *Diaphus holti*/*Lobianchia gemellarii*, *Mycteroperca rubra*/*Epinephelus costae* and *Dipturus batis*/*Dipturus oxyrhincus*. Among the latter pair, the species *Dipturus batis* is often confused with the congeneric *D. oxyrhinchus*, despite morphological and colour differences [43]. However, the validity of *D. batis* has not been confirmed in a recent study on *Dipturus* spp. of the North East Atlantic, using morphological and molecular phylogenetic analyses [44]; see also the annotation for *D. batis* in Carneiro et al. [21].

Cases of relatively high within-species distances

Among several of the species displaying relatively high within-species distances, possible mis-assignments of some specimens to a morphologically close sister species may have occurred. As opposed to mismatches described above, where DNA and morphology pointed to different species, species misidentifications were not obvious. Such discrepancies may derive from a lack of DNA barcode data for all established species within a given group. This could be the case of the small red scorpionfish *Scorpaena notata*, which occurs in two highly divergent branches (18.74%) in the BOLD-IDS tree, neither one matching with any other *Scorpaena* spp., and too divergent to be presumed to belong to the same species. Both branches comprise specimens from different origins, either from North East Atlantic or the Mediterranean. One of the clusters comprises several specimens from continental Portugal, Azores, Italy (Sicily and Liguria), Israel and Mediterranean Spain (Valencia). The other branch includes only 3 specimens, 2 from continental Portugal and 1 from Malta. According to the Food and Agricultural Organization [45], *S. notata* can be easily confused with the cadenat's rockfish *Scorpaena loppei*, and both species occur on the Mediterranean and the Portuguese coast [23], [21]. However, for *S. loppei* there is no sequence data currently available in BOLD nor in GenBank. Furthermore, the whole group is challenging from the viewpoint of identification [46]. Considering only the temperate North East Atlantic, there are 8 additional species reported [21] not included here (*Scorpaena azorica*, *Scorpaena canariensis*, *Scorpaena elongata*, *Scorpaena laevis*, *Scorpaena maderensis*, *Scorpaena plumieri*, *Scorpaena porcus*, *Scorpaena stephanica*), with only 3 of them having sequence data available in BOLD (data not shown). In the other scorpionfish investigated in our study – the red scorpionfish *Scorpaena scrofa* - we have also found intraspecific structure. Albeit in this case the divergence is not as high as with *S. notata*, it is high (6.3%) compared to variance typical of COI-5P divergence in marine fish [9], and it is likely to reflect the existence of 2 separate species. The clustering pattern for *S. scrofa* patent on the BOLD-IDS tree is somewhat intriguing. Our specimens from Malta (the only location where we collected this species) separated into two branches where they cluster together with specimens from South Africa. Still, a third, less divergent cluster, comprises specimens solely from Turkey [20]. Thus it appears that between 2 to 3 species may be included, where two of them have a wide, and at least partially overlapping, distribution range. Several synonyms are known for this species (e.g. *Scorpeana natalensis*, *Scorpaena lutea* [23]), warranting a taxonomic reconsideration of the genus. Alternatively to the above, high conspecific divergences may reflect actual cases of undescribed species or sub-species. The high divergence detected between North East Atlantic and Mediterranean specimens for the annular seabream *Diplodus annularis* (6.0%) appears to coincide with such hitherto undescribed species diversity. Specimens from this species form a clearly separated branch in the BOLD-IDS tree (data not shown), without any mismatching specimens. Further intraspecific genetic structure is visible in addition to our observations, where between 3 to 5 (or even 6) separate lineages can be discriminated. Notably, the *D. annularis* cluster includes several specimens from Turkey [20] that split in two branches, one closer to the Mediterranean specimens, and the other one to the North East Atlantic. There are DNA barcode sequences available for most *Diplodus* spp. occurring in the Mediterranean, and the only known synonym for *D. annularis* is a case of genus relocation (*Sparus annularis*), therefore reinforcing the case of hidden diversity within this species.

Sympatric populations from Portugal of the Atlantic bonito *Sarda sarda* display moderate COI-5P intraspecific distances. Specimens from Portugal match into two different groups separated by 2.2% genetic distance. There are no other species of *Sarda* spp. recorded for North East Atlantic and Mediterranean, and our sequences do not match any other *Sarda* spp. in BOLD. While we examine only a small sample size, some junior synonyms such as *Sarda mediterranea* (Bloch & Schneider, 1801) would benefit further attention.

Complete sorting between lineages was observed for the species *Nezumia sclerorhynchus*, displaying 7.1% genetic distance between 4 individuals analysed from the North East Atlantic and 2 from Malta, probably reflecting spatial genetic differentiation between the two areas. When verified in BOLD-IDS (data not shown), two clusters were observed. *N. sclerorhyncus* from the North East Atlantic did not match with specimens from other projects. Two *N. sclerorhyncus* specimens collected in Malta matched with conspecifics from Israel. The genetic distance between the two groups (North East Atlantic *versus* Malta) (7.1%) was considerably higher than the 2% observed between the *N. sclerorhynchus* cluster from North East Atlantic and specimens identified as *N. namatahi* from New Zealand.

On first inspection, our findings of high intraspecific distances within the Mediterranean rainbow wrasse *Coris julis* and the blotched picarel *Spicara maena* could be attributed to failure to recognise synonyms as valid species. According to FishBase [23], *C. julis* has a valid sister species *Coris atlantica*, known from the North East Atlantic from Cape Verde archipelago to the coast of Liberia. This species has been confirmed initially by 12S mtDNAs [47] and later using microsatellite analyses [48]. Notably, our specimens from Portugal and the central Mediterranean match

closely sequences of *C. atlantica* and *C. julis* respectively published by Kazancioglu et al. [49]. This could indicate that the *C. atlantica* specimens from Portugal were identified incorrectly as *C. julis*. However, the inspection of the BOLD-IDS tree reveals a second and distant branch containing two unpublished sequences of *C. atlantica*, amongst other clusters of *Coris* spp. Adding to these somewhat ambiguous findings, previous mtDNA and microsatellite studies [47], [48] have found some genetic differentiation between Atlantic and Mediterranean populations of *C. julis*, though results with the current work are not directly comparable because a different mtDNA gene was assayed, namely 12 s rDNA. Regarding the above mentioned picarel, our findings point to the existence of two separate species under the specimens identified as *S. maena* (10.9% within-species divergence). According to the BOLD-IDS tree, the specimen from Malta groups with another conspecific from France and two other published sequences [50], while specimens from Portugal, Israel and Italy cluster in a separate branch, where one single specimen assigned to *Spicara flexuosa* from Liguria, Italy, can also be found. There has been some dispute over the validity of *S. flexuosa*. Several sources consider it a junior synonym of *S. maena* [51], including FishBase [23]. Nevertheless, recent 16S rRNA sequence data appear to confirm very distinctly *S. maena* and *S. flexuosa* as two separate species [51]. Therefore, one of the lineages of *S. maena* we have detected could be possibly *S. flexuosa*.

Overall, we report a high level of congruence between current established taxonomic boundaries and the aggregation of DNA barcode sequences in NJ tree's branches, for most species investigated (96%). Still, the detection of several taxonomic discrepancies and of unusual levels of within-species divergence (2.02–18.74%), appear to reflect both the ongoing quest to secure robust and congruent morphology-based identifications in some species complexes, as well as possible overlooked species diversity. Mediterranean ichthyofauna diversity is likely to be especially vulnerable to such challenges. The lack of physical barriers, which have restricted migrations historically, i.e. from the cold North East Atlantic, via the Strait of Gibraltar [25], and from the warm waters of the Red Sea, through the Suez Canal [5], [6], allow greater dispersal and potential gene flow [2]. Furthermore, the complexity of population connectivity obstacles in the Mediterranean Sea, such as that documented between the two major Mediterranean sub-basins (western vs. eastern) probably contributes to high regional genetic differentiation [52], [53], [54].

Examining DNA barcode variation in multiple species across regions has allowed detection of significant population genetic differentiation, as well as highlighting possible cryptic species [28], [39]. Such information on genetic divergence of fish populations is particularly important in the case of economically relevant fish species. Patterns detected here warrant further scrutiny using additional specimens and populations, and detailed examination with complementary morphological and molecular approaches. Misidentifications frequently bias datasets due to various factors, such as lack of taxonomic expertise, operational errors within the DNA barcoding pipeline, as well as morphological ambiguities across different species.

Concluding Remarks

Here, the molecular identification of the marine ichthyofauna of the central basin of the Mediterranean Sea was verified by screening sequences against an accessible and curated reference library of DNA barcodes. Such a molecular approach, independent from morphology-based taxonomic identification, becomes crucial whenever the input from expert taxonomists is not possible, or when rare or invasive species occur [19]. The ability to identify species starting from tissue of unknown provenance, allows the employment of DNA barcodes in any application where the identification of the whole specimens is not possible. Annotated reference libraries of DNA barcodes provide a robust backbone for a variety of applications, from fish products authentication [55], [56], [57] to biosecurity [58] to the detection of the illegal use of protected or regulated species [59], and fisheries surveillance and management [11], [57]. Such molecular information is likely to be of key importance for improved stock delimitation of shared fishery resources by southern European countries. Moreover, a sustained analysis of trends in the geographic distribution of within- and among-species divergence will not only further elucidate the environmental and demographic factors impacting on marine biodiversity, but importantly also enable a consideration of threats and responses to ongoing anthropogenic change.

Currently, the main limitation appears to be the relative dearth of species' COI sequences in the reference database [60]. The occurrence of taxonomic ambiguities could be at least partially circumvented by the continuous revision and attribution of reliability grades (e.g. Knebelsberger et al. [36]). The expected expansion of the reference libraries of DNA barcodes for fish, as multiple contributions proceed [35], will likely disclose common taxonomic misidentifications and ambiguities, therefore providing opportunities for revision and clarification. The continuous growth of the reference datasets will also enable more among-region comparisons, and assist detection of unusual divergence levels among populations.

Supporting Information

Figure S1   Neighbour Joining tree of marine fishes from temperate North East Atlantic (NEA) and Mediterranean Sea (MED). NJ Tree resulting from 1278 sequences and obtained

using Kimura-2-parameter distance model. Branches are collapsed at species level and supported by bootstrap values based on 1000 replicates. For each species, the number of specimens analyzed in each region is reported within parenthesis. For congruence with Costa et al. (2012) [16], specimens identified at genus or family level are presented. In total, 218 species, 160 genera, and 91 families are here reported.

Table S1   List of 218 species analyzed and respective area of collection. NEA: North East Atlantic; MED: Mediterranean Sea; NEA and MED: species collected from both areas. † denotes the species newly analyzed in the reference dataset, compared to Costa *et al.* [16]. □ new addition to the global COI-5P library. * Species with distribution limited to the Archipelago of the Azores and to the Strait of Messina. ** Species occurring only in the Mediterranean Sea (www.fishbase.org).

Acknowledgments

We are thankful to Maria Judite Alves from the National Museum of Natural History of Portugal, for her considerable help in organizing the fish specimens' collection. We thank Jonathan Carabott who assisted in the sample collection, and Dr. Fabio Bertasi for doing the map of sampling sites. We wish to thank the constructive comments of this paper's reviewers, which greatly helped improving our manuscript.

Author Contributions

Conceived and designed the experiments: ML FOC GRC SLB MA. Performed the experiments: ML FOC SLB MA. Analyzed the data: ML FOC. Contributed reagents/materials/analysis tools: ML FOC GRC SLB MA RM MC. Wrote the paper: ML FOC. Read and approved the final manuscript: ML MD MA GB R MC GRC SLB FOC.

References

1. Coll M, Piroddi C, Steenbeek J, Kaschner K, Ben RaisLasram F, et al. (2010) The biodiversity of the Mediterranean Sea: Estimates, Patterns, and Threats. PLoS ONE 5: e11842.

2. Azzurro E, Moschella P, Maynou F (2011) Tracking Signals of Change in Mediterranean Fish Diversity Based on Local Ecological Knowledge. PLoS ONE 6: e24885.

3. Lo Brutto S, Arculeo M, Grant WS (2011) Climate change and population genetic structure of marine species. Chem Ecol 27: 107–119.

4. Bianchi CN, Morri C (2000) Marine biodiversity of the Mediterranean Sea: situation, problems and prospects for future research. Mar Pollut Bull 40: 367–376.

5. FAO (2003a) Trends in oceanic captures and clustering of large marine ecosystems—2 studies based on the FAO capture database. FAO fisheries technical paper 435. 71 p.

6. FAO (2003b) Fisheries Management-2. The Ecosystem Approach to Fisheries. FAO Technical Guidelines for Responsible Fisheries 4. 112 p.

7. Coll M, Piroddi C, Albouy C, Ben Rais Lasram F, Cheung WWL, et al. (2011) The Mediterranean Sea under siege: spatial overlap between marine biodiversity, cumulative threats and marine reserves. Glob Ecol Biogeogr 21: 465–480. doi: 10.1111/j.1466-8238.2011.00697.x

8. Hebert PDN, Cywinska A, Ball SL, de Waard JR (2003) Biological identification through DNA barcodes. Proc R Soc Lond B 270: 313–322.

9. Ward RD, Hanner R, Hebert PDN (2009) The campaign to DNA barcode all fishes, FISH-BOL. J Fish Biol 74: 329–356.

10. Joly S, Davies TJ, Archambault A, Bruneau A, Derry A, et al. (2014) Ecology in the age of DNA barcoding: the resource, the promise and the challenges ahead. Mol Ecol Res 14: 221–232.

11. Costa FO, Carvalho GR (2007) The Barcode of Life Initiative: synopsis and prospective societal impacts of DNA barcoding of fish. Genomics, Society and Policy 3: 29–40.

12. Costa FO, Antunes PM (2012) The contribution of the Barcode of Life initiative to the discovery and monitoring of Biodiversity. In: Mendonça, A., et al. (Ed.). Natural Resources, Sustainability and Humanity - A Comprehensive View. pp. 37–68.

13. Handy SM, Deeds JR, Ivanova NV, Hebert PD, Hanner RH, et al. (2011) A single-laboratory validated method for the generation of DNA barcodes for the identification of fish for regulatory compliance. J AOAC Int 94: 201–210.

14. Ratnasingham S, Hebert PDN (2007) BOLD: The Barcode of Life Data System (www.barcodinglife.org). Mol Ecol Notes 7: 355–364.

15. Steinke D, Hanner R (2010) The FISH-BOL collaborators' protocol. Mitochondrial DNA 21: 1–5.

16. Costa FO, Landi M, Martins R, Costa MH, Costa ME, et al. (2012) A ranking system for reference libraries of DNA barcodes: application to marine fish species from Portugal. PLoS ONE 7: e35858.

17. Ward RD, Zemlak TS, Innes BH, Last PR, Hebert PDN (2005) DNA barcoding Australia's fish species. Philos T Roy Soc B 360: 1847–1857.

18. Costa FO, Carvalho GR (2010) New insights into molecular evolution: prospects from the Barcode of Life Initiative (BOLI). Theor Biosci 129: 149–157.

19. Bergsten J, Bilton DT, Fujisawa T, Elliott M, Monaghan MT, et al. (2012) The effect of Geographical Scale of Sampling on DNA Barcoding. Syst Biol 61: 1–19.

20. Keskin E, Atar HH (2013) DNA barcoding commercially important fish species of Turkey. Mol Ecol Res 13: 788–797.

21. Carneiro M, Martins R, Landi M, Costa FO (2014) Updated checklist of marine fishes (Chordata: Craniata) from Portugal and the proposed extension of the Portuguese continental shelf. Eur J Taxon 73: 1–73.

22. Lago FC, Vieites JM, Espiñeira M (2012) Development of a FINS- based method for the identification of skates species of commercial interest. Food Control 24: 38–43.

23. Frose R, Pauly D (2013) FishBase. World Wide Web electronic publication. Available: http://www.fishbase.org. Version (04/2013).

24. Plan Bleu (2009) UNEP/MAP-Plan Bleu: State of the Environment and Development in the Mediterranean. UNEP/MAP-Plan Bleu, Athens, www.panbleu.org

25. Patarnello T, Volckaert FA, Castilho R (2007) Pillars of Hercules: is the Atlantic–Mediterranean transition a phylogeographical break? Mol Ecol 16: 4426–4444.

26. Ward RD (2009) DNA barcode divergence among species and genera of birds and fishes. Mol Ecol Resour 9: 1077–1085.

27. Kochzius M, Seidel C, Antoniou A, Botla SK, Campo D, et al. (2010) Identifying Fishes through DNA Barcodes and Microarrays. PLoS ONE 5: e12620.

28. Ward RD, Costa FO, Holmes BH, Steinke D (2008) DNA barcoding shared fish species from the North Atlantic and Australasia: minimal divergence for most taxa, but *Zeus faber* and *Lepidopus caudatus* each probably constitute two species. Aquat Biol 3: 71–78.

29. Martins R, Costa FO, Murta A, Carneiro M, Landi M (2012) First record of *Zenion hololepis* (Zenionidae) in Portuguese continental waters: the northernmost occurrence in the eastern Atlantic. Mar Biodivers Rec: 5: e30.

30. Tortonese E (1956) Fauna d'Italia (Vol II). Bologna: Calderini. 334 p.

31. Tortonese E (1970) Fauna d'Italia (Vol X). Bologna: Calderini. 551 p.

32. Tortonese E (1975) Fauna d'Italia (Vol XI). Bologna: Calderini. 636 p.

33. Whitehead PJP, Bauchot ML, Hureau JC, Nielsen J, Tortonese E (1984–86) Fishes of the North-eastern Atlantic and the Mediterranean. UNESCO, Paris. 1473 p.

34. Tamura K, Peterson D, Peterson N, Stecher G, Nei M, et al. (2011) MEGA5: Molecular Evolutionary Genetics Analysis using Maximum Likelihood, Evolutionary Distance, and Maximum Parsimony Methods. Mol Biol Evol 28: 2731–2739.

35. Ward RD (2012) FISH-BOL, A case study for DNA Barcodes. In: Kress JW, Erickson DL, editors. DNA Barcodes Methods and Protocols. pp. 423–439.

36. Knebelsberger T, Landi M, Neumann H, Kloppmann M, Sell AF, et al. (2014) A reliable DNA barcode reference library for the identification of the North European shelf fish fauna. Mol Ecol Res, online in advance of print. doi:10.1111/1755-0998.12238.

37. Kimura M (1980) A simple method of estimating evolutionary rate of base substitutions through comparative studies of nucleotide sequences. J Mol Evol 16:111–120.

38. Altschul SF, Gish W, Miller W, Myers EW, Lipman DJ (1990) Basic local alignment search tool. J Mol Biol 215: 403–410.

39. Zemlak TS, Ward RD, Connell AD, Holmes BH, Hebert PDN (2009) DNA barcoding reveals overlooked marine fishes, Mol Ecol Res 9: 237–242.

40. Mat Jaafar TM, Taylor MI, Nor SAM, Bruyn M, Carvalho GR (2012) DNA barcoding reveals cryptic diversity within commercially exploited Indo-Malay Carangidae (Teleosteii: Perciformes). PLoS ONE 7(11): e49623. doi:10.1371/journal.pone.0049623.

41. McCusker MR, Denti D, Van Guelpen L, Kenchington E, Bentzen P (2013) Barcoding Atlantic Canada's commonly encountered marine fishes. Mol Ecol Res 13: 177–188.

42. Cárdenas L, Hernández CE, Poulina E, Magoulas A, Kornfield I, et al. (2005) Origin, diversification, and historical biogeography of the genus *Trachurus* (Perciformes: Carangidae). Mol Phylogenet Evol 35: 496–507.

43. Cannas R, Follesa MC, Cabiddu S, Porcu C, Salvadori S, et al. (2012) Molecular and morphological evidence of the occurrence of the Norwegian skate *Dipturus nidarosiensis* (Storm, 1881) in the Mediterranean Sea. Mar Biol Res 6: 341–350.

44. Iglésias SP, Toulhoat L, Sellos DY (2009) Taxonomic confusion and market mislabelling of threatened skates: Important consequences for their conservation status. Aquat Conserv 20: 319–333.

45. Fischer W, Schneider M, Bauchot ML (1987) Fiches FAO d'identification des espèces pour les besoins de la pêche. FAO, Rome, 2: 761–1530.

46. Ordines F, Valls M, Gouraguine A (2012) Biology, Feeding, and Habitat preferences of cadenat's rockfish, *Scorpaena loppei* (Actinopterygii: Scorpaeniformes: Scorpaenidae), in the Balearic Island (Western Mediterranean). Acta Ichthyologica et Piscatoria 42: 21–30.

47. Guillemaud T, Cancela ML, Afonso P, Morato T, Santos RS, et al. (2000) Molecular insights into the taxonomic status of *Coris atlantica* (Pisces: Labridae). J Mar Biol Assoc U.K. 80: 929–933.

48. Aurelle D, Guillemaud T, Afonso P, Morato T, Wirtz P, et al. (2003) Genetic study of *Coris julis* (Osteichtyes, Perciformes, Labridae) evolutionary history and dispersal abilities. C R Biol 326: 771–785.

49. Kazancioglu E, Near TJ, Hanel R, Wainwright PC (2009) Influence of sexual selection and feeding functional morphology on diversification rate of parrot fishes (Scaridae). Proc. R. Soc. B, 276: 3439–3446.

50. Yamanoue Y, Miya M, Matsuura K, Yagishita N, Mabuchi K, et al. (2007) Phylogenetic position of tetraodontiform fishes within the higher teleosts: Bayesian inferences based on 44 whole mitochondrial genome sequences. Mol Phylogenet Evol, 45: 89–101.

51. Imsiridou A, Minos G, Gakopoulou A, Katsares V, Karids T, et al. (2011) Discrimination of two picarel species *Spicara flexuosa* and *Spicara maena* (Pisces: Centracanthidae) based on mitochondrial DNA sequences. J Fish Biol 78: 373–377.

52. Maggio T, Lo Brutto S, Garoia F, Tinti F, Arculeo M (2009) Microsatellite analysis of red mullet *Mullus barbatus* (Perciformes, Mullidae) reveals the isolation of the Adriatic Basin in the Mediterranean Sea. ICES J Mar Sci 66: 1883–1891.

53. Mejri R, Arculeo M, Hassine B, Lo Brutto S (2011) Genetic architecture of the marbled goby *Pomatoschistus marmoratus* (Perciformes, Gobiidae) in the Mediterranean Sea. Mol Phylogenet Evol 58: 395–403.

54. Mejri R, Lo Brutto S, Hassine B, Arculeo M (2009) A study on *Pomatoschistus tortonesei* Miller 1968 (Perciformes, Gobiidae) reveals the Siculo-Tunisian Strait (STS) as a breakpoint to gene flow in the Mediterranean basin. Mol Phylogenet Evol 53: 596–601.

55. Barbuto M, Galimberti A, Ferri E, Labra M, Malandra R, et al. (2009) DNA barcoding reveals fraudulent substitutions in shark seafood products: The Italian case of "palombo" (Mustelus spp.). Food Res Int 43: 376–381.

56. Hajibabaei M, Smith MA, Janzen DH, Rodriguez JJ, Whitfield JB, et al. (2006) A minimalist barcode can identify a specimen whose DNA is degraded. Mol Ecol Notes 6: 959–964.

57. Hanner R, Becker S, Ivanova NV, Steinke D (2011) FISH-BOL and seafood identification: Geographically dispersed case studies reveal systemic market substitution across Canada. Mitochondrial DNA 22: 106–122.

58. Armstrong KF, Ball SL (2005) DNA barcodes for biosecurity: invasive species identification. Philos T Roy Soc B 360: 1813–1823.

59. Rasmussen RS, Morrissey MT (2008) DNA-Based Methods for the Identification of Commercial Fish and Seafood Species. Compr Rev Food Sci Food Saf 7: 280–295.

60. Cerutti-Pereyra F, Meekan MG, Wei N-WV, O'Shea O, Bradshaw CJA, et al. (2012) Identification of Rays through DNA Barcoding: An Application for Ecologists. PLoS ONE 7: e36479. doi:10.1371/journal.pone.0036479.

14

A New Role for Carbonic Anhydrase 2 in the Response of Fish to Copper and Osmotic Stress: Implications for Multi-Stressor Studies

Anna de Polo[1]*, Luigi Margiotta-Casaluci[1,2], Anne E. Lockyer[1], Mark D. Scrimshaw[1]

1 Institute for the Environment, Brunel University, London, United Kingdom, **2** AstraZeneca, Global Health, Safety and Environment, Freshwater Quarry, Brixham, United Kingdom

Abstract

The majority of ecotoxicological studies are performed under stable and optimal conditions, whereas in reality the complexity of the natural environment faces organisms with multiple stressors of different type and origin, which can activate pathways of response often difficult to interpret. In particular, aquatic organisms living in estuarine zones already impacted by metal contamination can be exposed to more severe salinity variations under a forecasted scenario of global change. In this context, the present study aimed to investigate the effect of copper exposure on the response of fish to osmotic stress by mimicking in laboratory conditions the salinity changes occurring in natural estuaries. We hypothesized that copper-exposed individuals are more sensitive to osmotic stresses, as copper affects their osmoregulatory system by acting on a number of osmotic effector proteins, among which the isoform two of the enzyme carbonic anhydrase (CA2) was identified as a novel factor linking the physiological responses to both copper and osmotic stress. To test this hypothesis, two *in vivo* studies were performed using the euryhaline fish sheepshead minnow (*Cyprinodon variegatus*) as test species and applying different rates of salinity transition as a controlled way of dosing osmotic stress. Measured endpoints included plasma ions concentrations and gene expression of CA2 and the α1a-subunit of the enzyme Na^+/K^+ ATPase. Results showed that plasma ions concentrations changed after the salinity transition, but notably the magnitude of change was greater in the copper-exposed groups, suggesting a sensitizing effect of copper on the responses to osmotic stress. Gene expression results demonstrated that CA2 is affected by copper at the transcriptional level and that this enzyme might play a role in the observed combined effects of copper and osmotic stress on ion homeostasis.

Editor: Fanis Missirlis, CINVESTAV-IPN, Mexico

Funding: The first author was the recipient of an Isambard Scholarship grant. The funders had no role in study design, data collection and analysis, decision to publish, or preparation of the manuscript.

Competing Interests: The authors therefore have no competing or financial interests to declare. AstraZeneca had no role in the study design, the collection, analysis and interpretation of the data.

* Email: annadp84@gmail.com

Introduction

Ecotoxicological studies on environmental chemicals are usually performed under optimal exposure conditions, whereas aquatic organisms in their natural settings have to cope with additional stressors, such as variations in temperature, oxygen levels or salinity, which can affect the way they respond to chemical stressors [1]. Conversely, exposure to chemical stressors could impair organisms' responses to changes in environmental factors. In either cases, this two-way interaction may or may not lead to more adverse effects on the organisms and ultimately on the population, since the correlation between tolerance to chemical and non-chemical stressors can be positive or negative, depending on the type and level of stressors in question (e.g. [2,3]).

The potential interactions between toxic substances and environmental factors represent one of the main challenges for ecotoxicologists. The importance of refining toxicity studies and applying them to more complex scenarios has been recognized by both the EU and US EPA, who have expressed an increasing interest in studying the responses of biological systems to a combination of stressors, both chemical and environmental, rather than to single chemical in stable conditions [4,5]. An integrated examination of chemical and non-chemical stressors is especially pertinent when considering chemical pollution in the context of global change scenarios, where more fluctuating environmental variables are plausible to influence the responses of biological systems to chemical exposure [6,7]. This is again a two-way interaction, because global change can make organisms more sensitive to chemical stressors as well as exposure to chemical stressors can make organisms more sensitive to changes in environmental stressors caused by global change.

In this context, the challenge is to identify physiology-based interactions between non-chemical and chemical stressors affecting key physiological processes in an organism. As a first step to applying a multi-stressor approach into ecotoxicological studies, it is crucial to acquire an understanding of the mechanisms of action of the stressors in question, by dissecting the biological pathways through which they exert their effects [8]. One possible approach in this sense is the application of the adverse outcome pathway (AOP) concept, a unified framework that links molecular initiating events with the cascade of responses occurring across all levels of biological organization [9]. Such approach, despite being mainly qualitative [10], does highlight the importance of dissecting the physiological mechanisms and toxicodynamic processes underpinning the complexity of many biological responses, which is a key step in interpreting the stress-response dynamics displayed by any biological system in the complexity of the real world.

With this aim in mind, the present study examined the mechanisms of interaction between one chemical stressor, i.e. copper, and one environmental stressor, i.e. osmotic stress, which was applied in the form of salinity transitions from either freshwater to saltwater or from saltwater to freshwater. In order to put this laboratory-based study into an environmentally realistic context, we chose to investigate copper-salinity interactions by mimicking under laboratories conditions the salinity changes occurring in estuaries, which are environments where the combination of anthropogenic impacts and fluctuating abiotic factors represents an ideal context to study the interactions between chemical and non-chemical stressors. Considering future global change scenarios and their impacts on estuarine and coastal pollution [11], aquatic organisms inhabiting transitional environments are likely to be exposed to more severe and/or more frequent salinity fluctuations in the future, as a result of increased frequency of extreme events [12]. Therefore salinity changes, also defined as osmotic stress, were chosen as the non-chemical stressor to test in this study. As for the chemical stressor, we selected a metal (i.e. copper) because in general metal contamination is of particular concern in estuarine and coastal environments, where the historically high anthropogenic impacts, mainly due to shipping, urbanization and industrialization, often lead to elevated concentrations of metals both in the water column and in the sediments [13]. In particular, one of the main concerns about transitional zones is represented by copper, whose use as biocide in antifouling painting coatings has increased since TBT (tributyltin) and other organic biocides has been phased out [14]. We therefore focused on the interactions between copper and salinity changes, given their theoretical as well as applicative relevance for multi-stressor studies applied to transitional environments in a global change perspective. However, since this topic has been almost exclusively addressed from the point of view of the effects of different salinities on copper toxicity [15–18], we instead put more emphasis on the effects of copper exposure on the response of aquatic organisms, i.e. fish, to salinity transitions. In accordance with the mechanistic intent of the study, we hypothesized that, given that acute copper toxicity is mainly a consequence of osmotic disruption [19–21], copper-exposed fish are more sensitive to osmotic stresses, as copper can affect their osmoregulatory system by interacting with a number of osmotic effector proteins, such as the enzyme Na^+/K^+ ATPase (e.g. [22]). As we previously discussed in a review on copper toxicity in saline environments [23], a critical screening of the ecotoxicological data available in the literature on copper toxicity and salinity led to the hypothesis that one factor linking the main physiological responses to both copper and osmotic stress is the cytosolic isoform-2 of the enzyme carbonic anhydrase (CA2), potentially a copper target [24–26] with salinity dependent expression and activity [27–29]. Because CA2 displays its osmoregulatory functions not only in the gills, but also in the intestine of fish [30], we argued that in saline environments the intestine should be considered alongside the gills as a site of action for copper [23]. We hypothesized that the

combined effect of copper and osmotic stress is not the overall product of two independently acting factors, but rather the outcome of a mechanistic interaction between two stressors that act through similar pathways and affect similar effector proteins.

The *in vivo* experiments reported here were aimed at testing that hypothesis, using the euryhaline fish sheepshead minnow (*Cyprinodon variegatus*) as test species and applying different types and rates of salinity transitions as a way to administrate different doses of osmotic stress. Results of plasma sodium levels measured before and after the salinity transitions overall supported the hypothesis of a copper-disrupted response to osmotic stress, whilst gene expression results provided a mechanistic understanding of the copper-salinity interaction and supported the argument that the CA2 enzyme plays a role in the combined responses to copper and osmotic stress.

Materials and Methods

Ethics statement

These studies were carried out under project and personnel licences granted by the Home Office under the United Kingdom Animals Act (Scientific Procedures).

Experiment 1

Test species. Adult male and female sheepshead minnows (*Cyprinodon variegatus*) were obtained from the Brixham Environmental Laboratory (Brixham, UK) and acclimated to freshwater for approximately three weeks under flow-through conditions. Fish age was 4 month post-hatching and average wet weight was 4.7 ± 1.1 g.

Experimental design and protocol. The experiment was carried out using a continuous flow-through system. Thermostatically heated dechlorinated carbon filtered tap water, from a header tank, flowed through 6 flow-meters into 6 mixing chambers via silicon tubing (medical grade, VWR) at a rate of 120 ml/min. From each mixing chamber the water flowed into the fish tanks through silicone tubing. In total there were 6 glass fish tanks each with a working volume of 11 L. Each tank was aerated and received approximately 15 tank volume renewals per day. During the exposure period, copper was added as $CuSO_4 \cdot 5H_2O$ dissolved in 1% HNO_3 (Optima grade, Fisher Scientific) and dosed into each individual mixing chamber via a peristaltic pump at a rate of 0.12 ml/min. The stock solutions and mixing regime were designed to yield final nominal copper concentrations of 10 and 100 µg/L. Dilution water and chemical flow rates were checked twice per day and adjusted if necessary. Temperature, pH, oxygen, total alkalinity (see Table S1), nitrate and ammonia were monitored and recorded daily. Water samples for total organic carbon (TOC), copper and major constituents analyses were collected in 50 ml centrifuge tubes on days 3, 6, 8 and 9. Samples for TOC analysis were stored at $-20°C$ after collection, whereas samples for copper and major constituents analysis were acidified with 1% HNO_3 and stored at 4°C.

The 9-day exposure experiment consisted of 3 treatments: 0, 10 µg/L and 100 µg/L nominal copper concentrations. All treatments were in freshwater (FW) for the first 8 days of exposure. Each treatment was composed of 2 single-sex tanks, each tank containing 6 fish (12 fish per treatment, 6 males and 6 females). Fish were maintained in a photoperiod of 16 h of light followed by 8 h of dark, with 20 min dawn/dusk transition, and were fed *ad libitum* three times per day: twice with flake food (King British Tropical flake food, Lillico, Surrey) and once with brine shrimp (Tropical Marine Centre, gamma irradiated). Food was withheld 20 h prior to each sampling. After 8 days of exposure in FW, 3 fish per tank (6 fish per treatment) were humanely sacrificed according to UK Home Office procedures, using an overdose of buffered ethyl 3-aminobenzoate methanesulfonate (300 mg MS222/L, adjusted to pH 7.8, the average pH throughout the study). Blood samples were collected from the caudal peduncle using heparinised capillary tubes, transferred into Eppendorf tubes and kept on ice until plasma was separated by centrifugation at 14,000 g for 5 min, removed and stored at − 20°C. Fish were weighed and measured (fork length). Tissue samples (gills, liver and intestine) were collected by dissection. The intestine was divided into three segments (anterior, mid and posterior intestine) and sub-samples of each segment were kept for gene expression analysis. This procedure was chosen in order to account for potential differences in gene expression along the intestinal tract. All tissue samples were immediately frozen in liquid nitrogen after collection and stored at −80°C.

Salinity transition. After the first sampling on day 8, the salinity in all treatments was increased from 0 to 20 ppt over a time of 4 h by dosing hyper-concentrated (200 ppt) saltwater (SW) prepared by dissolving and mixing synthetic seasalts (Tropic Marin) in three glass tanks of 40 L volume each, filled with dechlorinated carbon filtered tap water. The brand of synthetic seasalts (Tropic Marin) used to prepare the saline stock solutions was selected as it has been shown to have a low content of trace elements, particularly copper, compared to other commercially available brands [31]. The hyper-saline stock solution was prepared 24 h in advance and left overnight with strong aeration to allow the pH to stabilize. SW was dosed from the stock solution into the mixing chambers at a rate of 12 ml/min to yield a final salinity of 20 ppt in all fish groups. Salinity was monitored every 15 min with a refractometer throughout the salinity transition period. After reaching 20 ppt, fish in all treatments were held in SW for 24 h, until the second sampling on day 9. Copper dosing was maintained constant during the entire study. On the 9th day of exposure, the remaining fish (6 per treatment) were sampled following the same procedure described above, with the exception of the MS222 solution, which was prepared in SW instead of FW and buffered at a pH of 8.4 (instead of 7.8), consistent with the salinity and average pH of the tanks after the salinity transition.

Experiment 2

Test species. Adult male and female sheepshead minnows (*Cyprinodon variegatus*) were obtained from a livestock bred and maintained at Brunel University (UK) in a SW re-circulated system supplied with UV-sterilizer, sand biofilter and protein skimmer (Marine Compact Filtration System, Tropical Marine centre). The synthetic seasalts (Tropic Marin) used for maintaining the system were the same used for the preparation of the artificial SW during the two experiments.

Approximately one month before the start of the experiment, a sub-stock of fish has been gradually moved to FW and then kept under these conditions for three weeks before starting the

exposure. Fish age at the beginning of the experiment was approximately 6 months post-hatching and average wet weight was 2.7 ± 0.7 g.

Experimental design and protocol. The experiment was carried out using a continuous flow-through system and artificial SW dosed from a concentrated stock for the groups held in SW. Thermostatically heated dechlorinated carbon filtered tap water, from a header tank, flowed through 16 flow-meters into 16 mixing chambers via silicon tubing (medical grade, VWR) at a rate of 60 ml/min. This final flow of 60 ml/min was half the rate of the one used in the Experiment 1 (120 ml/min). Thus, a full replacement was now reached within double the time, which implied that the salinity switch on day 20 would then be performed over 8 h instead of 4, hence representing a milder osmotic stress, but still within the time range of a tidal change in an average estuary. From each mixing chamber the water flowed into the fish tanks through silicone tubing. In total there were 16 glass fish tanks with a working volume of 11 L. Each tank was aerated and received approximately 7 tank volume renewals per day. During the exposure period, copper was added as $CuSO_4 \cdot 5H_2O$ dissolved in 1% HNO_3 (Optima grade, Fisher Scientific) and dosed into each individual mixing chamber via a peristaltic pump at a rate of 0.06 ml/min. The copper stock solutions and mixing regime were designed to yield final nominal copper concentrations of 32, 100 and 320 µg/L. Dilution water and chemical flow rates were checked twice per day and adjusted if necessary. Temperature, pH, oxygen, total alkalinity (see Table S2), nitrate and ammonia were measured and recorded daily. Water samples for TOC, copper and major constituents analysis were collected in 50 ml centrifuge tubes on days 1, 4, 8, 12, 16, 19 and 21. Water samples for TOC analysis were stored at −20°C after collection and water samples for copper and major constituents analysis were acidified with 1% HNO_3 and stored at 4°C.

The 21-day exposure experiment consisted of 4 treatments: 0, 32, 100 and 320 µg/L nominal copper concentrations (refer to Figure S2 for an outline of the experimental design and set-up). Each treatment consisted of 4 tanks, 2 in FW (0 ppt) and 2 in SW (20 ppt), of which one replicate tank contained 10 males and one 10 females ($n = 12$ in control tanks). SW conditions were maintained by dosing hyper-concentrated SW (100 ppt) from a dosing stock prepared daily by dissolving and mixing synthetic seasalts (Tropic Marin) in two fibreglass tanks of 80 L volume each, filled with dechlorinated carbon filtered tap water. The hyper-saline stock solution was made at least 24 h in advance and left overnight with strong aeration to allow the pH to stabilize. SW was transferred into two intermediate dosing stocks (40 L volume each) and from there dosed into the mixing chambers at a rate of 12 ml/min to yield a final salinity of 20 ppt in the SW groups (2 tanks per treatment). Throughout the experiment, fish were maintained under the same photoperiod and food regime as in the Experiment 1.

Table 1. Primers sequences.

Gene	Forward Primer	Reverse Primer	T (°C)
CA2	5′− GAAGGTTCTGGATGCTTTGG − 3′	5′− AGTTGGAGAAGGTGGTCTGC − 3′	59
NKA	5′− GCCACACAGCCTTCTTCAC − 3′	5′− ACAATAGAGTTCCTCCTGGTCTTG − 3′	59
18S	5′− GCTGAACGCCACTTGTCC − 3′	5′− CTCAGAGCAAGCAATAGCCTTA − 3′	57

Primers used for qPCR and respective optimal annealing temperatures.

Table 2. Inorganic copper speciation.

Inorganic copper speciation		
Species	% of total copper	
	FW	SW
Cu^{2+}	28.2	14.2
$Cu(OH)^+$	55.3	44.5
$Cu(OH)^2$	7.44	14.07
$Cu_2(OH)_2^{2+}$	4.44	4.53
$CuCl^+$	0.02	2.28
$CuCO_3$	3.75	18.26

Most relevant copper forms are reported (expressed as % of total dissolved copper concentrations), repectively in FW and SW, as calculated by Visual MINTEQ using measured water chemistry parameteres of Experiment 1.

After 19 days of copper exposure in either FW or SW, half of the fish (10 per treatment, 12 from controls) were humanely sacrificed according to UK Home Office procedures, using an overdose of ethyl 3-aminobenzoate methanesulfonate (300 mg MS222/L, buffered at pH 7.6, the average pH throughout the study, and adjusted to a salinity of either 0 or 20 ppt, consistently with the exposure conditions). Blood samples were collected and stored as indicated for Experiment 1. Fish were weighed and measured (fork length). Tissue samples of gills and intestine were collected by dissection. The intestine was divided into two segments (mid-anterior and posterior) and only sub-samples of the mid-anterior segment were kept for gene expression analysis, since preliminary gene expression analyses had shown that the mid-anterior segment displayed slightly higher expression levels of the measured genes, compared to the posterior segment of the intestinal tract (data not shown). All tissue samples were immediately frozen in liquid nitrogen after collection.

Salinity transition. On day 20 of exposure (after the first sampling), the salinity in all treatments was either increased from 0 to 20 ppt in the groups previously held in FW or decreased from 20 to 0 ppt in the SW groups, over a time of 8 h. This was achieved by swapping the final tubing between the mixing chambers and the fish tanks (Figure S2), with a time lag of two hours between treatments, in order to assure that each group would be held in the new conditions for the same time (24 h), assuming a sampling rate of 10 fish per hour on the following day.

Salinity was monitored at least every hour with a refractometer throughout all the salinity switch period (see Figure S1). After reaching either 0 or 20 ppt in all groups, fish were held in the new conditions for 24 h until the second sampling. Copper dosing was maintained constant during the entire study. On day 21 of exposure, the remaining fish (8 to 10 per treatment, 12 from controls) were sampled following the same procedure described for Experiment 1.

Chemical analyses and speciation modelling

Copper, total organic matter and major cations in water. Total copper concentrations in fish tank water were determined by either graphite furnace, GF-AAS (4100ZL Zeeman Atomic Absorption Spectrometer, Perkin Elmer) or flame atomic absorption spectroscopy, F-AAS (AAnalyst100 AAS, Perkin Elmer), depending on the expected range of concentrations, using standard operating conditions and dilutions when necessary. For analytical procedures, all acid was nitric acid (Optima grade, Fisher Scientific). In order to account for sodium interference in the SW series, calibration standards were prepared in acidified Milli-Q water, with the addition of artificial seasalts to yield a final salinity of 20 ppt (or lower, when samples were diluted). Copper standards were run every 6 samples to check measurement accuracy. Total organic carbon (TOC) was determined as the Non-Purgeable Organic Carbon fraction (NPOC) by high-temperature catalytic oxidation using a Shimadzu total organic

Table 3. Total copper speciation.

Total copper speciation		
Species	% of total copper	
	FW	SW
Cu^{+2}	3.08	2.46
$Cu(OH)^+$	6.02	7.71
$Cu(OH)^2$	0.81	2.44
$Cu_2(OH)_2^{2+}$	0.05	0.14
$CuCO_3$	0.41	3.17
Organic fraction	89.6	83.6

Most relevant copper forms are reported (expressed as % of total dissolved copper concentrations), repectively in FW and SW, as calculated by Visual MINTEQ using measured water chemistry parameteres of Experiment 1.

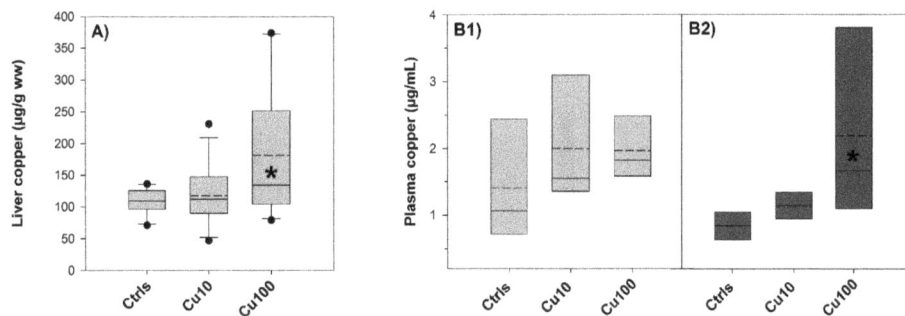

Figure 1. Copper in liver and plasma – Exp.1. (A) Copper liver burden (expressed as µg/g wet weight) in fish exposed to 0 (Ctrls), 10 and 100 µg/L copper for 9 days. Values are means ± SE ($n = 10/12$). The asterisk denotes statistically significant difference from the control value ($P<0.05$, Tukey t-test) (**B1–2**) Plasma copper levels (µg/ml) in fish exposed to 0, 10 and 100 µg/L copper and sampled either before (B1) or after (B2) the salinity transition. Values are means ± SE ($n = 4/7$). The asterisk denotes statistically significant difference from the control value post switch ($P<0.05$, Tukey t-test). Solid horizontal lines represent median values and dashed lines represent mean values.

carbon-V CPN Analyzer. Major cations (Na^+, Mg^{2+} and Ca^{2+}) concentrations in water samples were determined by F-AAS after appropriate dilution.

Copper speciation modelling. Based on water parameters, measured TOC, cations and total dissolved copper concentrations (Table S1, S3 and S4), inorganic and total speciation of copper in exposure conditions (both in FW and SW) was calculated using the Visual MINTEQ 3.0 software [32]. For the total speciation calculations, charges on dissolved organic matter were calculated based on speciation and the % of organic C was set at 50.

Copper burden in liver. Six individual 0.2 g aliquots of certified standard DOLT-4 (dogfish liver certified reference material for trace metals, NRC, Canada) were pre-digested overnight in 4 ml of 100% HNO_3, digested in a Microwave Accelerated Reaction System (MARS X, CEM Corporation) following the heating programme recommended by the manufacturer for fish tissue digestion (15 min ramp-to-temperature time and 15 min hold time at 200°C), diluted appropriately and analysed by F-AAS. The average wet weight of the liver samples was 0.1±0.06 g. For each batch of samples, one blank and one certified standard were analysed and the average percent recovery was 119±19% ($n = 5$).

Copper content in fish food. Copper content in fish food was analysed by GF-AAS after pre-digestion with HNO_3, microwave digestion (15 min ramp-to-temperature time and 15 min hold time at 200°C) and appropriate dilution with Milli-Q water.

Copper, chloride and major cations in plasma. Plasma copper concentrations were determined by GF-AAS after dilution in acidified Milli-Q water (1% HNO_3). Plasma Na^+, Mg^{2+} and Ca^{2+} concentrations were measured by F-AAS and plasma Cl^- concentrations by ion chromatography (DIONEX DX, Dionex Corp.), in both cases after appropriate dilution.

Bioinformatics and molecular analyses

RNA extraction and cDNA synthesis. Total ribonucleic acid (RNA) was isolated from individual gills and intestine samples (mean tissue weights were respectively 25.8±8.6 and 17.7±9.2 mg) using the RNeasy Fibrous Tissue Mini Kit (Qiagen), according to the manufacturer's instructions, which included tissue homogenization in buffer RLT using a TissueLyser II (Qiagen) for 3 min at maximum speed. The protocol also included a DNAse step to eliminate contaminating genomic DNA. The extracted RNA was resuspended in 50 µl of RNase-free water. Quantity and purity of each RNA sample were determined by spectrophotometry (Nanodrop, Fisher Scientific), and RNA integrity was visually checked by agarose gel electrophoresis. Complementary DNA (cDNA) was synthesised from 2 µg total RNA using Invitrogen SuperScript III First-Strand Synthesis System for reverse transcription-PCR kit according to the manufacturer's protocol, using random hexamers to prime synthesis. Diluted (1:5) cDNA samples were assessed for carbonic anhydrase isoform-2 (CA2) and Na^+/K^+ ATPase α1.a5-isoform (NKA) expression using quantitative real-time PCR (qPCR).

NKA and CA2 primer design. Sheepshead minnow specific qPCR primers were developed for both NKA and CA2 from partial *C. variegatus* mRNA sequences identified from GenBank using basic local alignment search tool (BLAST) searches. For NKA, no previously characterized sequence of *C. variegatus* NKA was available in the National Centre for Biotechnology Information (NCBI) databases (http://www.ncbi.nlm.nih.gov/). However, the screening of NCBI Expressed Sequence Tags (ESTs) database led to the identification of four highly similar *C. variegatus* ESTs (GenBank Acc. No. GE337281.1, GE337212.1, GE334919.1 and GE336240.1). GE337281.1 demonstrated 79% identity with the 3' region of the NKA sequence expressed in *Oncorhynchus mykiss*

Table 4. Degree of change in plasma ions concentrations – Exp.1.

	Sodium	Chloride	Magnesium	Calcium
Ctrls	20	38	11	19
Cu10	16	50	−15	13
Cu100	30	56	21	−12

Differences (expressed as percentage) in plasma sodium, chloride, magnesium and calcium levels before and after the salinity transition in fish exposed to 0 (Ctrls), 10 and 100 µg/L copper.

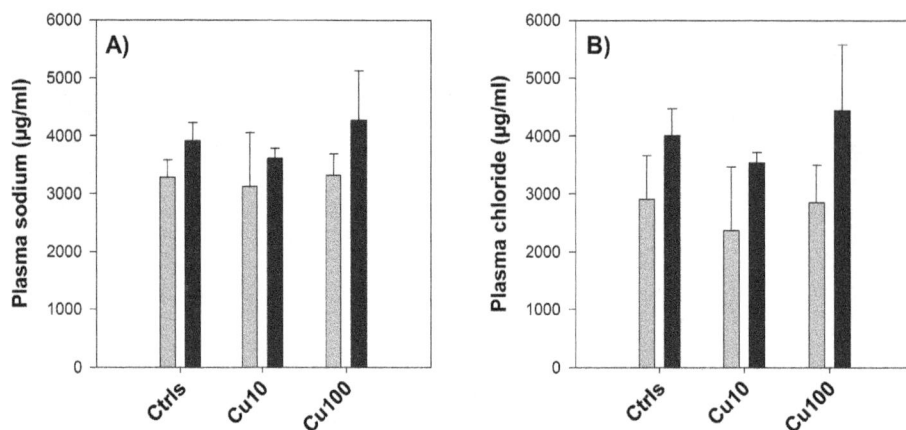

Figure 2. Plasma sodium and chloride – Exp.1. Plasma sodium (A) and chloride (B) levels (μg/ml) measured in fish exposed to 0 (Ctrls), 10 and 100 μg/L copper before (light grey bars) and after (dark grey bars) the salinity transition. Values are means ± SE (n = 6/7).

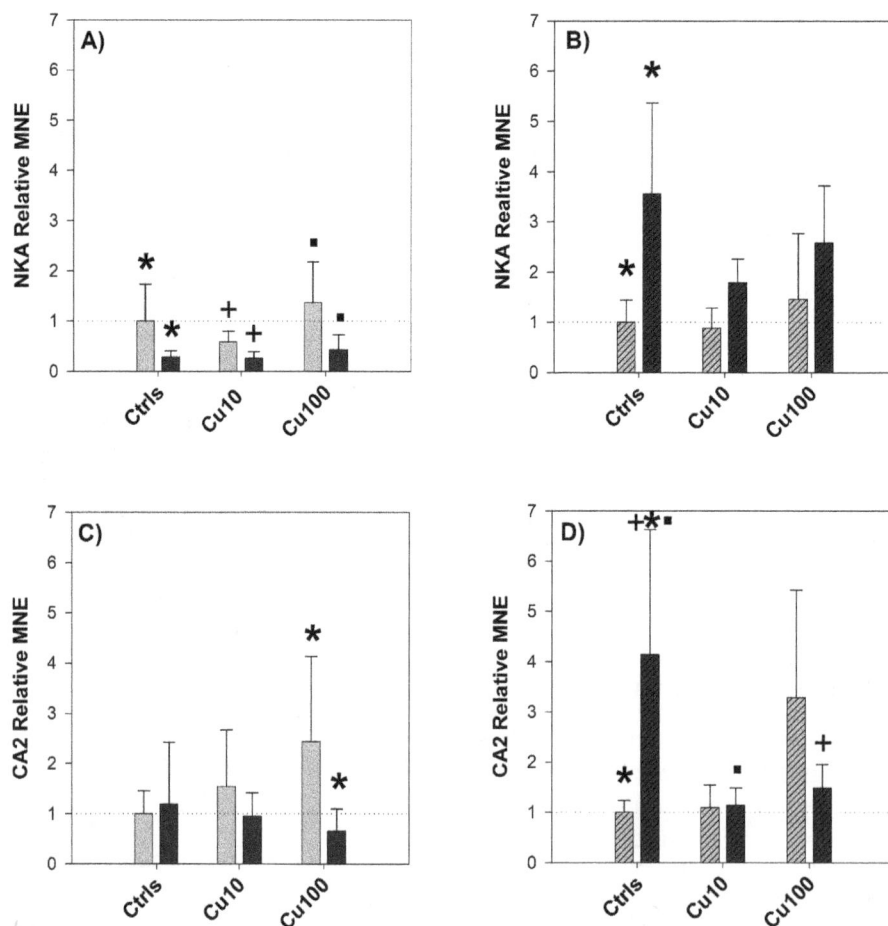

Figure 3. Pre-post change in NKA and CA2 expression – Exp.1. Relative Mean Normalized Expression (MNE) levels of NKA (graphs A and B), and CA2 (graphs C and D) measured in the gills (A and C) and in the mid-anterior tract of the intestine (B and D, striped bars) of sheepshead minnows exposed to 0 (Ctrls), 10 and 100 μg/L copper. Within each copper treatment, the pair of bars represents expression levels before (left side) and after (right side) the salinity transition. Relative MNE levels were determined by qPCR, normalized to control gene (18S) and expressed as fold change relative to pre-switch control value, which was set at 1 (dotted horizontal line). Values are means ± SE (n = 6/7). In each graph, bars sharing the same symbol (asterisk, cross or dot) are significantly different one from the other (P<0.05, Tukey t-test).

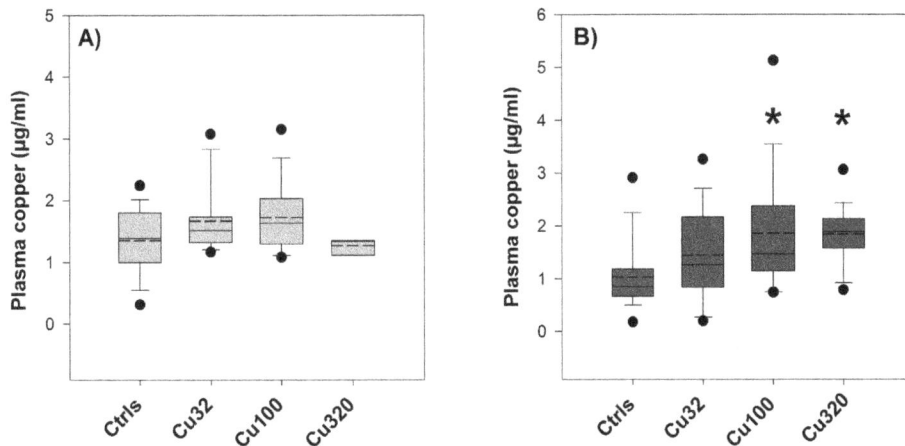

Figure 4. Plasma copper – Exp.2. Plasma copper levels (μg/ml) measured in fish exposed to 0 (Ctrls), 32, 100 and 320 μg/L copper in either FW (A) or SW (B). Values are means ± SE ($n = 18/20$, except for FW Cu320, where $n = 3$). The asterisk denotes statistically significant difference from the control value ($P < 0.05$, Tukey t-test). Solid horizontal lines represent median values and dashed lines represent mean values.

and in BlastX searches it identified NKAα1-isoforms from several species, including *Danio rerio* (Acc. No. AAH90285), which showed high similarity with *C. variegatus* EST and hence confirmed its identity. For CA2, a partial mRNA sequence of *C. variegatus* putative cytosolic carbonic anhydrase (cCA) (Acc. No. HM142344.1) had been previously identified. HM142344.1 showed 81% similarity with the 5′ region of the EST of CA2 expressed in *Pimephales promelas* (Acc. No. DT261041.1), which in turn identified *Oncorhynchus mykiss* carbonic anhydrase isoform-2 (NP_954685) in BlastX searches, confirming its identity and hence that of the original *C. variegatus* sequence. These identified sequences were used to design the primers for NKA and CA2 (Table 1), with the assistance of PRIMER3 web software (http://bioinfo.ut.ee/primer3-0.4.0/).

NKA and CA2 gene expression. The qPCR primers designed for *C. variegatus* CA2 and NKA were verified using PCR. Reactions contained diluted gill/intestine cDNA (1:5), 1× Buffer, 1.5 mM MgCl$_2$, 0.2 mM dNTPs mix, 0.5 μM forward and reverse primer and 0.1 μM *Taq* Polymerase. Cycling conditions were 2 min initial denaturation at 95°C, 35 cycles at 95°C for 30 sec, annealing at 58/62°C for NKA/CA2 respectively, then

72°C for 20 sec and final extension of 5 min at 72°C. PCR products were electrophoresed on 2% agarose gel containing GelRed (Biotium) to verify a single product and correct amplicon size. qPCRs for CA2, NKA and the reference gene 18S were performed in triplicate on cDNA from individual fish using a CFX96 Real-Time PCR detection system (Bio-Rad) and Fast SYBR Green Master Mix (Invitrogen) as per instructions of the manufacturer. Reactions were optimized for annealing temperature (see Table 1) and run for 2 min at 50°C followed by 10 sec at 95°C, then 40 cycles of 10 sec at 95°C, 30 sec at optimal annealing temperature and then 1 min ramp from 55 to 65°C. Finally a dissociation curve was obtained (melt curve: 65 to 95°C, increment 0.5°C for 0.05 sec) to confirm single products in each reaction. The relative expression of the target genes were normalized to the expression of the housekeeping gene 18S using the Excel-based software qGENE [33], which takes into account the amplification efficiency of both the target genes and the reference gene to calculate the mean normalized expression (MNE) of each target gene.

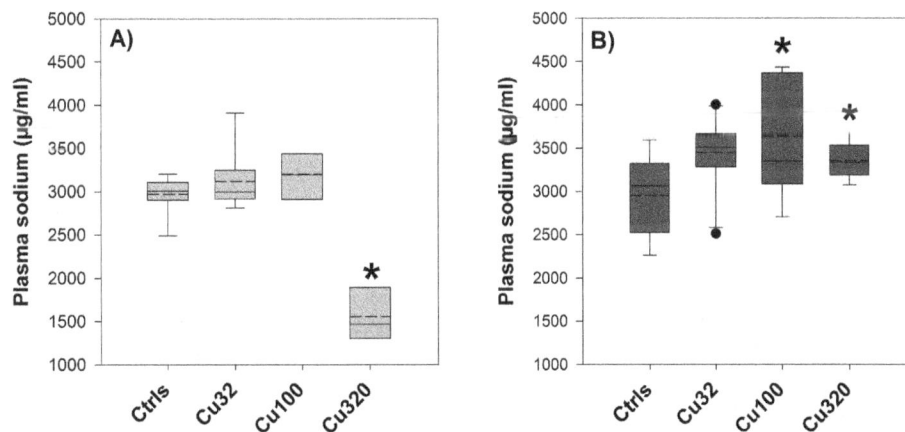

Figure 5. Plasma sodium – Exp.2. Plasma sodium concentrations (μg/ml) measured in fish exposed to 0 (Ctrls), 32, 100 and 320 μg/L copper in either FW (A) or SW (B). Values are means ± SE ($n = 10$, except for FW Cu320, where $n = 3$). The asterisk denotes statistically significant difference from the control value ($P < 0.05$, Tukey t-test). Solid horizontal lines represent median values and dashed lines represent mean values.

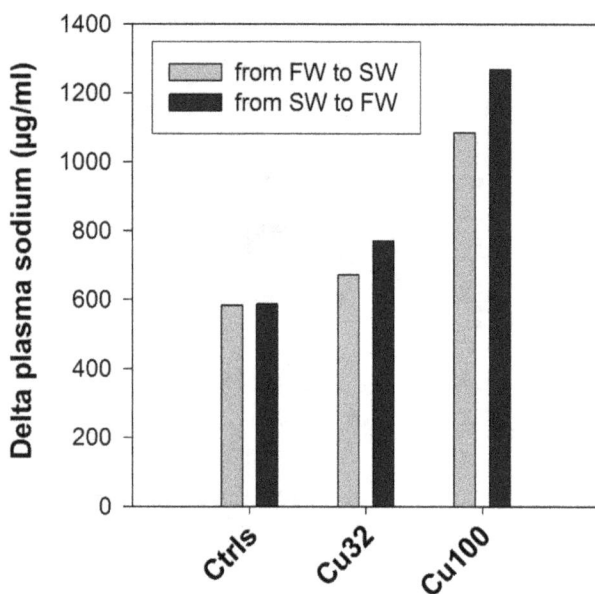

Figure 6. Degree of change in sodium levels. Difference (delta) in plasma sodium levels (µg/ml) before and after the salinity transition in fish exposed to 0 (Ctrls), 32 and 100 µg/L copper. Light grey and dark grey bars represent the delta change in the transition respectively from FW to SW and from SW to FW. No values are reported for the high-copper group as no fish survived after the salinity change. Values are calculated as difference between means and as such have no SE.

Statistical analysis

Numerical data are presented as means ± SD throughout. Statistical analyses were conducted using SigmaStat software (version 3.5 Systat Software Inc.). Data were analysed for normality (Kolmogorov-Smirnov test) and variance homogeneity (Levene's test). Where assumptions of normality and homogeneity were met, one-way analysis of variance (ANOVA) was performed, followed by all pairwise comparison using Dunn's *post-hoc* test. Where the assumptions were not met, data were analysed using a non-parametric test, Krustal-Wallis ANOVA on ranks, followed by Dunn's *post-hoc* test. Significant differences between pre- and post-salinity transition data were assessed by the Tukey *t*-test. In all cases, values were considered significantly different at $P<0.05$.

Results

Experiment 1

Water chemistry and copper speciation modelling. Average copper concentrations in the water during the exposure period were <1 µg/L in the controls, 10.04±0.77 in the low-copper and 115.5±9.97 µg/L in the high-copper treatment (Table S4).

The average concentration of organic matter in the water was 2.6±0.5 mg/L and remained stable throughout the exposure period, ranging between 1.8 and 3.9 mg/L. Among the water parameters (Table S1), alkalinity and pH underwent the most significant increase following the salinity switch, probably as a result of the increased amount of calcium carbonate in the water due to the dissolved seasalts used to prepare the SW stock solutions [31]. Higher calcium carbonate concentrations shifted the carbonate equilibria towards more basic conditions and this change, along with the increased water concentrations of cations and chloride (Table S3), affected the inorganic speciation of copper, as it can be observed in Table 2. The most relevant

difference in copper speciation between pre and post salinity switch was the decreased percentage fraction of free copper ions (from around 28% to 14% of the inorganic speciation), which are considered the most bioavailable and therefore toxic form. This fraction and the other percentage fractions of copper forms reported in Table 2 are calculated without including organic matter among the input parameters. When factoring it into the calculations, the model predicted that, given the same average amount of organic matter, around 90% and 84% of the total copper was bound to it, respectively in FW and SW (Table 3). Hence, although it decreased in SW, the organic fraction of copper remains the most represented fraction. As this form of copper is considered not bioavailable, internal copper concentrations, both in the liver and in the plasma, were measured in order to confirm that copper was indeed uptaken by the fish and its internal levels displayed a dose-response trend.

Liver copper burden and copper content in food. Hepatic copper levels in the high-copper treatment were significantly elevated after 9 days of exposure (Figure 1). However, the relatively high copper levels measured in controls fish (around 110 µg/g wet weight), given the low concentrations of copper in the water (<1 µg/L), led us to investigate fish food as a potential source of copper in control fish. Analysis showed that copper concentrations were 9.1±3.2 µg/g wet weight in the flakes and 1.5±0.8 µg/g wet weight in the brine shrimps, thus providing a possible explanation for the elevated hepatic copper levels in control fish.

Plasma copper. Despite model predictions indicating that most of the total dissolved copper was probably complexed by organic matter, 8 days of copper exposure resulted in an increase in plasma copper levels detectable at 10 µg/L and statistically significant at 100 µg/L post salinity switch (Figure 1B). Within the same treatment, plasma copper levels did not show a significant difference pre and post switch.

Plasma ions. Plasma sodium levels were around 3000 µg/ml in control fish after 8 days of exposure in FW and did not appear to be significantly affected by copper exposure in a concentration-dependant manner. However, the transition to SW resulted in an average 30% increase of sodium levels in the high-copper treatment, whilst only a 20% increase was detected in the controls. Plasma chloride levels followed a similar trend, displaying a 38, 50 and 56% increase after the salinity transition respectively in controls, Cu10 and Cu100 (Table 4). Plasma magnesium levels underwent a similar though smaller change, whereas calcium cations appeared to follow an opposite trend, showing a greater degree of change in the controls compared to the copper-exposed fish. However, none of the changes in plasma ions levels were statistically significant (Figure 2).

NKA and CA2 gene expression. NKA gene expression did not appear to be significantly affected by copper exposure in a concentration-dependant fashion, either in the gills or in the intestine. However, when comparing NKA gene expression in controls before and after the salinity switch (Figure 3A and B), the change in salinity resulted in a significant down-regulation of NKA expression in the gills and a significant up-regulation (~3.5-fold change) in the intestine. The same trend of down-regulation in the gills and up-regulation in the intestine was detected in the copper-exposed groups and in both tissues.

CA2 gene expression was affected by copper exposure in a concentration-dependant manner, both in gills and intestine, displaying in the high treatment a ~2-fold up-regulation in the gills and a ~3-fold up-regulation in the intestine. CA2 expression in the gills of the controls did not exhibit any substantial difference after the salinity switch, whereas in the intestine CA2 displayed a

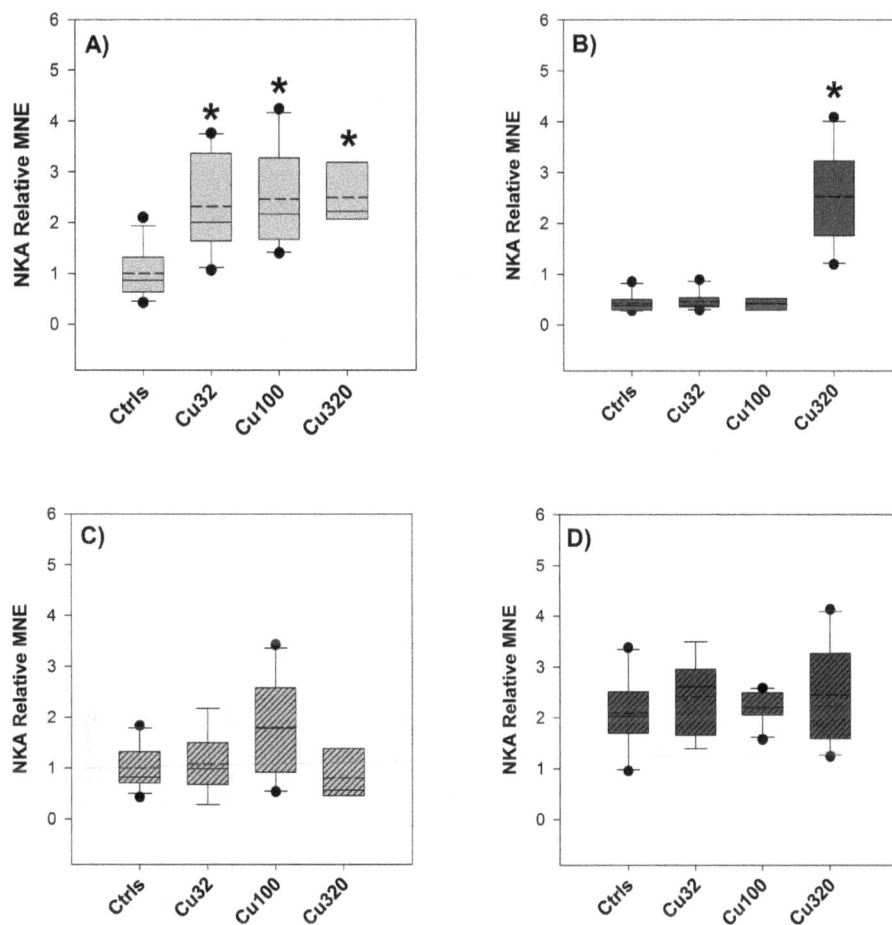

Figure 7. NKA expression – Exp.2. Relative Mean Normalized Expression (MNE) levels of NKA measured in the gills (A and B) and in the mid-anterior tract of the intestine (C and D, striped bars) of sheepshead minnows exposed to 0 (Ctrls), 32, 100 and 320 μg/L copper for 19 days. Light grey boxes (graphs A and C) and dark grey boxes (B and D) represent NKA levels respectively in FW and SW groups. Relative MNE levels were determined by qPCR, normalized to control gene (18S) and expressed as fold change relative to FW control value, which was set at 1. Values are means ± SE (n = 10/12, except for the FW Cu320 group, where n = 6). The asterisk denotes statistically significant difference from the control value (P<0.05, Tukey t-test). Solid horizontal lines represent median values and dashed lines represent mean values.

~4-fold up-regulation in the controls following the switch (Figure 3C and D).

Experiment 2

Water chemistry. Concentrations of copper in the water were in the expected range for all treatments. Mean measured concentrations in the 32, 100 and 320 μg/L groups were, respectively, 31.82±9.1, 116.6±6.6 and 375.2±21 μg/L in the FW groups, and 29.01±4.7, 81.20±16 and 317.4±52 μg/L in the SW groups. Copper concentrations were <1 μg/L in all the control groups. Water chemistry parameters (Table S2) were stable over the exposure period. In the SW groups, mean measured salinity was 19.9±0.3 ppt, mean pH was only slightly higher than in FW and mean alkalinity was 180/240 mg/L $CaCO_3$.

Plasma copper and sodium. The volume of plasma collected from the fish used in this experiment was lower than in the first experiment, due to the smaller fish size. Therefore, copper levels in plasma could be measured only where possible, given the dilution factor applied and the minimal sample volume required for the instrumental analysis. For this reason, results from fish sampled before and after the salinity switch within the same groups were pooled and reported graphically as one group (Figure 4), after verifying that there was no detectable difference

between pre and post values. In the FW groups, plasma copper levels were slightly but not significantly elevated by copper exposure, whereas in the SW groups the mid- and high-copper treatments (Cu100 and Cu320) displayed significantly higher copper levels than controls.

Given the limited amount of plasma sample available in the second experiment, we chose to measure only sodium as the other plasma endpoint besides copper. Plasma sodium levels in FW and SW controls were very similar, despite the very different osmotic conditions, measuring respectively 2932±343.8 and 2952±459.2 μg/ml (equivalent to, respectively, 127.5±14.9 and 128.3±21.5 μmol/ml). Both in FW and SW conditions, the low- and mid-copper treatments (Cu32 and Cu100) displayed elevated sodium levels before the salinity transition, though this increase was statistically significant only in the SW Cu100 group (Figure 5). In the Cu320 treatments, a markedly different type of response was observed between the two salinity groups: in the SW one, sodium levels were significantly higher than controls, even though not higher than the Cu100 group, whereas in the FW group a 50% drop in plasma sodium levels was detected, although here n equalled only 3, because in the last days of exposure 6 fish were left in that group (out of 10), of which just 3 provided sufficient volume of plasma for the analysis. The FW Cu320 is the only group in the

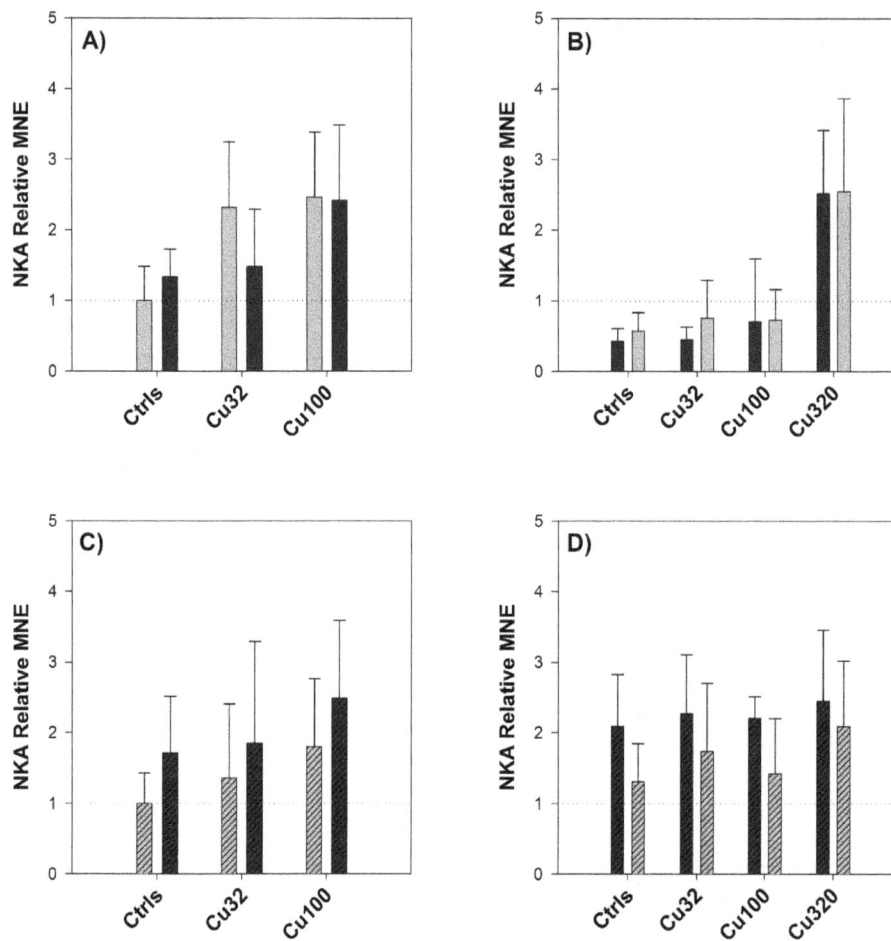

Figure 8. Pre-post change in NKA expression – Exp.2. Relative Mean Normalized Expression (MNE) levels of NKA measured in the gills (A and B) and in the mid-anterior tract of the intestine (C and D, striped bars) of sheepshead minnows exposed to 0 (Ctrls), 32, 100 and 320 µg/L copper. Within each copper treatment, the pair of bars represents NKA expression levels before (left side) and after (right side) the salinity transition. The colour of the bars represents the salinity conditions: light-grey for FW and dark-grey for SW. Relative MNE levels were determined by qPCR, normalized to control gene (18S) and expressed as fold change relative to pre-switch FW control value, which was set at 1 (dotted horizontal line). Values are means ± SE ($n = 10/12$). No values are reported for the high copper dose as no fish survived after the salinity change.

whole study were mortality was observed, presumably as a result of acute copper toxicity in freshwater conditions. However, despite the poor statistical power of the FW Cu320 group, the drastically lower levels measured in those samples were still significantly different from the control value. Unfortunately, in that treatment no fish survived the salinity transition, hence preventing any plasma analysis in that treatment following the switch.

As for the pre-post switch effect on plasma sodium levels in the other treatments, a trend of response in ion homeostasis was observed. In the FW control group, mean sodium levels exhibited a 584 µg/ml increased after the salinity change, and in the SW control group almost exactly the same delta of plasma sodium (587 µg/ml) was detected before and after the salinity change (Figure 6), although in this case it represented a decrease. Hence, the degree of change in the controls was +20% for the FW groups and −20% for the SW ones. An almost equally symmetrical delta change was observed in the low-copper groups (Cu32), where the degree of change was around 22% (of either increase from FW to SW or decrease from SW to FW), and in the mid-copper groups, with a 34% increase in one direction and 32% decrease in the other.

NKA gene expression. NKA gene expression was significantly up-regulated by copper exposure in the gills of both FW and SW groups, exhibiting a 2 to 3-fold change relative to control values (Figure 7A and B). In samples of intestine from FW groups, the low- and mid-copper groups displayed increased NKA expression, whilst in the high-copper group expression levels were down to control values (Figure 7C). No appreciable changes of NKA expression were detected in the intestine of SW groups (Figure 7D).

Considering the effects of the salinity change on NKA expression levels (Figure 8), a general trend of up-regulation from FW to SW and down-regulation from SW to FW was displayed in the intestine samples, whereas in the gills NKA expression did not change appreciably. However, when comparing NKA levels of expression in the gills of FW controls versus the SW controls, both of them before the salinity change, the FW gills exhibited a higher level of expression. The opposite was observed when comparing FW with SW controls in the intestine: in this tissue, the SW controls were the ones displaying higher expression levels.

CA2 gene expression. CA2 gene expression was up-regulated in response to copper exposure in the gills of both FW and SW groups, exhibiting a ~2-fold change in the FW Cu100 group and

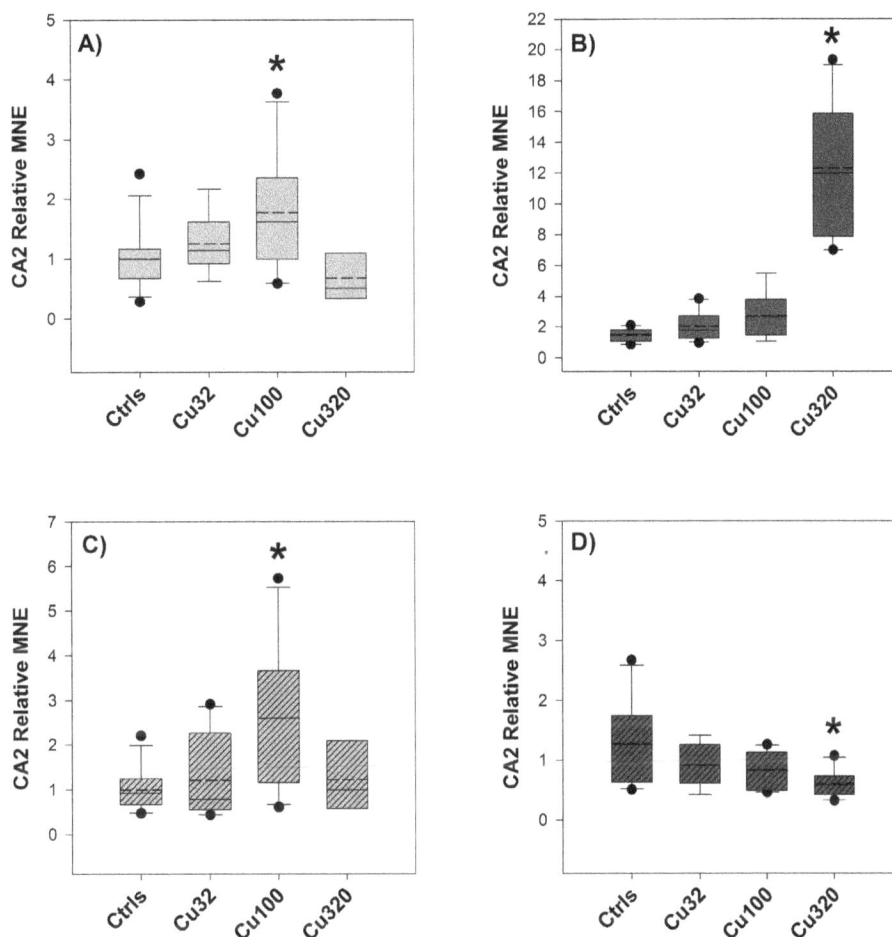

Figure 9. CA2 expression – Exp.2. Relative Mean Normalized Expression (MNE) levels of CA2 measured in the gills (A and B) and in the mid-anterior tract of the intestine (C and D, striped bars) of sheepshead minnows exposed to 0 (Ctrls), 32, 100 and 320 μg/L copper for 19 days. Light grey boxes (graphs A and C) and dark grey boxes (B and D) represent NKA levels respectively in FW and SW groups. Relative MNE levels were determined by qPCR, normalized to control gene (18S) and expressed as fold change relative to FW control value, which was set at 1. Values are means ± SE ($n = 10/12$, except for the FW Cu320 group, where $n = 6$). The asterisk denotes statistically significant difference from the control value ($P < 0.05$, Tukey t-test). Solid horizontal lines represent median values and dashed lines represent mean values.

a ~12-fold change in the SW Cu320 group (Figure 9A and B), whereas in the FW Cu320 group CA2 levels dropped down to control values. Similarly, CA2 expression in intestine samples of FW groups was affected by copper in a concentration-dependant manner up to the mid-copper treatment, whilst the high-copper group broke the trend, displaying similar expression levels to the low-copper group (Figure 9C). CA2 expression in intestine samples of the SW groups displayed a trend of down-regulation in response to copper exposure, in contrast with the general up-regulation observed in the gills of both groups and in the intestine of the FW groups (Figure 9D).

No statistically significant change in CA2 expression was detected either in the gills or in the intestine in response to the salinity change, nor was it possible to observe any appreciable effect of copper on the response to the osmotic stress (Figure 10), in contrast to what was detected in the first experiment. However, a comparison between CA2 levels in controls of FW and SW revealed that the SW controls had a significantly higher expression levels, compared to the FW ones, whereas no appreciable difference was seen between FW and SW controls in the intestine.

Discussion

Modelling of copper speciation in the exposure conditions revealed that a high proportion of the total dissolved copper was probably bound to organic matter and only a small percentage remained in the most bioavailable and hence toxic form of free metal ion (Table 2 and 3). Nevertheless, analyses of copper content in plasma showed that copper was still uptaken by fish in a concentration-related manner (Figure 1B and 4). One of the few studies that measured copper concentrations in the plasma reported only a small and transient increase of plasma copper levels after 24 hours of exposure, followed by a return to control values from day two onward [34]. This lack of a clear and consistent relationship between external concentrations and internal accumulation may appear to undermine the soundness of any biological effect attributed to water-borne metal exposure. However, metals are highly mobile chemical entities whose mode of action is intrinsically dynamic, in accordance with the high degree of complexation and compartmentalisation they undergo when entering the organism [35]. It is therefore their passing through some biological compartments, such as gills, rather than their accumulation, that bears most of their toxicological meaning and potential adverse effect, especially in short/mid-term

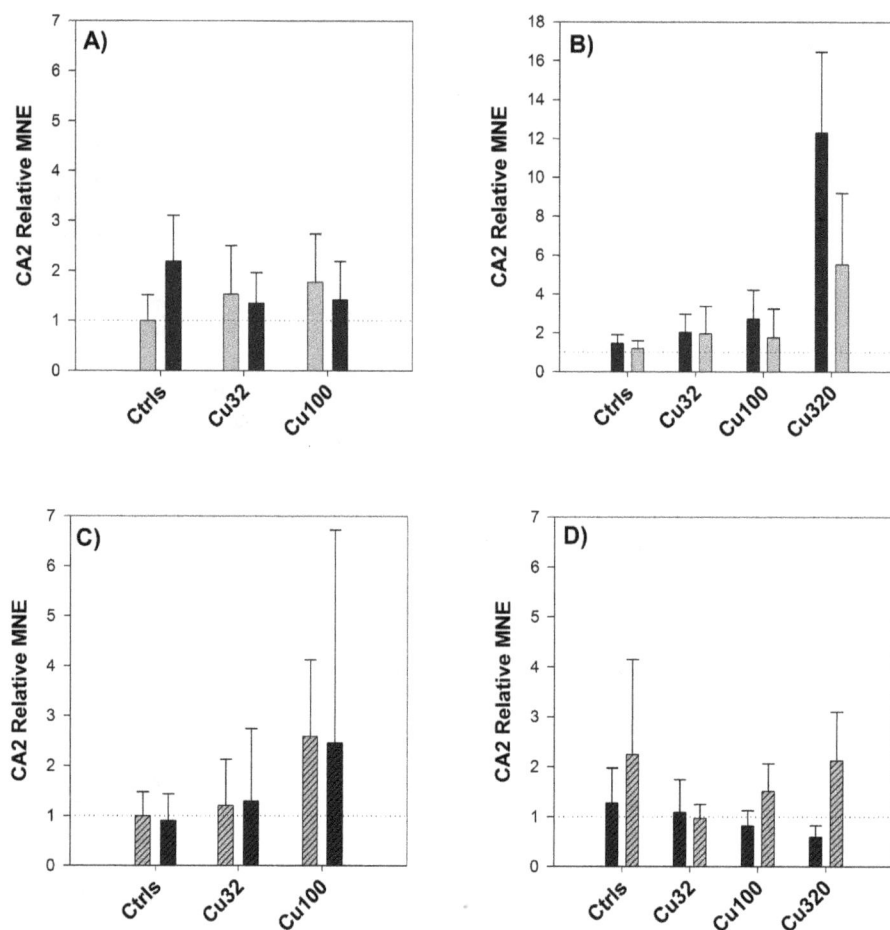

Figure 10. Pre-post change in CA2 expression – Exp.2. Relative Mean Normalized Expression (MNE) levels of CA2 measured in the gills (A and B) and in the mid-anterior tract of the intestine (C and D, striped bars) of sheepshead minnows exposed to 0 (Ctrls), 32, 100 and 320 µg/L copper. Within each copper treatment, the pair of bars represents CA2 expression levels before (left side) and after (right side) the salinity transition. The colour of the bars represents the salinity conditions: light-grey for FW and dark-grey for SW. Relative MNE levels were determined by qPCR, normalized to control gene (18S) and expressed as fold change relative to pre-switch FW control value, which was set at 1 (dotted horizontal line). Values are means ± SE ($n = 10/12$). No values are reported for the high copper dose as no fish survived after the salinity change.

exposures at low concentrations. They do accumulate in some tissues, mainly liver and bile [19,36], but often in complexed and hence relatively inactive forms [37], whose toxicological relevance depends to the type of study and the endpoints of interest. This observation is supported by the high hepatic copper levels we detected in our study, possibly as a result of a copper-rich diet rather than of water-borne exposure (Figure 1A).

In the context of this study and its aims, the anchor point from where to start dissecting the results and build up their discussion is constituted by plasma ion homeostasis, as represented by plasma levels of sodium, chloride, calcium and magnesium, and their perturbation in response to copper and osmotic stress. All these four parameters responded to the salinity transition in the first experiment (Table 4); therefore, in the second experiment we chose to focus our attention only on sodium (Figure 5 and 6), which was used as a proxy of plasma ion homeostasis (or internal osmotic pressure). First of all, it is interesting to note that in both experiments and regardless of the exposure conditions (either FW or SW), control values of sodium in plasma before the salinity transition were all extremely close, around 3000 µg/ml, confirming that this parameter is very tightly and actively regulated by fish.

According to the literature, the iso-osmotic point of sheepshead minnow, meaning the salinity point at which the internal osmotic pressure equals that of the external medium, is around 10 ppt [38]. Based on this information, we should expect that both the fish held in FW and those held in SW (20 ppt) actively regulate their internal plasma concentrations against the same gradient of 10 ppt. However, we must also consider that our stocks of fish were bred and kept at 20 ppt for most of their life, before being adapted to FW conditions for a month. If we take into account this aspect, it is plausible to assume that strict FW conditions require a higher energetic cost to maintain the internal plasma ion homeostasis. Bearing this in mind, the first question we want to ask is whether the plasma sodium parameter was affected by copper exposure, in FW and in SW, irrespective of the salinity switch. Since copper is commonly considered an osmoregulatory disrupter [19,39], displaying its osmoregulatory effects mainly through the disruption of sodium excretion mechanisms, we should expect that this impairment results in either increased sodium levels, when the gradient between internal and external plasma ion levels is negative (i.e. SW), or decreased sodium levels, when the gradient is positive (i.e. FW) [40]. And indeed, SW copper-exposed fish had higher levels of sodium in their plasma, compared to control ones, in so proving that copper exerted its osmoregulatory disruption, whereas in FW sodium levels displayed an increase in response to copper exposure, contrary to what we would expect given the positive osmotic gradient between internal and external osmotic pressure in these conditions. However, this small increase was observed only in the low- and mid-copper treatments, whereas in the high-copper treatment, Cu320, a dramatic drop of ~50% in sodium levels was detected. It is important to stress here that in that particular group only 6 fish had survived after three weeks of copper exposure in FW: this in itself showed how those fish, exposed to the "double stress" of copper and FW conditions, were physiologically struggling, as confirmed by the 50% decrease in sodium levels. Given the tight regulation of this parameter under normal conditions, such a change can be considered as the effect of acute toxicity, similarly to other plasma parameters, such as human plasma glucose concentration [41]. This is an important element to consider when analysing this plasma sodium dataset with the aim of answering question *(1)*, which is at the heart of the hypothesis we are testing in this study. To address this question we should look at the degree of change in plasma sodium levels before and after the salinity transition (Figure 6). If the degrees of change (or delta sodium) calculated for the controls are equal or similar to those calculated for the copper-exposed groups, then copper did not affect fish response to the osmotic stress. Alternatively, if the degrees of change, from FW to SW and from SW to FW, calculated for the copper-exposed groups are higher than those of the controls, we could conclude that copper did have an effect on the fish response to the osmotic stress and impaired their adaptation to the new conditions [20–21]. Results presented in Figure 6 not only support the latter option by showing a higher delta change in the copper treatments, but also display a concentration-dependent trend of increasing degrees of change at increasing copper concentrations, further supporting the hypothesis of a copper-disrupted osmoregulatory response to the salinity change. This was true in both directions of salinity transitions, from FW to SW and from SW to FW. One possible weakness of this set of results is the narrow range of change in plasma levels, i.e. few percentage points. However, this observation should be put in the context of this particular endpoint and its physiological regulation: as on the one side a 20% change was the "normal" degree of change observed in the controls, we know that on the other side a 50% change in plasma sodium is highly toxic for the organisms (see results in the Cu320 FW group). Framed in such context, we can consider a 22 or a 34% delta change as very physiologically significant.

Having addressed the question of whether or not copper affected the physiological response of fish to osmotic stress, the next step is to understand if the observed interaction is the combined effect of two independent effects, or if it is the outcome of a mechanistic interaction between the two stressors, i.e. copper and osmotic stress. According to the reasoning presented in the 2012 review [23], we hypothesize that the interaction between copper exposure and osmotic stress is indeed of a mechanistic nature. We also speculate that it is indirect, rather than direct, and that, in the exposure conditions applied in this study, it takes place mainly at the transcriptional level. In brief and according to our hypothesis, it should be an interaction rather than a combined effect because the pathways activated by both of them are similar (i.e. osmotic stress pathways) and they all share some common elements, one of which is the enzyme CA2. Given the numerous and diverse functional roles of CA2 [42–43], it is more plausible to think that copper affects the transcription of CA2 not directly, such as through the binding of some metal-responsive elements upstream of the gene coding for CA2, but rather indirectly,

through the activation of osmotic-stress-related factors, which in turn regulate CA2 transcription. This indirect effect can be a form of compensatory response caused by impaired ionoregulation by copper exposure.

In order to interpret the results of the molecular analyses, it can be useful to apply the same set of questions used to interpret the results of plasma sodium levels, particularly *(a)* and *(c)*. The first question can be re-formulated as to whether copper exposure affected the transcriptional levels of CA2 and NKA, whereas the second one addresses the effect of copper on the transcriptional response to the osmotic stress. The expression of NKA was not significantly affected by copper exposure in the first experiment, in contrast with the second one, where there was a significant induction in response to copper in both FW and SW (Figure 7). CA2 expression was affected by copper in a dose-dependent manner, in both experiments and in both gills and intestine (Figure 9), demonstrating a maximum 12-fold up-regulation in the gills in response to copper (320 µg/L) in SW. As for the second question, i.e. whether copper affected the transcriptional response of fish to the salinity challenge, the two experiments provided similar results for NKA, whereas CA2 responses were somehow different between experiments. Considering the effect of copper on the regulation of NKA in response to the salinity transition, according to the results of both experiments (Figure 3 and 8) copper did not affect the response of NKA to the salinity change, as its regulation was not impaired by copper exposure. However, if we consider the results of the first experiment and compare the response of NKA in the two tissues, gills and intestine, it is interesting to note that the down-regulation of NKA caused by the salinity change in the gills was accompanied by an almost symmetrical up-regulation in the intestine, a pattern that might suggest a "deactivation" of NKA in one tissue and a parallel "activation" in the intestine as a result of the transition to SW conditions. This is in line with the expression patterns displayed by this isoform of NKA in rainbow trout during salinity transfer [43]. Considering the effect of copper on CA2 and its regulation in response to the salinity change, this is where some disagreements between the two datasets emerge. In the first experiment, CA2 expression in the intestine displayed a clear up-regulation in the controls following the salinity change (Figure 3C and D) and, notably, this up-regulation in the controls was not detected in the low-copper treatment and was even replaced by a down-regulation in the high-copper treatment. This opposite trend of response suggests that, under normal conditions, an osmotic stress induces the expression of CA2, at least in the intestine, but this response is disrupted by copper. Overall, these results support the original hypothesis that copper-exposed fish struggle to adapt to new salinity conditions because copper affects their osmoregulatory response to osmotic stress, and it does so by interfering with the regulation of osmotic effector proteins such as CA2. However, contrary to what was observed in the first experiment, the results from the second experiment did not show any significant effect of copper on the regulation of CA2 either in the gills or in the intestine (Figure 10).

We speculate that the disagreement between the two datasets can be explained by addressing the interaction of copper and salinity from a multi-stressor perspective, where copper exposure and osmotic stress are regarded as a chemical and an environmental stressor that disrupt the homeostasis of the system, i.e. the fish. In this systemic context, it is plausible to assume that the organism responds to the perturbation of its homeostasis by activating responses that are of a magnitude and complexity appropriate to the degree of perturbation, i.e. level of stress. Put another way, if the organism responded to a mild stress with a disproportionally complex response, it would waste its cellular resources, whereas underestimating the level of stress may compromise its cellular function. Therefore, it is energetically sensible for the organism to size its adaptive response to the severity of the perturbation. Applying this concept to the case of osmotic stress responses, we should expect that different doses of osmotic stress elicit different magnitudes of response, as was indeed shown by our results on CA2 transcriptional levels in the two experiments. Since osmoregulatory mechanisms are finely modulated according to varying degrees of osmotic stress [44], the different rate at which the salinity transition was performed in the two exposures (4 hours in the first and 8 hours in the second one) could explain the activation of different transcriptional responses, resulting in a significant induction of CA2 in the first experiment and in an overall unaffected response of the same enzyme in the second one. Additionally, the responses of some endpoints at high copper doses were completely different than those displayed at low and mid doses, suggesting that somewhere between the mid and the high dose a threshold of different pathways activation was passed, as the dose of chemical stress applied went from mild to severe. Of course we are aware that this is another way to formulate the classical concept of chronic and acute dose, but this may actually need to be reformulated, when multiple stressors of different source and nature are studied in atypical combinations. Although these arguments are admittedly speculative, they in any case hint at the complexity of multi-stressor studies. Since such a complexity lies at the level of biological responses to different stressors, it cannot be overlooked when it comes to modelling metal toxicity in multi-stressor scenarios.

Supporting Information

Figure S1 Salinity switch – Exp.2. Salinity measurements (ppt) throughout the salinity transition period (8 h) of Experiment 2. The graph on the left side represents the FW groups in the 4 treatments (Controls, Cu32, Cu100 and Cu320) that were moved towards SW conditions, whilst the graph on the right represents the SW groups moved towards FW. Control groups (white circles) were started at 10 am, Cu32 (grey triangles) at 12 am, Cu100 (dark grey squares) at 14 pm and Cu320 (black diamonds) at 16 pm. A 2 h-lag was chosen to ensure that all treatments were held in the new conditions for exactly 24 h prior to sampling on the following day.

Figure S2 Exposure set-ups – Exp.2. Exposure set-ups one on one side of the exposure room (the other was symmetrical), respectively before (top figure) and after (bottom figure) the salinity switch. Blue colour represents the saltwater (SW) supply and dark green colour the dilution water (DW) supply. The SW stock was mixed with a RZR 2052 overhead stirrer (Heidolph) in two 80 L tanks (one per side) and then transferred into two 40 L tanks, from which the SW was dosed at a rate of 12 ml/min via a peristaltic pump into the mixing chambers of the SW groups. Copper stock solutions were dosed at a rate of 0.06 ml/min via one single peristaltic pump that fed both sides of the room.

Table S1 Water parameters in fish tanks – Exp.1. Reported values are means ± SD of the measurements taken daily in all 6 groups ($n = 6$) respectively before (PRE) and after (POST) the salinity switch.

Table S2 Water parameters in fish tanks – Exp.2. Reported values are means ± SD of the measurements taken daily over the exposure period, respectively in the freshwater (FW) groups and in the saltwater (SW) groups.

Table S3 TOC and major cations concentrations in the water – Exp.1. Experiment 1 water concentrations (µg/mL) of Total Organic Carbon (TOC) and Na^+, Ca^{2+} and Mg^{2+} analysed respectively by Shimadzu total organic carbon-V CPN Analyzer and F-AAS. Reported FW and SW values are means ± SD of all 6 groups ($n = 6$) respectively before (PRE) and after (POST) the salinity switch.

Table S4 Copper concentrations in the water – Exp.1. Experiment 1 water concentrations (µg/L) of copper during the exposure period, analysed by GF-AAS (controls and 10 µg/L) and F-AAS (100 µg/L). Reported values are means ± SD ($n = 3$ for each given time point and concentration).

Acknowledgments

The authors are grateful to Professor John Sumpter for his valuable input in the design of the *in vivo* experiments.

Author Contributions

Conceived and designed the experiments: AdP MDS. Performed the experiments: AdP LMC. Analyzed the data: AdP AEL. Contributed reagents/materials/analysis tools: MDS. Contributed to the writing of the manuscript: AdP.

References

1. Holmstrup M, Bindesbøl A, Oostingh GJ, Duschl A, Scheil V, et al. (2010) Interactions between effects of environmental chemicals and natural stressors: A review. Sci Total Environ 408: 3746–3762.
2. Leitão J, Ribeiro R, Soares A, Lopes I (2013) Tolerance to copper and to salinity in *Daphnia longispina*: implications within a climate change scenario. PLoS ONE 8: e68702.
3. Lopes I, Baird D, Ribeiro R (2009) Resistance to metal contamination by hystorically-stressed populations of *Ceriodaphnia pulchella*: environmental influence versus genetic determination. Chemosphere 61: 1189–1197.
4. Løkke H (2010) Novel methods for integrated risk assesment of cumulative stressrs-results from the *NoMiracle* project. Total Environ 408: 3719–3724.
5. Rider CV, Boekelheide K, Catlin N, Gordon CJ, Morata T, et al. (2014) Cumulative risk: toxicity and interactions of physical and chemical stressors. Toxicol Sci 137: 3–11.
6. Noyes PD, McElwee MK, Miller HD, Clark BW, Van Tiem LA, et al. (2009) The toxicology of climate change: environmental contaminants in a warming world. Environ Int 35: 971–986.
7. Schiedek D, Sundelin B, Readman J, Macdonald R (2007) Interactions between climate change and contaminants. Mar Pollut Bull 54: 1845–1856.
8. Hooper M, Ankley G, Cristol D, Maryoung L, Noyes P, Pinkerton K (2013) Interactions between chemical and climate stressors: a role for mechanistic toxicology in assessing climate change risks. Environ Toxicol Chem 32: 32–48.
9. Ankley GT, Bennett RS, Erickson RJ, Hoff DJ, Hornung MW, et al. (2010) Adverse outcome pathways: a conceptual framework to support ecotoxicology research and risk assessment. Environ Toxicol Chem 29: 730–741.
10. Caldwell D, Mastrocco F, Margiotta-Casaluci L, Brooks B (2014) An integrated approach for prioritizing pharmaceuticals found in the environment for risk assessment, monitoring and advanced research. Chemosphere S0045-6535(14)00087-3: in press.
11. Sheahan D, Maud A, Wither C, Moffat C, Engelke C (2013) Impacts of climate change on pollution (estuarine and coastal). MCCIP Science Review 4: 1-xxx.
12. IPPC (United Nations Intergovernmental Panel on Climate Change) (2007) Climate change, 2007: climate change impacts, adaptation and vulnerability. Cambridge University Press, Cambridge, UK.
13. Lotze HK, Lenihan HS, Bourque BJ, Bradbury RH, Cooke RG, et al. (2006) Depletion, degradation, and recovery potential of estuaries and coastal seas. Science 312: 1806–1809.

14. Brooks S, Waldock M (2009) Copper biocides in marine environment. In Arai T, Harino H, Ohji M, Langston WJ, eds, Ecotoxicology of antifouling biocides, Vol 3. Springer, Japan, pp 413–428.

15. Lee JA, Marsden ID, Glover CN (2010) The influence of salinity on copper accumulation and its toxic effects in estuarine animals with differing osmoregulatory strategies. Aquat Toxicol 99: 65–72.

16. Miller TG, Mackay WC (1980) The effects of hardness, alkalinity and pH of test water on the toxicity of copper to rainbow trout (*Salmo gairdneri*). Water Res 14: 129–133.

17. Arnold WR, Diamond RL, Smith DS (2010) The effects of salinity, pH, and dissolved organic matter on acute copper toxicity to the rotifer, *Brachionus plicatilis* ("L" Strain). Arch Environ Contam Toxicol 59: 225–234.

18. Blanchard J, Grosell M (2005) Effects of salinity on copper accumulation in the common killifish (*Fundulus heteroclitus*). Environ Toxicol Chem 24: 1403–1413.

19. Stagg RM, Shuttleworth TJ. (1982) The effects of copper on ionic regulation by the gills of the seawater-adapted flounder (*Platichthys flesus* L.). J Comp Physiol B 149: 83–90.

20. Grosell M (2011) Copper. In *Fish Physiology*, Homeostasis and Toxicology of Essential Metals: Volume 31A. Elsevier Inc., pp 53–133.

21. Laurén D, McDonald D (1985) Effects of copper on branchial ionoregulation in the rainbow trout, *Salmo gairdneri* (Richardson). J Comp Physiol B 155: 635–644.

22. Atili G, Canli M (2011) Essential metal (Cu, Zn) exposures alter the activity of ATPases in gill, kidney and muscle of tilapia Oreochromis niloticus. Ecotox 20: 1861–1869.

23. de Polo A, Scrimshaw MD (2012) Challenges for the development of a biotic ligand model predicting copper toxicity in estuaries and seas. Environ Toxicol Chem 31: 230–238.

24. Skaggs HS, Henry RP (2002) Inhibition of carbonic anhydrase in the gills of two euryhaline crabs, *Callinectes sapidus* and *Carcinus maenas*, by heavy metals. Comp Biochem Physiol C 133: 605–612.

25. Vitale AM, Monserrat JM, Castilho P, Rodriguez EM (1999) Inhibitory effects of cadmium on carbonic anhydrase activity and ion regulation of the estuarine crab *Chasmagnathus granulata* (Decapoda, Grapsidae). Comp Biochem Physiol C C122: 121–129.

26. Soyut H, Beydemir Ş, Hisar O (2008) Effects of some metals on carbonic anhydrase from brains of rainbow trout. Biol Trace Elem Res 123: 179–190.

27. Sattin G, Mager EM, Beltramini M, Grosell M (2010) Cytosolic carbonic anhydrase in the Gulf toadfish is important for tolerance to hypersalinity. Comp Biochem Physiol A 156: 169–175.

28. Scott GR, Claiborne JB, Edwards SL, Schulte PM, Wood CM (2005) Gene expression after freshwater transfer in gills and opercular epithelia of killifish: insight into divergent mechanisms of ion transport. J Exp Biol 208: 2719–2729.

29. Kultz D, Bastrop R SD (1992) Mitochondria-rich (MR) cells and the activities of Na$^+$/K$^+$-ATPase and carbonic anhydrase in the gill and opercular epithelium of *Orechromis mossambicus* adapted to various salinities. Comp Biochem Physiol 102B: 293–301.

30. Grosell M, Gilmour KM, Perry SF (2007) Intestinal carbonic anhydrase, bicarbonate, and proton carriers play a role in the acclimation of rainbow trout to seawater. Am J Physiol Regul Integr Comp Physiol 293: 2099–2111.

31. Atkinson MJ, Bingman C (1997) Elemental composition of commercial seasalts. Journal of Aquaculture and Aquatic Science VIII, no. 2: 39–43.

32. KTH, Department of Sustainable Development, Environmental Science and Engineering (2010) Visula MINTEQ - A free equilibrium speciation model. 2012.

33. Muller PY, Janoviak H, Miserez AR, Dobbie Z (2002) Processing of gene expression data generated by quantitative real-time RT-PCR. 32. Short Technical Report. BioTechniques.

34. Grosell M, McDonald MD, Walsh PJ, Wood CM (2004) Effects of prolonged copper exposure in the marine gulf toadfish (*Opsanus beta*) II: copper accumulation, drinking rate and Na+/K+-ATPase activity in osmoregulatory tissues. Aquat Toxicol 68: 263–275.

35. Harrison MD, Dameron CT (1999) Molecular mechanisms of copper metabolism and the role of the Menkes Disease protein. J Biochem Mol Toxicol 13: 93–105.

36. Adeyemi JA, Deaton LE, Pesacreta TC, Klerks PL (2012) Effects of copper on osmoregulation in sheepshead minnow, *Cyprinodon variegatus*. Aquat Toxicol 109: 111–117.

37. Laurén DJ, McDonald DG (1987b) Acclimation to copper by rainbow trout, *Salmo gairdneri*: biochemistry. Can J Fish Aquat Sci 44: 105–111.

38. Grosell M, Blanchard J, Brix KV, Gerdes R (2007) Physiology is pivotal for interactions between salinity and acute copper toxicity to fish and invertebrates. Aquat Toxicol 84: 162–172.

39. Yeo R, Sawdon D (2013) Hormonal control of metabolism: regulation of plasma glucose. Anaest Intens Care Med 14: 296–300.

40. Henry RP (1996) Multiple roles of carbonic anhydrase in cellular transport and metabolism. Annu Rev Physiol 58: 523–538.

41. Lionetto MG, Caricato R GM, Erroi E, Schettino T (2012) Carbonic anhydrase as pollution biomarker: an ancient enzyme with a new use. Int J Environ Public Health 9: 3965–3977.

42. Supuran CT, Scozzafava A (2007) Carbonic anhydrases as targets for medical chemistry. Bioorg Med Chem 15: 4336–4350.

43. Richards JG, Semple JW, Shulte PM (2003) Pattern of Na$^+$/K$^-$ ATPase isoform expression in rainbow trout (*Oncorhynchus mykiss*). Integr Comp Biol 42: 1300.

44. Evans TG (2010) Co-ordination of osmotic stress responses through osmosensing and signal transduction events in fishes. J Fish Biol 76: 1903–1925.

15

Environmental and Biotic Correlates to Lionfish Invasion Success in Bahamian Coral Reefs

Andrea Anton[1]*[¤], Michael S. Simpson[2], Ivana Vu[2]

1 Curriculum for the Environment and Ecology, University of North Carolina, Chapel Hill, North Carolina, United States of America, 2 Department of Biology, University of North Carolina, Chapel Hill, North Carolina, United States of America

Abstract

Lionfish (*Pterois volitans*), venomous predators from the Indo-Pacific, are recent invaders of the Caribbean Basin and southeastern coast of North America. Quantification of invasive lionfish abundances, along with potentially important physical and biological environmental characteristics, permitted inferences about the invasion process of reefs on the island of San Salvador in the Bahamas. Environmental wave-exposure had a large influence on lionfish abundance, which was more than 20 and 120 times greater for density and biomass respectively at sheltered sites as compared with wave-exposed environments. Our measurements of topographic complexity of the reefs revealed that lionfish abundance was not driven by habitat rugosity. Lionfish abundance was not negatively affected by the abundance of large native predators (or large native groupers) and was also unrelated to the abundance of medium prey fishes (total length of 5–10 cm). These relationships suggest that (1) higher-energy environments may impose intrinsic resistance against lionfish invasion, (2) habitat complexity may not facilitate the lionfish invasion process, (3) predation or competition by native fishes may not provide biotic resistance against lionfish invasion, and (4) abundant prey fish might not facilitate lionfish invasion success. The relatively low biomass of large grouper on this island could explain our failure to detect suppression of lionfish abundance and we encourage continuing the preservation and restoration of potential lionfish predators in the Caribbean. In addition, energetic environments might exert direct or indirect resistance to the lionfish proliferation, providing native fish populations with essential refuges.

Editor: Howard I. Browman, Institute of Marine Research, Norway

Funding: This project was funded by a PADI Foundation Grant (http://www.padifoundation.org/), a UNC-Off Campus Fellowship (http://gradschool.unc.edu/), and a C.V. STARR Scholarship (http://cgi.unc.edu/awards/cv-starr) to AA, a Watts Hill Jr. Research Fellowship (http://our.unc.edu/) to MSS, and an NSF OCE #0746164 grant (http://www.nsf.gov/) to JF Bruno. The funders had no role in study design, data collection and analysis, decision to publish, or preparation of the manuscript.

Competing Interests: The authors have declared that no competing interests exist.

* Email: androide@live.unc.edu

¤ Current address: Queen's University Belfast, School of Biological Sciences, Belfast, United Kingdom

Introduction

Establishment of non-native species in new biogeographic regions can have serious consequences on biodiversity [1] and is now recognized as one the world's most critical conservation challenges [2]. Both physical and biological characteristics of the new environment affect the fate and success of exotic species [3–5]. Clearly, the physical environment must be physiologically tolerable: harsh environments such as deserts have been shown to be the least invaded worldwide [6], perhaps because the suite of non-native species pre-adapted to those extreme conditions is limited. Alternatively, when environmental conditions are tolerable, biotic resistance may inhibit local invasion success [7]. Biotic resistance stems from community diversity [8] or from the effects of strong local enemies (e.g. predators, competitors, or pathogens), affecting the fate of the exotic species in the new range. For instance, the native blue crab (*Callinectes sapidus*) provides biotic resistance against invasion by green crabs (*Carcinus maenas*) through direct predation in eastern North America [9]. Similarly, communities are more susceptible to invasion if they provide essential resources [10] or if the exotic species outcompetes native species in resource acquisition. For instance, invasive Argentine ants (*Linepithema humile*) outcompete native ants for food sources, depressing native ant abundance in northern California [11].

Invasive lionfish (*Pterois volitans*), a native species from the Indo-Pacific, was first detected in Florida in 1985 [12] and spread rapidly throughout the tropical Caribbean, subtropical southeast Atlantic coast [13] and has been recently spotted in the Mediterranean Sea [14]. This particular invasion is now ranked as one of the top-ten most serious emerging environmental issues in the world [15]. Densities of lionfish in their new biogeographic region are up to 15 times those in their native environment [16]. On reefs in the Bahamas, lionfish consume small fish and are thereby capable of reducing native fish abundance [17], biomass [18], and richness [19]. These findings are consistent with a meta-analysis that reveals that some novel predators can exert impacts on prey populations roughly double that of native predators [20]. Possible explanations for the successful lionfish invasion of the Atlantic include its diet breadth, comprising dozens of species of native fishes [see Table S1 for a list of fish species that are lionfish current prey in the Atlantic and Caribbean], naiveté of prey towards exotic lionfish [21,22], and the possibility of a geographic escape from control by natural enemies [22,23] (although see [24,25]). Threats posed by invading lionfish are particularly serious because of the high ecological and economic values of coral reefs in the Caribbean [26]. Similarly, lionfish are a threat to reefs in southeastern North America [27,28] which are habitat for valuable reef fishes of the snapper-grouper complex already seriously stressed by overfishing [29].

Here we utilize the invasion of lionfish in reefs of San Salvador, Bahamas, first documented in 2005 [13] to quantify potentially important physical and biological environmental characteristics to determine which factors contribute to the success of the lionfish invasion. We operationally define invasion success by lionfish abundance, either density or biomass. By assessing the effect of wave exposure on lionfish abundance, we test how physical energy relates to lionfish invasion success on coral reefs. By evaluating how the rugosity of the reefs affects lionfish abundance, we measure the role of structural complexity on the invasion process. By exploring how lionfish abundance relates to abundances of large native groupers and other predatory fishes, we infer whether biotic resistance to invasion may be provided by natural predators or competitors. Finally, by relating lionfish abundance to abundance of small and medium fishes, we infer whether prey availability may be facilitating lionfish invasion success.

Methods

Ethics statement

No protected or endangered species were involved in this field study. Surveys were performed through visual census and no vertebrates were handled or collected. Approvals by the Department of Marine Resources of The Bahamas were obtained to perform the surveys.

Field Sampling

We conducted field surveys at 18 sites around the island of San Salvador, Bahamas in July-August 2009, in coral reef habitat at depths between 13–17 m (Fig. 1). Lionfish were detected in San Salvador in 2005 [13]. Sites were separated by more than 1.5 km. Buoys, which are used to moor boats over the reefs, were avoided when selecting sites to minimize possible influences of spearfishermen on lionfish. Three replicate, haphazardly placed 50 m long transects, separated by approximately 20 m gaps, were deployed to perform surveys of benthic habitat cover and fish abundances at each site. Transects were oriented parallel to shore and surveys were conducted between 9:00–16:00 h. On each transect, divers working together but on different sections along the transect followed a sequence of sampling protocols (Fig. S1). Fish surveys were performed using standard underwater visual belt transect methods [30], which were conducted by two divers for safety reasons. One diver quantified lionfish and large (>30 cm in total length, TL) native predatory fish abundances by species (listed in Table S2), and estimated TL of each individual within 500 m² (50×10 m; large quadrat) area along the transect (Fig. S1). Care was taken to examine cryptic habitats by thoroughly inspecting reef crevices and overhangs, to avoid underestimating lionfish densities [31]. Simultaneously another diver quantified potentially suitable prey fishes of two different sizes: Fishes of less than 5 cm total length (TL), termed small fishes, were quantified in 30 m² (15×2 m; small quadrat) area and prey fishes of 5–10 cm TL, termed medium fishes, were counted in 120 m² (30×4 m; medium quadrat) area (Fig. S1). At each transect, the large quadrat contained the medium and small quadrats and the medium quadrat did not overlap with the small quadrat (Fig. S1). To reduce the effect of one diver on the observations of a second diver, the two divers that performed the fish surveys advanced simultaneously along the transect line, with the diver examining

the large quadrat performing 1-way ziz zag swims centered on the transect, while the second diver progressed in a straight forward motion along the transect. Small quadrats were surveyed after the large and medium quadrats to minimize any influence on large fish. Fishes were recorded by species except nocturnal (such as Apogonidae) and highly cryptic (such as Blenniidae and Gobiidae) fish species, that were not quantified, following other comparable previous studies [32]. Only small and medium fish species identified in the literature as lionfish prey were included in the statistical analysis (Table S1 and Table S2).

Fish counts from the three same-size quadrats surveyed per site were pulled together to calculate fish counts per site per unit of area surveyed to avoid pseudoreplication and spatial autocorrelation (transects in the same site are more similar that transects from other sites). Small and medium fish counts were assessed per unit of area (small fish of 0–5 cm TL in 90 m^{-2} area and medium fish of 5–10 cm TL in 360 m^{-2} area) and then extrapolated to individual 1500 m^{-2} to aid comparison with the fish counts in the larger transect. Fish densities for each species were transformed to biomass by using the allometric length-weight conversion formula $W = aTL^{b}$, where W is the weight of each individual fish in grams, TL is the total length recorded for each fish in cm and the parameters a and b are species-specific constants. The parameters a and b for each species were obtained from FishBase ([33]; Table S2). Lionfish lengths (cm) were converted to biomass (g) using empirically fitted, allometric scaling parameters (a = 0.00492 and b = 3.31016) obtained from the weight and length of 137 lionfish from Abaco Island, Bahamas [22].

Environmental predictors: Wave exposure and habitat complexity

We estimated the average bottom velocity (i.e. velocity of the water near the sea floor) at each site as a metric of the degree of wave exposure to demersal and semi-demersal fish species. Land masses can modify the wave energy near the bottom and wave exposure was calculated as follows. First, we determined vectors of the oceanic waves that could strike each site (all directions from which the waves could reach a site) in San Salvador using geographic maps. Bottom velocity depends on wave direction, dominant wave period, wave height and depth [34]. Depth was measured *in situ* on each study site using dive computers. Wave direction, dominant wave period, and wave height were obtained from data available online (National Data Buoy Center website. Available: http://www.ndbc.noaa.gov/. Accessed 2014 August 7) from two permanent moored buoys (41047-NE Bahamas and 41046-East Bahamas) owned and maintained by the National Oceanic and Atmospheric Administration. We assumed that the same waves that were reaching these buoys also reached our study sites. The historical public record of wave data from buoys is intermittent but included data from May to December 2009, January to July 2010 and January to December 2011 from the NE Bahamas buoy and data from August to December 2010 from East Bahamas buoy. Buoys collect data hourly from which we estimated bottom velocity [34] hourly for each site for all waves that directly reached that site: otherwise bottom velocity was recorded as zero. We then computed monthly average bottom velocities for every study site over all the time periods (above) for which these buoys recorded wave data. We used estimated site means of bottom velocities from May through August ("summer" months) to construct box plots of the hourly bottom velocity for each month, allowing visual comparison of wave exposure between sites (Fig. 2). This time period includes the field sampling months of July and August plus the two preceding months, which could also have strong influences on biotic patterns.

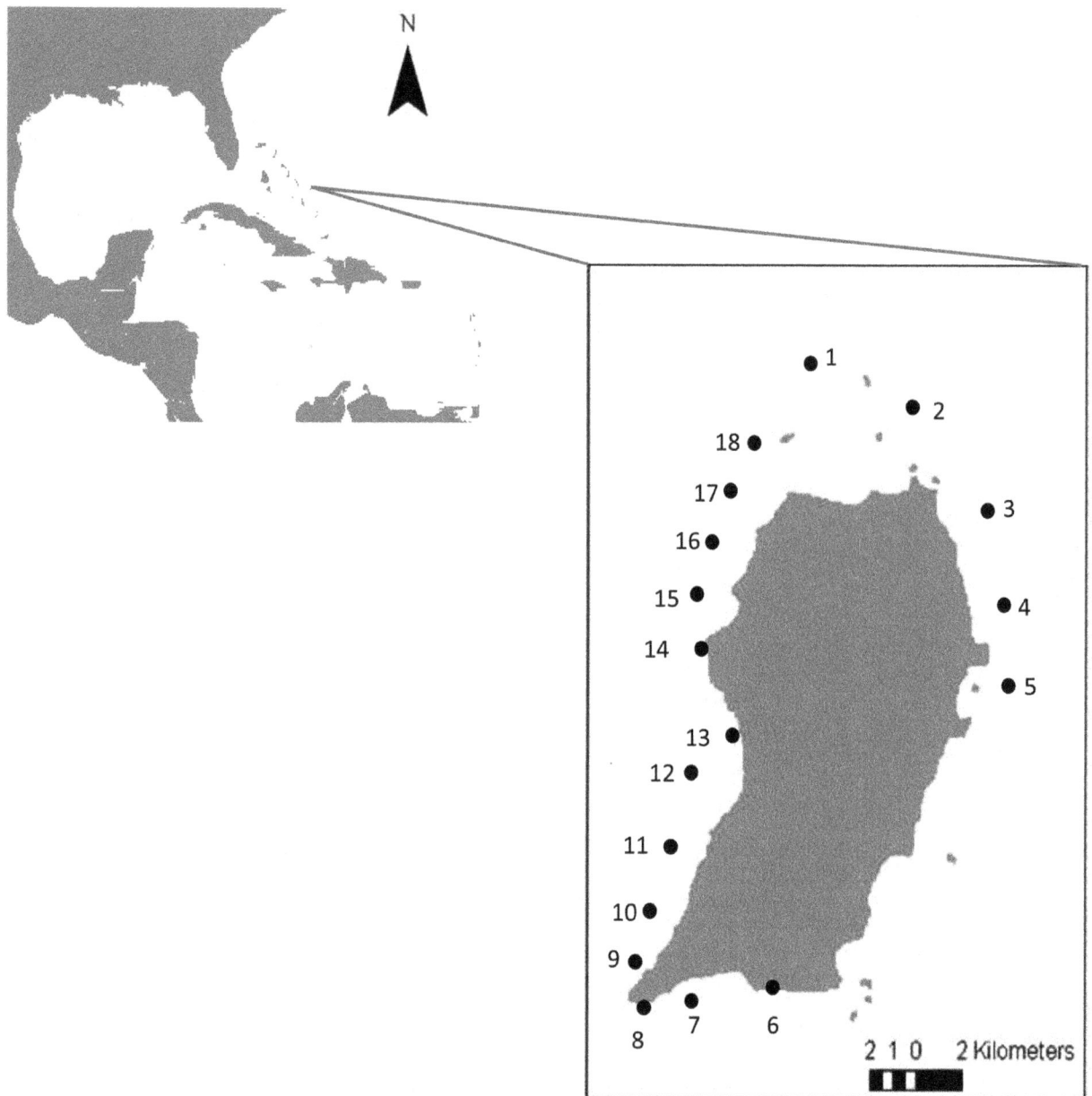

Figure 1. Map of study sites in the island of San Salvador (The Bahamas). Circles indicate the study sites. Numbers indicate study sites as follows: (1) White Island, (2) Catto Cay, (3) Light House, (4) Baptism, (5) Crab Cay, (6) La Crevasse, (7) Danger Point, (8) Double Caves, (9) Grotto, (10) Great Cut, (11) Red House, (12) Gardness, (13) Sangrila, (14) Runway, (15) Club Med, (16) Yellow House, (17) Rocky Point, and (18) Green Cay. This map was generated using a publicly available shapefile from the World Resources Institute (World Resources Institute website. Available: http://www.wri. org. Accessed 2014 August 7).

Topographic complexity (e.g., the rugosity of the reef) was measured on each transect (3 times per site) by carefully laying a 30 m steel chain (2 cm long links) to the reef surface. The chain was deployed following the length of the measuring tape used in the fish surveys. A rugosity index (C) was calculated per site as $C = 1 - d/l$, where d is the horizontal distance covered by the chain when following the contour of the reef and l is the length of the chain when fully extended (30 m; [35,36]).

In addition, we quantified benthic habitat cover along 30 m of the transect line placed on the bottom at each site to investigate the effects of environmental predictors of benthic habitat. We classified benthic habitat type as coral cover (including a subsection of important reef-building corals-the *Orbicella* habitat, which comprised *Montastraea annularis*, *M. flaveolata* and *M. franksi* species), macroalgae cover, turf algae cover, sponge cover, and gorgonian cover. Other benthic habitat types (sand, cyanobacteria and crustose coralline algae) were also identified but not used in this study because they provide no emergent habitat structure for fish. We identified habitat category at 50 cm intervals directly below marked points on the transect line. Benthic cover measurements were pooled by transect and then averaged across transects to produce site means for each habitat type.

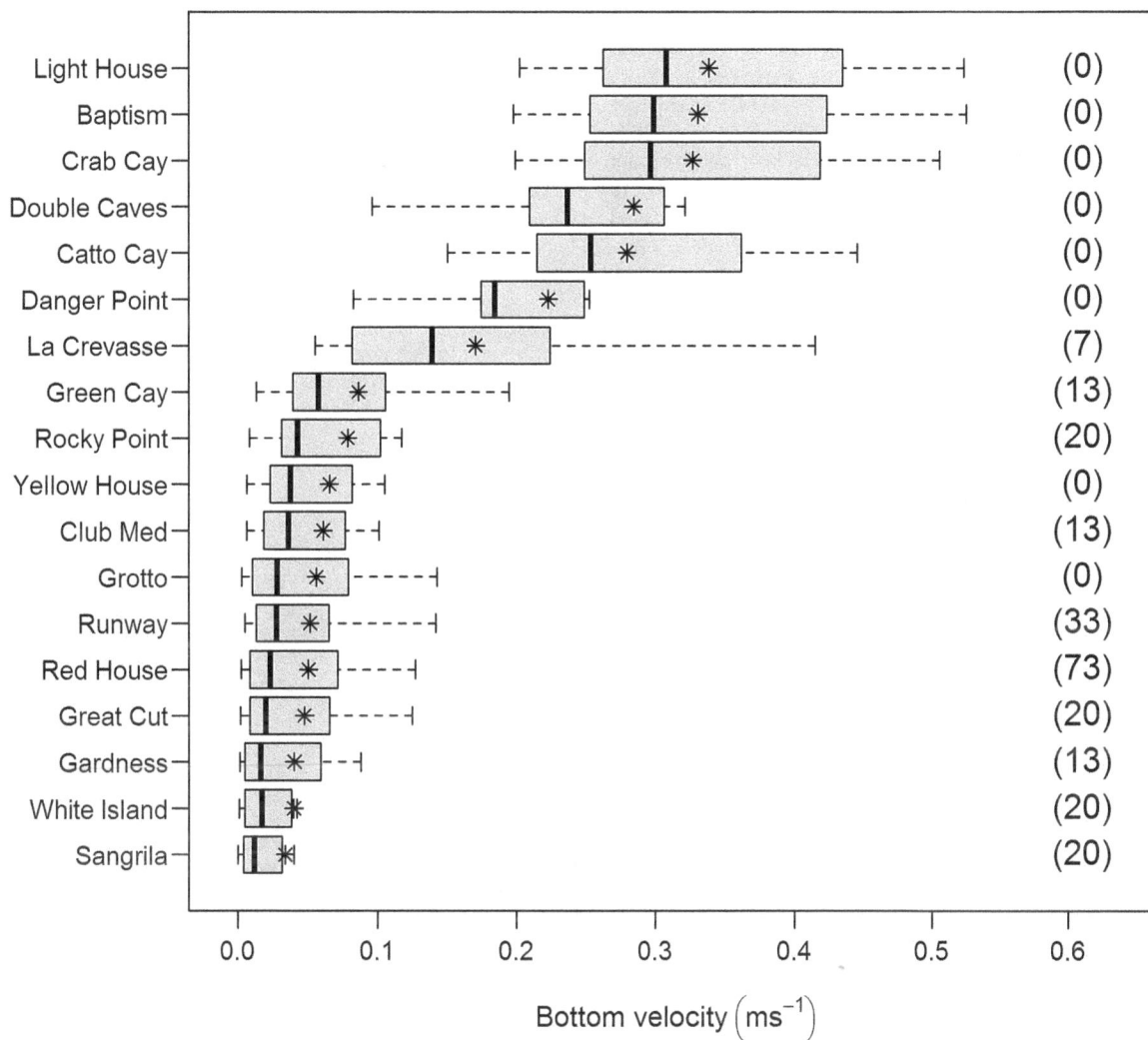

Figure 2. Boxplots of bottom velocity averaged over summer months as a function of site. Mean bottom velocity (m s^{-1}) at 18 sites in San Salvador Island (Bahamas) for the summer months (May, June, July, and August). Sites were ordered, listed with increasing mean bottom velocities. The wave parameters used to calculate bottom velocity were collected hourly by oceanic buoys. Lionfish mean density (individuals ha^{-1}) per site was indicated in parenthesis.

Statistical analyses

A subset of large, native predatory groupers (so forth collectively termed "large grouper"), consisted of Nassau grouper (*Epinephelus striatus*) and tiger grouper (*Mycteroperca tigris*), were selected to aid comparison of findings with previous studies ([24]; Table S2).

First, to corroborate the independence between model predictors, we tested for correlation among our abiotic (wave exposure and structural complexity) and biotic (small fish-, medium fish-, large grouper-, large predatory fish- density or biomass) independent variables by creating a Pearson's product-moment correlation coefficients table (Table S3). The correlation matrix indicated no correlation among preditors with the exception of 1) wave exposure and small fish density and biomas and 2) large predatory fish density and large grouper density (Table S3). To avoid data missinterpredation arising from models with correlated predictors, we dropped small fish density and biomass from the analyses while large predatory fish abundance and large grouper abundance were not included together as predictors in the same statistical analyses.

We employed generalized linear models to determine the effect of biotic (medium fish abundance, large predator abundance, and grouper abundance) and abiotic factors (wave exposure and rugosity) on lionfish abundance (density and biomass). Four independent generalized linear models were run to test the effect of the environmental and fish variables on lionfish abundance: 2 models had lionfish density as dependent variable and the other 2 models had lionfish biomass as dependent variable (Table 1). Fish independent variables were either medium fish abundance and large predatory fish abundance or medium fish abundance and large grouper abundance (Table 1). We ran separate analysis with grouper abundance (instead of predator abundance) as a predictor because 1) grouper abundance was contained within large predator abundance, 2) large predatory fish density and large grouper density were significantly correlated (Table S3) and 3) to allow comparisons with previous studies [24]. Models with were fitted by negative binomial distributions on zero-inflated models (ZINB) because lionfish data distributions were heavily sckewed towards zero, containing more zeros than expected based on a negative binomial distribution [37]. A ZINB model is a mixture model consisting of two parts: A binomial model (zero-inflation model) that accounts for the excess zeros and a count model that includes the counts and the expected zeros modeled with a negative binomial distribution [37]. Both parts of the mixture model can include independent variables. We did not include predictors in the zero-inflated part of the model because we do not suspect that the probability of false zeros was a function of any of our predictors [37]. To validate the models we plotted the model residuals against the fitted values and no patterns were detected in any of the four models. No interaction terms were included in the models to avoid overfitting. The range of large predator and large grouper biomass was 100 and 9 times larger than lionfish biomass respectively (Table 2) and these predictors were standardized (centered and scaled by sustracting the mean and dividing by the standard deviation) to improve model convergence.

To elucidate potential indirect bottom-up effects of wave exposure or reef rugosity on lionfish abundance, we performed correlations among each habitat type (the % cover of corals, the *Orbicella* habitat, macroalgae, turf algae, sponges and gorgonias) and the abiotic variables (wave exposure and reef rugosity) by creating a Pearson's product-moment correlation coefficients table (Table 3).

Finally, a clear separation existed between sites with relatively low and high wave exposure (7 versus 11 sites respectively; see Fig. 2): To corroborate our results on the effect of wave exposure on lionfish abundance (density and biomass) and reef rugosity, we performed three additional statistical analyses. First, we ran two independent zero-inflated generalized linear models to test for

differences in lionfish density or biomass across sites with low and high wave exposure. Second, we used a two sample one-tailed t-test to assess differences in reef rugosity between low and high wave exposure sites. All statistical analyses were performed with R version 3.1.0 (R project for Statistical Computing website. Available: http://www.r-project.org. Accessed 2014 August 7.) in RStudio (RStdio website. Available: https://www.rstudio.com/. Accessed 2014 August 7.) with packages MASS [38] and pscl [39].

Results

To facilitate the comparison of fish abundances with previous studies, we built a table (Table 2) that includes the fish counts and their calculated biomass per site (individuals 1500 m^{-2} and grams 1500 m^{-2} respectively) and also a conversion of the same variables into the common units used in the literature to report fish abundance (density as individuals ha^{-1} and biomass as g 100 m^{-2}).

Summer-time estimated near-bottom velocities for the 4 years of buoy wave data (mean per site) ranged from 0.033 to 0.337 with a mean of 0.14 (± 0.12) m s^{-1} (Fig. 2) and rugosity index (C) ranged from 0.021 to 0.57 with a mean of 0.37 (± 0.13). Lionfish density in our study ranged from 0 (in eight sites) to 73 with a mean ($\pm$SD) of 13 (± 18) individuals ha^{-1} (Table 2 and Fig. 2) and lionfish biomass ranged from 0 to 173 with a mean of 27 (± 45) g 100 m^{-2} (Table 2). Medium fish, large predatory fish, and large grouper mean densities were 540 (± 320), 39 (± 45), and 10 (± 11) individuals ha^{-1} respectively (Table 2).

Table 1. Statistical zero-inflated negative binomial models for the effects of environment (wave exposure and structural complexity), and fish (small and medium fishes, and large predatory fishes) abundance (density and biomass) on lionfish abundance (density and biomass).

Dependent variable	Independent variable	Coefficient Estimate	SE*	z-value	p-value
Lionfish density	Intercept	1.445	1.598	0.904	0.365
(individuals 1500 m^{-2})	Medium fish density	−0.007	0.003	−1.958	0.05
	Large predator density	0.076	0.031	2.382	**0.017**
	Rugosity	1.406	4.406	0.319	0.75
	Wave exposure	−14.847	5.4	−2.749	**0.006**
Lionfish density	Intercept	1.127	1.7163	0.657	0.511
(individuals 1500 m^{-2})	Medium fish density	−0.002	0.004	−0.45	0.653
	Large grouper density	0.197	0.136	1.444	0.149
	Rugosity	1.603	5.228	0.307	0.759
	Wave exposure	−15.636	6.531	−2.394	**0.016**
Lionfish biomass	Intercept	5.983	2.209	2.368	0.008
(g 1500 m^{-2})	Medium fish biomass	−9.586	<−0.001	−1.098	0.272
	Large predator biomass†	−0.276	0.301	−0.891	0.373
	Rugosity	10.41	6.361	1.636	0.102
	Wave exposure	−42.56	8.95	−4.755	**<0.001**
Lionfish biomass	Intercept	5.681	2.437	2.331	0.019
(g 1500 m^{-2})	Medium fish biomass	−0.001	<−0.001	−1.763	0.078
	Large grouper biomass†	−0.235	0.368	−0.639	0.523
	Rugosity	11.79	7.725	1.527	0.126
	Wave exposure	0.059	11.32	−3.872	**<0.001**

The four models had 7 degrees of freedom.
†This variable was centered and scaled
Bolded values denote significant differences at p<0.05

Table 2. Conversion table with the density and biomass of lionfish, medium fish, large predatory fish and large grouper.

	Density, individuals 1500 m^{-2}	Biomass, g 1500 m^{-2}
Lionfish	1.94±2.7 (0–11)	409±674 (0–2591)
Medium fish	81±48 (8–162)	646±406 (102–1434)
Large predatory fish	5.9±6.7 (0–22)	42471±80070 (0–292213)
Large grouper	1.55±1.61 (0–6)	3580±3912 (0–13221)
	Density, individuals ha^{-1}	Biomass, g 100 m^{-2}
Lionfish	13±18 (0–73)	27.3±44.9 (0–173)
Medium fish	540±320 (53–1080)	4306±2706 (680–9560)
Large predatory fish	39±45 (0–147)	2831±5338 (0–19481)
Large grouper	10±11 (0–40)	238±261 (0–881)

Values are presented as mean ± standard deviation (minimum value-maximum value).

Our statistical models relating lionfish abundance across sites to abundances of various groupings of fishes and two environmental predictors help uncover possible functional relationships affecting lionfish invasion success. Lionfish density was negatively related to wave exposure, positively related to large predator abundance and did not exhibit any response to reef rugosity, medium fish- or large grouper- density (Table 1, Fig. 3, and Fig. 4). Lionfish biomass was also negatively related to wave exposure (Table 1 and Fig. 4) and did not exhibit any response to reef rugosity or the biomass of any of the fish groups (medium fish-, large predatory fish-, and large grouper- biomass; Table 1 and Fig. 3).

Correlations of the environmental predictors (rugosity and wave exposure) and benthic habitat type (the % cover of corals, the *Orbicella* habitat, macroalgae, turf algae, sponges and gorgonias) helped to identify direct effects of the environment on the benthos and the associated, potential indirect effects on lionfish invasion success. Only two benthic habitat types were correlated to one environmental predictor: macroalgal cover was negatively correlated to wave exposure in contrast with turf algae that was positively correlated to wave exposure (Table 3).

Additional analyses performed after segregating wave exposure into two categories, low and high, confirmed the results obtained in our previous statistical analysis, where wave exposure was included as a continuous predictor. Lionfish abundance (density and biomass) differed significantly between low and high wave-exposure environments (z = −2.906, p = 0.0036, df = 4 and z = −3.141, p = 0.0016, df = 4 for lionfish density and biomass

respectively; Fig. 4). Average lionfish density was more than 22 times higher on low than on high wave exposure sites (20.6 versus 0.93 ha^{-1}) (Fig. 4). Lionfish biomass exhibited an even more extreme pattern of a 122-fold higher average level in high versus low wave exposure environments (44.45 versus 0.36 g 100 m^{-2}; Fig. 4), a consequence of finding larger lionfish at the sheltered sites as compared with the high wave exposure sites. We did not detect differences in reef rugosity between low and high wave exposure environments (t = −0.0131, p = 0.5051, df = 16).

Discussion

Environmental wave exposure had a large influence on lionfish density and biomass (Fig. 4). Sheltered sites had a 22-fold higher density and a more than 120-fold greater biomass of lionfish than the exposed sites (Fig. 4). The apparent inhibition of lionfish invasion success in the sites with the highest wave exposure may reflect direct impacts of physical stresses on the lionfish. Some other fishes are also scarce in wave-exposed environments, perhaps because the energetic costs of locomotion may be a considerable barrier to occupation [40,41]. In addition, when lionfish are hunting, they hover over or near their potential prey, and usually flare and spread their oversized, interconnected pectoral fins before striking (*personal observation*). They often blow jets of water at their prey, presumably to disorient them before striking [42]. These complex and sophisticated hunting behaviors could be difficult under conditions of high water velocities or strong oscillatory water motion. Our results agree with a recent

Table 3. Table of the Pearson's product-moment correlation coefficients between benthic habitat type (coral cover, macroalgae cover, turf algae cover, *Orbicella* habitat cover, sponge cover and gorgonian cover) and abiotic predictors (wave exposure and rugosity).

Variable	Rugosity (C)	Wave exposure (m s^{-1})
Coral cover (%) [10.5±5.1]	0.129	0.044
Macroalgae cover (%) [61.1±21.1]	−0.116	*−0.83
Turf cover (%) [10.9±10.4]	0.233	*0.874
Orbicella habitat cover (%) [3.1±2.8]	0.096	0.353
Sponge cover (%) [1.4±1.7]	−0.212	−0.211
Gorgonian cover (%) [1.2±1.2]	0.309	0.040

The asterisk (*) indicates significant differences at p-values <0.05.

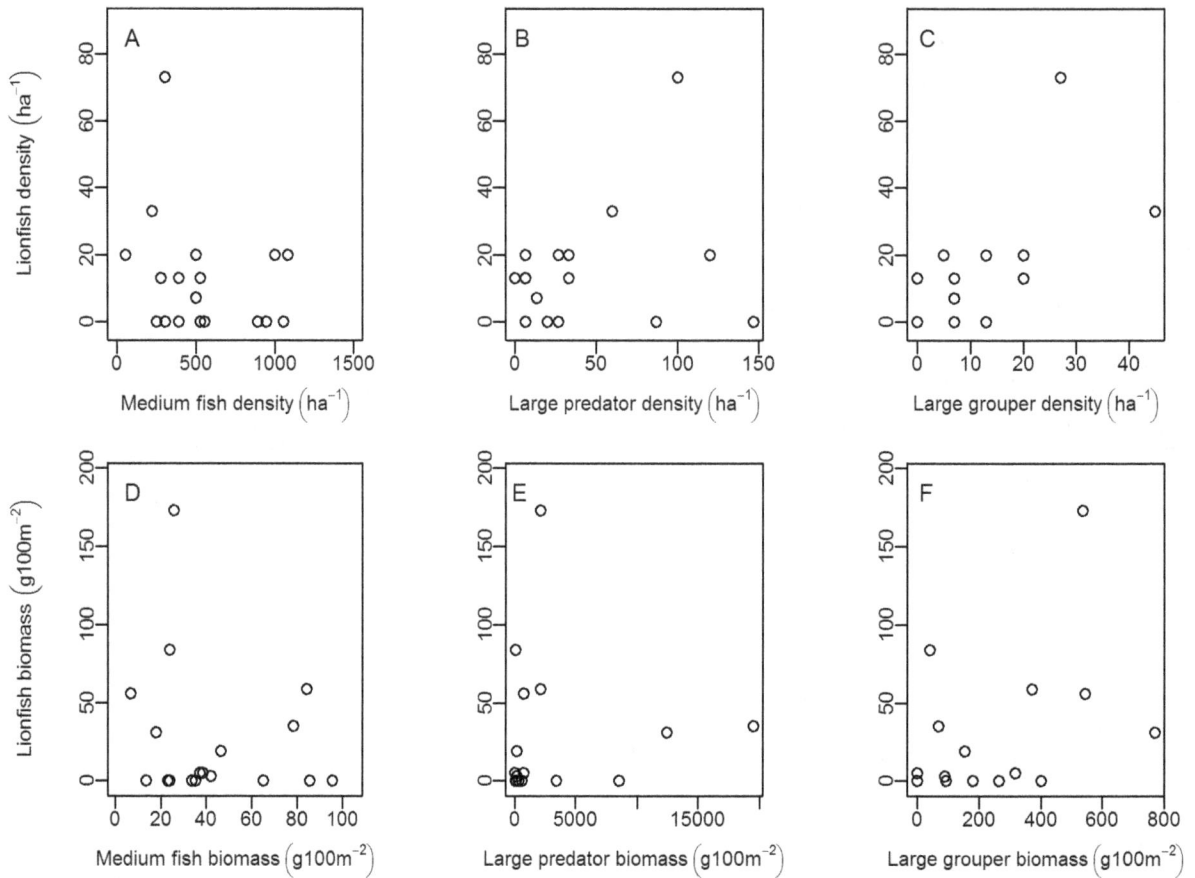

Figure 3. Effect of fish abundance (density or biomass) on lionfish abundance (density or biomass). Relationships between lionfish density and (A) medium fish density (individuals ha^{-1}), (B) large predator density (individuals ha^{-1}), and (C) large grouper density (individuals ha^{-1}). Also relationships between lionfish biomass (g 100 m^{-2}) and (D) medium fish biomass (g 100 m^{-2}), (E) large predator biomass (g 100 m^{-2}), and (F) large grouper biomass (g 100 m^{-2}).

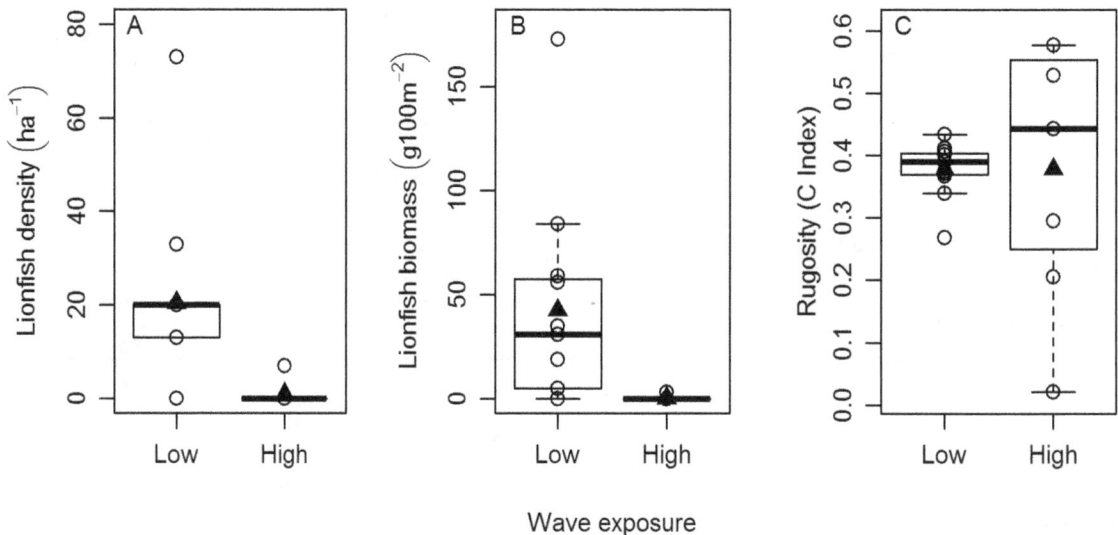

Figure 4. Effect of wave exposure (low/high) on lionfish abundance (density or biomass) and rugosity. Box plot of the relationships between wave exposure (categorized as low or high) and (A) lionfish density (individuals ha^{-1}), (B) lionfish biomass (g 100 m^{-2}), and (C) reef rugosity (C Index). Triangles denote mean values and open circles represent the mean value on each study site.

study [43] that reported that "wind exposure has a weak negative effect on lionfish abundance". However, we found that wave exposure significantly suppressed lionfish abundance and biomass, indicating that lionfish invasion success can be strongly affected by physically energetic conditions. If confirmed by subsequent experiments that lionfish are sensitive to hydrodynamic perturbations, it would imply some optimism that energetic environments may serve as refuges for coral reef fish populations even as lionfish may fundamentally modify fish communities in more protected environments [18].

Our findings of low lionfish abundances in the sites with relatively high wave exposure can not be explained by more effective lionfish removals by divers. To our knowledge, no lionfish derbies, like the ones organized by the Reef Environmental Education Foundation (REEF), have been held in San Salvador. [23] reported lower abundances of lionfish inside marine reserves than in several control-fished sites in the Caribbean and attributed the low lionfish densities within the marine parks to targeted and regular culling of the invasive fish by managers, dive operators, and/or tourist. However, none of our study sites were within marine reserves or regularly visited by dive operations, fishermen, and/or tourist (local fishermen and divers, *personal communication*). Therefore, at least in this Bahamian island, we can not attribute the low lionfish densities found in the wave-exposed sites to sustained lionfish removals.

The rugosity of coral reefs can be shaped by wave exposure, where coral reefs in sheltered locations are usually more structurally complex than reefs on wave-exposed environments [44–47]. However, we found that the rugosity on the reefs around San Salvador island appeared to be unaffected by wave exposure (Fig. 4). Two non-mutually exclusive hypotheses might explain these results. First, the relationship between habitat complexity and wave exposure might have been undetected due to our marginal sample size (n = 3). Second, benthic structural complexity is affected by wave exposure but also by depth, which can have a positive effect on reef rugosity [48]. The depth of our sites in San Salvador ranged from 13 m to 17 m, which might have been sufficient to buffer the differences in reef rugosity between exposed and sheltered environments ([48], Fig. 4). In fact, while swimming and diving around some of our wave-exposed study sites we noticed that the reef structure was visibly flatter in shallower areas than at the depth range where we were performing our surveys. Nonetheless, reef rugosity was unrelated to lionfish density and biomass, implying that structural complexity does not seem to facilitate lionfish invasion success (Table 1 and Fig. 4). These results agree with a recent study [43] that shows no correlation between habitat structural complexity and lionfish density. Reef rugosity in the [43] study was determined by visually assessing a substratum complexity category [49], which provides a limited qualitative estimation of habitat complexity. Our measurements of topographic complexity using the "chain and tape" method represent a fine scale quantification of reef rugosity [50–52]. Hence, our observations in San Salvador support the hypothesis that structural complexity is not a proximate driver of lionfish abundance.

The apparent inhibition of lionfish at the sites with high wave exposure could also occur indirectly through environmentaly driven biotic variables. Macroalgae were more abundant on sites with low wave-exposure which was related with more small fish (Table 3). The greater abundance of macroalgae in reefs previously built by *Orbicella* (when it was living) in areas of low wave exposure has been previously shown [53]. If benthic habitat plays a significant role in affecting the success of lionfish with regard to wave exposure, it would also likely influence lionfish invasion success by indirectly providing refuge and prey for small fishes, which are themselves prey

for lionfish [12]. Macroalgae may also provide invertebrate prey to lionfish directly and thus represent an alternative indirect effect facilitating greater lionfish invasion success on the protected sites of this Bahamian island. Small fish (<5 cm total length) density and biomass were negatively correlated to wave exposure (Table S3). Because wave exposure and small fish abundance were autocorrelated (Table S3), we were unable to examine the effect of small prey fish on lionfish invasion success. However, we found that lionfish abundance was not associated with medium fish abundance, which are also prey of lionfish, suggesting a potential lack of an indirect bottom-up control of lionfish invasion success on this island.

Medium fish abundance was included in our models as a predictor under the assumption that the medium fish community has not yet been affected by lionfish presence. We did not conduct prey fish data counts before the lionfish invasion of San Salvador and we can not discern if lionfish had already influenced their prey fish community. The mean lionfish density in our study was relatively low when compared to other locations in the Bahamas (mean±SD: 13±18 individuals ha^{-1} in our study in San Salvador versus 101±103 individuals ha^{-1} in New Providence Island [54]). In fact, our lionfish densities in San Salvador were comparable to just a few sites with the lowest abundance of lionfish in New Providence. Another study in New Providence Island [18] reported that the detrimental effect of lionfish on native fish communities in coral reefs can be pronounced and quick: a 65% decline in the biomass of lionfish's prey fishes was quantified six years after the first lionfish sighting on nine coral reefs. On these reefs in New Providence, lionfish reached an abundance of nearly 40% of the total predator biomass in the system [18]. However, lionfish biomass in our sites in San Salvador represented only 1% of the large predator abundance in the reefs (Table 2) and these lionfish abundances appear to have been fairly low since the lionfish invasion of the island in 2005 (*personal communication* with local divers). Hence, it is likely that the effect of predatory lionfish on the medium fish communities of San Salvador 4 years after their arrival was limited.

Lionfish density was positively related to density of large native predatory fishes, but lionfish biomass was not associated with either large predatory fish density or biomass. The relationship between lionfish abundance and abundance of large predatory fishes implies a limited impact of competition and perhaps also predation on lionfish invasion success on this island. Instead, this positive relationship may arise indirectly through joint influences of some other variable on both lionfish and large native predatory fishes. For instance, predatory fish are often more abundant in sheltered environments [40,48], which also seems to be the case for invasive lionfish. It is interesting that this positive relationship between the abundance of lionfish and native large predatory fishes exists even if fishing effects on larger predators may be higher in sheltered environments (because of better accessesability for fishermen) than on wave-exposed habitats. Therefore, the quantified abundances of large predators in the sheltered sites might have been relatively low compared to historic densities.

The lack of an effect of grouper abundance (density or biomass), which included only those fish >30 cm in total length, on lionfish abundance suggests that on San Salvador native predatory groupers are not providing biotic resistance against lionfish invasion, as shown in [24]. The limited top-down effects on lionfish found in San Salvador may not be surprising given the potent venom that lionfish carry in their dorsal, anal, and pelvic spines [55]. Although the act of any predation on healthy, free-roaming lionfish has not yet been reported, numerous studies of another successful toxic invader, the cane toad invading Australia, show low predation in the newly established range [56,57].

The lack of a negative relationship between grouper abundance and lionfish abundance in our study contrasts with the conclusions in [24] from their study of lionfish and grouper biomass at sites along a chain of the Exuma Cays, also in the Bahamas. The Exuma reef sites included two sets: one in the Exuma Cays Land and Sea Park (ECLSP), where native grouper biomass is now high after protection from fishing, and another set to the north, where fishing continues and grouper biomass is lower. The authors concluded that when protected from fishing for long enough to rebuild grouper population biomass, predation by these native groupers can suppress the proliferation of lionfish on Exuma reefs. Grouper biomass in the Exuma protected area was on average approximately 9 times what we documented in San Salvador, so our failure to detect suppression of lionfish proliferation on this island could be explained by the relatively low biomass of native groupers. In addition, a recent study in Little Cayman Island reports predation by two native predatory fish species, Nassau grouper (*Epinephelus striatus*) and nurse shark (*Ginglymostoma cirratum*), on tethered but healthy lionfish [58], suggesting that predation of lionfish in the Caribbean might already be occurring. While the question of whether Atlantic native fish predators might exert a top-down control on invasive lionfish deserves further empirical investigation, the restoration and preservation of potential lionfish predators, in combination with selected removals of this invader [59,60], are useful conservation efforts to manage the lionfish invasion of the Caribbean. In addition, energetic environments might impose direct or indirect resistance to the lionfish invasion, serving as fundamental refuge for native coral reef fish populations.

Supporting Information

Figure S1 Diagram of fish field surveys. Diagram depicting the area of the different quadrats used on the fish surveys performed in the field.

Table S1 Prey fish species of *Pterois volitans* in the Atlantic reported in the scientific literature.

Table S2 List of the fish species include in each of the fish categories: small fish (*a*), medium fish (*b*), large predatory fish (*c*) and large grouper (*d*). a and b are the allometric length-weight parameters used to convert fish lenth into biomass.

Table S3 Table of the Pearson's product-moment correlation coefficients between the biotic (density and biomass of small fish, medium fish, large predatory fish and large grouper) and environmental (wave exposure and rugosity) model predictors. *d* indicates density in individual 1500 m^{-2} and *b* indicates biomass in g 100 m^{-2}. The asterisk (*) indicates significant differences at p-values <0.05.

Acknowledgments

We are grateful to the Gerace Research Center in San Salvador, P. Peterson, N. Geraldi, P. Mumby, J. Rossman, P. Whitfield, J. Weiss, B. VanDusen, R. Puntila, L. Lerea, J. Cable, C. Cox, and one anonymous reviewer.

Author Contributions

Conceived and designed the experiments: AA. Performed the experiments: AA MSS IV. Analyzed the data: AA. Contributed reagents/materials/analysis tools: AA MSS IV. Wrote the paper: AA. Contributed ideas, comments, and editing: MSS IV.

References

1. Fritts T, Rodda G (1998) The role of introduced species in the degradation of island ecosystems: A case history of Guam. Annu Rev Ecol Evol Syst 29:113–140.
2. Pejchar L, Mooney HA (2009) Invasive species, ecosystem services and human well-being. Trends Ecol Evol 24: 497–504.
3. Madrigal J, Kelt DA, Meserve PL, Gutierrez JR, Squeo FA (2011) Bottom-up control of consumers leads to top-down indirect facilitation of invasive annual herbs in semiarid Chile. Ecology 92: 282–288.
4. Martin CW, Valentine JF (2012) Eurasian milfoil invasion in estuaries: physical disturbance can reduce the proliferation of an aquatic nuisance species. Mar Ecol Prog Ser 449:109–119.
5. Geraldi NR, Smyth AR, Piehler MF, Peterson CH (2014) Artificial substrates enhance non-native macroalga and N₂ production. Biol Invasions: 16: 1819–1831.
6. Lonsdale WM (1999) Global patterns of plant invasions and the concept of invasibility. Ecology 80: 1522–1536.
7. Elton CS (1958) The ecology of invasions by animals and plants. Methuen, London.
8. Stachowicz JJ, Whitlatch RB, Osman RW (1999) Species diversity and invasion resistance in a marine ecosystem. Science 286: 1577–1579.
9. deRivera CE, Ruiz GM, Hines AH, Jivoff P (2005) Biotic resistance to invasion: Native predator limits abundance and distribution of an introduced crab. Ecology 86:3364–3376.
10. Davis MA, Grime JP, Thompson J (2000) Fluctuating resources in plant communities: a general theory of invasibility. J Ecol 88: 528–534.
11. Human KG, Gordon DM (1996) Exploitation and interference competition between the invasive Argentine ant, *Linepithema humile*, and native ant species. Oecologia 105:405–412.
12. Morris JA Jr, Akins JL (2009) Feeding ecology of invasive lionfish (*Pterois volitans*) in the Bahamian archipelago. Environ Biol Fishes 86: 389–398.
13. Schofield PJ (2010) Update on geographic spread of invasive lionfishes (*Pterois volitans* [Linnaeus, 1758] and *P. miles* [Bennett, 1828]) in the Western North Atlantic Ocean, Caribbean Sea and Gulf of Mexico. Aquat Invasions 5: 117–122.
14. Bariche M, Torres M, Azzurro E (2013) The presence of the invasive lionfish *Pterois miles* in the Mediterranean Sea. Mediterr Mar Science 14: 292–294.
15. Sutherland WJ, Clout M, Cote IM, Daszak P, Depledge MH, et al. (2010) A horizon scan of global conservation issues for 2010. Trends Ecol Evol 25: 1–7.
16. Kulbicki MM (2012) Distributions of Indo-Pacific lionfishes *Pterois* spp. in their native ranges: Implications for the Atlantic invasion. Mar Ecol Prog Ser 446: 189–205.
17. Albins MA, Hixon MA (2008) Invasive Indo-Pacific lionfish *Pterois volitans* reduce recruitment of Atlantic coral-reef fishes. Mar Ecol Prog Ser 367: 233–238.
18. Green SJ, Akins JL, Maljkovic A, Cote IM (2012) Invasive lionfish drive native Atlantic coral reef fish declines. PLoS One 7(3): e32596.
19. Albins MA (2013) Effects of invasive Pacific red lionfish *Pterois volitans* versus a native predator on Bahamian coral-reef fish communities. Biol Invasions 15: 29–43.
20. Salo P, Banks PB, Dickman CR, Korpimaki E (2010) Predator manipulation experiments: Impacts on populations of terrestrial vertebrate prey. Ecol Monogr 80: 531–546.
21. Cure KC, Benkwitt E, Kindinger TL, Pickering EA, Pusack TJ, et al. (2012) Comparative behavior of red lionfish *Pterois volitans* on native Pacific versus invaded Atlantic coral reefs. Mar Ecol Prog Ser 467: 181–192.
22. Anton A (2013) Ecology and evolution of the lionfish invasion of Caribbean coral reefs: Resistance, adaptation, and impacts. The University of North Carolina at Chapel Hill. PhD Dissertation. ProQuest, UMI Dissertation Publishing. 3594121.
23. Hackerott S, Valdivia A, Green SJ, Côté IM, Cox CE, et al. (2013) Native predators do not influence invasion success of Pacific lionfish on Caribbean reefs. PLoS One 8(7): e68259 doi:10.1371
24. Mumby PJ, Harborne AR, Brumbaugh DR (2011) Grouper as a natural biocontrol of invasive lionfish. PLoS One 6(6): e21510.
25. Mumby PJ, Brumbaugh DR, Harborne AR, Roff G (2013) On the relationship between native grouper and invasive lionfish in the Caribbean. PeerJ PrePrints 1:e45v1
26. Barbier EB, Hacker SD, Kennedy C, Koch EW, Stier AC (2011) The value of estuarine and coastal ecosystem services. Ecol Monogr 81: 169–193.
27. Whitfield PE, Gardner T, Vives SP, Gilligan MR, Courtenay WR, et al. (2002) Biological invasion of the Indo-Pacific lionfish *Pterois volitans* along the Atlantic coast of North America. Mar Ecol Prog Ser 235: 289–297.
28. Whitfield P, Hare J, David A, Harter S, Munoz RC, et al. (2007) Abundance estimates of the Indo-Pacific lionfish *Pterois volitans/miles* complex in the Western North Atlantic. Biol Invasions 12: 53–64.

29. National Marine Fisheries Service (2008) 2008 Status of U.S. Fisheries. National Marine Fisheries Service report.

30. Sandin SA, Smith JE, DeMartini EE, Dinsdale EA, Donner SD, et al (2008) Baselines and Degradation of Coral Reefs in the Northern Line Islands. PLoS ONE 3(2): e1548. doi:10.1371/journal.pone.0001548.

31. Green SJ, Tamburello N, Miller SE, Akins JL, Cote IM (2013) Habitat complexity and fish size affect the detection of Indo-Pacific lionfish on invaded coral reefs. Coral Reefs 32: 413–421.

32. Mumby PJ, Edwards AJ, Arias-Gonzalez JE, Lindeman KC, Blackwell PG, et al. (2004) Mangroves enhance the biomass of coral reef fish communities in the Caribbean. Nature 427: 533–536.

33. Froese R, Pauly D (2011). FishBase website. Available: https://www.fishbase.org. Accessed 2014 Aug 7.

34. Kundu PK (1990) Fluid mechanics. Academic Press, San Diego.

35. Risk MJ (1972) Fish diversity on a coral reef in the Virgin Islands. Atll Res Bull 153: 1–6

36. Aronson RB, Precht WF (1995) Landscape patterns of reef coral diversity: A test of the intermediate disturbance hypothesis. J Exp Mar Bio Eco 192: 1–14.

37. Zuur AF, Ieno EN, Walker NJ, Saveliev AA, Smith GM (2009) Mixed effects models and extensions in ecology with R. Springer, New York.

38. Venables WN, Ripley BD (2002) Modern Applied Statistics with S (Fourth Edition). Springer, New York.

39. Zeileis A, Kleiber C, Jackman S (2008) Regression Models for Count Data in R. J Stat Softw 27(8):1–25.

40. Krajewski JP, Floeter SR (2011) Reef fish community structure of the Fernando de Noronha Archipelago (Equatorial Western Atlantic): the influence of exposure and benthic composition. Environ Biol Fishes 92: 25–40.

41. Bellwood DR, Wainwright PC, Fulton CJ, Hoey A (2002) Assembly rules and functional groups at global biogeographical scales. Funct Ecol 16: 557–562.

42. Albins MA, Lyons PJ (2012) Invasive Indo-Pacific lionfish *Pterois volitans* direct jets of water at prey fish. Mar Ecol Prog Ser 448: 1–5.

43. Valdivia A, Bruno JF, Cox CE, Hackerott S, Green SJ (2014) Re-examining the relationship between invasive lionfish and native grouper in the Caribbean. PeerJ 2:e348; DOI 10.7717/peerj.348.

44. Goreau TF (1956) The ecology of Jamaican coral reefs. 1. Species composition and zonation. Ecology 40: 67–90.

45. Wolff N, Grober-Dunsmore R, Rogers CS, Beets J (1997) Management implications of fish trap effectiveness in adjacent coral reef and gorgonian habitats. Environ Biol Fishes 55: 81–90.

46. Torres R, Chiappone M, Geraldes F, Rodriguez Y, Vega M (2001) Sedimentation as an important environmental influence on Dominican Republic reefs. Bull Mar Science 69: 805–818.

47. Harborne AR, Mumby PJ, Zychaluk K, Hedley JD, Blackwell PG (2006) Modeling the beta diversity of coral reefs. Ecology 87: 2871–2881.

48. Karkarey R, Kelkar N, Lobo A, Alcoverro T, Arthur R (2014) Long-lived groupers require structurally stable reefs in the face of repeated climate change disturbances. Coral Reefs 33: 289–302.

49. Polunin NVC, Roberts CM (1993) Greater biomass and value of target coral-reef fishes in two small Caribbean marine reserves. Mar Ecol Prog Ser 100: 167–176.

50. Brokovich E, Baranes A, Goren M (2006) Habitat structure determines coral reef fish assemblages at the northern tip of the Red Sea. Ecol Indic 6: 494–507.

51. Mumby P, Wabnitz CCC (2002) Spatial patterns of aggression, territory size, and harem size in five sympatric Caribbean parrotfish species. Environ Biol Fishes 63: 265–279.

52. Bejarano S, Mumby PJ, Sotheran I (2011) Predicting structural complexity of reefs and fish abundance using acoustic remote sensing (RoxAnn). Mar Biol 158: 489–504.

53. Mumby PJ (2014) Stratifying herbivore fisheries by habitat to avoid ecosystem overfishing of coral reefs. Fish and Fisheries. doi: 10.1111/faf.12078.

54. Darling ES, Green SJ, O'Leary JK, Cote IM (2011) Indo-Pacific lionfish are larger and more abundant on invaded reefs: a comparison of kenian and Bahamian lionfish populations. Biol Invasions 13: 2045–2051.

55. Balasubashini MS, Karthigayan S, Somasundaram ST (2006) *In vivo* and *in vitro* characterization of the biochemical and pathological changes induced by lionfish (*Pterois volitans*) venom in mice. Toxicol Mech Methods 16: 525–531.

56. Llewelyn J, Schwarzkopt L, Alford R, Shine R (2010) Something different for dinner? Responses of a native Australian predator (the keelback snake) to an invasive prey species (the cane toad). Biol Invasions 12: 1045–1051.

57. Shine R (2010) The ecological impact of invasive cane toads (*Bufo marinus*) in Australia. Q Rev Biol 85: 253–291.

58. Diller JL, Frazer TK, Jacoby CA (2014) Coping with the lionfish invasion: Evidence that naïve, native predators can learn to help. J Exp Mar Biol Ecol 455: 45–49.

59. Barbour AB, Allen MS, Frazer TK, Sherman KD (2011) Evaluating the Potential Efficacy of Invasive Lionfish (*Pterois volitans*) Removals. PLoS ONE 6(5): e19666. doi:10.1371/journal.pone.0019666.

60. Green SJ, Dulvy NK, Brooks AL, Akins JL, Cooper AB, et al. (2014) Linking removal targets to the ecological effects of invaders: a predictive model and field test. Ecol Appl. http://dx.doi.org/10.1890/13-0979.1 (In press)

16

Occupancy Models for Monitoring Marine Fish: A Bayesian Hierarchical Approach to Model Imperfect Detection with a Novel Gear Combination

Lewis G. Coggins Jr.[1,2]*, **Nathan M. Bacheler**[1], **Daniel C. Gwinn**[3]

1 National Marine Fisheries Service, Southeast Fisheries Science Center, Beaufort, North Carolina, United States of America, **2** United States Fish and Wildlife Service, Yukon Delta National Wildlife Refuge, Bethel, Alaska, United States of America, **3** Biometric Research, and Program for Fisheries and Aquatic Sciences, School of Forest Resources and Conservation, University of Florida, Gainesville, Florida, United States of America

Abstract

Occupancy models using incidence data collected repeatedly at sites across the range of a population are increasingly employed to infer patterns and processes influencing population distribution and dynamics. While such work is common in terrestrial systems, fewer examples exist in marine applications. This disparity likely exists because the replicate samples required by these models to account for imperfect detection are often impractical to obtain when surveying aquatic organisms, particularly fishes. We employ simultaneous sampling using fish traps and novel underwater camera observations to generate the requisite replicate samples for occupancy models of red snapper, a reef fish species. Since the replicate samples are collected simultaneously by multiple sampling devices, many typical problems encountered when obtaining replicate observations are avoided. Our results suggest that augmenting traditional fish trap sampling with camera observations not only doubled the probability of detecting red snapper in reef habitats off the Southeast coast of the United States, but supplied the necessary observations to infer factors influencing population distribution and abundance while accounting for imperfect detection. We found that detection probabilities tended to be higher for camera traps than traditional fish traps. Furthermore, camera trap detections were influenced by the current direction and turbidity of the water, indicating that collecting data on these variables is important for future monitoring. These models indicate that the distribution and abundance of this species is more heavily influenced by latitude and depth than by micro-scale reef characteristics lending credence to previous characterizations of red snapper as a reef habitat generalist. This study demonstrates the utility of simultaneous sampling devices, including camera traps, in aquatic environments to inform occupancy models and account for imperfect detection when describing factors influencing fish population distribution and dynamics.

Editor: Ilaria Corsi, University of Siena, Italy

Funding: The authors have no support or funding to report.

Competing Interests: The authors have declared that no competing interests exist.

* Email: lewis_coggins@fws.gov

Introduction

Ecological surveys are important for understanding spatial and temporal variability in plant and animal populations, as well as providing the necessary feedback to guide policy options in the context of state-dependent and adaptive-management programs [1]. Individuals of a population are often counted directly in order to draw inferences about their abundance and distribution, but rarely are all individuals observed [2]. When the detection of individuals is imperfect, some portion of the population will be missed leading to erroneous conclusions and possibly erroneous management [3,4,5,6]. Many ecological surveys instead use count or capture-rate data to index abundance. The implicit, and often violated, assumption of abundance indices is that capture probability does not vary systematically across space, time, habitat types, or environmental conditions [7,8]. An alternative approach is to explicitly account for imperfect detection in sampling methodologies. Plot, distance, capture-recapture, and removal methods have all been used in terrestrial and aquatic environments to estimate animal abundance while accounting for imperfect capture probabilities [9], but these approaches are often impractical or expensive for many species [10].

The use of occupancy models to describe the distribution of populations while accounting for imperfect detection has increased in popularity over the last decade. These models require repeated sampling at spatially replicated sites to simultaneously estimate occupancy and detection probability, thereby correcting for imperfect detection [11,4,12]. Although the occurrence of a species at a site describes a different population process than abundance, occupancy models can be structured to estimate abundance directly by making some structural assumptions about the relationship between detection and abundance [3]. One major advantage of occupancy models for estimating population distribution and abundance is the use of incidence data that are often less costly to collect than data to estimate abundance directly (e.g., tagging information). Thus, occupancy models are gaining popularity in the wildlife literature as a monitoring tool.

Examples of occupancy modeling to index abundance or distribution are currently sparser in the fisheries literature than the wildlife literature. One reason for this discrepancy is the difficulties in sampling fish populations in ways that meet the assumptions of the model, particularly in marine systems, but see [13,14,9]. Because sampling fish is often invasive (e.g., electrofishing, trawling), replicate samples may not meet the assumption of independence [15] as the first sample may affect the detection of fish in the following samples. Temporal replicates that allow enough time between samples for the fish to recover from previous handling can be employed, but increase the risk of violating the population closure assumptions [11,16]. Spatial replication of fish sampling at a site is often employed to ameliorate this issue; however, substituting spatial for temporal replicates can cause bias in parameters estimates under common sampling schemes [17,18]. These sampling issues put fisheries managers at a great disadvantage because fisheries indices of abundance that inform policy choices are known to be plagued with issues of detection [19,20,21]. The lack of account for these detection issues has, in some cases, lead to inappropriate management choices and devastating ecological and economic losses, e.g., [22,23]. For example, the collapse of the North Atlantic cod stock in 1992 is cited as one the greatest social and economic tragedies in Canada's history and is partially attributed to a systematic increase in detection probability that caused abundance indices to remain stable as the stock declined [24]. Thus, methods that allow fisheries researchers to account for incomplete or variable detection when estimating the abundance and distribution of stocks are paramount.

An alternative sampling scheme to achieve temporal and spatial replication is the use of multiple sampling gears simultaneously. The use of multiple gears with occupancy models is uncommon in the ecological literature, but has been utilized to expand detection opportunities across species and individuals with variable vulnerability to different detection methods [25,26]. Additionally, multiple gears employed in a nested design have been used to estimate occupancy probability at different spatial scales [27]. The use of non-invasive sampling gears such as camera traps in combination with traditional fish sampling [28,29] may resolve some issues associated with replicate sampling described above. With replicate observations from simultaneous deployment of different gears in time and space, issues of bias induced by closure violations and non-independence of individual detections may be avoided. Thus, sampling with multiple gears combined with occupancy models may represent a powerful tool for monitoring fish populations.

Red snapper *Lutjanus campechanus* along the southeast USA coast (SEUS) is an economically important marine fish species and their management would benefit greatly from basic knowledge regarding their abundance and distributional patterns [30]. Since 2010, the SEUS red snapper fishery has been closed due to overfishing, and the only long-term survey data that exist (i.e., chevron trapping) have been deemed unusable in recent stock assessments due to overdispersed catches, as well as the perceived low detection probability [31,32]. Beginning in 2010, high-definition video cameras have been attached to chevron traps to presumably increase gear detection probability [28]. Thus, the addition of cameras to the existing monitoring program is deemed critical for recovering and sustainably managing the red snapper fishery.

Here we develop occupancy and abundance models that employ incidence data collected with a combination of traditional

invasive (chevron trap) and non-invasive (camera trap) sampling gears. This gear combination is novel to marine fisheries research and allows for a gear-for-time substitution for generating replicate samples that we expect will better meet the required assumptions of binomial sampling. Our specific objectives are to demonstrate the utility of these sampling methods by, 1) evaluating the relative fit of models that obtain sample replication from a combination of chevron trap and aggregated and disaggregated camera trap data, and 2) apply these models to evaluate how time and habitat influence red snapper occupancy probability and abundance as well as how gear and habitat influence detection.

Materials and Methods

Ethics Statement

Data collection for this study was authorized in a 5-year Scientific Research Permit (that commenced in 2010), issued by the Administrator of Southeast Regional Office of the National Marine Fisheries Service, National Oceanic and Atmospheric Administration, United States Government. This Scientific Research Permit covered all areas sampled in the study. All research followed the guidelines of the U.S. Government Principles for the Utilization and Care of Vertebrate Animals Used in Testing, Research, and Training (http://grants.nih.gov/grants/olaw/references/phspol.htm#USGovPrinciples). Red snapper collected in fish traps were euthanized by being placed on ice, after which a variety of biological samples were extracted per the guidelines of the Scientific Research Permit.

Sampling Program

Sampling occurred in Atlantic Ocean continental shelf waters off Georgia and Florida, USA, which encompass the historical center of the red snapper fishery in the SEUS [31] (Figure 1). Sampling targeted red snapper and other reef fishes that typically associate with hard substrates, which occur as scattered patches within the dominant sand and mud substrate of the region [33]. Patches of hard substrates in the SEUS are diverse and consist of flat limestone pavement, ledges, rocky outcroppings, or reefs, and are often colonized by various types of attached biota [34,35]. The major oceanographic feature of the SEUS is the Gulf Stream, which influences outer sections of the continental shelf as it flows northward.

Sampling was conducted by the Southeast Fishery-Independent Survey (SEFIS), a fishery-independent sampling program created by the National Marine Fisheries Service in 2010 to increase fishery-independent sampling in the SEUS. Hard bottom sampling sites were selected for sampling in one of three ways. First, most sites were randomly selected from a sampling frame of hard bottom sampling points developed by SEFIS or the Marine Resources Monitoring, Assessment, and Prediction program of the South Carolina Department of Natural Resources. Second, some sites were sampled opportunistically even though they were not randomly selected for sampling in a given year. Third, new sites were added during the study period using information from fishermen, charts, and historical survey information. These locations were investigated using the vessel echo sounder and sampled if hard bottom was suspected to be present. Overall, less than 10% of the sampled sites included in the study were selected non-randomly via the second and third methods above. All sampling for this study occurred in 2010–2011 during daylight hours aboard the R/V *Savannah*, NOAA Ship *Nancy Foster*, or NOAA Ship *Pisces*. Depths ranged from 16 to 83 m.

Chevron/camera traps were deployed at each selected site and consisted of a chevron trap outfitted with an outward-looking

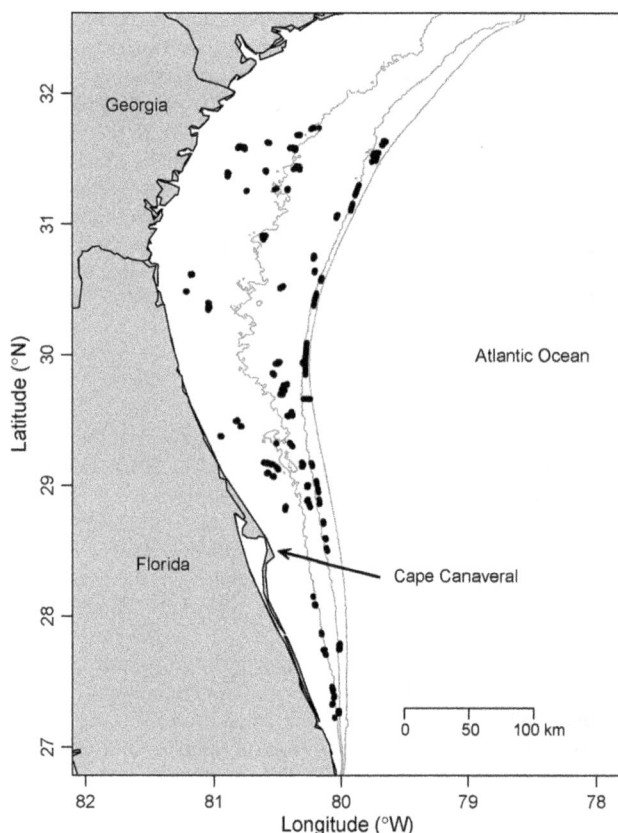

Figure 1. Study area with dots marking sampling locations. The offshore contours are depth isoclines at 30 m, 50 m, and 100 m.

high-definition video camera (Figure 2). Eighteen to 24 trap combinations were deployed for approximately 90 min each day during April-October of each year. Traps were always spaced more than 200 m apart and each chevron trap was baited with 24 menhaden *Brevoortia* spp. Chevron traps have been used widely to index the abundance of reef fish and invertebrate species [36,37,38]. Chevron traps were constructed from plastic-coated galvanized 12.5-ga wire (mesh size = 3.4 cm^2), and were shaped like an arrowhead measuring 1.7 m×1.5 m×0.6 m, with a total volume of 0.91 m^3 (Figure 2).

A GoPro Hero (2010) or Canon Vixia HFS200 (2011) camera was attached over the mouth of each chevron trap. This camera positioning allowed for the potential detection of fish that are available to be caught by the chevron trap whether they actually enter or do not enter the trap. These cameras have similar viewing areas and resolution, and we assumed that camera type did not influence detection probability. We examined 1-second "snapshots" every 30 seconds beginning 10 minutes after the chevron trap was deployed and continuing for 20 minutes (for a total of 41 snapshots; [39]). If red snapper were seen in any of the 41 snapshots, they were considered present in the camera trap. This sampling strategy allowed us to gain replication at each sampling site from the simultaneous use of the chevron trap sample and the aggregate sample of the camera, or gain additional replication by using the disaggregated 41 1-second snapshots from the camera.

Hypothesized Predictors of Occurrence and Detection

Red snapper site occurrence is likely influenced by latitude, depth, temperature, and various localized reef characteristics

associated with substrate type and bottom topography [40,41]. Habitat features such as substrate relief (i.e., the amount of topographic variation), percent hard bottom (i.e., substrate consisting of consolidated sediments), and percent attached biota (i.e., substrate with attached coral or sponges) are potentially important drivers of habitat suitability. Latitude, depth, and temperature are also likely to influence occurrence at broader spatial scales via processes potentially related to the native range of the species, the probability of influence by warm Gulf Stream currents, or fishery exploitation patterns.

Covariates that influence the observation process may be shared or unique to each gear type. Chevron trap detection probability is possibly related to deployment time (influencing the probability of fish encountering the trap), current speed (influencing the intensity and area of the bait plume scent), temperature (influencing fish activity or metabolism), and current direction (influencing how fish might orient to the trap). The camera trap detection process is possibly influenced by water clarity (limiting the detection range) and current direction and speed, which may affect the orientation or staging location of fish relative to the camera orientation.

We only analyzed sites that contained both valid chevron and camera trap samples. Sites were excluded if the chevron trap bounced or drifted, the chevron trap mouth opening was blocked by rocks, the video was dark, or any video files were missing. Year, latitude, longitude, depth, and bottom temperature were recorded at each sampling station. Bottom temperature (°C) was determined using a Seabird "Conductivity, Temperature, and Depth" instrument package (CTD; model SBE 25, Bellevue, Washington, USA). CTD casts were conducted near the middle of each sampling period, and the instrument was lowered to within 2 m of the bottom. Underwater videos were used to determine micro-habitat features, water clarity, and current direction and magnitude around the chevron/camera trap (Table 1). The data analyzed in this study are reported in Table S1.

Modeling Overview

We modeled the occurrence of red snapper within a Bayesian hierarchical framework using a distribution sub-model that described how organisms are distributed among sites and a detection sub-model that described the data-generating process [42]. We also extended the occupancy model to account for variation in detection probability due to variation in fish abundance among sites using methods developed by Royle and Nichols [3]. This model formulation assumes a relationship between the probability of detecting a species and the number of individuals of that species at a site. This is accomplished by estimating the probability of detecting a single individual (individual-based detection). The advantage of modeling abundance is that variation in the probability of detecting a species due to abundance variation among sites is explicitly accounted for and covariates are evaluated for effects on the abundance of fish. We evaluated performance of six candidate models including both the basic occupancy model [11] and the abundance model (Royle-Nichols [3]) formulation with both distribution and detection model covariates, with and without random effects on the detection process, and with aggregated and disaggregated detection histories for the camera trap (Table 2). Additionally, we fit the Royle-Nichols formulation both with and without random effects on mean site abundance.

Basic Occupancy Model Structure

We defined occurrence as z_i where z is a binary variable indicating the latent occupancy state of red snapper at site i with $z = 1$ indicating presence and $z = 0$ indicating absence. We

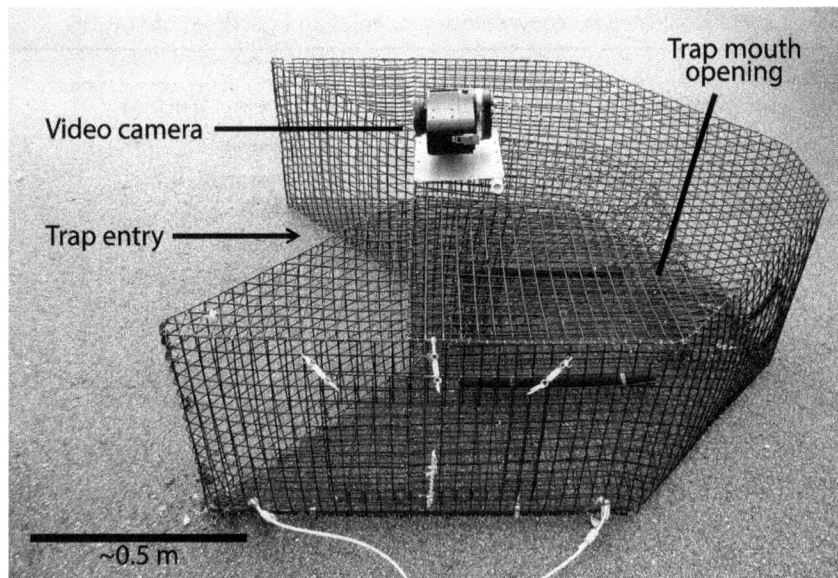

Figure 2. Chevron fish trap outfitted with an outward-looking Canon high-definition video camera over the mouth of the trap.

assumed that z_i was the result of a Bernoulli trial represented by z_i ~Bernoulli(ψ_i), where ψ_i represents the probability of occurrence of red snapper at site i. Because the true occupancy state is observed imperfectly, we modeled the probability of detection of red snapper as a separate Binomial process for each gear, where the unconditional probability of detection at site i with gear j is z_i p_{ij}. Thus, the number of observations of red snapper is represented as $y_{ij} \sim \text{Binomial}(k_j, z_i p_{ij})$, where p_{ij} is the detection probability that is conditional on $z_i = 1$, and k indicates the number of replicate samples collected by each gear at each site.

We incorporated potential covariate effects in the distribution sub-model using a logit link [43] specified as:

Table 1. Variables evaluated as potential covariates influencing patterns in occurrence, abundance, and/or detection of red snapper *Lutjanus campechanus*.

Covariate	Description
Yr2011	Data collected in 2011
depth	Depth of water in meters
depth²	Depth squared
lat	Latitude of sample site in decimal degrees
lat²	Latitude squared
temp	Bottom temperature (°C) of water at sample location
temp²	Bottom temperature squared
livebot.l	0–10% of substrate covered by live bottom (e.g., corals, sponges)
livebot.m	11–40% of substrate covered by live bottom
livebot.h	>40% of substrate covered by live bottom
hardsub.l	0–10% of substrate is hard bottom (e.g., rocks, boulders, ledges)
hardsub.m	11–40% of substrate is hard bottom
hardsub.h	>40% of substrate is hard bottom
relief.m	Maximum topographical relief of substrate is 0.3–1.0 m
relief.h	Maximum topographical relief of substrate is >1.0 m
soak	The total amount of time the trap was deployed (minutes)
cdir.p	Current direction is perpendicular to the trap mouth opening
cdir.a	Current direction is away from the trap mouth opening
cspeed	Speed of the current (low or high)
turb.h	Indicates high turbidity (i.e., cannot see bottom habitat)

Covariates not listed here (e.g., *yr2010*) inform the models' intercept predictions.

Table 2. General model structures evaluated for convergence properties and goodness of fit (GOF).

Model Type	Model Structure	Camera Trap Data	G-R	GOF
Basic Occupancy	**Bin(ψ) Bin($p_{chevron}$) Bin(p_{camera})**	**pooled**	**1.0**	**0.99**
Basic Occupancy	Bin(ψ) Bin($p_{chevron}$) Bin(p_{camera})	disaggregated	1.0	0.00
Basic Occupancy	Bin(ψ) Bin($p_{chevron}$) Bin-logNorm(p_{camera})	disaggregated	1.0	0.00
Royle-Nichols	**Pois(λ) Bin($p_{chevron}$) Bin(p_{camera})**	**pooled**	**1.0**	**0.71**
Royle-Nichols	Pois(λ) Bin($p_{chevron}$) Bin(p_{camera})	disaggregated	1.0	0.00
Royle-Nichols	Pois-logNorm (λ) Bin($p_{chevron}$) Bin(p_{camera})	disaggregated	1.0	0.00

The camera trap data column indicates models using pooled detections versus disaggregated detections. Values of GOF approaching zero indicate lack of fit while values approaching one indicate no evidence of lack of fit. Values of the Gelman-Rubin statistic (G-R) close to one indicate model convergence while values greater than one indicate lack of model convergence. Bolded models converged and displayed no evidence of lack of fit.

$$\text{logit}(\psi_i) =$$

$$\beta_1 + \beta_2 yr2011_i + \beta_3 depth_i + \beta_4 depth_i^2 + \beta_5 lat_i + \beta_6 lat_i^2 +$$

$$\beta_7 temp_i + \beta_8 temp_i^2 + \beta_9 livebot.l_i + \beta_{10} livebot.m_i + \quad (1)$$

$$\beta_{11} livebot.h_i + \beta_{12} hardsub.l_i + \beta_{13} hardsub.m_i +$$

$$\beta_{14} hardsub.h_i + \beta_{15} relief.m_i + \beta_{16} relief.h_i,$$

where β_1 represents the intercept of the distribution sub-model and β_2, β_3,....β_{16} represent the logit-scale effects of each variable on the probability of the occurrence. Variables and their abbreviations are defined in Table 1.

A detection probability sub-model was developed for each gear type. We specified the chevron trap detection probability $p_{ij=1}$ as:

$$\text{logit}(p_{ij=1}) = \alpha_1 + \alpha_2 temp_i + \alpha_3 temp_i^2 + \alpha_4 soak_i + \alpha_5 cdir.p_i + \\ \alpha_6 cdir.a_i + \alpha_7 cspeed_i, \quad (2)$$

where α_1 represents the intercept and α_2, α_3,......α_7 represent the logit-scale effects of each covariate on detection probability (Table 1). We specified camera trap detection probability $p_{ij=2}$ as:

$$\text{logit}(p_{ij=2}) = \varphi_1 + \varphi_2 turb_i + \varphi_3 cdir.p_i + \\ \varphi_4 cdir.a_i + \varphi_5 cspeed_i + \varepsilon_i, \quad (3)$$

where φ_1 represents the intercept and φ_2 through φ_5 represent the logit-scale effects of each variable on the probability of detection by the camera trap (Table 1). The parameter ε_i is a site-specific random effect to account for possible variation in detection probability among sites not explained with covariates (see Table 2). We modeled ε_i as a normally distributed random variable with mean equal to zero and standard deviation σ_ε. The random effects model structure was only used when considering the disaggregated camera trap data containing sufficient information to assure all model parameters were identifiable.

Royle-Nichols Model Structure

We modeled the abundance of red snapper based on methods first proposed by [3] and then extended into a hierarchical framework by [42]. Because we cannot observe the abundance at sites directly, we specified site abundance N_i as a latent random effect with variation that is explained by a Poisson distribution across i sites as $N_i \sim \text{Poisson}(\lambda_i)$, where λ_i represents the mean site

abundance. Our data y_{ij} are the frequencies of red snapper detections at site i with gear j. We assume that y_{ij} is the result of binomial outcomes as $y_{ij} \sim \text{Binomial}(k_j, p_{ij})$, where p_{ij} is the probability of detecting at least one individual at site i with gear j and k_j is the number of replicate samples collected with gear j. We linked the distribution sub-model to the detection sub-model by specifying the relationship between p_{ij} and N_i, per [3], as $p_{ij} = 1 - (1 - r_{ij})^{N_i}$ where r_{ij} is individual-based detection probability, as opposed to p_{ij}, which is the probability of detecting at least one individual at site i with gear j. This formulation essentially models the detection probability p_{ij} as a random effect due to variation in fish abundance among sites.

We utilized a similar covariate structure for the of the Royle-Nichols distribution sub-model as we did for the basic occupancy model with log of mean site abundance (λ_i) specified as:

$$\log(\lambda_i) =$$

$$\beta_1 + \beta_2 yr2011_i + \beta_3 depth_i + \beta_4 depth_i^2 + \beta_5 lat_i +$$

$$\beta_6 lat_i^2 + \beta_7 temp_i + \beta_8 temp_i^2 + \beta_9 livebot.l_i + \beta_{10} livebot.m_i \quad (4)$$

$$+ \beta_{11} livebot.h_i + \beta_{12} hardsub.l_i + \beta_{13} hardsub.m_i +$$

$$\beta_{14} hardsub.h_i + \beta_{15} relief.m_i + \beta_{16} relief.h_i + \varepsilon_i.$$

The parameter ε_i was a site-specific random effect drawn from a Normal distribution with mean of zero and standard deviation (σ_ε) to account for extra-Poisson variation in abundance across sites. Similar to the basic occupancy model structure, the random effects model was only fit to the disaggregated camera trap data to assure all parameters were identifiable. Covariates were incorporated into the detection sub-models with a logit link as:

$$\text{logit}(r_{ij=1}) = \alpha_1 + \alpha_2 temp_i + \alpha_3 temp_i^2 + \alpha_4 soak_i + \alpha_5 cdir.p_i + \\ \alpha_6 cdir.a_i + \alpha_7 cspeed_i, \quad (5)$$

$$\text{logit}(r_{ij=2}) = \varphi_1 + \varphi_2 turb_i + \varphi_3 cdir.p_i + \varphi_4 cdir.a_i + \varphi_5 cspeed_i, \quad (6)$$

where $r_{ij=1}$ is the individual detection probability for chevron traps and $r_{ij=2}$ is the individual detection probability for camera traps. All continuous variables were centered on the mean and scaled by one standard deviation to help with fitting and allow for unambiguous comparison of parameter estimates.

Model Evaluation

We evaluated the performance of six candidate models to determine the model structure that best described the data (Table 2). These models included both the basic occupancy model and the Royle-Nichols model fit with camera trap detection models employing both aggregated and disaggregated detection data. The aggregated camera trap detections used the entire 41 snap-shots as one sample while the disaggregated camera trap detections considered each snap-shot as an independent sample to determine if there was a trade-off between the number of replicates and effort per replicate that may optimize model fit. We hypothesized that more replicates would better inform estimates of the detection process than a single aggregated sample. However, disaggregated detections may sacrifice model fit because of auto-correlation among the snap shots. For the basic occupancy models and considering the disaggregated camera trap data, we evaluated model structures with and without site-specific random effects on the detection process of the camera traps to account for variation in detection that was not explained by the site-specific covariates (3). For the Royle-Nichols models and considering the disaggregated camera trap data, we evaluated model structures with and without site-specific random effects on abundance in the distribution sub-model to account for extra-Poisson variation in abundance (4). This model extension, in turn, modeled additional variation in the detection process of both the cheveron trap and the camera trap through the relationship between detection probability and abundance.

We performed our model evaluations at two different levels. First we evaluated model fit with a Chi-squared goodness of fit (GOF) test developed by [44]. Although lack of fit can result for many different reasons, adequate fit of a model would indicate that the assumptions of the model such as independent detections among camera snap shots or between the camera and the chevron trap were adequately met. The GOF test is a form of posterior predictive check that assumes the frequency of detection histories conform to a Chi-square distribution. Point estimates of all parameter values from the full model are used to perform a parametric bootstrap to determine the expected distribution of model deviances. The deviance of the observed detection frequencies is then compared to the distribution of expected deviances to determine model fit. Goodness of fit values range from zero to one with values approaching zero indicating lack of fit and values approaching one indicating adequate fit.

For all models judged to fit adequately by the GOF [44] procedure, we further evaluated the support of each covariate as our second level of model evaluation. We considered all combinations of covariates plausible and estimated the inclusion probability of each covariate using a mixture modelling approach where each covariate is multiplied by an "inclusion parameter" ([12], pages 72–79). The inclusion parameters (w_v for all variables in the model) were latent binary variables with uninformative prior probabilities of 0.5 (i.e., equal probability model inclusion or exclusion). The posterior probabilities of the inclusion parameters correspond to the probability that the given variable is included in the model, and parameter summaries and predictions are averaged across all covariate combinations. Following [45], we assumed that parameters with inclusion probability greater than 50% were adequately supported by the data.

Posterior probability distributions of model parameters were estimated using a Monte Carlo-Markov chain (MCMC) algorithm implemented in program JAGS [46]. We called JAGS from within program R [47] with the library RJAGS (http://mcmc-jags. sourceforge.net). All prior distributions were uninformative distributions specified to have little influence on the posterior

probability distributions. The prior distributions of all logit-scale parameters were specified as t-distributions with $\sigma = 1.566$ and $\nu = 7.763$ per [48] such that back-transformed values assigned equal probability for all values between zero and one. Priors of standard deviation parameters were specified as uniform distributions with equal probability between zero and 100 and were verified to not influence the range of the posterior distributions. Inference was drawn from 30,000 posterior samples taken from 3 chains of 100,000 samples thinned to every 10. We allowed a burn in of 10,000 samples to remove the effects of initial values. Convergence was diagnosed for the full model by visual inspection of the MCMC chains for adequate mixing and stationarity and by using the Gelman-Rubin statistic (with values <1.1 indicating convergence; [49]).

Results

The Gelman-Rubin statistic indicated that all of the models converged to stable posterior distributions (Table 2). Based on the [40] GOF statistic, all of models considering disaggregated camera trap data exhibited poor fit. Inspection of the raw incidence data suggests that the poor fit was largely influenced by a small number of sites with much higher camera trap detection frequencies than predicted by the model structure and covariates considered. The remaining two models exhibiting good fit considered the pooled data and did not include log-normal random effects on either abundance or camera trap detection. We present detailed results from both of these models, Bin(ψ) Bin($p_{chevron}$) Bin(p_{camera}) and Pois(λ) Bin($p_{chevron}$) Bin(p_{camera}), and compare and contrast the basic occupancy versus the Royle-Nichols models. The computer code for these models is reported in Appendix S1 and S2.

The occupancy probability across years was estimated as 0.45 (95% CI = [0.30, 0.57]) for the basic occupancy model and 0.48 (95% CI = [0.32, 0.61]) for the Royle-Nichols model. As expected, the observed occupancy rates, uncorrected for imperfect detection, were lower for both the chevron trap (0.14) and camera trap (0.31) data. The estimated mean abundance (λ) across sites from the Royle-Nichols model ranged between 0.01 and 2.08 fish site^{-1}. In contrast, the observed chevron trap catch across sites ranged between 0 and 13 fish site^{-1} and the observed maximum camera trap count ranged between 0 and 14 fish site^{-1}. While maximum numbers of fish observed per site are larger for the chevron and camera trap than the maximum estimated λ, the proportion of sites where the observed numbers of fish was greater than 2 was small for both the chevron (0.04) and the camera (0.07) trap data.

Distribution

The posterior inclusion probabilities for the basic occupancy model indicated that the distribution sub-model best supported by the data included the covariates: water depth (depth, Pr = 0.87; Table 3), latitude (lat, Pr = 0.95), and its squared term (lat², Pr = 0.99). This model predicted that occupancy probability should vary between approximately 0.6 and 0.3 with a depth change from 20 m to 60 m (Figure 3). The model also predicted that occupancy probability should vary quadratically with latitude peaking (~0.48) at approximately 29.5°N and declining to near 0.2 at 27.2°N and 31.3°N. Notably, the year covariate (yr2011) was not supported by the data, suggesting no temporal trend in the occurrence probability of red snapper between 2010 and 2011.

The distribution sub-model of the Royle-Nichols model produced similar results to the basic occupancy model likely due to both low site-specific true abundances and little variation in true abundance among sites. Water depth (depth, Pr = 0.96; Table 4) and both latitude (lat, Pr = 1.00) and its squared term (lat²,

Table 3. Posterior probability summaries of parameters evaluated in the red snapper *Lutjanus campechanus* basic occupancy model using pooled camera trap data and without a site-specific random effect on camera trap detection probability (Bin(ψ) Bin($p_{chevron}$) Bin(p_{camera})).

| Parameter | Mean | SD | Credible interval | | Inclusion |
			2.5%	97.5%	probability
Distribution model					
intercept	−0.188	0.281	−0.844	0.279	–
yr2011	−0.003	0.075	−0.166	0.082	8%
depth	**−0.330**	**0.167**	**−0.602**	**0.000**	**87%**
depth²	0.013	0.055	0.000	0.215	8%
lat	**−0.401**	**0.155**	**−0.672**	**0.000**	**95%**
lat²	**−0.421**	**0.105**	**−0.624**	**−0.220**	**99%**
temp	0.007	0.047	0.000	0.158	6%
temp²	0.000	0.018	0.000	0.000	3%
livebot.l	0.153	0.288	0.000	0.957	30%
livebot.m	0.044	0.178	−0.045	0.684	14%
livebot.h	−0.307	0.401	−1.202	0.000	46%
hardsub.l	0.070	0.216	0.000	0.799	17%
hardsub.m	0.126	0.276	0.000	0.938	25%
hardsub.h	0.005	0.163	−0.333	0.365	11%
relief.m	0.008	0.080	−0.023	0.235	8%
relief.h	−0.055	0.194	−0.726	0.011	16%
Chevron trap detection model					
intercept	−0.499	0.184	−0.876	−0.155	–
temp	−0.055	0.130	−0.455	0.000	20%
temp²	0.003	0.036	0.000	0.056	5%
soak	−0.005	0.043	−0.106	0.000	6%
cdir.p	0.008	0.103	−0.105	0.280	9%
cdir.a	0.014	0.123	−0.130	0.392	10%
cspeed	0.051	0.410	−0.769	1.249	21%
Camera trap detection model					
intercept	1.126	0.533	0.132	2.213	–
turb.h	**0.559**	**0.639**	**0.000**	**1.920**	**55%**
cdir.p	−0.067	0.332	−1.082	0.477	20%
cdir.a	**0.652**	**0.783**	**0.000**	**2.405**	**54%**
cspeed	−0.034	0.642	−1.595	1.571	28%

All metrics were calculated from model averaged posterior distributions using the Bayesian mixture modeling approach. Bolded parameters had inclusion probability greater than 50%.

Pr = 1.00) were the most important covariates influencing abundance in the model (Table 4). On average, predicted abundance ranged between 1.0 and 0.5 individuals per site as depth changed from 20 m to 60 m (Figure 3). Abundance was also predicted to have a quadratic relationship to latitude, with a predicted abundance of about 0.20 individuals per site at 27°N and 31.5°N and peaking at approximately 0.75 individuals per site at 29.5°N (Figure 3). The Royle-Nichols model indicated that high coverage of the substrate with biota had a negative influence on abundance (*livebot.h*, Pr = 0.55; Table 4). This effect predicts a decrease of about 0.66 individuals per site between low (0–10%) and high levels (>40%) of live bottom coverage. The basic occupancy model suggested a similar effect, though the inclusion probability was slightly less than the 0.5 standard for inclusion. As

with results of the basic occupancy model, mean abundance did not appear to differ between 2010 and 2011.

We used the distribution sub-models from both the basic occupancy and Royle-Nichols models to predict both the occupancy probability and the mean abundance across a grid from approximately latitude 27°N to 32°N and from near-shore to a depth of 65 m (Figure 4). The model predicts that occupancy of reefs is highest in shallow waters (~20 m) and between approximately Cape Canaveral (28.4°N) and the Georgia-Florida state boundary line. The Royle-Nichols model predicts highest abundance at reefs in similar locations. Both models predict the center of distribution to be located at reefs offshore from Cape Canaveral.

Basic Occupancy Model

Royle–Nichols Occupancy Model

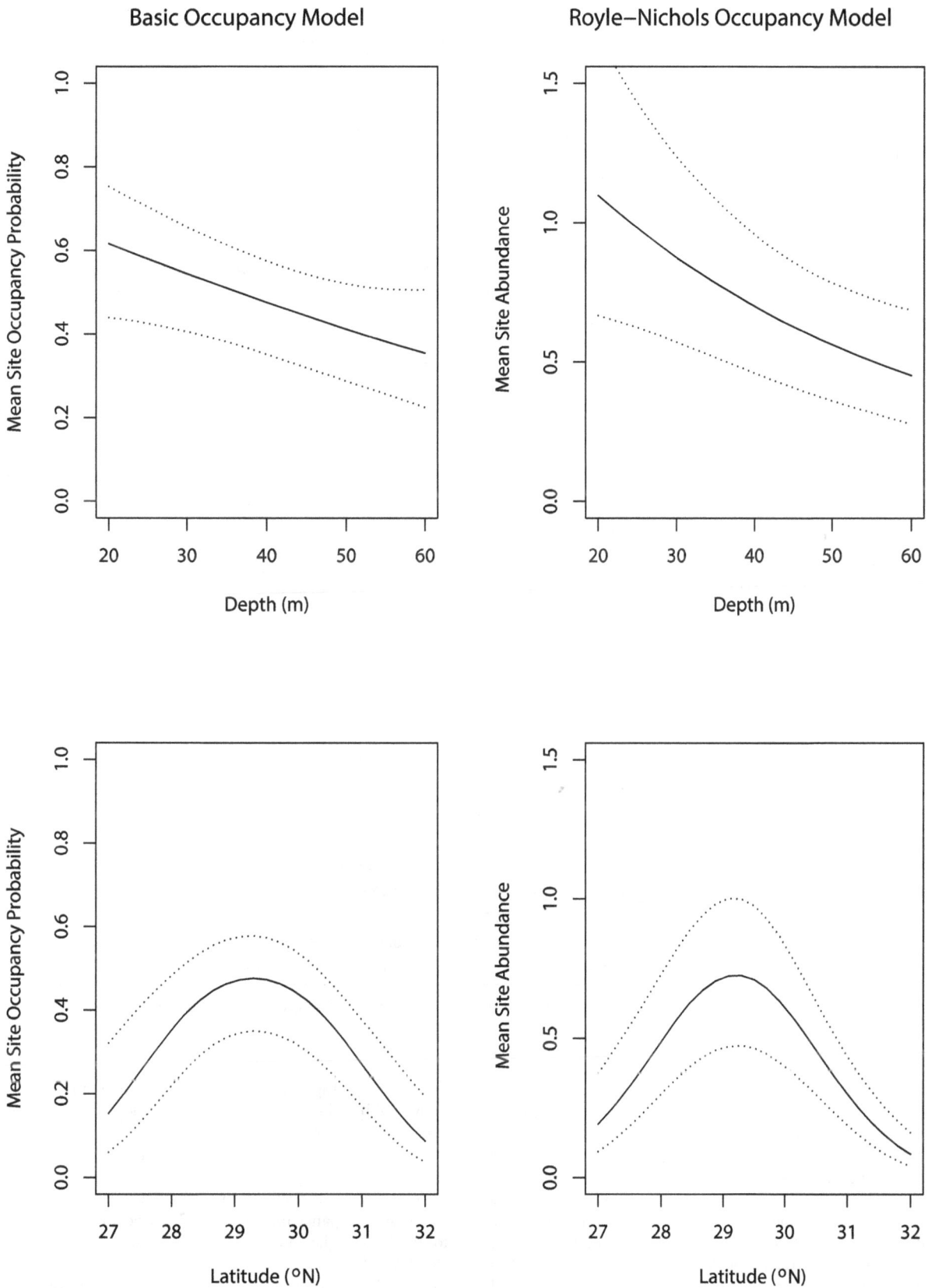

Figure 3. Mean site occupancy probability and abundance of red snapper as a function of depth and latitude. The left column contains the estimated mean site occupancy probability as a function of depth (top panel) and latitude (bottom panel) from the basic occupancy model. The right column contains the estimated mean site abundance as a function of depth (top panel) and latitude (bottom panel) from the Royle-Nichols occupancy model. The intervals represent 95% credible intervals.

Table 4. Posterior probability summaries of parameters evaluated in the red snapper *Lutjanus campechanus* Royle-Nichols occupancy model using pooled camera trap data and without a site-specific random effect on abundance (Pois(λ) Bin($p_{chevron}$) Bin(p_{camera})).

Parameter	Mean	SD	Credible interval		Inclusion
			2.5%	97.5%	probability
Distribution model					
intercept	−0.433	0.222	−0.954	−0.066	-
yr2011	−0.006	0.060	−0.183	0.070	10%
depth	**−0.277**	**0.101**	**−0.453**	**0.000**	**96%**
depth²	0.002	0.024	0.000	0.063	6%
lat	**−0.414**	**0.106**	**−0.627**	**−0.214**	**100%**
lat²	**−0.375**	**0.074**	**−0.525**	**−0.234**	**100%**
temp	0.006	0.036	0.000	0.127	8%
temp²	0.001	0.016	0.000	0.007	4%
livebot.l	0.140	0.228	0.000	0.744	37%
livebot.m	0.054	0.165	−0.081	0.589	21%
livebot.h	**−0.281**	**0.318**	**−0.941**	**0.000**	**55%**
hardsub.l	0.037	0.134	−0.022	0.485	16%
hardsub.m	0.098	0.196	0.000	0.648	30%
hardsub.h	0.013	0.125	−0.199	0.373	15%
relief.m	0.027	0.094	−0.005	0.350	15%
relief.h	−0.042	0.147	−0.532	0.077	19%
Chevron trap detection model					
intercept	−0.940	0.217	−1.391	−0.542	-
temp	−0.083	0.150	−0.481	0.000	31%
temp²	0.005	0.049	−0.039	0.149	9%
soak	−0.011	0.058	−0.207	0.000	10%
cdir.p	0.012	0.136	−0.262	0.407	16%
cdir.a	0.037	0.170	−0.204	0.569	19%
cspeed	0.062	0.445	−0.899	1.252	32%
Camera trap detection model					
intercept	0.512	0.483	−0.388	1.515	-
turb.h	**0.470**	**0.525**	**0.000**	**1.595**	**59%**
cdir.p	−0.025	0.294	−0.840	0.646	26%
cdir.a	**0.817**	**0.710**	**0.000**	**2.259**	**73%**
cspeed	−0.039	0.563	−1.397	1.322	36%

All metrics were calculated from model averaged posterior distributions using the Bayesian mixture modeling approach. Bolded parameters had inclusion probability greater than 50%.

Detection

Our mixture model procedure indicated support for covariates of camera trap detection, but did not for covariates of chevron trap detection as all covariate inclusion probabilities were <0.50 (Tables 3 & 4). The largest effect on detection was between the chevron traps and the camera traps. Camera trap detection probabilities were predicted to be about twice the detection probabilities predicted for chevron traps for both the basic occupancy model and the Royle-Nichols model (Figure 5). Mean detection probabilities estimated with the basic occupancy models were similar in magnitude to mean individual-based detection probabilities estimated with the Royle-Nichols occupancy model, likely due to low mean site abundances. Mean species detection probability and individual-based detection probability values for the chevron traps were 0.39 and 0.30, respectively (Figure 5). For

the camera traps, mean species detection probability and individual-based detection probability values were 0.75 and 0.61, respectively (Figure 5).

Covariates found to influence camera trap detection were water turbidity and current direction and were consistent between the basic occupancy model and the Royle-Nichols occupancy model. The basic occupancy model predicted species detection probability to be 0.09 higher in turbid water than clear water (*turb.h*, Pr = 0.55, Table 3, Figure 6) and that mean detection probability to be at least 0.10 higher when the current direction was away from the camera lens (*dir.a*, Pr = 0.54). The results for individual-based detection probability (r) from the Royle-Nichols model were similar, but slightly stronger. The model predicted mean individual detection probability to be 0.10 higher in turbid water than clear water (*turb.h*, Pr = 0.59, Table 4) and mean individual-

Basic Occupancy Model

Royle–Nichols Occupancy Model

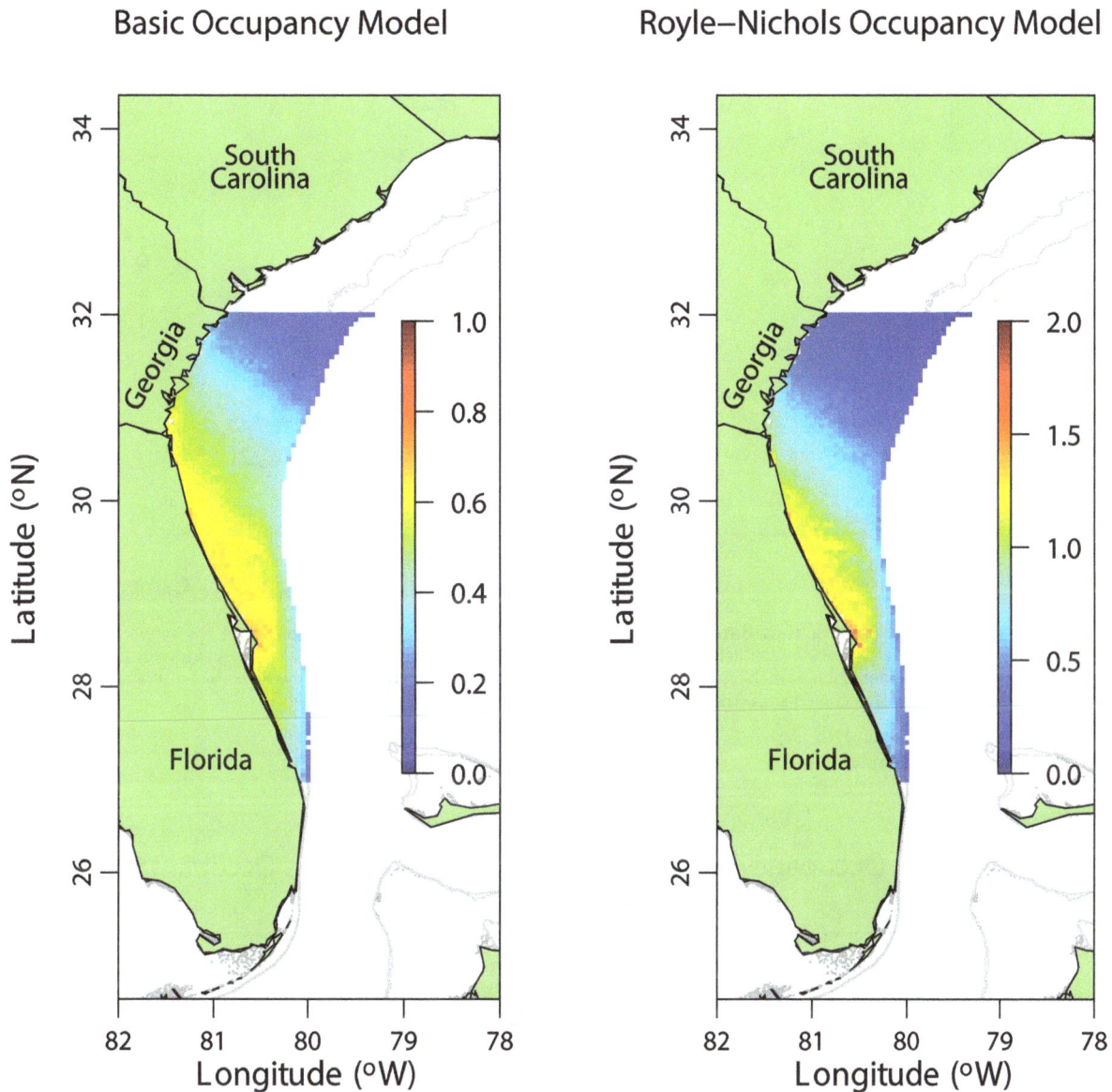

Figure 4. Mean site occupancy probability and abundance of red snapper in reef habitat off the coasts of Georgia and Florida. Estimated mean occupancy probability of reef sites from the basic occupancy model (left panel) and abundance from the Royle-Nichols model (right panel) as a function of depth and latitude.

based detection probability to be at least 0.17 higher when the current direction was away from the camera lens ($dir.a$, $\mathrm{Pr} = 0.73$).

Discussion

We explored the utility of combining observations from novel camera trap methods with traditional chevron traps to inform occupancy models that characterize temporal and spatial patterns in reef fish distribution. While camera traps are occasionally combined with other detection methods in terrestrial studies [e.g., 50, 25], we are unaware of studies utilizing camera and fish traps to provide the replicate observations required to account for incomplete detection of marine fish. The use of multiple gears sampling simultaneously assures closure among replicate samples, which is a major advantage for informing occupancy models. This is particularly true in aquatic systems where the need to use invasive sampling methodologies can violate the closure assumption among temporal replications.

We initially hypothesized that the disaggregated camera trap data would be more informative than the pooled camera trap data because of larger numbers of replications. However, the models considering the disaggregated camera trap data exhibited poor fit and simple examination of the raw data suggested the presence of extra-binomial variation possibly caused by non-independence in detections among individuals at the same site. This potential violation of the binomial sampling assumption appears to limit the utility of the disaggregated data for both the basic and Royle-Nichols models. In particular, the Royle-Nichols model assumes detections are independent among individuals in the relationship between detection and abundance ($p = 1 - (1 - r)^N$; [3,51]). Violations of this assumption could occur if replicate samples are correlated or demonstrate extra-binomial variation. Correlation among samples could occur if camera trap snap shots were taken

Basic Occupancy Model

Royle–Nichols Occupancy Model

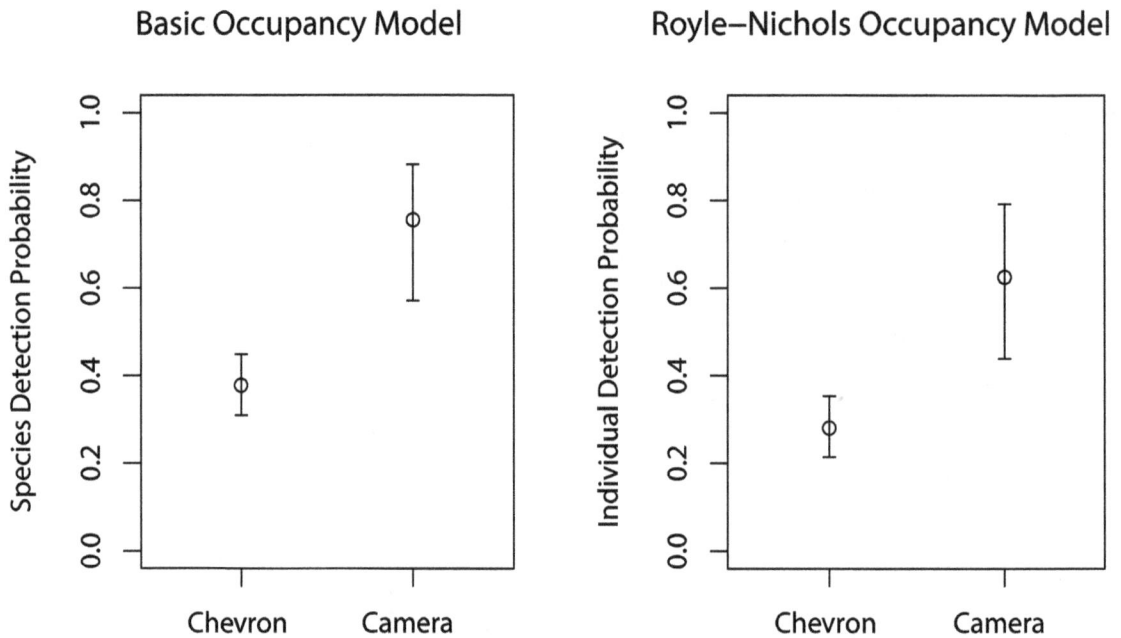

Figure 5. Mean chevron and camera trap detection probabilities. Estimated detection probabilities at the reference covariate values (intercept) from both the basic occupancy model (left panel) and the Royle-Nichols model (right panel) for red snapper. Note that for the basic occupancy model framework the detection probability is for the species at occupied sites. In contrast, for the Royle-Nichols framework the detection probability is for each individual at occupied sites. The intervals represent 95% credible intervals.

Basic Occupancy Model

Royle–Nichols Occupancy Model

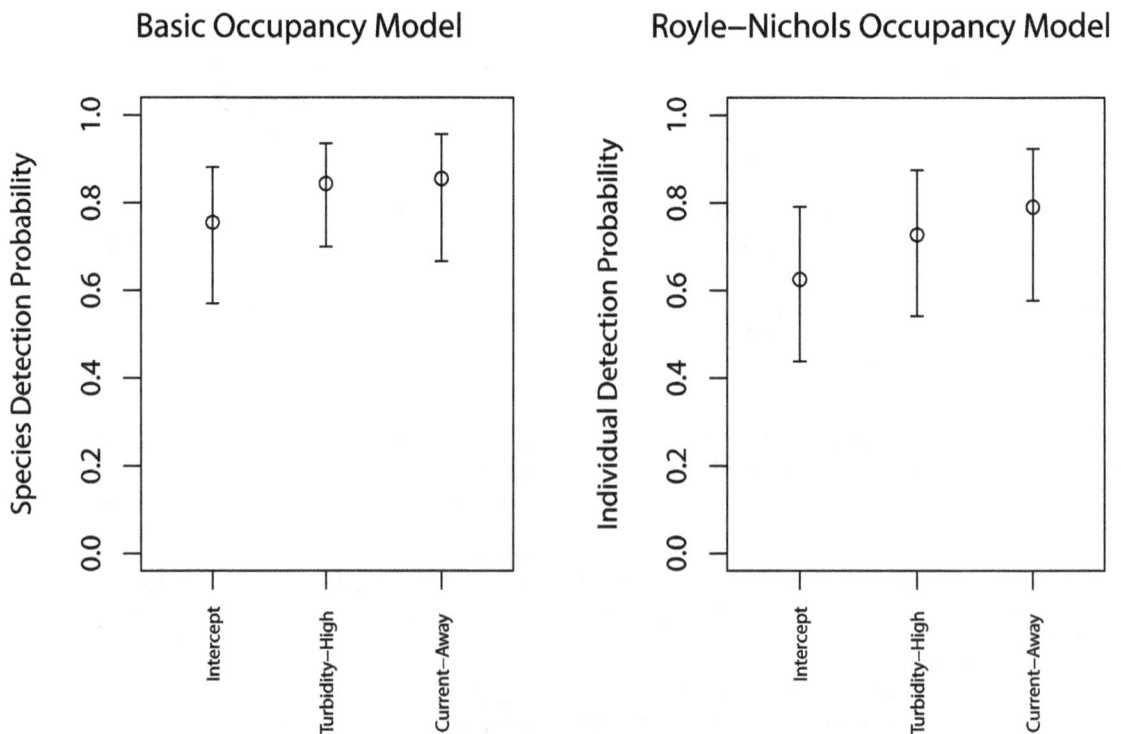

Figure 6. Camera trap detection probabilities as influenced by turbidity and current direction. The estimated detection probabilities at reference covariate values (intercept) and at high turbidity and with current direction away from the camera lens from both the basic occupancy model (left panel) and the Royle-Nichols model (right panel) for red snapper. Note that for the basic occupancy model framework the detection probability is for the species at occupied sites. In contrast, for the Royle-Nichols framework the detection probability is for each individual at occupied sites. The intervals represent 95% credible intervals.

at too short of a time interval leading to serial auto-correlation. Alternatively, extra-binomial variation can occur when animals demonstrate characteristics that result in clustering as would be expected from schooling behavior in fish. Red snapper exhibit weak schooling behavior [52], which corroborates the evidence of lack of independence observed in our data.

The consequence of violating the binomial sampling assumption is a divergence from the relationship of occupancy and abundance [51] because groups of fish are perceived as individuals by the model. This violation results in an underestimation of abundance. Our analysis showed evidence of negative bias because the estimate of mean site abundance was less than the mean maximum count across sites. However, we expect little impact on the accuracy of occupancy probability estimates because occurrence is defined by the presence of one or more individuals. Furthermore, we expect the divergence of patterns in occurrence from patterns in abundance to be low because of low numbers of individuals per site, likely due to their overfished status in the SEUS [31].

Our findings suggested that depth and latitude explained most of the variation in red snapper occupancy rate and mean abundance among hard bottom sites. While this finding is clearly supported by the data and analysis, it is important to recognize that this does not imply that red snapper are not associated with reef habitat. Because nearly all samples considered in the analyses were collected on or near some amount of reef habitat, the analysis is incapable of describing the difference in occupancy probability or site abundance between hard bottom reef sites and non-reef sites composed entirely of unconsolidated sediments with little or no bottom relief. Instead, our analysis attempts to uncover differences among reefs with different amounts of hard bottom, attached biota, and topographic complexity. Our data supported only a weak and negative relationship between red snapper abundance and high incidence of attached biota. Overall this supports the existing evidence that red snapper may not strongly require specific reef characteristics and instead may be reef habitat generalists [40,41]. However, caution must be practiced when extrapolating these results outside of our sample region, as a subset of our sites was selected non-randomly.

Both the basic occupancy and the Royle-Nichols models estimated the inclusion probability of the year effect ($yr2011$) to be much smaller than the 50% threshold. As such, the data do not support a positive or negative difference in either occupancy probability or site abundance between 2010 and 2011. Red snapper is currently under a management rebuilding plan to recover from a designated "overfished" status [31]. While our results do not imply a change in population between 2010 and 2011, we recognize that a longer time-series of data may be needed to observe any extant population trend [32]. However, evaluation of a temporal effect from occupancy models is potentially useful to both management and future stock assessments, particularly as more years of data become available.

A major benefit of analyzing the camera and chevron trap data in the occupancy modeling framework is to estimate both the species and individual detection probabilities for each gear. The model estimated that the camera-trap detection probability increased with higher turbidity and if the current direction was away from the camera lens. While we were not surprised that a current direction that tends to cause fish to orient themselves in the camera field of view would increase detection probability, we do not have a good explanation as to why high turbidity would favor detection over low turbidity. Perhaps red snapper tend to stage closer to the chevron and camera traps when the water is more turbid because they are visual predators and the chevron trap is perceived as structure; however, the effect was small indicating this would not be a pronounced behavior. The original motivation to collect video information during chevron trap deployments was to evaluate whether red snapper were frequently present near chevron traps but not captured. Our work suggests that camera traps are approximately twice as likely as chevron traps to detect red snapper. Additionally, at least one red snapper will be detected with >95% probability with the camera trap so long as ≥ 4 individuals are present. In contrast, ≥ 9 individuals must be present to have a 95% probability of detecting the species with the chevron trap.

We evaluated the use of occupancy style models that estimate patterns in occurrence and abundance [3] with incidence data. One limitation of these analyses for evaluating distributional patterns occurs for species that inhabit most sampling sites and therefore demonstrate little variation in occurrence rates. Under these conditions, occupancy probability is not an informative metric of population change. A further limitation occurs for species with high abundance at occupied sites such that the species is detected in most or all replicate samples. When this occurs, the Royle-Nichols model can produce negatively biased estimates of abundance. Although these were not the conditions for red snapper in the SEUS, they can be the condition for more ubiquitous or abundant species. For example, black sea bass *Centropristis striata* demonstrate high catches in the SEFIS data and would likely limit the utility of the Royle-Nichols model for generating unbiased estimates of abundance. Thus, occupancy models may perform best for species with some level of rarity such as red snapper, but may be a less useful monitoring tool for more ubiquitous or abundant species.

A natural extension of occupancy models not limited by high occurrence rates and high abundance is a class of models referred to as binomial-mixture models or *N*-mixture models [53]. Similar to the Royle-Nichols occupancy model, binomial-mixture models assume that abundance is distributed across sites according to a distribution (e.g. Poisson) and that catches are the result of replicated binomial processes. However, binomial-mixture models fit count data as opposed to incidence data, which can be more informative of the abundance and detection process. Research on the use of binomial-mixture models for monitoring fish is more limited than for occupancy models (but see [54]); however, some research has been done on the use of multiple sampling methods with binomial-mixture models of terrestrial species. For example, [24] applied multiple sampling methods in a binomial-mixture model of grizzly bears in Glacier National Park. [55] then investigated the conditions when it is appropriate to combine multiple sampling methods into a single binomial-mixture model for grizzly bears. Evaluating these methods for marine fish catch data generated with the combined chevron trap and camera trap described here would be a valuable contribution to the ecological literature.

Accounting for incomplete and variable detection of fish is rare in the ecological literature because invasive sampling methods inhibit appropriate replication. Our analysis demonstrates how the use of non-invasive camera traps can be paired with invasive fish traps to generate replicate samples needed to account for incomplete and variable detection for marine fish. This work has broad implications for fisheries management because fisheries data are notoriously expensive and plagued with issues of variable detection. Furthermore, the costs of inappropriate management of fish stocks are high for commercially and recreationally valued species. Thus, the methods we demonstrate here could improve the management of many exploited fish stocks and reduce the risks of economic and biodiversity loss.

Supporting Information

Table S1 Data analyzed in this study. Note that the first two columns contain the numbers of red snapper detections (det) from the chevron and camera traps. Refer to Table 1 for a description of the covariate abbreviations.

Appendix S1 JAGS code for basic occupancy model of red snapper occurrence and detection with pooled camera trap detections. The "#" symbol precedes annotation remarks. Refer to the Methods section of the main text for symbol equations and parameter definitions.

Appendix S2 JAGS code for Royle and Nichols (2003) occupancy model of red snapper abundance and detection with pooled camera trap detections. The "#" symbol precedes

annotation remarks. Refer to the Methods section of the main text for symbol equations and parameter definitions.

Acknowledgments

We are grateful to D. Berrane, T. Kellison, W. Mitchell, M. Reichert, C. Schobernd, Z. Schobernd, and many others for assistance in the field and reading videos, and the Captain and crew of the R/V *Savannah*, NOAA Ship *Pisces*, and NOAA Ship *Nancy Foster* for carrying out field work. We also thank K. Shertzer, K. Purcell, E. Williams, A. Hohn, A. Chester, J.T. Fisher, and one anonymous reviewer who provided helpful and insightful comments that improved this manuscript.

Author Contributions

Conceived and designed the experiments: LGC NMB DCG. Performed the experiments: NMB. Analyzed the data: LGC NMB DCG. Wrote the paper: LGC NMB DCG.

References

1. Walters CJ (1986) Adaptive Management of Renewable Resources. New York: MacMillan Pub. Co.
2. Williams BK, Nichols JD, Conroy MJ (2002) Analysis and management of animal populations. San Diego: Academic Press.
3. Royle JA, Nichols JD (2003) Estimating abundance from repeated presence-absence data or point counts. Ecology, 84, 777–790.
4. Tyre AJ, Tenhumberg B, Field SA, Niejalke D, Parris K, et al. (2003) Improving precision and reducing bias in biological surveys: estimate false-negative error rates. Ecological Applications 13: 1790–1801.
5. Mazerolle MJ, Desrochers A, Rochefort L (2005) Landscape characteristics influence pond occupancy by frogs after accounting for detectability. Ecological Applications 15: 824–834.
6. Archaux F, Henry P, Gimenez O (2011) When can we ignore the problem of imperfect detection in comparative studies? Methods in Ecology and Evolution 3: 188–194.
7. Conroy MJ, Nichols JD (1996) Designing a study to assess mammalian diversity. In: Wilson DE, Cole FR, Nichols JD, Rudran R, Foster MS, editors. Measuring and Monitoring Biological Diversity: Standard Methods for Mammals. Washington D.C.: Smithsonian Institution Press. pp. 41–49.
8. Pollock KH, Nichols JD, Simons TR, Farnsworth GL, Bailey LL, et al. (2002) Large scale wildlife monitoring studies: statistical methods for design and analysis. Environmetrics 13: 105–119.
9. Katsanevakis S, Weber A, Pipiton C, Leopold M, Cronin M, et al. (2012) Monitoring marine populations and communities: methods dealing with imperfect detectability. Aquatic Biology 16: 31–52.
10. Yoccoz NG, Nichols JD, Boulinier T (2001) Monitoring of biological diversity in space and time. Trends in Ecology and Evolution 16: 446–453.
11. MacKenzie DI, Nichols JD, Lachman GB, Droege S, Royle JA, et al. (2002) Estimating site occupancy rates when detection probabilities are less than one. Ecology 83: 2248–2255.
12. MacKenzie DI, Nichols JD, Royle JA, Pollock KH, Bailey LL, et al. (2006) Occupancy Estimation and Modeling: Inferring Patterns of Dynamics of Species Occurrence. Amsterdam: Academic Press.
13. MacNeil MA, Fonnesbeck CJ, McClanahan TR (2010) Occupancy models for estimating the size of reef fish communities. In: Riegle B, Dodge R, editors. Proceedings of the 11th International Coral Reef Symposium. Nova Southeastern University: National Coral Reef Institute. pp. 785–789.
14. Issaris Y, Katsanevakis S, Salomidi M, Tsiamis K, Katsiaras N, et al. (2012) Occupancy estimation of marine species: dealing with imperfect detectability. Marine Ecology Progress Series 453: 95–106.
15. Mesa MG, Schreck CB (1989) Electrofishing, mark-recapture, and depletion methodologies evoke behavioral and physiological changes in cutthroat trout. Transactions of the American Fisheries Society 118: 644–658.
16. Rota CT, Fletcher Jr RJ, Dorazio RM, Betts MG (2013) Occupancy estimation and the closure assumption. Journal of Applied Ecology 46: 1173–1181.
17. Hines JE, Nichols JD, Royle JA, MacKenzie DI, Gopalaswamy AM, et al. (2010) Tigers on trails: occupancy modeling for cluster sampling. Ecological Applications 20: 1456–1466.
18. Guillera-Arroita G (2011) Impact of sampling with replacement in occupancy studies with spatial replication. Methods in Ecology and Evolution 2: 401–406.
19. Walters C (2003) Folly and fantasy in the analysis of spatial catch rate data. Canadian Journal of Fisheries and Aquatic Science 60: 1433–1436.
20. Martell SJD, Walters CJ (2002) Implementing harvest rate objectives by directly monitoring exploitation rates and estimating changes in catchability. Bulletin of Marine Science 70: 695–713.
21. Walters C (1998) Evaluation of quota management policies for developing fisheries. Canadian Journal of Fisheries and Aquatic Science 55: 2691–2705.
22. Rose GA, Kulka DW (1999) Hyperaggregation of fish and fisheries: how catch-per-unit-effort increased as the northern cod (*Gadus morhua*) declined. Canadian Journal of Fisheries and Aquatic Science 56: 118–127.
23. Erisman BE, Allen LG, Claisse JT, Pondella DJ, Miller EF, et al. (2011) The illusion of plenty: hyperstability masks collapses in two recreational fisheries that target fish spawning aggregations. Canadian Journal of Fisheries and Aquatic Sciences 68: 1705–1716.
24. Walters C, Maguire J (1996) Lesson for stock assessment from northern cod collapse. Reviews in Fish Biology and Fisheries 6: 125–137.
25. Graves TA, Kendall KC, Royle JA, Stetz JB, Macleod AC (2011) Linking landscape characteristics to local grizzly bear abundance using multiple detection methods in a hierarchical model. Animal Conservation 14: 652–664.
26. Haynes TB, Rosenberger AE, Linberg MS, Whitman M, Schmutz JA (2013) Method- and species-specific detection probabilities of fish occupancy in arctic lakes: implications for design and management. Canadian Journal of Fisheries and Aquatic Sciences 70: 1055–1062.
27. Nichols JD, Bailey LL, O'Connell AF, Talancy NW, Campbell Grant EH, et al. (2008) Multi-scale occupancy estimation and modelling using multiple detection methods. Journal of Applied Ecology 45: 1321–1329.
28. Bacheler NM, Schobernd CM, Schobernd ZH, Mitchell WA, Berrane DJ, et al. (2013) Comparison of trap and underwater video gears for indexing reef fish presence and abundance in the southeast United States. Fisheries Research 143: 81–88.
29. Bischof R, Hameed S, Ali H, Kabir M, Younas M, et al. (2014) Using time-to-event analysis to complement hierarchical methods when assessing determinants of photographic detectability during camera trapping. Methods in Ecology and Evolution 5: 44–53.
30. Cowan JH Jr (2011) Red snapper in the Gulf of Mexico and U.S. South Atlantic: data, doubt, and debate. Fisheries 36: 319–331.
31. SEDAR (2009) SEDAR-15: Stock assessment report 1: South Atlantic red snapper. South Atlantic Fisheries Management Council, Charleston. Available: http://www.sefsc.noaa.gov/sedar/. Accessed 17 January 2014.
32. Conn PB (2011) An evaluation and power analysis of fishery independent reef fish sampling in the Gulf of Mexico and U.S. South Atlantic. NOAA Technical Memorandum. NMFS-SEFSC-610.
33. Fautin D, Dalton P, Incze LS, Leong JC, Pautzke C, et al. (2010) An overview of marine biodiversity in United States waters. PLoS ONE 5(8): e11914. doi:10.1371/journal.pone.0011914.
34. Kendall MS, Bauer LJ, Jeffrey CJG (2008) Influence of benthic features and fishing pressure on size and distribution of three exploited reef fishes from the southeastern United States. Transactions of the American Fisheries Society 137: 1134–1146.
35. Schobernd CM, Sedberry GR (2009) Shelf-edge and upper-slope reef fish assemblages in the South Atlantic Bight: habitat characteristics, spatial variation, and reproductive behavior. Bulletin of Marine Science 84: 67–92.
36. Recksiek CW, Appeldoorn RS, Turingan RG (1991) Studies of fish traps as stock assessment devices on a shallow reef in southwestern Puerto Rico. Fisheries Research 10: 177–197.
37. Evans CR, Evans AJ (1996) A practical field technique for the assessment of spiny lobster resources of tropical islands. Fisheries Research 26: 149–169.
38. Rudershausen PJ, Mitchell WA, Buckel JA, Williams EH, Hazen E (2010) Developing a two-step fishery-independent design to estimate the relative abundance of deepwater reef fish: application to a marine protected area off the southeastern United States coast. Fisheries Research 105: 254–260.
39. Schobernd ZH, Bacheler NM, Conn P (2014) Examining the utility of alternative video monitoring metrics for indexing reef fish abundance. Canadian Journal of Fisheries and Aquatic Sciences 71: 464–471.

40. Szedlmayer ST (2007) An evaluation of the benefits of artificial habitats for red snapper, *Lutjanus campechanus*, in the northeast Gulf of Mexico. Proceedings of the Gulf and Caribbean Research Institute 59: 223–230.

41. Gallaway BJ, Szedlmayer ST, Gazey WJ (2009) A life history review of red snapper in the Gulf of Mexico with an evaluation of the importance of offshore petroleum platforms and other artificial reefs. Reviews in Fisheries Science 17: 48–67.

42. Royle JA, Dorazio RM (2008). Hierarchical Modeling and Inference in Ecology: The Analysis of Data from Populations, Metapopulations, and Communities. San Diego: Academic Press. 444 p.

43. McCullagh P, Nelder JA (1989) Generalized Linear Models, 2nd edition. London: Chapman and Hall.

44. MacKenzie DI, Bailey LL (2004) Assessing the fit of site-occupancy models. Journal of Agricultural, Biological, and Environmental Statistics 9: 300–318.

45. Barbieri MM, Berger JO (2004) Optimal predictive model selection. Annals of Statistics 32: 870–897.

46. Plummer M (2009) JAGS: A program for analysis of Bayesian graphical models using Gibbs samping. In: Leisch F, Zeileis A, editors. Proceedings of the 3rd International Workshop on Distributed Statistical Computing. Vienna, Austria.

47. R Development Core Team (2014) R: A language and environment for statistical computing. R Foundation for Statistical Computing, Vienna. Available: http://www.R-project.org/. Accessed 17 January 2014.

48. Dorazio R, Gotelli NJ, Ellison AM (2011) Modern methods of estimating biodiversity from presence absence surveys. In: Grillo O, editor. Biodiversity Loss in a Changing Planet. Rijeka Croatia: InTech. pp. 1–27.

49. Gelman A, Carlin JB, Stern HS, Rubin DB (2004) Bayesian Data Analysis, 2nd edn. Boca Raton: Chapman and Hall.

50. O'Connell AF, Talancy NW, Bailey LL, Sauer JR, Cook R, et al. (2006) Estimating site occupancy and detection probability parameters for meso and large mammals in a coastal ecosystem. Journal of Wildlife Management 70: 1625–1633.

51. McCarthy MA, Moore JL, Morris WK, Parris KM, Garrard GE, et al. (2012) The influence of abundance on detectability. Oikos 122: 717–726.

52. McDonough M, Cowan J (2007) Tracking red snapper movements around an oil platform with an automated acoustic telemetry system. Proceedings of the Gulf and Caribbean Fisheries Institute 59: 159–163.

53. Royle JA (2004) *N*-Mixture models for estimating population size from spatially replicated counts. Biometrics, 60, 108–115.

54. Wenger SJ, Freeman MC (2008) Estimating species occurrence, abundance, and detection probability using zero-inflated distributions. Ecology 89: 2953–2959.

55. Graves TA, Royle JA, Kendall KC, Beier P, Stetz JB, et al. (2012) Balancing precision and risk: should multiple detection methods be analyzed separately in *N*-Mixture models? PLoS ONE 7(12): e49410. doi:10.1371/journal.pone.0049410.

17

A Resolution to the Blue Whiting (*Micromesistius poutassou*) Population Paradox?

Fabien Pointin, Mark R. Payne*

Centre for Ocean Life, National Institute of Aquatic Resources (DTU-Aqua), Technical University of Denmark, Charlottenlund, Denmark

Abstract

We provide the strongest evidence to date supporting the existence of two independent blue whiting (*Micromesistius poutassou* (Risso, 1827)) populations in the North Atlantic. In spite of extensive data collected in conjunction with the fishery, the population structure of blue whiting is poorly understood. On one hand, genetic, morphometric, otolith and drift modelling studies point towards the existence of two populations, but, on the other hand, observations of adult distributions point towards a single population. A paradox therefore arises in attempting to reconcile these two sets of information. Here we analyse 1100 observations of blue whiting larvae from the Continuous Plankton Recorder (CPR) from 1948–2005 using modern statistical techniques. We show a clear spatial separation between a northern spawning area, in the Rockall Trough, and a southern one, off the Porcupine Seabight. We further show a difference in the timing of spawning between these sites of at least a month, and meaningful differences in interannual variability. The results therefore support the two-population hypothesis. Furthermore, we resolve the paradox by showing that the acoustic observations cited in support of the single-population model are not capable of resolving both populations, as they occur too late in the year and do not extend sufficiently far south to cover the southern population: the confusion is the result of a simple observational artefact. We conclude that blue whiting in the North Atlantic comprises two populations.

Editor: Athanassios C. Tsikliras, Aristotle University of Thessaloniki, Greece

Funding: The research leading to these results has received funding from the European Union 7th Framework Programme (FP7 2007–2013) under grant agreement numbers 264933 (Euro-Basin project) and 308299 (NACLIM). Analysis of the CPR fish larval samples from 1979 to 2005 was funded by the United Kingdom Department of Environment, Fisheries and Rural Affairs (Defra) through project MF1101. The funders had no role in study design, data collection and analysis, decision to publish, or preparation of the manuscript.

Competing Interests: The authors have declared that no competing interests exist.

* Email: mpay@aqua.dtu.dk

Introduction

Blue whiting (*Micromesistius poutassou* (Risso, 1827)) is a small mesopelagic planktivorous gadoid found throughout the North-East Atlantic. The species has been the subject of a large but highly variable commercial fishery since the late 1970s. Fisheries surveys and formal stock assessments have been in place since the early 1980s, and management agreements in more recent times. The first scientific reports date back more than a century [1] and the species is generally regarded as playing an important role in the ecology of the North-East Atlantic [2].

In recent decades the stock (and the associated fishery) has undergone dramatic changes. From moderate levels in the early 1990s, the stock and fishery swelled during the late 1990s and early 2000s: in 2004, landings reached 2.4 millions tonnes, making it the third largest marine fishery in the world [3]. The stock has since reduced dramatically in size [4], however, and at one point, scientific advice recommended the closure of the fishery altogether [5]. The most recent stock assessments suggest that the decline has stabilised and that the population may be increasing again [6]. Yet, in spite of the relative importance of this fish population, and the wealth of information and studies that normally are associated with an assessed species, there are still important gaps in our understanding.

One such outstanding question is that of population structure. The species is widely distributed throughout the North-East Atlantic. The core of the distributional range is from the Bay of Biscay along the continental shelf edge to the Norwegian Sea (Figure 1). The edges of the distribution include the southern Iberian Peninsula and the Mediterranean Sea, the Barents Sea, the North Sea (although not the Baltic) and the Mid-Atlantic ridge, East-Greenland and the east coast of North America [7,8]. The Mediterranean population is typically considered as a separate population that is isolated from the rest of the Atlantic population and is not considered further here.

However, the Atlantic population structure, if any, is the subject of some controversy. One long-running line of argument (see *e.g.* [7] for early references) proposes the existence of two separate Atlantic populations. According to this hypothesis, one population (hereafter the northerly population) spawns in spring to the west of Great Britain and the Outer Hebrides along the continental shelf edge, in the Rockall Trough and around the Rockall Plateau and Hatton Bank: this population then migrates northwards into the Norwegian and Barents Seas where it feeds during summer, and possibly overwinters. The second (southerly) population is thought to spawn around Porcupine Bank and the Porcupine Seabight, and possibly further to the south in the Bay of Biscay. This population may migrate southwards to the Bay of Biscay to feed during summer, although the understanding of the migrations and distributions in this region is limited.

A variety of different studies support this hypothesis. Early genetic studies based on allozyme markers were able to show

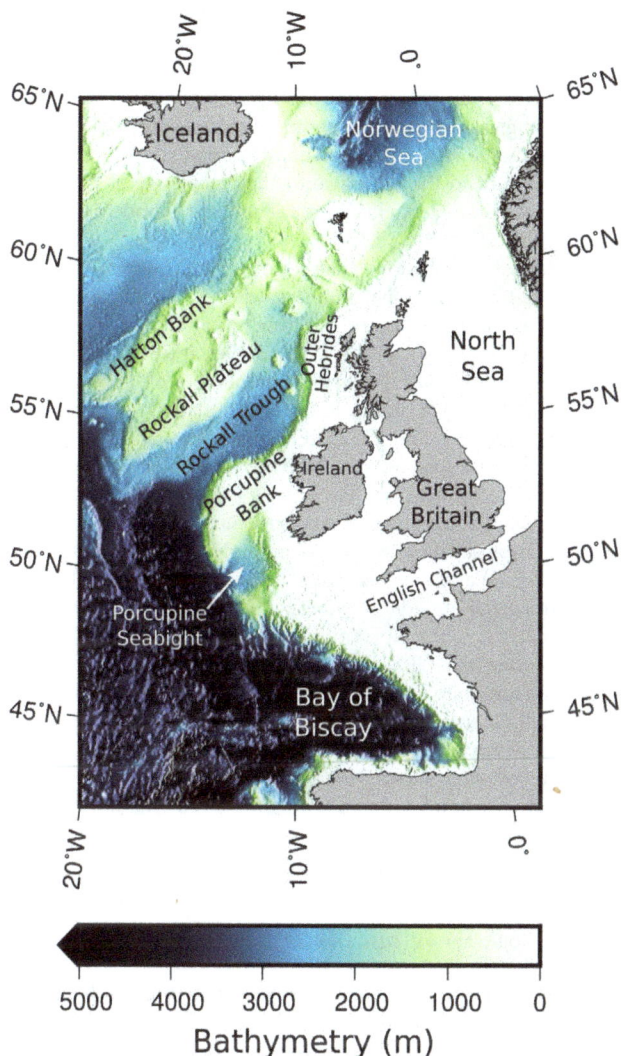

Figure 1. Bathymetric relief map of the study area. Features mentioned in the text are labelled.

differences between individuals caught at the edges of the distribution [9,10] (*e.g.* between the Mediterranean and Barents Seas). More modern studies based on microsatellite loci [11,12] have provided more detail, with differences exhibited between individuals from the Hebrides and Porcupine Bank. Growth studies based on the larval region of otoliths captured from adults suggested that individuals captured in southern areas (Porcupine Bank and Bay of Biscay) grew significantly faster during their larval stage than those from northern areas (the Hebrides and Norwegian sea), suggesting that fish from these regions do not mix randomly [13]. Otolith shape analysis [14] suggests systematic differences between the Celtic Sea and the Norwegian Sea. Morphometric and meristic data also support a separation between the Hebrides and Porcupine Bank [15]. Circulation studies lend further support to this idea by providing a mechanism that can maintain the separation: larvae spawned north of 53–55 °N are advected northwards, while those south of this region drift southwards [16–18].

The current management structure, however, does not reflect this evidence. Blue whiting in the North-East Atlantic is managed as a single stock, with one quota to cover the entire domain. This was not always the case: the initial management structure upon

establishment of the ICES Blue Whiting Assessment Working Group in the early 1980s was a two-population construct. Surveys performed during this time were often reported in terms of southern and northern populations, and separate abundance estimates were generated for each population (*e.g.* [19]). However, the two populations were merged into a single stock in 1993, due to reasons of convenience and the absence of data to the contrary [18].

During the intervening two decades, the single-stock paradigm has come to dominate both the management of this stock and the science performed upon it. Most modern publications on this topic (*e.g.* [4,20–22]) start from this assumption and interpret their results in terms of a single population. Recent management advice even goes so far as to deny any evidence to the contrary, stating "...*there is no scientific evidence in support of multiple stocks with distinct spawning locations or timings.*" [6]. On the other hand, the steady accumulation of results undermining the single-stock paradigm has lead to blue whiting being cited as an example of the mismatch between genetic studies and management [23].

Part of the reason for the dominance of the single-stock approach lies in the observations of blue whiting on the spawning grounds. Acoustic fisheries surveys have covered the spawning grounds since the early 1980s, and are generally regarded as one of the best sources of information about the spatial distribution of this species. Such surveys, however, generally show a continuum of fish running from the Hebrides all the way to Porcupine Bank (Figure 2). The question can therefore be raised: if, as the two-population hypothesis suggests, there are truly two populations with separate spawning grounds, why can we not see them in the surveys? Alternatively, if, as the acoustic observations suggest, there is mixing at spawning time, how can the genetic and morphometric separations observed be maintained? It is this paradox, with a conflict between two conceptual models, both of which seem reasonable when viewed individually but are nevertheless mutually exclusive, that is at the core of the conflict between the two models of blue whiting population structure.

Resolving this controversy requires a fresh approach. One potential data source that could shed new light on this issue is the Continuous Plankton Recorder (CPR). The CPR is a sampling device that is towed behind ships of opportunity throughout European waters (and more recently on a global scale) and captures both phytoplankton and zooplankton together with fish eggs and larvae [24]. Starting in 1931, it is one of the longest running biological sampling programs in the world, and provides a unique and invaluable insight into the dynamics of marine systems. The CPR is especially closely linked to the history of blue whiting: the species is one of the most commonly occurring fish species in the CPR record, comprising approximately 10% of all fish ichthyoplankton identified [25] and 75% of all larvae west of the British Isles [26]. The broad spatial and temporal coverage of the CPR, and its penchant for blue whiting, lead to the identification of large concentrations of blue whiting larvae around Rockall Trough and Rockall Plateau in the 1950s [27,28], and the CPR is therefore frequently credited as playing a crucial role in the identification and development of the fishery [29]. The same broad coverage can potentially shed fresh light on the population structure of this species.

In this work, we aim to investigate the population structure of blue whiting using the CPR larval observations. In particular, we will apply modern statistical modelling techniques to this unique dataset to develop a comprehensive overview of the spatial and temporal distribution of the spawning products. These results can then be used to assess support for the various conceptual models of blue whiting population structure in the North-East Atlantic.

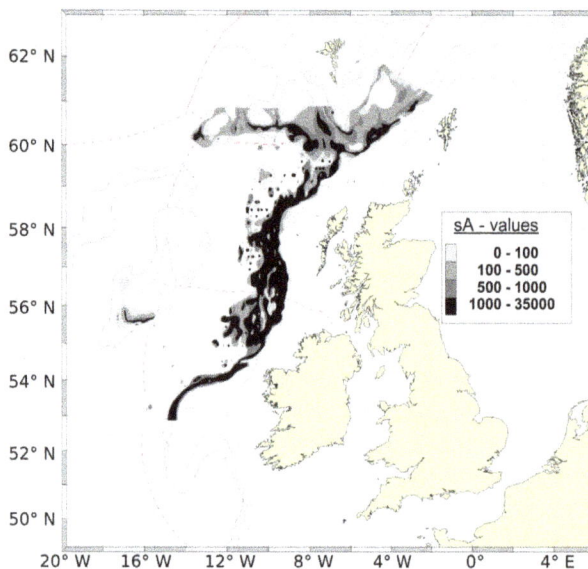

Figure 2. Distribution of the blue whiting spawning stock from a fisheries acoustic survey. The acoustic intensity of blue whiting (sA, which is directly related to abundance) from the International Blue Whiting Spawning Stock Survey (IBWSS) is shown for the 2013 acoustic survey [52]. Isobaths are plotted as grey lines. International maritime boundaries are plotted as red dotted lines. Note the continuous distribution along the shelf edge and limited southern extension of the survey.

Materials and Methods

Continuous Plankton Recorder (CPR) Data

The CPR is towed behind ships of opportunity at depths of 7 m to 10 m. Sea water enters the recorder through a small opening in the front of the device, and is filtered through a silk screen with a mesh size of approximately 270 μm. The silk cloth is stored on a roll and replaced continuously as the recorder is towed through the water: after being exposed to sea water, the cloth is covered with a second layer of unexposed silk and then enters a tank of formalin to preserve the samples. On shore, the silk is divided into squares that correspond to approximately 10 nautical miles of towing distance, and analysed under a microscope by a taxonomist. Details of the sampling and analysis procedure are published elsewhere [24].

Initially, all fish larvae were identified to species level on all samples. Reductions in funding in the late 1970s lead to the cessation of species-level identification from the early 1980s onwards: fish larvae after this point were noted but not identified. However, a new initiative was commenced in the late 2000s and with funding from the UK government the archived fish larvae were reanalysed to species level in a restricted region around the British Isles [30].

CPR blue whiting larval observations were provided upon request by the Sir Alister Hardy Foundation for Ocean Science (SAHFOS), Plymouth, UK. In addition to the spatial domain incorporated in the modern reanalysis project (from 20 °W to 10 °E and 44 °N to 64 °N), we also obtained observations back to 1948 over the entire North Atlantic domain. Both presence and absence observations were incorporated in the data obtained.

Modelling approach

The goal of our data analysis was to find a model that synthesizes the data available and accounts for the complex spatial-temporal distribution of the samples. We apply an Information Theoretic approach to the development of this model [31], defining an ensemble of candidate model structures in advance and fitting them to the observations. We then choose the model that gives the most parsimonious representation of the data, as judged by the Akaike Information Criteria (AIC), a metric that balances the fit to the data against the complexity of the model (number of parameters employed). The "best" model is the one with the lowest AIC score.

We differ from previous analyses of CPR fish larval data (e.g. [30,32,33]) by disregarding abundance data. CPR fish data are not recorded as true abundances, but rather as abundance categories. Beyond the first categories (0, 1, 2, and 3 larvae), where there is an unambiguous relationship between the number of larvae and the category, there is a rapid loss of information e.g. the next categories are 4–11, and 12–25. The approaches applied by other authors, typically assuming a Gaussian or Poisson observation model, are therefore not valid in this case. A statistically valid model to handle this observational structure would require a high degree of sophistication, based, for example, on continuation ratio logits [34]. We choose instead to simplify the problem by disregarding the abundance information and instead focusing on the presence/absence aspect of the data.

Considering the CPR data as presence/absence observations lends itself naturally to Generalised Additive Models (GAMs) with a Bernoulli observational structure. We employ a GAM using the metadata of each observation (spatial position, year, day of year and time of day) as the basis for the explanatory variables. Specifically, we employ the following model structure:

$$P(X_i = \text{TRUE} \mid \pi_i) \sim \text{Bernoulli}(\pi_i) \qquad (1a)$$

$$\text{logit}(\pi_i) = g(east_i, north_i, doy_i, year_i) + DN_i \qquad (1b)$$

where X_i is presence/absence observation i, π_i is the probability of X_i being true (present), and doy_i, and $year_i$ are the day of year and year of the observation. The spatial domain is represented in the Universal Transverse Mercator (UTM) projection (Zone 28) to minimise the effect of coordinate distortions due to the curvature of the earth. The spatial position is thus represented by the eastings, $east_i$, and northings, $north_i$, in Equation 1b above.

The variable DN_i is a categorical factor indicating whether the sample was taken during the day or night. The ability of the CPR to capture fish larvae may change with the light environment, due to either active avoidance of the gear or diel vertical migrations of the larvae. The DN_i variable was therefore incorporated to account for such effects and was based on the solar-elevation at the time and position of each observation, as calculated using the solarpos() function in the "maptools" package in R [35]. Sunrise/sunset were defined following the "civil dawn" convention i.e. night is where the sun is six degrees or more below the horizon. The DN_i term was used in all models considered.

The function g() in Equation 1b is the main unknown element. We consider an ensemble of different terms for g(), ranging from a fully separable model, where each space-time dimension influences the probability of occurrence independently, to full three-dimensional interactions between space and day of year. We do not consider four dimensional interactions (i.e. space - day-of-year - year interactions), due to the limited number of presence observations.

Two different structures are considered for the year term. The first, and simplest model does not consider a year term, and simply assumes the abundance of larvae in each year to be the same.

Alternatively, interannual variations in adult abundance (and therefore of the probability of observing larvae) were accounted for as smoothly varying covariates of time (denoted by an $s(year)$ term in $g()$).

The full list of models considered is given in Table 1.

Models were fitted using the mgcv package in R [36,37]. Following the recommendation of [36], each model is fitted with a "gamma" parameter set to 1.4, to avoid overfitting. Cyclic cubic regression splines were used as smoothers for day of year: standard cubic regression splines were used for all other terms. Two and three dimensional tensor-product interaction smoothers [36,38] were used for interaction terms, where appropriate.

Model validation and evaluation

Model validation for models with non-Gaussian responses is somewhat more challenging than for standard linear modelling, where an array of diagnostic plots exist to assess the validity of the fit. This is particularly the case for a binary response variable, such as the presence/absence observations used here, where the concept of a residual becomes difficult to interpret. Binary response variables are, by definition, Bernoulli distributed, so there are no distributional assumptions to check.

Our model validation is therefore limited to checking that the smoothers are neither over-constrained nor are overfitting. We follow the guidelines described in [36] and in the internal documentation of the mgcv package in this regard, relying heavily on the gam.check() function.

We assess model goodness of fit using two standard measures. The area-under-the-curve (AUC) of a receiver-operator curve (ROC) is a commonly employed measure of the ability of a model to distinguish between binary outcomes. A value of 1 indicates perfect discrimination between presence and absence, whilst a value of 0.5 is that expected from a random number generator: models with values in excess of 0.75 are typically regarded as having a "useful" ability to discriminate between absences and presences [39,40]. Although the validity of this metric has been questioned [41], we present these results here for consistency with other analyses. The AUC for each model was calculated using the verification package in R [42]. We also considered the "explained deviance" as a second metric of the model goodness of fit [36]: this metric can be considered as an analogue of the coefficient of determination, R^2, for generalised linear and generalised additive models.

Model fits were visualised by evaluating the fitted model on a regular three dimensional grid (*east*, *north*, *doy*) for a given year. The annual distributions were then normalised and the mean marginal distributions determined. Interannual variability in spawning was visualised by integrating the probability of larval occurrence across these grids for each year, with confidence intervals generated by resampling from the posterior distribution of the fit [36].

Results

Data exploration

In total, 134 260 CPR observations that had been checked for blue whiting larvae were obtained in the North Atlantic region. The spatial distribution of these samples clearly shows a high concentration of samples in the North Sea and to the west of Great Britain and Ireland, from the continental shelf out to approximately 20°W, north of the Iberian peninsula, and south of Iceland (Figure 3). Discontinuities and inhomogeneities arise in the spatial distribution of samples due to both the pattern of shipping routes employed by the CPR, and the boundaries imposed by the

modern reanalysis project (which is focused on the North Sea, and the waters to the west of Great Britain and Ireland).

The domain covered by the modern reanalysis is, fortunately, also the region that clearly contains the most blue whiting larval observations. A few presences are seen outside of this region, particularly towards the Mid-Atlantic ridge, and are consistent with other reports [43]. However, the presence of blue whiting larvae in the North Sea and English Channel has not been reported previously, and is not consistent with existing knowledge. We have therefore interpreted these observations as misidentifications or errors in data entry.

In order to simplify the analysis, we focus the modelling efforts on the region of highest sampling density and most frequent larval-presence, as denoted by the region in Figure 3. The region-of-interest polygon is drawn to follow the boundaries of the modern reanalysis to the west of Great Britain and Ireland. Regions in the Norwegian Sea and Bay of Biscay are also excluded, due to sparse sampling coverage. 34 out of 1161 presence observations are excluded by this spatial filtering, an acceptably low number (3%) that highlights the peripheral nature of these regions. The final data set consisted of 59 042 observations, of which 1127 were presences (1.9%).

The interannual distribution of the samples and the presences in the study region show a number of systematic patterns (Figure 4). Although the annual distribution of samples is relatively constant (Figure 4a), the number of presences reported varies over time (Figure 4b), and is markedly reduced from 1975 onwards. This reduction can be explained in part by a closer examination of the spatial distribution of samples in each year (Figure S1). Sampling intensity in the Rockall region in particular was reduced during this time and is associated with the close of the ocean weather ships in this region (and their associated CPR routes) and may account for the changes in the frequency of presence observations.

The distribution of samples with respect to the day of year immediately reveals the spawning period of blue whiting. The CPR samples are uniformly spread throughout the year, although there is a clear monthly sampling cycle, with the greatest sampling intensity in the middle of each month (Figure 4c). However blue whiting larvae are predominately found in the months of March, April and May, with two outliers occurring in November (Figure 4d). These observations may be erroneous but in the absence of other information, are retained in the analysis.

The distribution of larval abundances supports the choice of presence/absence modelling (Figure 5). Of the approximately 1100 presence observations, 60% are of abundance category 1, 2 or 3, and can therefore be directly related to their actual abundance. However, the remaining 40% are reported as abundance ranges which are not readily modelled using standard statistical techniques. Based on these results, the decision to employ presence/absence modelling appears justified.

Model fitting and validation

The quality of the fits from the initial model ensemble (Models 1–6 in Table 1) showed a strong dependence on the space-time formulation, $g()$, employed. Increasing the degree of interaction between space and time increased the quality of the model fit to a degree that outweighed the penalties associated with the addition of extra fitting parameters (as judged by the AIC criteria). The quality of the fit also improved, as judged by both the deviance explained and the area under the receiver-operator curve (AUC) statistics. Models that were fully separable, with no interaction terms were the worst, whilst those with full three-dimensional interactions between eastings, northings and day of year were the best according to both of these criteria. Year effects were clearly

Table 1. Model fitting results.

Model	g() function	Comps.	Dev. Expl.	AUC	AIC	ΔAIC₁	ΔAIC₂	DN
1	east + north + doy	1	0.384	0.945	6930	1127	1199	0.20
2	east + north + doy + syear	1	0.424	0.953	6558	755	827	0.18
3	east*north + doy	1	0.388	0.947	6918	1116	1187	0.21
4	east*north + doy + syear	1	0.432	0.956	6515	712	784	0.22
5	east*north*doy	1	0.465	0.966	6127	324	396	0.18
6	east*north*doy + syear	1	0.501	0.970	5803	0	72	0.18
7	east*north + doy + syear|comp	2	0.445	0.959	6396		665	0.22
8	east*north + doy|comp + syear	2	0.464	0.964	6165		434	0.20
9	east*north + doy|comp + syear|comp	2	0.471	0.964	6108		377	0.20
10	east*north*doy + syear|comp	2	0.510	0.972	5731		0	0.17

The g() function is the representation of space and time in the model: "*" indicates the use of 2D or 3D interaction smoothers, whereas "+" indicates additive terms. "|comp" indicates that the term is conditional on the component (northern or southern component). "syear" indicates a spline smoother year-term. Comps: The number of components fitted in the model. AIC: Akaike Information Criteria. AUC: Area under the Curve (of a Receiver Operator Curve). Dev. Expl.: the fraction of deviance explained by the model. ΔAIC₁: Difference in the AIC from the minimum AIC of the first model ensemble. ΔAIC₂: Difference in the AIC from the minimum AIC of the second model ensemble (multicomponent models). DN: The value of the day-night coefficient in the model.

required. In each of the three cases, for a constant space and day-of-year formulation, adding the year effect lead to a better quality model. Model 6 is clearly the best of these candidate models, with the next best model (Model 5) having an AIC value more than 300 units greater: ΔAIC values of more than 20 are typically characterised as a model having "essentially no empirical support" [31,44].

However, initial model evaluation suggested a further refinement to the model ensemble that had hereto been overlooked. All models showed a clear local minimum in the density of blue whiting larvae on Porcupine Bank (Figure S2), between approximately 52 and 54°N, with spawning centres to the north and south of this feature. This result is clearly in line with other published results suggesting the presence of two-populations. Furthermore, these two spawning regions also appear to have distinct spawning times that are separated by a month or more (Figure S3). There is thus a clear suggestion of two distinct spawning-grounds in these results.

A second ensemble of models was therefore generated by expanding the first to include this alternative structure. Specifically, we drew a dividing line at 53°N based on the results of Model 6 (see Figures S2 and S3). Larvae observed north of this line are associated with the "northern component", and those south of the line are associated with the "southern component". Models allowing for component-dependent interannual abundance variations (Model 7), component-dependent spawning times (Model 8) or both (Models 9–10) were created and fitted.

The two-component models are systematically better than their corresponding single-spawning-ground models (Table 1). The addition of the two-component feature leads to a substantial reduction in the AIC and increase in the AUC in models where there is no interaction between space and season (day of year) (i.e. Model 4 compared with model 7). Model 10 which incorporates full space-season interaction with component-dependent interannual variations in abundance, is clearly an improvement on its one-component counterpart (Model 6), and is now the best model overall.

All models appear to fit the data well. Model validity checks performed as part of the fitting procedure suggest that the smoothers are capturing the variability. The models also capture the majority of the deviance (Model 10 captures 51%). The AUC scores are particularly impressive, and exceed 95% for nearly all models, suggesting a high degree of skill in discriminating between the presence and absence of larvae, although may be unrealistically high due to the low number of presences. The model fits therefore appear valid representations of the data, and the best fitting model, Model 10, is therefore adopted as the basis for the remainder of this study.

Model visualisation

The spatial patterns apparent in the simpler one-component models, are also clearly apparent in the best-fitting two-component model, Model 10. There appear to be two main centres of larval density (Figure 6). The first is in the Rockall Trough in the deep water off the continental shelf-edge to the north-west of Ireland and west of the Outer Hebrides. A second high-density region is centred south of the Porcupine Bank and south-west of Ireland, offshore from the Porcupine Seabight. Importantly, there appears to be a clear minimum between these two regions, hinting at their independence (Figure 6).

The two centres also clearly exhibit different distributions in the timing of spawning. The timing of the local maximum in larval density (Figure 7) is strongly dependent on space, exhibiting a systematic increase from the south to the north. The core of the

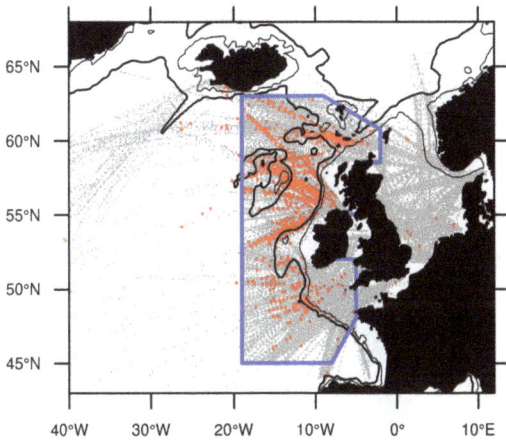

Figure 3. Spatial distribution of CPR samples. Grey points are locations where CPR samples have been checked for fish larvae. Red circles are where these samples were found to contain blue whiting. The blue box denotes the spatial region of interest used in further model-based analyses.

two centres appear to differ substantially in the timing of maximum larval-density.

The zonal dependence of the larval temporal distribution is clearly apparent when the meridional dimension is integrated out (Figure 8). The temporal distribution of larval from the southern component appears to lead the northern component by at least 30–45 days. Furthermore, the temporal distribution of the northern component appears more protracted than that in the south, with appreciable larval densities into mid- and late-May.

The overall abundance of the two components also appear to show different interannual dynamics (Figure 9). However, the confidence intervals about the median estimate are large, a result that is unsurprising given the poor sampling coverage in some years. The high uncertainty means that it is not appropriate to draw inference about the trends, nor to make comparisons with, for example, the spawning stock biomass from the stock assessment. Nevertheless, incorporating different interannual dynamics for the two components (from Model 6 to Model 10) resulted in a greatly improved fit to the data *i.e.* the abundance trends in each component are statistically different. Furthermore, although we have not tested it explicitly, the results clearly suggest that the southern component typically has an integrated abundance that is smaller, on average, than the northern component.

Finally, the day-night (DN) factor for the best fitting model, Model 10, was 0.17 (with a 95% confidence interval of [0.05,0.35]). All models showed comparable values for this factor. When translated into actual catchability, this results suggests that

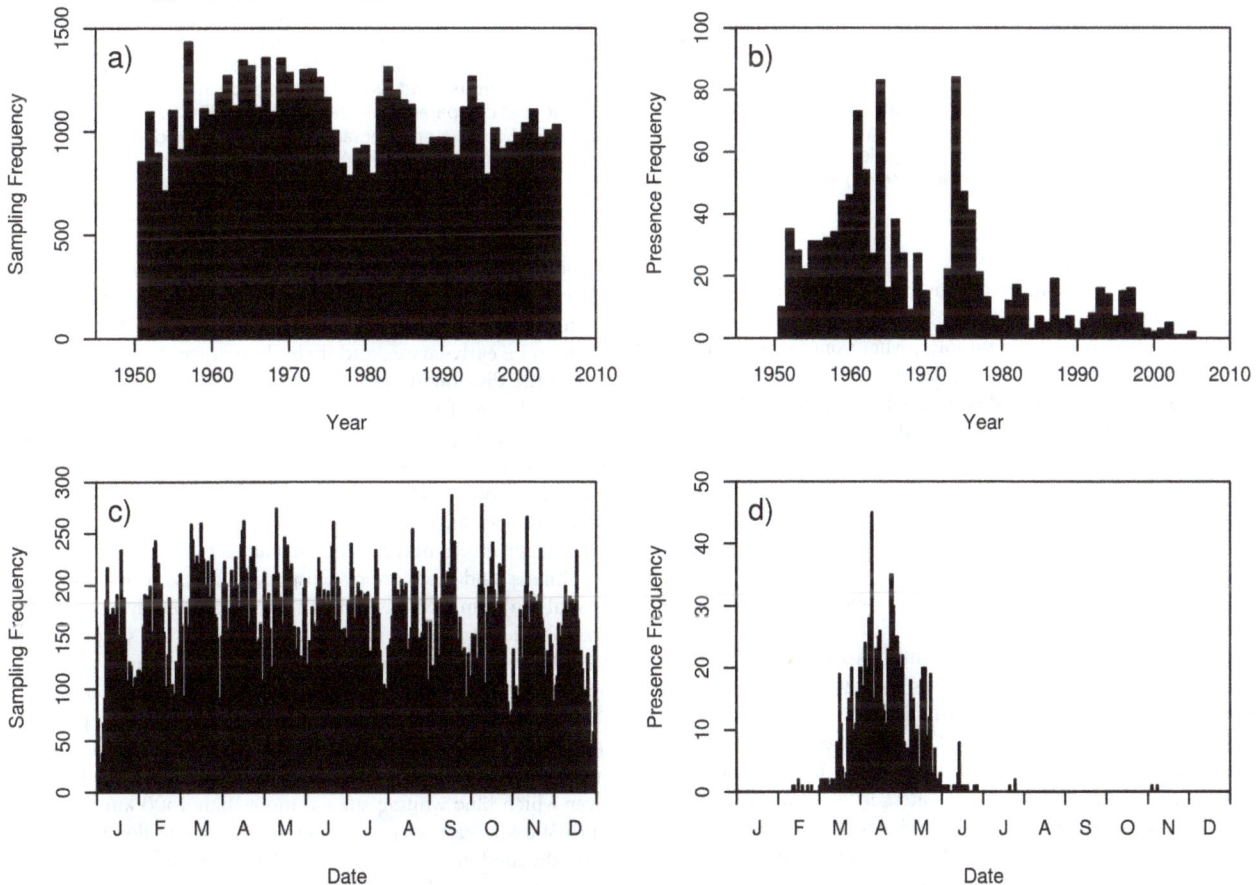

Figure 4. Temporal distribution of CPR samples. Temporal distribution of samples checked for blue whiting larvae obtained from the CPR in the region of interest outlined in Figure 3. a) Sampling frequency in each year b) Presence frequency in each year c) Sampling frequency as a function of date in the year d) Presence frequency as a function of date in the year. In a) and b), each bar corresponds to a single year, whilst in c) and d) it corresponds to a day of year.

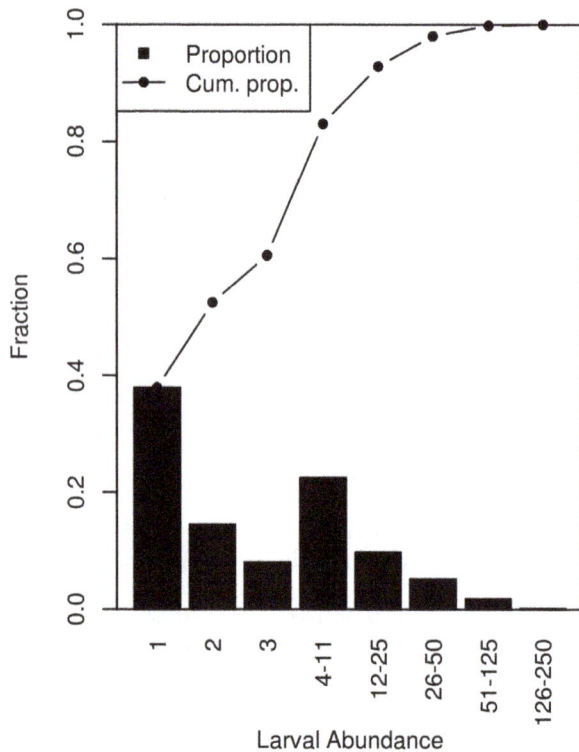

Figure 5. Distribution of larval abundances reported in the CPR. The relative proportion of each non-zero abundance category reported (bars) and the cumulative proportion (line) are show. Cumulative proportion is defined here as the proportion of presences with an abundance less than or equal to the given category. Note that the abundances are the abundance categories reported by the CPR survey [24]. Observations of zero larvae (absence) are omitted from this distribution.

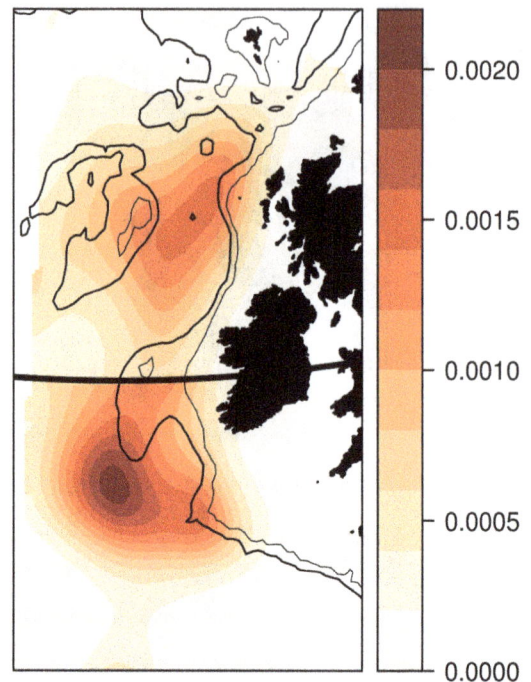

Figure 6. Spatial larval-presence probability distribution. Results predicted from Model 10 ($g()$ = east * north * doy + s(year)| comp) are plotted as a probability density function for each population (*i.e.* the spatial integral over the domain of each of the two populations is 1). The black horizontal line indicates the location of the arbitrary division between a northern and southern population at 53 °N. Note abundances cannot be compared between the domains, as each domain is normalised to give an integral of 1. Isobaths are draw at 200 m (thin line) and 1000 m (thicker line) depths for reference. Map projection is UTM Zone 28.

the CPR is marginally more effective at capturing larvae during the day than it is during the night. Such a result is not consistent with active avoidance of the sampler, where one would expect a reduced probability of capturing larvae during daylight hours. Instead, the result suggests diel vertical migration, where the larvae migrate close to the surface during the day and are therefore more readily captured by the CPR sampler.

Discussion

Reliability of the CPR data

In this study we infer the spatial, seasonal and interannual variability in the spawning of blue whiting from the presence and absence of blue whiting larvae in Continuous Plankton Recorder (CPR) samples. We argue that this is a valid proxy for the distribution of blue whiting spawning. Blue whiting spawn at depths of between 300 m and 600 m and, once hatched, rise to surface waters over the course of the first two-three weeks of life: larval length upon reaching these waters is 2–5 mm [45]. These field observations agree with the larval length distributions of blue whiting in the CPR reported by [25], who found all but a small minority of the larvae (approximately 5–10%) to be smaller than 6 mm. For contrast, while the length-at-metamorphosis of blue whiting is unknown [7], 15 mm larvae have been observed in other studies (*e.g.* [46,47]) and there is a single report of a 42 mm larvae [48]. Similarly, Coombs *et al.* [49] performed detailed studies of blue whiting egg and larval development in the

laboratory and demonstrated that yolk-sack absorption is complete after two weeks, at which point the larvae are approximately 5 mm in length. The blue whiting larvae in the CPR are highly-likely to be early-larvae, and their abundance therefore is likely to reflect the distribution of the adults that spawned them.

The choice of a presence/absence model, rather than a fully-developed abundance model, could potentially provide problems in interpretation. However, we note that single larvae are the most frequently observed class, and thus will have the strongest influence on an abundance-based model anyway. A reliable abundance model may also be difficult to develop due to the likely patchiness (and therefore overdispersion and zero-inflation) in the spatial and temporal distribution, and could easily be dominated by a few large catches. Nevertheless, future work should examine the use of the abundance categories in more detail.

This work provides another example of the utility of the CPR for investigating the characteristics of fish populations [33,50,51]. The study of the spawning distribution of this species in this region using fisheries surveys is made extremely difficult by the large areas over which blue whiting spawn: more than 1500 km north-south and 500 km east-west. In spite of the small flotilla of vessels typically used to cover this region, developing a synoptic picture of the distribution of this fish is challenging: multiple snapshots, enabling the dynamics of the spawning process to be tracked throughout the season, are simply not feasible. On the other hand, at least prior to the 1980s, the CPR provides observations with broad spatial and temporal coverage. Furthermore, the long time-series and consistency of the method allow insights into both the

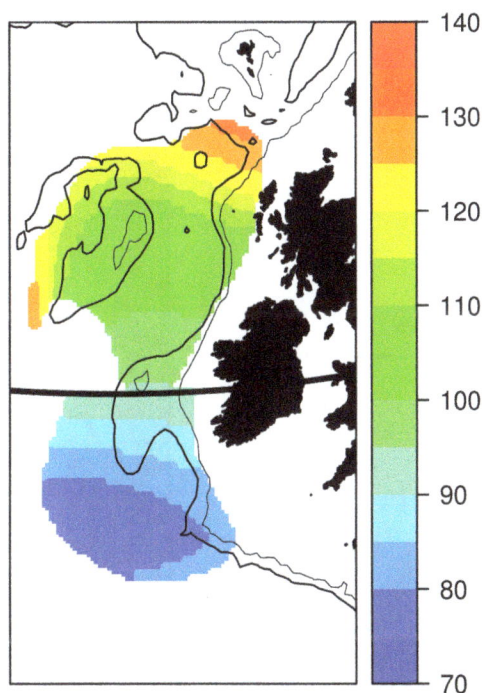

Figure 7. Timing of peak probability of occurrence. Results predicted from Model 10 (g() = east * north * doy + s(year)|comp). The day of year (colour scale) when the local maximum in probability of larval presence occurs is plotted as a function of space. The black horizontal line indicates the location of the arbitrary division between a northern and southern population (at 53 °N). The spatial distribution in Figure 6 is used to mask the output so that only the core 75% of the larval distribution in each region is plotted: regions where there are few larvae, and the estimated timing of spawning is therefore imprecise, are thus omitted. Isobaths are draw at 200 m (thin line) and 1000 m (thicker line) depths for reference. Map projection is UTM Zone 28.

history and population structure of this species that would not otherwise be possible.

This study, however, also highlights some of the limitations of CPR data. The irregular, and varying sampling pattern, with many gaps in coverage, the low frequency of larval occurrence, and the use of categorical abundances make the analysis of this data challenging. Nevertheless, the development of modern statistical tools, combined with ready access to powerful computers, have opened up many new possibilities. In particular, the development of Generalised Additive Models, and their packaging in a user-friendly form (e.g. [36]) allow for non-Gaussian responses (presence-absence) to be modelled with complex predictors (e.g. the eastings-northings-day-of-year tensor-product smoothers employed here). Such tools were not available even a decade ago, and offer great potential for the future use of CPR data.

However, although these technical challenges can be solved, the most important limitation of the CPR for this study, the reduction in the sampling coverage in the Rockall region during recent times, cannot. Routes through the Rockall region have been reduced in frequency since the 1980s, and have been virtually eliminated since the 2000s (e.g. Figure S1), at least during the spring spawning-period of this species. These changes are unfortunate as these are the time periods that coincide with the modern fishery, the advent of scientific surveys, and the interesting scientific questions concerning population dynamics and the influence of the physical environment on this stock [4,21,32]. The current CPR spatial distribution is inadequate for monitoring

this stock in this region: the reintroduction of regular haul lines through this area would be of great benefit to both the blue whiting community and all pelagic science performed in this region.

The reduced sampling also prevents extraction of useful measures of interannual variability from this data. Other studies have shown that the spatial distribution of blue whiting varies from year to year in concert with the sub-polar gyre [21,32]. Unfortunately, the poor coverage means that it is probably not possible to study these processes based on CPR observations, at least during the post-1990s changes described elsewhere. Similarly, the poor precision in the modelled abundance estimates means that direct comparisons against the stock assessment, for example, are not practical. Analyses of interannual variability in both abundance and spatial distribution prior to 1980, where the spatial coverage is much greater, may be feasible, but are made more challenging by the lack of other data during this time. Instead, focus should be placed here upon the spatial (Figure 6) and seasonal (Figure 8) distributions of larvae. Disregarding the interannual processes, these results therefore become a form of climatological distribution averaged over the entire 55-year period for which CPR observations are available.

One potential weak point of our analysis is the post-hoc modification of the model ensemble to include two-component models, which represents a form of data-dredging [31]. However, this modification has a solid and independent scientific basis to support it and two-component models could therefore have been included in the original ensemble. Furthermore, we have chosen to be transparent about where this step fits in the modelling process, and we present results from both the original and expanded ensembles. Importantly, we note that the separation of the spawning-grounds in both time and space is clear in models from both the original and expanded ensembles. Thus, although a small amount of data-dredging has occurred in this work, we feel it is to an acceptable degree and do not believe that the validity of our results are unduly affected by it.

A Resolution to the Paradox?

Our results suggest the presence of two unique spawning components. There is a clear separation between the two spawning centres, with a minimum in spawning activity occurring between 52 and 54 °N. Furthermore, we have also shown a difference in the timing of spawning of around a month between the two populations: in particular, spawning on the southern spawning ground appears to be nearly finished before it starts on the northern ground (c.f. Figure 8). Finally, we have shown a difference in the interannual abundances of these two components: although there is a large amount of noise in the interannual abundance estimates, a model (Model 10) with different interannual variations between the components is statistically superior to one (Model 6) assuming a common trend (c.f. Table 1).

Furthermore, the spatial separation into two spawning components closely mirrors the results obtained elsewhere, particularly from particle tracking studies. Bartsch et al. [16] suggested a separation between the populations at around 53/54 °N, whilst based on a different oceanographic model Svendsen et al. [17] and Skogen et al. [18] suggested a similar line at 54.5 °N. Here we chose a separation line at 53 °N, but the choice is essentially arbitrary and there appears to be a clear region of zero or minimal spawning between the components that also encompasses the aforementioned separation lines. Our direct observations of blue whiting larval distributions are therefore in line with these results.

Most importantly, our results suggest a resolution to the blue whiting population paradox. The crux of the problem is the

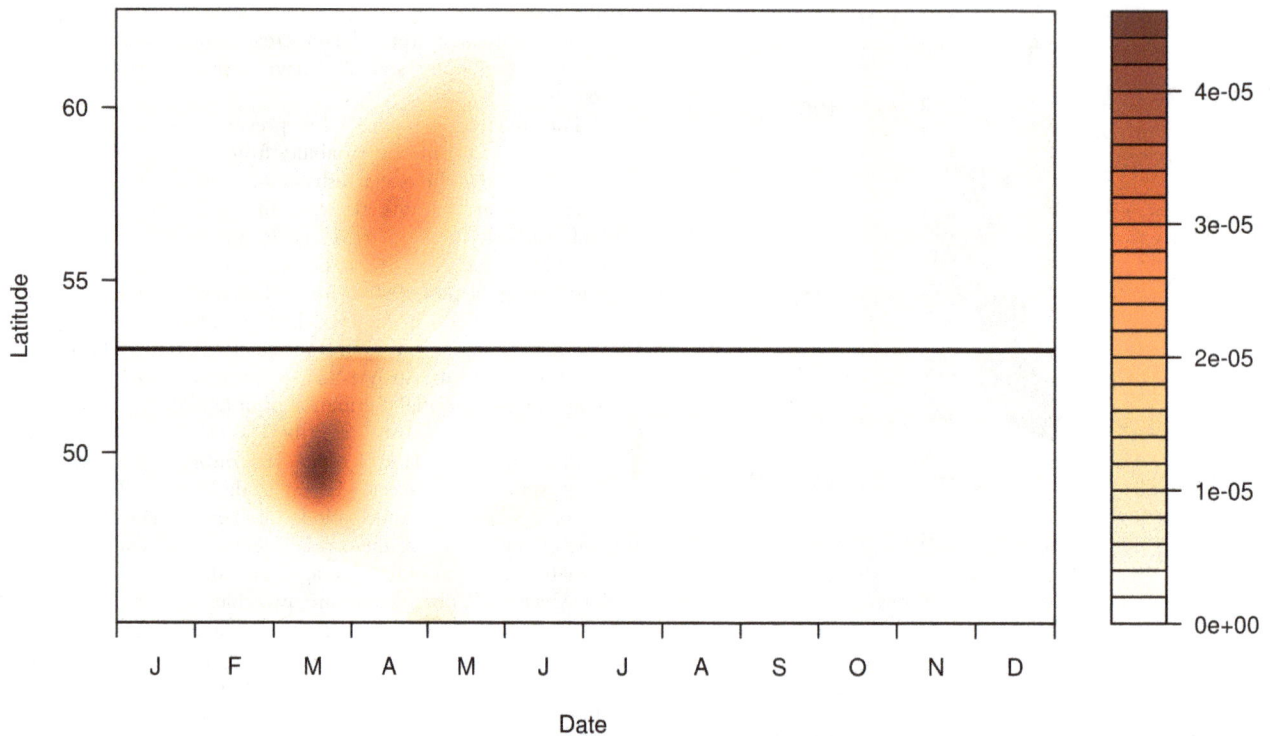

Figure 8. Zonally integrated larval-presence probability distribution. Results from Model 10 ($g()$ = east * north * doy + s(year)|comp), plotting the probability distribution of larval-presence as a function of latitude and day of year. The probability of larval-presence is expressed as a density function for each population (*i.e.* the integral over each of the two populations is 1). The black horizontal line indicates the location of a hypothesised division between a northern and southern spawning population (at 53 °N). Note that because this model allows the relative abundances of the two populations to vary from year to year, abundances cannot be compared between the domains. The projected UTM coordinates used in the fitted model have been reprojected back to longitude here for ease of interpretation.

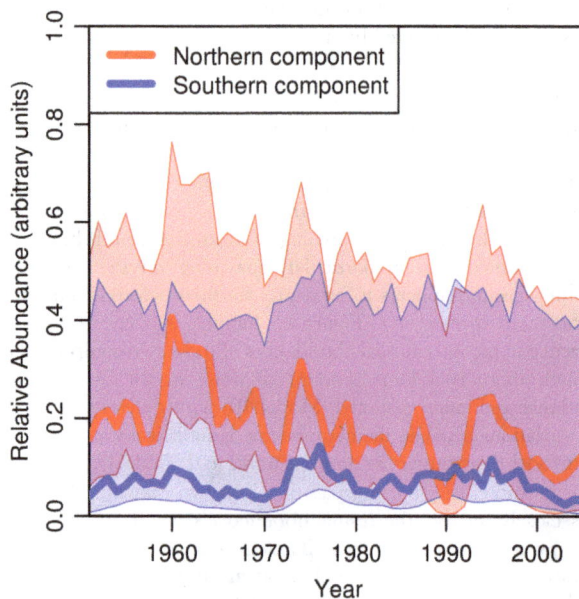

Figure 9. Annually integrated larval occurrence-probability. Results from Model 10 ($g()$ = east * north * doy + s(year)|comp). The probability of observing larvae integrated over the spatial domain and day of year is a measure of larval abundance in that year and is plotted as a function of the year for the northern (red) and southern (blue) populations, with the associated 67% (*i.e.* corresponding to 1 standard deviation) confidence intervals. The units of larval abundance plotted here are arbitrary but scale linearly.

supposed lack of evidence supporting the separation of the two hypothesised populations on the spawning grounds. However, we propose that this picture is simply an artefact of survey design. For example, the most recent (2013) survey took place over two weeks at the very end of March and the beginning of April and stretched from 53 °N to 62 °N (Figure 2) [52]. Such a survey will not capture spawning in the southern population for two reasons. Firstly, it occurs too late: the abundance of larvae in the southern population is essentially zero by the end of March (*c.f.* Figure 8), and therefore spawning, occurring approximately two weeks earlier than the larvae that we observed, peaked at least one month prior. Noting the highly migratory nature of blue whiting, it is not unreasonable to expect that the fish may have left the southern spawning grounds by late March.

Secondly, the survey does not extend sufficiently far south. The current survey design stops at Porcupine Bank (53 °N: Figure 2), whereas the southern population spawns offshore from the Porcupine Seabight, between 48 and 52 °N (Figure 6). Such an omission is not unique to modern times: a review of all acoustic surveys [32] shows regular coverage of Porcupine Bank, but not further south into the seabight where we suggest the southern population spawns.

We therefore conclude that the blue whiting population paradox is simply an observational artefact. While the distribution of the spawning products is clearly and cleanly separated in space and time, the acoustic observations of adult fish are not capable of resolving the southern population due to their restricted temporal and spatial coverage. Confusion therefore arises because the observations are only capable of capturing the northern population, rather than both populations, creating the (false) appearance of a continuous distribution on the spawning grounds.

With the insight afforded by these new results, an important inconsistency in the literature becomes apparent. Many authors have previously considered the Porcupine Bank to be the spawning ground of the southern population and designed their studies accordingly: however, these results suggest that the north-side of the Porcupine Bank should be considered as northern "territory" (e.g. Figure 6). This revelation suggests that the interpretation of many existing studies need to be reconsidered. For example, the results of a microsatellite genetic study on blue whiting population structure [12] lumped the north-side of the Porcupine Bank together with the Outer Hebrides and Rockall Plateau, whilst samples taken from the Porcupine Seabight were genetically distinct. To a researcher working under the (previous) assumption that the Porcupine Bank is the "southern" component, these results are confusing. However, when combined with the results presented here, where Porcupine Bank is part of the northern population, they are consistent. Similar reinterpretations occur when re-examining the otolith juvenile growth [13] and shape [14] studies. Furthermore, observational studies reporting spawning fish off the Porcupine Seabight [53], which made little sense in the previous conceptual model, now give both meaning and lend their support. There is therefore a need for a comprehensive re-examination of the published literature on this topic: however, that is clearly beyond the scope of this work.

Nevertheless, and other studies not withstanding, there is now clear evidence that the North Atlantic blue whiting population should be considered as two independent stocks. Studies based on both genetics [12] and otoliths [13,14] support this separation, while circulation studies [16–18] provide a physical mechanism that maintains the separation between the larvae spawned in these two locations. In this study, we have shown a clear physical separation between the two populations, and that there is at least a month difference in the timing of peak spawning. Furthermore, the interannual variations in the abundances of each population are also statistically different. With the lack of structure in the adult observations now explained as an observational artefact, the case for two-populations already appears irresistibly strong.

The current management paradigm, however, is based on a single stock approach and is likely to be so for some time to come. In contrast to early assessments (e.g. [19]), little attention is paid to quantifying the southern population and there is therefore a risk of inadvertently fishing it to collapse. Studies in other small pelagic species (e.g. herring, Clupea harengus) suggest that maintaining stock/population diversity provides resilience against both natural and anthropogenic stresses and helps maintain productivity [54–56]. However, even in the absence of separating these two populations into unique management units, improvements in the monitoring of these populations are possible. The most obvious is the extension of the spawning acoustic survey both in space and time to cover the spawning of the southern population. Secondly,

the re-establishment of CPR haul lines through the Rockall region would allow direct comparison with modern observations, and therefore aid the interpretation of the historical CPR observations. Such changes should be considered as critical steps towards the precautionary management of blue whiting in the North Atlantic.

Supporting Information

Figure S1 Annual spring distribution of CPR samples. Samples checked for fish larvae obtained from the CPR. Grey points are locations where CPR samples have been checked for fish larvae. Red circles are where these samples were found to contain blue whiting. As blue whiting larvae are predominately captured in the first half of the year, only observations from January to June (inclusive) are plotted here. Map projection is UTM Zone 28.

Figure S2 Spatial larval-presence probability distribution. Results from Model 6 ($g()$ = east * north * doy + s(year)), plotted as a probability density function (i.e. the spatial integral over the domain is 1). Isobaths are draw at 200 m (thin line) and 1000 m (thicker line) depths for reference. Map projection is UTM Zone 28.

Figure S3 Zonally integrated probability distribution. Results from Model 6 ($g()$ = east * north * doy + s(year)), plotting larval occurrence probability as a function of latitude and day of year. The probability of larval-occurrence is expressed as a probability density function (i.e. the integral over the domain is 1). The UTM coordinates used in the fitted model have been reprojected back to longitude for ease of interpretation.

Acknowledgments

The authors wish to thank Sophie Pitois (Cefas) and David Johns (SAHFOS) for providing the CPR data and answering questions about its genesis. We thank the members of the ICES Working Group of International Pelagic Surveys (WGIPS) for providing Figure 2. Ana Sofia Ferreira (DTU Aqua) and the members of the ICES Working Group on Widely Distributed Stocks (WGWIDE) provided valuable feedback on an early version of this work. We would like to express our thanks to Georg H. Engelhard and a second, anonymous, reviewer, who provided thorough and valuable critiques of this manuscript during the review stage.

Disclaimer
Modelled distributions of blue whiting larvae are available via the panagea data repository (doi:10.1594/PANGAEA.834212), or upon request from the corresponding author.

Author Contributions

Conceived and designed the experiments: MRP. Analyzed the data: FP MRP. Wrote the paper: MRP.

References

1. Schmidt S (1909) The distribution of the pelagic fry and the spawning regions of the gadoids in the North Atlantic from Iceland to Spain. Rapports et Proces-verbaux des Réunions Conseil International pour l'Éxploration de la Mer 10: 1–229.
2. Trenkel VM, Huse G, MacKenzie BR, Alvarez P, Arrizabalaga H, et al. (2014) Comparative ecology of widely distributed pelagic fish species in the North Atlantic: implications for modelling climate and fisheries impacts. Progress in Oceanography In Press.
3. FAO (2007) The state of world fisheries and aquaculture 2012. Technical report, FAO, Rome.
4. Payne MR, Egan A, Fässler SMM, Hátún H, Holst JC, et al. (2012) The rise and fall of the NE Atlantic blue whiting (Micromesistius poutassou). Marine Biology Research 8: 475–487.
5. ICES (2010) Report of the ICES Advisory Committee, 2010. Book 9. Technical report.
6. ICES (2013) Report of the ICES Advisory Committee, 2013. Book 9. Technical report.
7. Bailey RS (1982) The population biology of blue whiting in the North Atlantic. Advances in Marine Biology 19: 257–355.
8. Monstad T (2004) Blue Whiting. In: Skjoldal H, editor, The Norwegian Sea Ecosystem, Trondheim: Tapir Academic Press. pp. 263–288.

9. Giaever M, Stien J (1998) Population genetic substructure in blue whiting based on allozyme data. Journal of Fish Biology 52: 782–795.

10. Mork J, Giaever M (1995) Genetic variation at isozyme loci in blue whiting from the north-east Atlantic. Journal of Fish Biology 46: 462–468.

11. Ryan AW, Mattiangeli V, Mork J (2005) Genetic differentiation of blue whiting (*Micromesistius poutassou* Risso) populations at the extremes of the species range and at the Hebrides–Porcupine Bank spawning grounds. ICES Journal of Marine Science 62: 948–955.

12. Was A, Gosling E, McCrann K, Mork J (2008) Evidence for population structuring of blue whiting (*Micromesistius poutassou*) in the Northeast Atlantic. ICES Journal of Marine Science 65: 216–225.

13. Brophy D, King P (2007) Larval otolith growth histories show evidence of stock structure in Northeast Atlantic blue whiting (*Micromesistius poutassou*). ICES Journal of Marine Science 64: 1136–1144.

14. ICES (2012) Report of the Benchmark Workshop on Pelagic Stocks (WKPELA 2012), 13–17 February 2012, Copenhagen, Denmark. ICES CM 2012/ACOM:47. Technical report.

15. Isaev NA, Seliverstov AS (1991) Population structure of the Hebridean-Norwegian school of blue whiting, *Micromesistius poutassou*. Journal of Ichthyology 31: 45–58.

16. Bartsch J, Coombs S (1997) A numerical model of the dispersion of blue whiting larvae, *Micromesistius poutassou* (Risso), in the eastern North Atlantic. Fisheries Oceanography 6: 141–154.

17. Svendsen E, Skogen S, Monstad T, Coombs SH (1996) Modelling the variability of the drift of blue whiting larvae and its possible importance for recruitment. ICES CM 1996/S:31.

18. Skogen MD, Monstad T, Svendsen E (1999) A possible separation between a northern and a southern stock of the northeast Atlantic blue whiting. Fisheries Research 41: 119–131.

19. ICES (1986) Report of the Blue Whiting Assessment Working Group. Copenhagen, 25 September - 2 October 1985. ICES CM 1986/Assess:3. Technical report.

20. Heino M, Engelhard GH, Godø OR (2008) Migrations and hydrography determine the abundance fluctuations of blue whiting (*Micromesistius poutassou*) in the Barents Sea. Fisheries Oceanography 17: 153–163.

21. Hátún H, Payne MR, Beaugrand G, Reid PC, Sando AB, et al. (2009) Large bio-geographical shifts in the north-eastern Atlantic Ocean: From the subpolar gyre, via plankton, to blue whiting and pilot whales. Progress In Oceanography 80: 149–162.

22. Huse G, Utne KR, Fernö A (2012) Vertical distribution of herring and blue whiting in the Norwegian Sea. Marine Biology Research 8: 488–501.

23. Reiss H, Hoarau G, Dickey-Collas M, Wolff WJ (2009) Genetic population structure of marine fish: mismatch between biological and fisheries management units. Fish and Fisheries 10: 361–395.

24. Richardson AJ, Walne AW, John AWG, Jonas TD, Lindley JA, et al. (2006) Using continuous plankton recorder data. Progress In Oceanography 68: 27–74.

25. Bainbridge V, Cooper G (1973) The distribution and abundance of the larvae of the blue whiting, *Micromesistius poutassou* (Risso), in the North-east Atlantic, 1948–1970. Bulletins of Marine Ecology 8: 99–114.

26. Coombs S (1980) Continuous Plankton Records: A plankton atlas of the North Atlantic and North Sea: Supplement 5 - Young Fish, 1948–1972. Bulletins of Marine Ecology 8: 229–281.

27. Henderson GTD (1957) The distribution of young *Gadus poutassou* (Risso). Bulletins of Marine Ecology 4: 179–202.

28. Henderson GTD (1961) Continuous plankton records: Contributions towards a plankton atlas of the north-eastern Atlantic and the North Sea. Part 5. Young Fish. Bulletins of Marine Ecology 5: 105–111.

29. Corten A, Lindley JA (2003) The use of CPR data in fisheries research. Progress in Oceanography 58: 285–300.

30. Edwards M, Healouet P, Halliday N, Beaugrand G, Fox C, et al. (2011) Fish larvae atlas of the NE Atlantic. Results from the Continuous Plankton Recorder survey 1948-2005. Plymouth, U.K.: Sir Alister Hardy Foundation for Ocean Science, 22 pp.

31. Burnham KP, Anderson DR (2002) Model selection and multimodel inference: a practical information-theoretic approach. Springer, 2nd edition, 488 pp. doi: 10.1016/j.ecolmodel.2003.11.004.

32. Hátún H, Payne MR, Jacobsen JA (2009) The North Atlantic subpolar gyre regulates the spawning distribution of blue whiting (*Micromesistius poutassou*). Canadian Journal of Fisheries and Aquatic Sciences 66: 759–770.

33. Jansen T, Campbell A, Kelly C, Hátún H, Payne MR (2012) Migration and Fisheries of North East Atlantic Mackerel (*Scomber scombrus*) in Autumn and Winter. PLoS ONE 7: e51541.

34. Agresti A (2010) Analysis of ordinal categorical data. Hoboken, New Jersey: John Wiley and Sons, 2nd edition.

35. Bivand R, Lewin-Koh N (2013) maptools: Tools for reading and handling spatial objects. URL http://cran.r-project.org/package=maptools.

36. Wood SN (2006) Generalized additive models: an introduction with R. Boca Raton, FL: Chapman Hall/CRC Press.

37. Wood SN (2011) Fast stable restricted maximum likelihood and marginal likelihood estimation of semiparametric generalized linear models. Journal of the Royal Statistical Society: Series B (Statistical Methodology) 73: 3–36.

38. Wood SN (2006) Low-rank scale-invariant tensor product smooths for generalized additive mixed models. Biometrics 62: 1025–36.

39. Elith J, Graham CH, Ferrier S, Guisan A, Anderson RP, et al. (2006) Novel methods improve prediction of species' distributions from occurrence data. Ecography 29: 129–151.

40. Jones MC, Dye SR, Pinnegar JK, Warren R, Cheung WWL (2012) Modelling commercial fish distributions: Prediction and assessment using different approaches. Ecological Modelling 225: 133–145.

41. Lobo JM, Jiménez-Valverde A, Real R (2008) AUC: a misleading measure of the performance of predictive distribution models. Global Ecology and Biogeography 17: 145–151.

42. NCAR (2012) verification: Forecast verification utilities. URL http://cran.r-project.org/package=verification.

43. Gerber E (1993) Some data on the distribution and biology of the blue whiting, *Micromesistius poutassou*, at the Mid-Atlantic Ridge. Journal of Ichthyology 33: 26–34.

44. Burnham KP, Anderson DR, Huyvaert KP (2011) AIC model selection and multimodel inference in behavioral ecology: some background, observations, and comparisons. Behavioral Ecology and Sociobiology 65: 23–35.

45. Ådlandsvik B, Coombs S, Sundby S, Temple G (2001) Buoyancy and vertical distribution of eggs and larvae of blue whiting (*Micromesistius poutassou*): observations and modelling. Fisheries Research 50: 59–72.

46. Kloppmann M, Mohn C, Bartsch J (2001) The distribution of blue whiting eggs and larvae on Porcupine Bank in relation to hydrography and currents. Fisheries Research 50: 89–109.

47. Bailey M, Heath M (2001) Spatial variability in the growth rate of blue whiting (*Micromesistius poutassou*) larvae at the shelf edge west of the UK. Fisheries Research 50: 73–87.

48. Conway DVP (1980) The food of larval blue whiting, *Micromesistius poutassou* (Risso), in the Rockall area. Journal of Fish Biology 16: 709–723.

49. Coombs SH, Hiby AR (1979) The development of the eggs and early larvae of blue whiting, *Micromesistius poutassou* and the effect of temperature on development. Journal of Fish Biology 14: 111–123.

50. Pitois SG, Lynam CP, Jansen T, Halliday N, Edwards M (2012) Bottom-up effects of climate on fish populations: data from the Continuous Plankton Recorder. Marine Ecology Progress Series 456: 169–186.

51. Lynam C, Halliday N, Höffle H, Wright PJ, Van Damme CJ, et al. (2013) Spatial patterns and trends in abundance of larval sandeels in the North Sea: 1950–2005. ICES Journal of Marine Science 70: 540–553.

52. ICES (2014) Report of the Working Group of International Pelagic Surveys (WGIPS), 20–24 January 2014. ICES CM 2014/SSGESST:01. Technical report.

53. Gerber YM, Demenin AA (1993) On Spawning of Blue Whiting in the West European basin. Journal of Ichthyology 33: 77–86.

54. McPherson AA, Stephenson RL, O'Reilly PT, Jones MW, Taggart CT (2001) Genetic diversity of coastal Northwest Atlantic herring populations: implications for management. Journal of Fish Biology 59: 356–370.

55. Secor DH, Kerr LA, Cadrin SX (2009) Connectivity effects on productivity, stability, and persistence in a herring metapopulation model. ICES Journal of Marine Science 66: 1726–1732.

56. Kell LT, Dickey-Collas M, Hintzen NT, Nash RDM, Pilling GM, et al. (2009) Lumpers or splitters? Evaluating recovery and management plans for metapopulations of herring. ICES Journal of Marine Science 66: 1776–1783.

18

Coral Reef Disturbance and Recovery Dynamics Differ across Gradients of Localized Stressors in the Mariana Islands

Peter Houk[1,2]*, David Benavente[3], John Iguel[3], Steven Johnson[3], Ryan Okano[3]

1 University of Guam Marine Laboratory, UOG Station, Mangilao, Guam, **2** Pacific Marine Resources Institute, Saipan, Northern Mariana Islands, **3** CNMI Bureau of Environmental and Coastal Quality, Saipan, Northern Mariana Islands

Abstract

The individual contribution of natural disturbances, localized stressors, and environmental regimes upon longer-term reef dynamics remains poorly resolved for many locales despite its significance for management. This study examined coral reefs in the Commonwealth of the Northern Mariana Islands across a 12-year period that included elevated Crown-of-Thorns Starfish densities (COTS) and tropical storms that were drivers of spatially-inconsistent disturbance and recovery patterns. At the island scale, disturbance impacts were highest on Saipan with reduced fish sizes, grazing urchins, and water quality, despite having a more favorable geological foundation for coral growth compared with Rota. However, individual drivers of reef dynamics were better quantified through site-level investigations that built upon island generalizations. While COTS densities were the strongest predictors of coral decline as expected, interactive terms that included wave exposure and size of the overall fish assemblages improved models (R^2 and AIC values). Both wave exposure and fish size diminished disturbance impacts and had negative associations with COTS. However, contrasting findings emerged when examining net ecological change across the 12-year period. Wave exposure had a ubiquitous, positive influence upon the net change in favorable benthic substrates (i.e. corals and other heavily calcifying substrates, $R^2 = 0.17$ for all reeftypes grouped), yet including interactive terms for herbivore size and grazing urchin densities, as well as stratifying by major reeftypes, substantially improved models ($R^2 = 0.21$ to 0.89, lower AIC scores). Net changes in coral assemblages (i.e., coral ordination scores) were more sensitive to herbivore size or the water quality proxy acting independently ($R^2 = 0.28$ to 0.44). We conclude that COTS densities were the strongest drivers of coral decline, however, net ecological change was most influenced by localized stressors, especially herbivore sizes and grazing urchin densities. Interestingly, fish size, rather than biomass, was consistently a better predictor, supporting allometric, size-and-function relationships of fish assemblages. Management implications are discussed.

Editor: Benjamin Ruttenberg, California Polytechnic State University, United States of America

Funding: This study was funded by the National Oceanic and Atmospheric Administration (NOAA) coral-reef monitoring grants administered to the CNMI Coastal Resources Management Office, and USEPA Region IX funding administered to the CNMI Division of Environmental Quality. Financial support for PH was provided through a NOAA domestic coral reef grant award (NA11-NOS4820015). The funders had no role in study design, data collection and analysis, decision to publish, or preparation of the manuscript.

Competing Interests: The authors have declared that no competing interests exist.

* Email: houkp@uguam.uog.edu

Introduction

A vast array of acute disturbances and chronic stressors threaten coral reefs [1–3]. Yet, our knowledge of the role that individual disturbances and stressors play in determining reef dynamics through time remains limited for most locales [4]. One cause of this uncertainty stems from the rarity of ecological datasets that span across sufficient timeframes to capture disturbance and recovery periods, which can help partition the variance associated with population dynamics, and attribute cause, proportionally, to individual stressors. For coral reefs, two putative, localized stressors of primary concern are unsustainable fishing and pollution that may act independently or in combination with disturbance cycles to diminish the growth of reef accreting organisms such as corals [5,6]. Many studies have compared reefs where high and low human influences existed to generalize how water quality and herbivory have impacted coral growth and

resulted in macroalgal replacement [7,8]. However, less attention has been given to understanding their individual contributions towards growth dynamics outside of laboratory or manipulative settings [9–11], and thus, our ability to upscale evidence from manipulative studies to entire reefscapes remains limited. In turn, context-dependent roles of localized stressors continue to be the focus of much research, and varying findings provide support for many ideologies [12–14]. When used out of context, or when limited information exists to develop an appropriate context, uncertainty and improperly informed decision making can result.

Acute disturbances such as typhoons, *Acanthaster planci* outbreaks, and climate-induced bleaching are well appreciated for their role in driving coral population dynamics. Disturbance and recovery cycles have traditionally been investigated using coral cover trends, integrated across both local and regional scales, and through time [15,16]. Yet, coral cover can be an inconsistent metric of recovery due to varying natural environmental regimes, such as wave exposure, that dicate coral growth capacity [17–19], and thus the time needed for recovery. Improved assessments of reef condition and calcification potential have emerged from benthic-substrate datasets by simultaneously considering the abundances of macroalgae and other less-calcifying organisms in comparison to corals and other heavily-calcifying organisms [9,20,21]. Such integrated metrics can better account for the inherent environmental variation that drives coral cover trajectories, and represents one useful metric of overall recovery potential when comparing across reefs that is furthered within the present study.

In addition to coverage estimates, shifting species abundances [22] and colony-size distributions [23,24] have also proven to be sensitive indicators of disturbance-and-recovery cycles for coral populations. Studies show that faster growing corals (*Acropora*, *Pocillopora*, *Stylophora*) have lower tolerances to disturbance events, while others (*Porites* and numerous faviids) have a prolonged ability to deal with both acute disturbance and chronic stress [5,25–27]. However, many of the same faster-growing corals may be more resilient (or adaptable) to repeated climate-induced disturbance in the absence of chronic stress [28,29]. Thus, size distributions and abundance patterns of corals that have low thresholds, fast recovery, and the potential for adaptability provides an additional means of evaluating coral population dynamics [20,30].

The premise for the present study is that disturbance and recovery are not necessarily co-dependent processes on coral reefs, and by examining them independently with the aid of refined metrics described above, an improved understanding of causation and predicted resilience can emerge [3,31]. While many studies have described the nature of acute disturbances to reefs from a variety of agents, relatively few have followed recovery dynamics with respect to individual factors. In the Caribbean, Mumby and Harborne [10] reported significantly higher recovery of coral coverage and colony sizes in response to a no-take fishery closure over a 2.5 year period. Similarly, the recovery of coral colony sizes [22], but not recruitment [32], were found to be most heavily dependent upon fishing pressure in Kenya. In American Samoa, Houk et al. [9] found differential recovery of calcifying benthic substrates and coral species evenness to be interactively driven by herbivore biomass and water quality, with their hierarchical influence shifting based upon geological reef settings (i.e., a potential proxy for connectivity with groundwater discharge). In contrast, a meta-analysis of coral reef recovery across all major oceanic basins provided some evidence for unintuitive, reduced recovery rates within fisheries closures following disturbance events [3]. These findings were a perceived artifact of higher pre-disturbance coral coverage within no-take closures, and not attributed to management status. Interestingly, the causes behind recovery trajectories were not consistent across major geographic regions. Clearly recovery is context-dependent with respect to physical settings as well as management regimes [33–37]. Considering that predicting reef futures is becoming more of a priority for both local and global management efforts, further study is needed to interpret when and if generalizations may exist, and at what spatial scales they might operate at.

We examined 12-year trends in coral-reef assemblages across fore-reefs in the Commonwealth of the Northern Mariana Islands (CNMI) during a time period when significant disturbance impacts and differential recovery occurred. We first described the nature of *Acanthaster planci* (i.e., Crown-of-Thorn Starfish, COTS) population dynamics during the study period. COTS population trends were augmented with typhoon records to define pre-, during-, and post-disturbance time periods across our 12-year study. Coral

cover and colony sizes were then evaluated for two study islands that differed in human population, development, and geological setting. Island-scale coral decline and recovery trajectories provided an initial framework for interpreting the nature of coral growth cycles across the study period, and along with previous studies, highlighted primary factors influencing site-specific reef dynamics. Ensuing site-level regressions were performed between coral decline and COTS densities (i.e., the hypothesized, primary disturbance agent), while including key factors that quantified localized stressors and environmental regimes as interactive covariates. Similar site-level regressions were also performed to examine net ecological change across the study period. Combined, our study provided a descriptive framework for disentangling the differences between 'disturbance impacts' and 'net ecological change', and attributed cause, proportionally, to a suite of factors that help to inform management.

Methods

Study location

The Commonwealth of the Northern Mariana Islands (CNMI) represents a series of active volcanic and inactive raised limestone islands located in the Western Pacific Ocean (Figure 1). The present study focused upon the southernmost, inactive limestone islands where the majority of the human population resides. From north to south, the study islands were Saipan (48,220 people, capital island), Tinian (3,136 people), Aguijan (no inhabitants but the nearshore resources are accessed from Saipan and Tinian), and Rota (2,527 people) (Census 2010 statistics, http://commerce. gov.mp/divisions/central-statistics/) (Figure 1).

CNMI's local monitoring program has been collecting standardized benthic, coral, macroinvertebrate, and fish assemblage data since 2000 [38,39]. Monitoring designs were stratified based upon geological reeftypes, management status, and watershed development. Geological reeftypes have previously been described with respect to wave exposure and submarine groundwater discharge through karst watersheds that were attributed to specific coral assemblages and reef growth through time [18]. Cumulatively, Houk and van Woesik [18] described four distinct geological reeftypes in the CNMI: 1) optimal spur-and-groove structures, 2) constructional, high-relief interstitial framework, 3) low-relief framework with limited Holocene deposition found only on Rota, and 4) incipient coral assemblages with little to no deposition. Coral coverage, diversity, and evenness peak on the first two, with spur-and-groove structures being the most optimal settings for modern coral assemblage growth.

The most significant acute disturbance since the inception of monitoring efforts has been high *Acanthaster planci* populations (COTS herein) evident between mid-2003 and 2006. In complement to COTS disturbances, several tropical storms passed nearby the study islands during a similar timeframe, with the strongest being Pongsona (passing 40 km south of Rota, 95 knots maximum sustained windspeed) and Chaba (passing over Rota, 118 knot sustained windspeed) in December 2002 and August 2004, respectively. While large wave events were recorded, observations by the authors before and after these storms suggested negligible direct impacts compared with the onset of COTS, examined further within. COTS events became widespread throughout the North Pacific during this timeframe [40]. However, no information exists to understand any potential linkages between the storm events and the subsequent emergence of COTS. Initial examinations were undertaken by coupling COTS abundance and tropical storm histories to define three time periods within the present study *a priori*: pre-, during-, and post-disturbance. Subsequently,

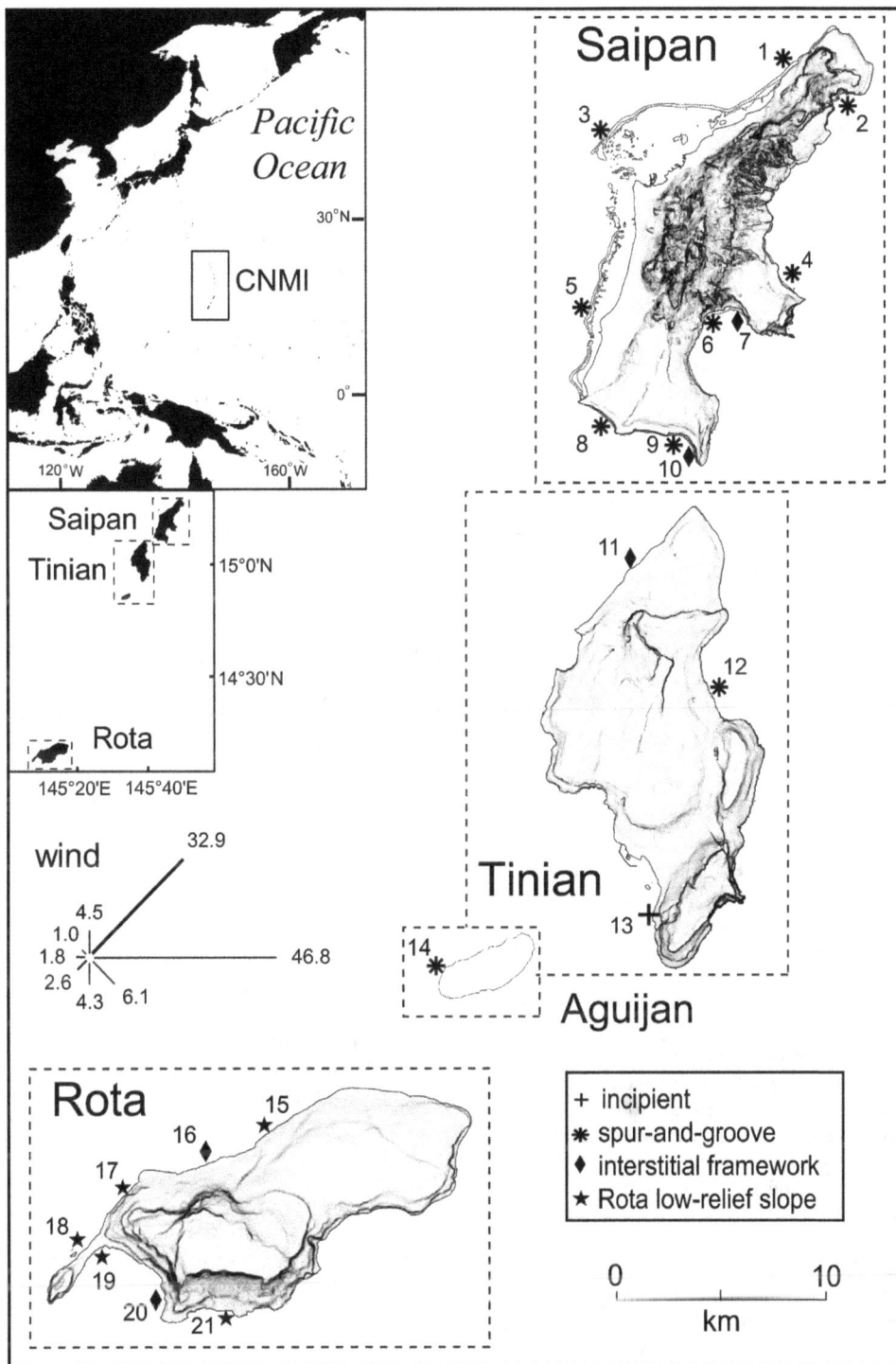

Figure 1. A map of the Western Pacific Ocean and the Commonwealth of the Northern Mariana Islands study islands. Distance between islands is not drawn to scale as dashed boxes indicate individual island entities. Wind vectors show the percent of time that winds originated from each of 8 quadrants (length and corresponding number) as well as the mean annual intensity (thickness). Reeftypes are indicated by symbols referred to in the legend. Topographic lines infer the steepness and size of watersheds.

we quantified how much of the site-level coral decline during the 'disturbance' period could be accounted for by COTS.

Ecological data collection

Data were collected as part of the CNMI Division of Environmental Quality and the CNMI Coastal Resources Management Office coral-reef monitoring program. These programs have the legislative authority to conduct monitoring activities, and given the non-invasive nature of this research, no further permits were required.

Long-term monitoring data have been collected at 1 to 3 year intervals for 25 monitoring stations across the CNMI in most cases, however, longer intervals between repeated site-visits existed in a few instances (Appendix 1). The prerequisite for inclusion in the present study was that coral, benthic, and/or macroinvertebrate data were available for all three timeframes: before, during, and after the disturbance period. Among the 25 monitoring sites, data from 21 met these criteria (Table S1, Figure 1).

Sites were identified by global positioning system coordinates coupled with directional bearings to indicate transect placement. During each survey event, five, 50 m transects were placed at the 8 m depth contour to guide fieldwork. In three instances, homogeneous substrates were not consistently available, and three, 50 m transects were used instead (sites 2, 16, and 18, Figure 1). Benthic substrate abundances were estimated from photographs of 0.5×0.5 m quadrats. In 2000 and 2001, 25 photographs were taken along each transect line, and the substrate under each of 16 data points was identified and recorded. Since 2002, 50 photographs were taken from each transect line, and the substrate under each of 5 data points was recorded. Methods were shifted in order to improve detection limits for temporal change [38]. With respect to the present study, the changes in benthic substrate abundances associated with COTS disturbance were a magnitude of order higher than the expected detection limits based upon 2002 datasets [38]. In all instances, benthic categories chosen for analysis were corals (to genus level), turf algae (non-identifiable turfs typically less than 2 cm), macroalgae (readily identifiable alga typically greater than 2 cm, to genus level if abundant), calcareous encrusting algae known to actively shed epithelial layers or inhibit the survival of juvenile corals (*Peyssonnelia, Pneophyllum*) [41–44], crustose coralline algae (CCA) known to promote reef accretion and juvenile coral settlement, sand, and other invertebrates (genus level if abundant). Using these substrate categories, we defined a benthic substrate ratio by the percent cover of heavily calcifying (corals and CCA) versus non-or-less calcifying (turf, shedding-calcareous, macroalgae) substrates.

Coral assemblage data have been collected since 2003 by the same observer using a standard quadrat-based technique. During each survey, 16 replicate 0.5×0.5 m quadrats were haphazardly tossed at equal intervals along the transect lines. All corals whose centerpoint resided within the quadrat boundary were identified, and the maximum diameter, and the diameter perpendicular to the maximum, were recorded. Surface area was calculated from these measurements assuming colonies were elliptical in nature. For three-dimensional corals, measurements were extended along the colony surface area. Coral taxonomy followed Veron [45]. Species-level data were used for analyses of richness and evenness, while species were grouped by 'sub-genus' (i.e., genus and growth form, digitate *Acropora*, massive *Porites*) to examine multivariate trends and subsequent species weightings described below.

Macroinvertebrate densities were estimated along individual transect lines (noted above) using 50×4 m observation belts. All sea cucumbers, sea urchins, shellfish, and other conspicuous macroinvertebrates were identified to species level and recorded. Count data derived from macroinvertebrate belt transects were used to examine COTS and grazing urchin populations through time.

Fish census protocols were recently added into CNMI long-term monitoring program. The present study used data collected between 2011 and 2012 by a single observer. Fish assemblages were estimated from 12 stationary-point-counts (SPC's) conducted at equal intervals along the transect lines, following a modified version of [46]. During each SPC, the observer recorded the name (species) and size of all food-fish within a 5 m circular radius for a period of 3-minutes. Food fish were defined by acanthurids, scarids, serranids, carangids, labrids, lethrinids, lutjanids, balistids, kyphosids, mullids, and holocentrids that are a known to be harvested. Sharks were also included. Size data were binned into 5 cm categories (i.e., 12.5–17.5 cm = 15 cm, inferring the size estimates of 13, 14, 15, 16, an 17 were considered as 15 cm), and converted to biomass using coefficients reported in fishbase (www.fishbase.org). Species data were binned into functional categories based upon maximum adult sizes and family level taxonomy for some graphical interpretations and analyses, such as large/small bodied parrotfishes and surgeonfishes, and large/small bodied snappers and groupers. We defined large-and-small-bodied species based upon estimated mean reproductive sizes greater or less than 30 cm, respectively, based upon fishbase records, or the life-history wizard. Planktivores were excluded from the present analyses due to their low abundance in comparison to other trophic groups, and low sample sizes for defining confidence intervals. Last, fish recruits less than 7.5 cm (i.e., 8–12 cm size class bin) were omitted from all analyses to avoid potential bias from differential recruitment with unknown post-settlement mortality dynamics. Recruits comprised less than 3% of the surveyed population.

Environmental data collection

A proxy for water quality was developed from geographic information system (GIS) layers pertaining to topography, landuse, and human population. Digital elevation models (i.e., topographic data) were used to define watershed boundaries. Landuse data were then overlaid upon the watersheds, and a measure of disturbed land was calculated by combining the coverage of barren land, urbanized vegetation, and developed infrastructure within each watershed (United States Forest Service, http://www.fs.usda.gov/r5). Equal weighting was given to each category because variation in pollution contribution was expected within each. For example, urbanized lands were associated with both septic systems and sewer collection systems that differ with respect to their pollution contribution. Human population density in the watershed adjacent to each site was derived from the 2010 CNMI census. Landuse and human population data were also standardized to weight them equally, and averaged to establish an overall proxy to watershed pollution that represents the simplest assumptions of equal contributions from two known sources of watershed-based pollution. In order to match the low-to-high scale of other localized stressors (i.e., herbivore size), the inverse of the proxy was used in regression modeling.

Wave energy was derived from long-term wind datasets and estimates of fetch [47]. For each site, fetch (i.e., distance of unobstructed open water) was first calculated for 16 radiating lines equally distributed between 20 to 360 degrees. Fully develop sea conditions were considered if unobstructed exposure existed for 20 km or greater. Ten-year windspeed averages were calculated from Saipan airport data (http://www7.ncdc.noaa.gov/), and used as inputs to calculate wave height following Ekebom et al. [47]. Mean height was calculated by:

$$Hm = 0.018 \cdot U^{1.1} \cdot F^{0.45} \qquad (1)$$

Where Hm is the wave height (m) for each quadrant, U is the windspeed at an elevation of 10 m, and F is the fetch (km). Windspeed corrections for varying elevations were made following Ekebom et al. [47]. Last, wave height was converted to energy following:

$$E = (1/8) \cdot \rho \cdot g \cdot H^2 \qquad (2)$$

Where ρ is the water density (kg/m^3), g is the acceleration due to gravity (9.81 m/s^2), and H is the wave height (m).

Data Analyses

Island comparisons

We first generalized disturbance and recovery trajectories for two islands that differed with respect to human population, development, and geological setting, Rota and Saipan. These islands were selected because they both had sufficient sampling effort and spatial coverage over the study period to provide a generalization of temporal reef dynamics (Figure 1, Table S1). Initial, island-scale examinations were conducted for coral cover and colony sizes. Due to varying geological settings, coral growth capacity is inherently lower on Rota compared with Saipan [18], and thus we focused investigations upon the nature and relative rate of decline and recovery, and not absolute values of change. Coral cover data were site-averaged within each of the study time periods, and repeat measures ANOVA tests with post-hoc Tukey's pairwise tests were conducted between the study periods for each island. Repeat measures ANOVA tests were similarly used to examine coral population density across the disturbance time-frames. Last, differences in coral colony-size distributions were examined using Kolmogorov-Smirnov (K-S) tests that provided a P-value that is generated from the distances between cumulative frequency plots [48].

We next examined sea cucumber and grazing urchin densities across the study periods. Macroinvertebrate count data had zero-inflated, negative binomial distributions that best conformed to zero-inflated, repeat-measures ANOVA models [49]. Zero-inflated models have two distinct parts, one that describes the probability of obtaining a zero count, and one that describes the expected density given a non-zero count. Tests of significance took both simultaneously into account (i.e., hurdle models).

Fish biomass and size were compared between Rota and Saipan within each trophic category noted above: herbivores/detritivores, secondary consumers, and tertiary consumers. Standard comparative tests were used to examine differences in fish size and biomass between the two islands (t-tests, Mann-Whitnet U-test if normality assumptions were not met).

Site-level analyses

Deeper investigations of disturbance and recovery trajectories were conducted at the site-level, in order to match the spatial scale that localized stressors and key environmental regimes such as wave exposure operate at. A stepwise regression modeling process was performed to determine the likelihoods and magnitudes of influence for COTS, pertinent localized stressors, and wave exposure. These factors were considered based upon both initial island-scale analyses and previous studies describing their driving influence upon CNMI's reefs [18], with more details provided below to describe localized stressors. Stepwise modeling to understand the impacts during the disturbance time frame utilized coral cover as the primary, sensitive dependent variable. Modeling to understand net ecological change across the 12-year period used two dependent variables that better described coral reef 'condition' and relied less upon coral cover alone, which is known to differ across species and environments [18]. These were: 1) the net change in the benthic substrate ratio, and 2) net change in multivariate coral ordination scores (Figure 2). Benthic and coral

ordination metrics were weakly correlated (r<0.4, all comparisons). Coral ordination metrics were calculated based upon two-dimensional principle component ordination plots. Coral species abundance data were log-transformed to produce a Bray-Curtis similarity matrix that quantified multivariate similarities between all sites and years. Similarity matrices were projected in two-dimensional space using principle component ordination (PCO) plots that depicted multivariate differences using axes (i.e., eigenvectors) that hierarchically account for the variation in multi-species datasets (PCO) [50]. Corals that were the strongest drivers of multivariate dissimilarities were overlaid on the plot to indicate the nature of shifting assemblages over the years. The percent net change for coral assemblages represented the two-dimensional PCO-movement (i.e., vector magnitude) away from *Montipora, Acropora, Stylophora,* and *Pocillopora* assemblages to faviid and *Porites* assemblages, and the magnitude of return proportional to its starting position (Figure 2). These corals cumulatively accounted for the top 30% of the variation in the PCO plot. Percent change values were calculated because they are comparable across sites, whereas absolute values would not be. Second, we used mean values of the benthic substrate ratio defined above. Percent net change was again defined by differences between the post- and pre-disturbance period, divided by the initial pre-disturbance period value (Figure 2). Cumulatively, larger reductions in percent change values indicated a shift to reefs with non-calcifying substrates and smaller, tolerant coral assemblages.

A forward, stepwise regression modeling process was conducted to evaluate the likelihood and magnitude of independent factors in determining the observed biological changes. This process began by searching for factors with a ubiquitous influence upon biological change across all reeftypes. These included COTS in the disturbance timeframe models and wave exposure in the benthic net-change models. No ubiquitous factor was revealed for models describing the net change in coral ordinations, so only individual terms were used. Forward steps were only taken when/if interactive models improved the fit (R^2-values) and likelihood (AIC-values, described below) of predictions. Interactive combinations were considered because independent variables were all scaled in a consistent, low-to-high manner describing a gradient of weak to strong influences, and are known to have contextual effects that are not independent of one another [9,33]. If stepwise terms did not improve the model fit and likelihood they were dropped, not presented in the results, and not further considered. This process continued until all interaction terms were examined. We first examined regression models including sites from all reeftypes grouped together, and subsequently examined subsets of the most favorable reef settings for coral growth to determine if herbivory, water quality, or wave exposure might have context-dependent roles that differed in accordance with reeftype. Comparisons of the explanatory power and likelihood of independent variables were only made across models examining the same reeftypes. Our goal was not to suggest that a single "best-fit" model existed, but rather to highlight that several plausible models existed, and provide the details of each. The influence of each factor was assessed based upon: 1) the (added) model fit, 2) overall presence across the suite of models, and 3) the likelihood scores and influence of outliers.

Regression modeling was performed using R [51]. Independent variables were standardized to provide equal scaling, and a constant value was added to make all numbers positive. Positive numbers were required when log-transformations were needed to ensure residual normality. Models were examined for normality using Shapiro-Wilk tests, and ranked based upon their explanatory

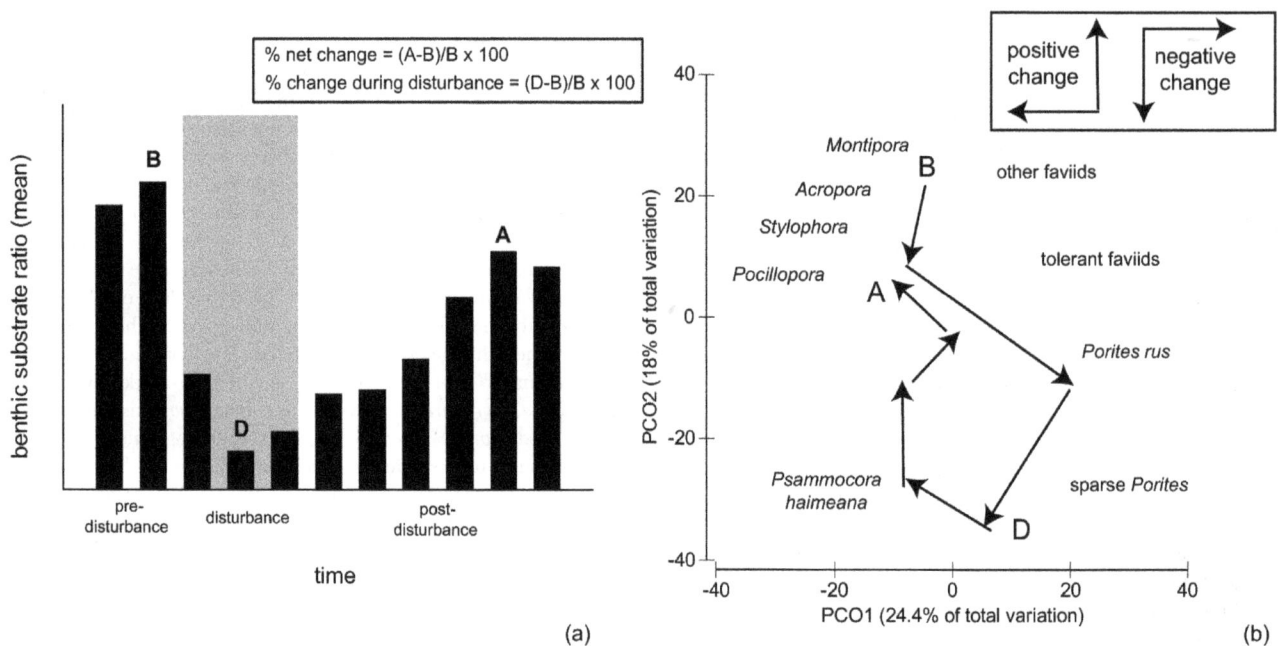

Figure 2. Methods used to calculate the percent decline and net change in (a) the benthic substrate ratio and (b) multivariate measures of the coral assemblages. B, D, A – before, during, and after the disturbance period, respectively. Percent declines were calculated by taking the difference between the minimum and maximum values during and before the disturbance period, and dividing by the pre-disturbance values. Net change values were calculated by taking the difference between maximum values before and after the disturbance period. Values for coral assemblages were calculated based upon their vector magnitude from the origin (0,0), with positive values given for PCO movement towards *Acropora*, *Montipora*, and *Pocillopora* assemblages, and negative values for PCO movement towards tolerant faviids and sparse *Porites*. See methods and Figure 6 for a better description of the corals depicted on the PCO plot.

power (R2-value) and likelihood as measured by Akaike's Information Criterion (AIC). Lower AIC scores indicated a better fit, based upon the least number of parameters and greatest residual normality that together maximize the probability of given outcomes based upon independent predictors. In all instances, the independent predictors used in interactive regression models were not highly correlated (r<0.30, P>0.05).

Independent variables were selected to represent factors that have a strong influence on CNMI's coral-reef assemblages and dictate disturbance and recovery dynamics in general. These were COTS densities (i.e., as a disturbance agent), wave exposure, the water quality proxy, mean fish size and biomass, and grazing urchin density (defined by the sum of *Echinothrix*, *Echinometra*, and *Diadema* urchins). When examining models dealing with net ecological change, we used mean herbivore/detritivore fish size and biomass. When examining models dealing with coral cover decline during the disturbance period, we used overall fish size and biomass. This was done in accordance with: 1) the literature describing herbivores as key ecosystem engineers facilitating net recovery dynamics [6,22], and 2) existing/emerging relationships between enhanced overall fish size/biomass and reduced COTS impact [12,30,52].

Last, the present study placed a reliance upon 2010/2011 fish data to predict disturbance and recovery trajectories over the past decade. We do not assume that fish populations were static across the disturbance period, as long-term studies across disturbance events elsewhere clearly define fish assemblage dynamics [53,54,55]. However, this approach does assume that site-level relationships were preserved, and relative differences with respect to environmental regimes and/or human exploitation maintained.

This was supported by several lines of evidence. Fishery-dependent studies conducted in the CNMI over the past 20 years that showed consistent trends in fish sizes across numerous geographic sectors in the CNMI based upon wave exposure and proximity to human population centers [56,57,58], despite evidence for an overall shift (i.e., decline) in several measures of the fishery resource. These studies also highlighted a shift towards increased herbivore dominance and smaller herbivores over the years, suggesting the relevance of herbivore assemblage metrics. In addition, a recent coral reef resilience assessment conducted for 35 sites around Saipan found that wave exposure and MPA-status together were highly correlated with a fishing pressure metric, and fish abundance trends in the present dataset, which were derived from an independent opinion survey taken by CNMI's resource managers [59]. Beyond CNMI, long-term studies show responses of fish assemblages to disturbances whereby a decline in overall fish diversity, and a decline in the abundance of coral-associated species are most often noted with coral loss [53,54,60]. In response to algal growth following disturbances, these studies also show increases in some acanthurids and/or scarids that can respond rapidly to algal substrate availability, with their abundances gradually decreasing to pre-disturbance levels after several, typically 4 to 5, years. Thus, the present regression models were based upon: 1) environmental gradients and management factors known to persist over longer time periods and more extensive spatial gradients than the present study, and 2) fish abundance datasets that were collected 4 to 5 years following the disturbance period when any pulses of herbivores were expected to diminish, and 3) utilized metrics of food-fish size and biomass that were not influenced by smaller, coral-associated species.

Results

High COTS densities were evident across the CNMI between 2003 and 2006 (Figure 3a, Table S2), concomitant with two tropical storms that passed by the study islands. Together, these findings formed the basis for determining 'before', 'during', and 'after' study periods. Transect-based densities of COTS were largest during the disturbance years on Saipan, however, increased densities were noted across all islands. Following disturbance years, COTS densities declined to pre-disturbance levels for Saipan and Rota, while one unique site on Tinian had an anomalous high density become emergent during 2011 (i.e., site 11 was driving the trend in Figure 3a for Tinian in 2011, Table S2). Cumulatively, COTS densities agreed with 2003 to 2006 as peak disturbance years, but also suggest that Rota and Saipan had similar temporal dynamics, with higher absolute abundances being found on Saipan.

Island comparisons

The disturbance period had a negative impact on coral cover throughout the CNMI, however rates of change were not uniform across the study islands. The largest decline was evident on Saipan, where high human populations and development existed, and smallest on Rota, where human presence and geological foundations for optimal reef growth were lower (Figure 3b). Coral cover declined from 34% to 23% on Saipan (32% decline, F-Statistic = 3.7, P = 0.05, repeat measures ANOVA, pairwise Tukey's post-hoc q-statistic = 3.6, P = 0.05), with no significant recovery since the disturbance years. Conversely, on Rota where coral cover is naturally lower than Saipan as a consequence of island geology, cover had a non-significant decrease from 11.5% to 9.7%, with recovery back to 12%. Yet, Rota coral cover trends were heavily influenced by a single locality where the coral *Porites rus* is abundant and dominant (Site 20 coral cover 30% whereas all other sites have 10%, Figure 1, Table S2). Re-analysis of the coral cover trends with this site omitted indicated that coral cover had a more substantial decline (8.5% to 4.6%) and recovery back to 7.7% (F-Statistic = 4.0, P = 0.06, repeat measures ANOVA). The impact of disturbances to coral-colony sizes and population densities were also markedly different (Figure 4). Mean coral colony-size on Saipan declined from 7.5 to 5.1cm (P<0.001, K-S test), with no change in the years after disturbance (Figure 3). Demographic changes were complemented by significant increases in coral population density during the disturbance years (F-Statistic = 5.96, P = 0.01, repeat measures ANOVA, pairwise Tukey's post-hoc q-statistic = 4.8, P<0.05), mainly due to small faviids and *Porites*. In contrast, mean colony size on Rota had a non-significant decline and recovery (4.3 to 3.7cm, followed by a recovery to 4.1cm). Population density comparisons were also non-significant.

Grazing urchin densities declined across the disturbance period on both islands, yet the magnitude of decline was much larger for Rota, where *Echinothrix* densities were reduced from over 6 individuals per 100 m^2 to less than 2 during the onset of disturbances (Figure 3c, z-statistics = 4.6 and 1.8, respectively, for the increased probability of obtaining zero counts and reduction in densities where non-zero counts existed, P<0.01 for both, repeat-measure, zero-inflated hurdle comparisons before and during). Urchin densities were initially lower on Saipan and had a gradual decrease across the study period that was most pronounced in the years following disturbance (z-statistics = 2.8 and 1.4, P<0.05, for greater zero counts and reduced densities, respectively, zero-inflated comparisons during and after), representing a timeframe when more abundant COTS existed compared to Rota (Fig-

ure 3a). Models indicated that declines were attributed to both higher probabilities of sites with no urchins being present (30% and 38% decline in significant modeled estimates, respectively for Rota and Saipan), and sites with urchins present at lower densities (48% and 58% decline, respectively). No significant differences in sea cucumber densities existed across the study periods.

Fish biomass and density were consistently greater for large-bodied primary and secondary consumers on Rota compared with Saipan (i.e., species that attain larger reproductive sizes, Figure 5). In contrast, smaller-bodied counterparts were higher in biomass and density on Saipan. These trends were most pronounced for large-bodied groupers, snappers, and parrotfish on Rota, and small-bodied acanthurids and parrotfish on Saipan (P<0.05, comparative t-tests and Mann-Whitney U-tests). The findings support that fewer, large-bodied fish with varying functional roles existed on Saipan, with more numerous, small-bodied species comprising a majority of the biomass.

Stepwise regression modeling at the site level

Focusing investigations at the site level revealed that disturbance and recovery cycles varied markedly, with differences attributable, in part, to varying reeftypes and islands that dictated site geomorphology (Figures 6 and 7, Table S3). Percent decline in coral cover during the disturbance period was primarily driven by the observed, transect-based COTS densities regardless of reeftype or island (Table 1). When considering all reeftypes grouped together, only 18% of the variance in coral decline was attributable to COTS (P = 0.06). Yet, the amount of variance explained increased to over 50% when examining the subset of reeftypes with the highest coral growth capacity (52% for interstitial and spur-and-groove reefs combined, and 67% for spur-and-groove reefs alone, P<0.01 for both). The addition of an interaction term describing the mean size of the fish assemblages enhanced the amount of variance accounted for (R^2 = 0.60 and 0.86, P = 0.009 and 0.005, for interstitial and spur-and-groove reefs combined, and spur-and-groove reefs alone, respectively), while wave exposure improve the fit for spur-and-groove reefs alone (R^2 = 0.83, P = 0.003, Table 1). Fish size had a negative association with COTS densities for coral-dominated reefs (r = 0.28, Pearson correlation coefficient, interstitial and spur-and-groove reefs combined), while an interactive term of *fish size ×wave exposure* had a stronger association with COTS (r = 0.48), highlighting the inter-dependence among these three factors. In sum, COTS were primary predictors of coral decline (i.e., the disturbance period), while COTS densities and coral impacts were diminished with higher wave exposure and larger fish assemblages, especially on reefs with a high capacity for coral growth.

The general response of coral assemblages to disturbance was a decline in *Acropora*, *Montipora*, *Stylophora*, and *Pocillopora* corals, with faviid and *Porites* assemblages becoming emergent, and differential rates of return in the post-disturbance years. Two-dimensional PCO-plots accounted for 40 to 50% of the variance in coral assemblage similarities, and provided a quantitative basis to assess disturbance and net change across individual sites (Figures 2 and 6). Similarly, photo-quadrat data revealed that benthic substrate ratios were sensitive to disturbance (i.e., the ratio of heavily calcifying corals and crustose coralline algae divided by less-calcifying turf, fleshy-coralline, and macroalgae, Figure 7), with differential levels of decline and net change.

In contrast to regression models describing spatial patterns in coral cover declines, COTS densities did not emerge as a significant predictor of net ecological change. This was the case when considering COTS densities during the disturbance time-

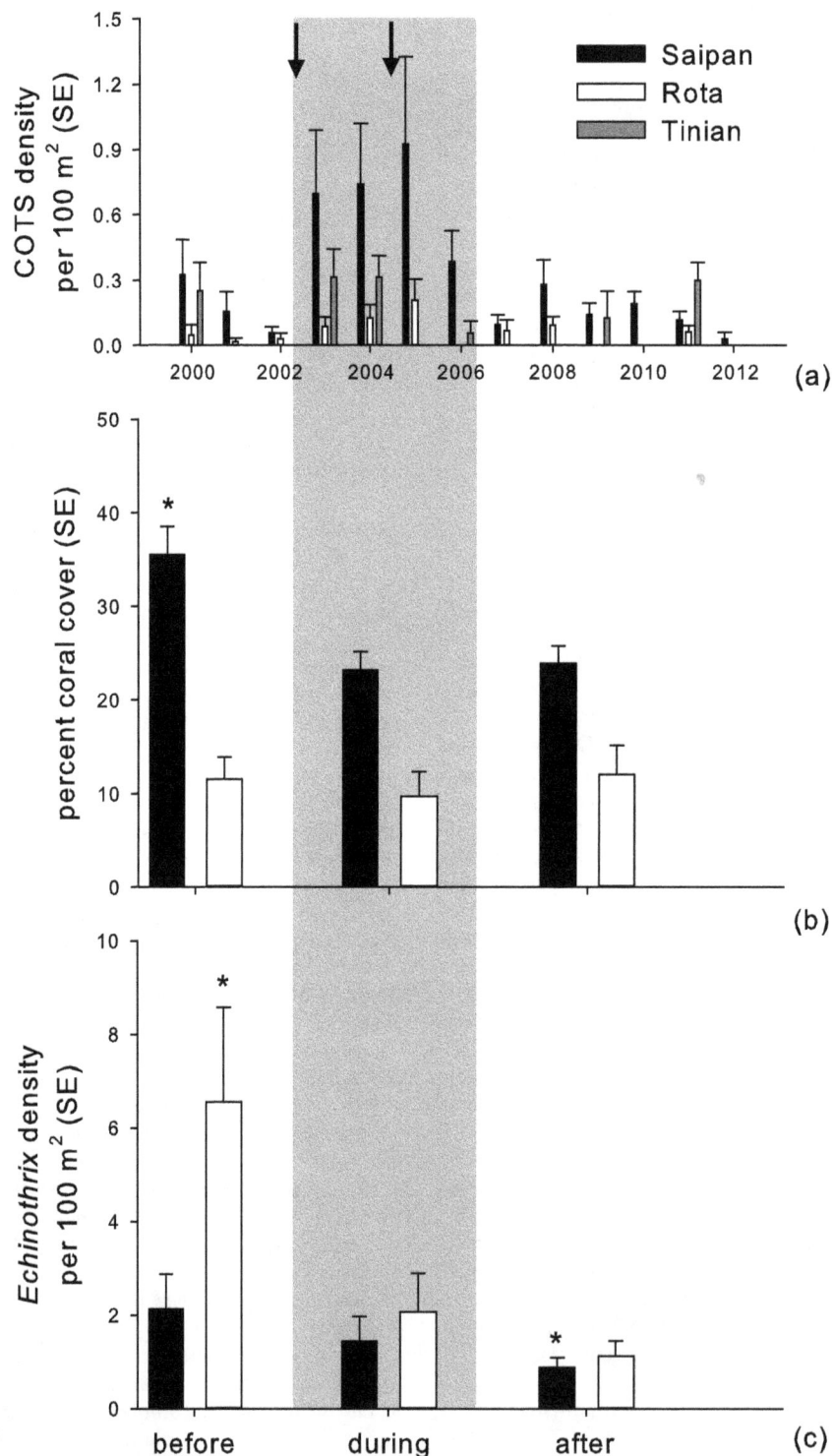

Figure 3. Density of Crown-of-Thorn starfish (COTS) based upon belt-transect data across Saipan, Tinian, and Rota from 2000 to 2012. (a). Densities represent island-based averages that diminish the highest and lowest observations in order to establish patterns across study years. Disturbance reduced coral cover on all islands, but recover trajectories, or the net rate of change, differed by island (b). Coral cover declined on Saipan with no significant recovery (*indicates P<0.05, repeat measures ANOVA and post-hoc tests), while a non-significant decline and recovery was noted on Rota. *Echinothrix* urchins also declined (c) in density during the disturbance period, with a further decline in the years after disturbance (*indicates P<0.05, zero-inflated hurdle models). Black arrows indicate tropical storms that passed by the study islands during the disturbance timeframe (grey rectangle box indicates the disturbance timeframe).

frame only, or integrated across the entire study period. Stepwise regression modeling for the net change in favorable benthic substrates reported that wave exposure was the strongest individual factor across all reeftypes, accounting for 20% of the

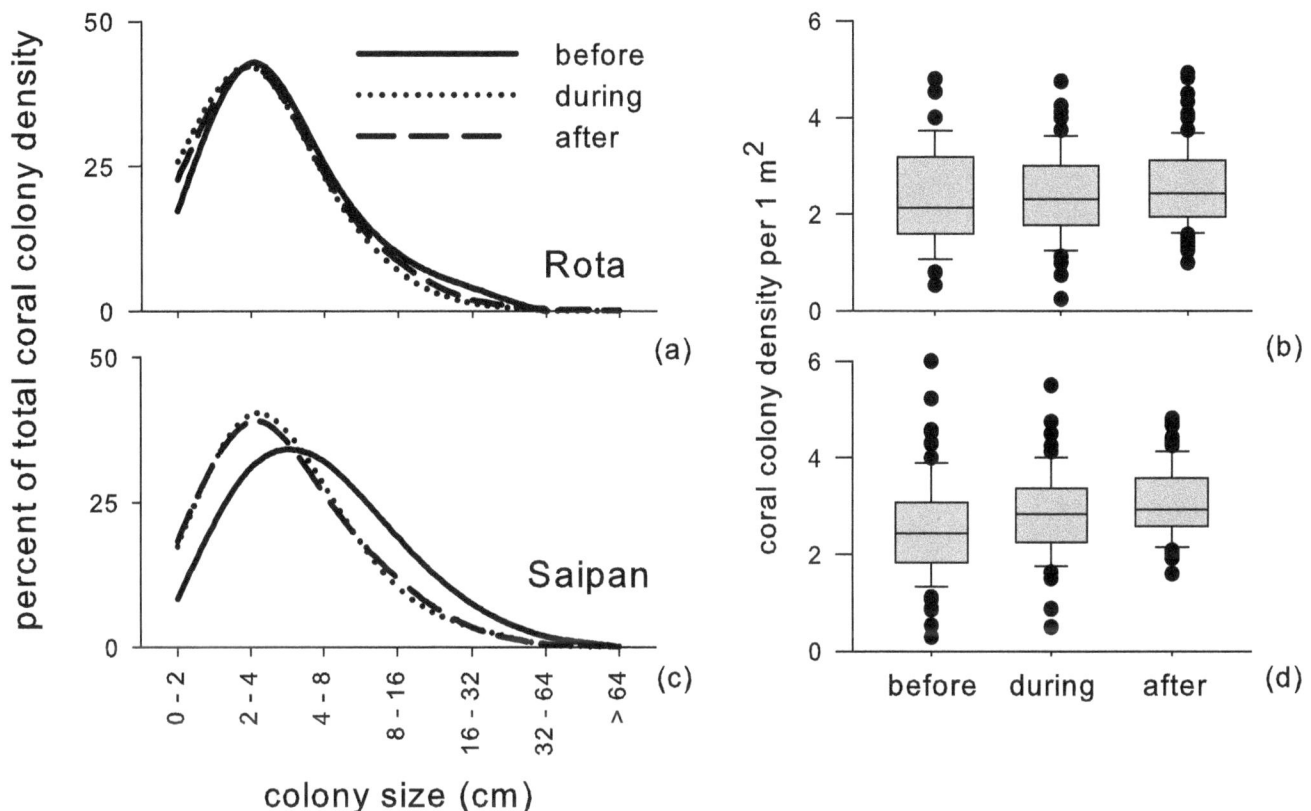

Figure 4. Dynamics of coral colony-size distributions and population densities across the disturbance periods. Rota had a non-significant decline and recovery in colony-size across the study periods (a), as well as a non-significant, sequential increase in population density (b). Reductions in colony-size were evident for Saipan (c) during the COTS period, accompanied by increases in population densities (d), attributed to the emergence of numerous small faviid and *Porites* corals (*indicates P<0.05, repeat measures ANOVA with post-hoc tests, see also Fig. 6).

variance (P = 0.03, Table 2). The influence of wave exposure grew when considering the subset of reefs with highest coral growth capacity ($R^2 = 0.22$, interstitial and spur-and-groove combined, $R^2 = 0.5$, spur-and-groove only). Yet, improved model fitting consistently required the inclusion of mean herbivore size and/or grazing urchin density that both increased the explanatory power and likelihood of resultant models. Explanatory power increased by 3%, 16%, and 14% when including herbivore size for all reeftypes, interstitial and spur-and-groove reefs combined, and spur-and-groove reefs alone, respectively (Table 2). Explanatory power increased by 0%, 20%, and 33% when including grazing urchin densities for these same reeftypes, respectively. Last, for coral-dominated reeftypes, including the interaction between herbivore size and grazing urchin density increased model fit by 33%. In all instances, interaction models lowered AIC scores suggesting their greater likelihoods.

No single factor consistently emerged as the primary driver of net change in the coral assemblages. Net change was predicted individually and not interactively by both herbivore size and the water quality proxy. When examining sites across all reeftypes, the water quality proxy had a slightly greater explanatory power ($R^2 = 0.28$ versus 0.38, herbivore size and water quality proxy, respectively), however the AIC-based likelihood was lower for water quality due to the relatively strong influence of 1 to 2 sites that diminished the normality of residuals (i.e., residuals still met the requirements of normality, but less so as compared with the herbivore size model, reducing the AIC score). The only other notable models emerged when examining spur-and-groove reefs, whereby herbivore size ($R^2 = 0.44$, AIC = 18.9) was a slightly

better predictor of net coral assemblage change as compared to the water quality proxy ($R^2 = 0.39$, AIC = 27.1, Table 1).

Discussion

Significant coral loss occurred in the CNMI between 2003 and 2006 concurrent with high COTS densities and several typhoons that passed through CNMI. Given that tropical storm paths during these years were consistently in closer proximity to Rota, passing between 60 and 100 km from Saipan, we purport that COTS activity was the primary driver of disturbance that simultaneously impacted the study islands during the mid-2000's. In support, COTS densities were ubiquitous in models explaining the spatial patterns in coral decline during the disturbance years. However, potential synergies and/or linkages between storm activity and COTS were not approached, and deeper mechanisms may have existed.

While starfish densities were similarly elevated during the disturbance years for two islands that differed in human presence and geology, Rota and Saipan, abundances were highest and most persistent on Saipan. These findings help to explain the greater impacts to coral decline on Saipan. Yet, COTS abundances provided no significant explanation of net recovery patterns to the favorable benthic substrates or coral assemblage dynamics across the entire 12 year study period. We hypothesized the island-scale differences in resistance and recovery were attributable to a suite of factors. First, due to varying island geomorphology, Rota naturally had less coral to begin with (i.e., less prey), including a reduction in preferential prey, *Acropora* and *Montipora* [18]. This

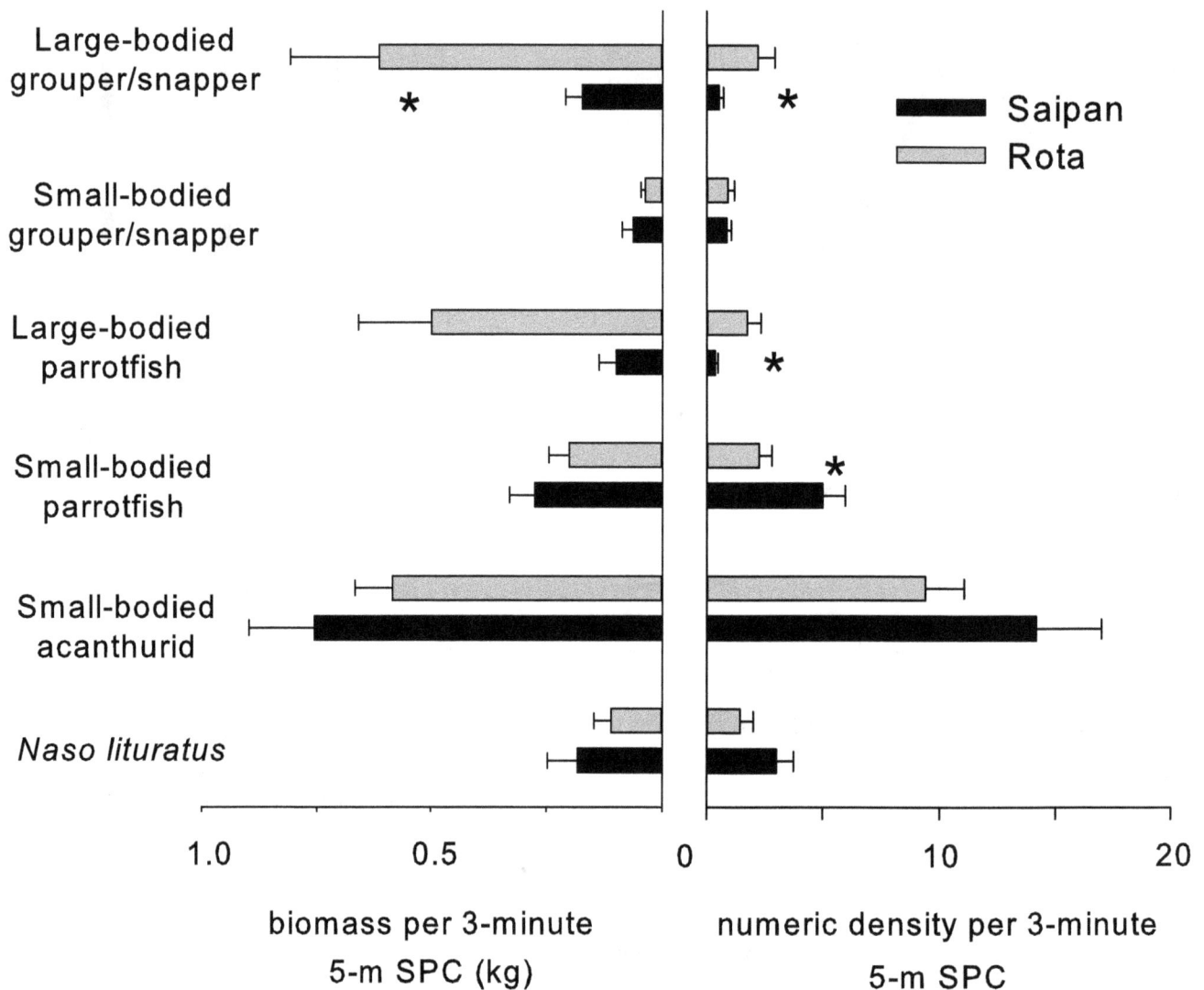

Figure 5. Comparisons of fish biomass and numeric density for several functional fish groups on Saipan and Rota. (*indicates P<0.05, comparative tests described in methods).

situation seems most relevant when interpreting why COTS persistence and coral impacts were diminished on Rota. Second, numerous studies continue to support that fish assemblages comprised of larger individuals across all trophic levels are associated with reduced impacts from COTS disturbances [12,30,51], and may help to explain COTS persistence and coral recovery dynamics [10,22]. Larger biomass and body-size of fish assemblages on Rota compared with Saipan supported this notion. Third, nutrient enrichment from watershed runoff is known to contribute to persistent, localized COTS populations [61,62]. CNMI water quality reports have consistently found better water quality on Rota over the past decade, with a 50% reduction in bacteria violations on Rota compared with Saipan in 2012 [63]. When considering where persistent populations existed (site 16 on Rota, sites 5, 6, 7, 8, and 10 on Saipan all had densities of 0.2 individuals per 100 m^2 across several post-disturbance years, Table S2), support for these combined hypotheses grows. Sites 7, 16, and 10 were associated with interstitial framework reefs, posited to have high connections with the karst aquifers compared to others [18] (Figure 1). These sites, in addition to others (5, 6, and 8), all had the smallest overall fish sizes. We synthesize that

inherent geological difference as well as localized stressors were influential in describing disturbance dynamics across Rota and Saipan, and utilized site-based analyses to better approach the individual and interactive roles of a suite of factors.

Site-level drivers of change on CNMI's reefs

Wave exposure has long been considered to shape modern coral assemblages and reef growth through time in CNMI [64], whereby full exposure to prevailing northeast trade winds has selected against geological reef development through time (i.e., reeftype 4 noted in the methods). Yet, beside the incipient reef development that exists along much of CNMI's eastern shoreline, significant variation in wave exposure remained among the subset of reeftypes where high coral growth capacity existed, despite having a lower overall magnitude. In fact, this secondary gradient in wave exposure (i.e, low to moderate levels) was the most influential, positive determinant of net benthic substrate change in all reef settings. This may be a result of greater flushing, nutrient transfer rates, and/or the removal of detrital build-up with wave energy [65]. After accounting for wave exposure, the process of grazing, as represented by herbivore/detritivore size and grazing

Table 1. Stepwise regression models predicting coral decline.

Independent variables	Slope	SE	Intercept	R^2	P-Value	AIC
All reeftypes (n = 16)						
COTS^{-1}	**6.57**	**3.20**	**0.68**	**0.18**	**0.06**	**46.2**
Interstitial and spur-and-groove reefs (n = 12)						
COTS^{-1}	7.84	2.18	0.47	0.52	0.005	25.7
COTS^{-1} × fish_size	1.55	0.43	1.62	0.60	0.009	18.4
Spur-and-groove reefs (n = 7)						
COTS^{-1}	8.88	2.43	0.23	0.67	0.01	16.8
COTS^{-1} × log(exposure)	8.21	1.50	0.44	0.83	0.003	12.2
COTS^{-1} × fish_size	2.17	0.38	1.30	0.86	0.005	8.6

Summary of forward, stepwise regression models that examined the drivers of coral decline during disturbance years. Methods describe the suite of independent variables examined and the basis for their selection. Significant independent variables presented below include mean *Acanthaster planci* densities during disturbance years (COTS), wave exposure, and mean fish size for all trophic groups combined. COTS densities were inversely scaled for consistency with other localized stressors (i.e., low-bad/high-good). AIC-scores were used to indicate the relative likelihood of models being able to predict outcomes, and are only comparable within each reeftype grouping.

urchin densities, was the strongest and most reliable predictor of favorable benthic substrates based upon their presence across the suite of models, the added variance accounted for, and the improved model likelihood scores (AIC values). These findings were amplified when stratifying by reeftype, and including only reefs with the most favorable geological foundations for coral growth. Water quality emerged as a significant predictor of change

for coral assemblages ordinations along with mean herbivore size, however, we purport contextual roles of water quality that are furthered below.

Synthesizing the findings reveals that localized stressors were most influential to reefs with low to moderate wave exposure, and high inherent capacity for coral growth. Given the distribution of geological reef settings in the CNMI (Figure 1), this means that

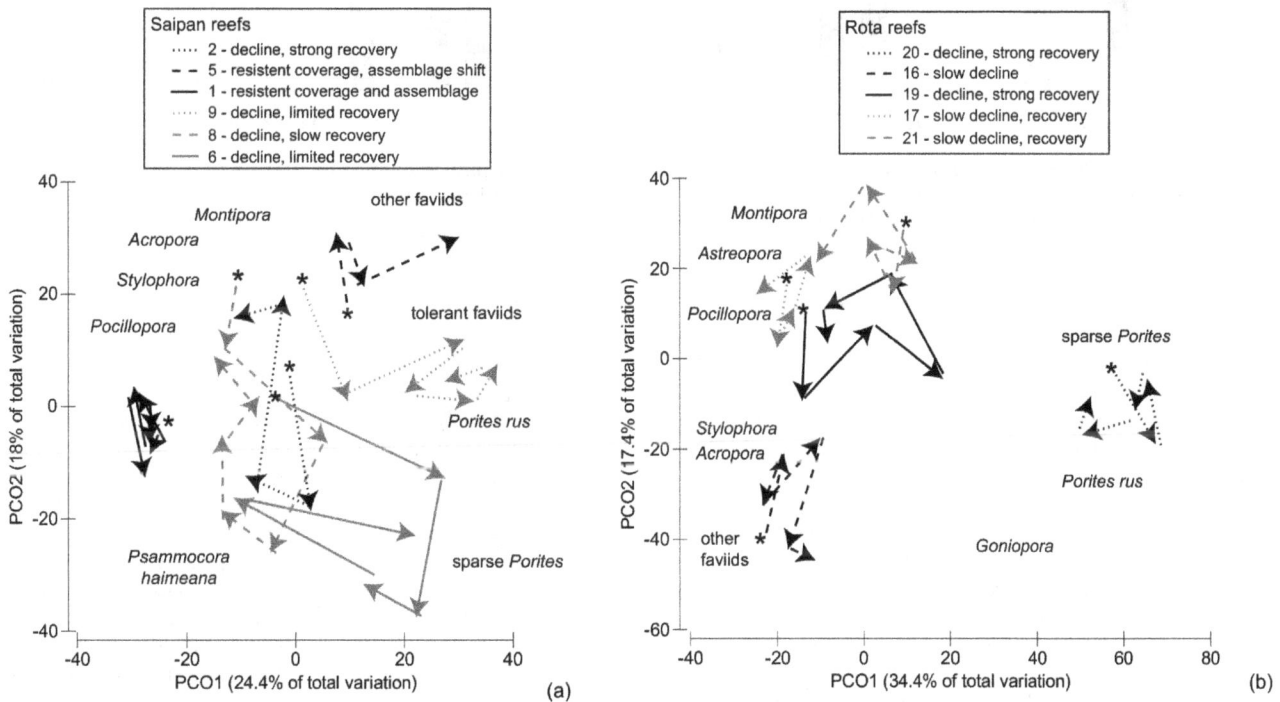

Figure 6. Principle components ordination of coral assemblages for six representative monitoring sites around Saipan (a) and Rota (b). See Figure 1 for site identification and Table S3 for summary statistics. Pre-disturbance assemblages are indicated with an asterisk (*), while vectors depict directional change through time. Sparse *Porites* refers to a dominance of *P. lichen*, *P. vaughani*, and small colonies of other massive species. Tolerant faviids consisted of *Leptastrea purpurea*, *Goniastrea retiformis*, *G. edwardsi*, *Favia matthaii*, *F. pallida*, and *F. favus*. Other faviids consisted of *Favia stelligera*, *Platygyra* spp., *Cyphastrea* spp., and *Favites abdita*.

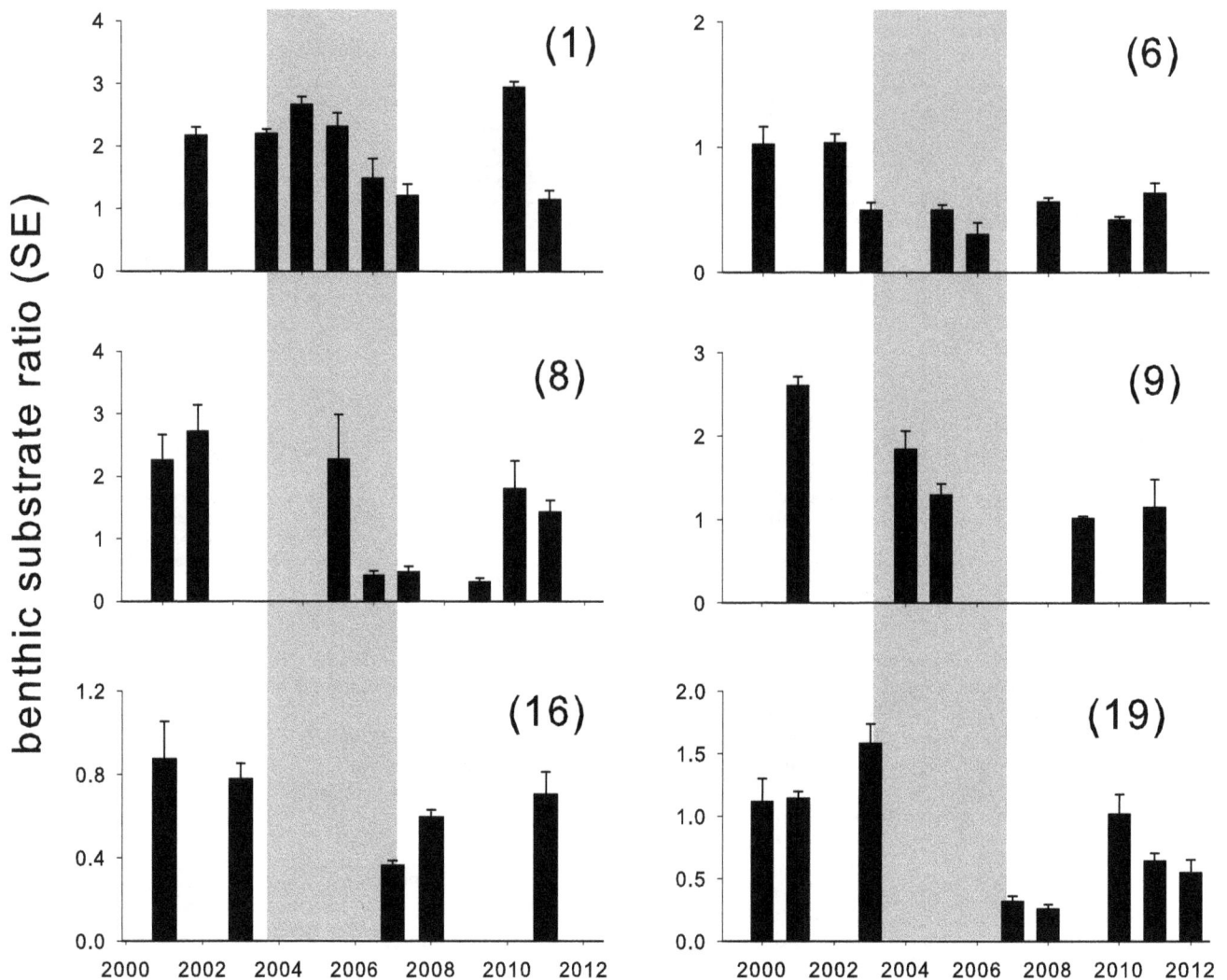

Figure 7. Benthic substrate ratio dynamics for representative monitoring sites around Saipan (sites 1, 9, 8, and 6) and Rota (sites 19 and 16). See Figure 1 for site identification and Table S3 for trends from all sites. Grey bars indicate the disturbance period. Benthic substrate ratios indicate the proportion of heavily-calcifying versus less-or-non-calcifying substrates (*see methods*).

Saipan reefs were weighted disproportionally within the subset of regression models associated with localized stressors, representing 63% of the sites with favorable geological settings (i.e., interstitial or spur-and-groove reefs). Hence, localized stressors were most influential for Saipan, the most populated island where reef-based tourism is centralized, and constitutes a key component of the economy [66].

The present results further an interesting and emerging association between high COTS impacts, smaller-bodied fish assemblages, and low wave exposure. While mechanisms remain unclear and of interest, similar patterns describing diminished COTS impacts with higher fish abundances in successful, no-take marine protected areas have been observed elsewhere [12,30,51]. Fish sizes (i.e., herbivores) were also influential to net change metrics, and the collective findings pertaining to the mean sizes of the fish assemblages rather than their biomass resonated well with power laws that describe relationships between body-size, physiology, and function in ecology [67]. Specific to the present study, power laws have been shown to govern numerous physiological traits such as grazing efficiency [68] and reproductive potential [69], whereby a doubling in fish size equates to an exponential

increase in function. Thus, even if similar biomass exists, fish assemblages comprised of mainly small-bodied species are expected to have a reduced ecological function within coral-reef food webs [70,71]. Rasher et al. [72] described that a subset of larger, functionally-dissimilar herbivores play a disproportional role in macroalgal grazing (key constituents include *Chlorurus* spp., *Siganus* spp., and two *Naso* spp., *N. lituratus* and *N. unicornis*), while a suite of other species were more reliant upon generalized detritus and turf grazing from the reef substrate. Species within these functional groups represent highly desirable food fish in CNMI. Market studies have revealed their declining sizes and abundance over the past two decades [55,56,57], and also reported a disproportionally small reef-area-per-person, and mean fish size-at-capture, compared to other Micronesia jurisdictions [57]. We reconcile that both size and functional diversity appear to be key attributes of CNMI herbivore assemblages that are sensitive to harvesting pressure, and influential to coral reef recovery patterns, making their improved management desirable for reef futures.

The reduction in *Echinothrix* urchin densities concomitant with COTS disturbances was a novel association to our knowledge.

Table 2. Stepwise regression models predicting net ecological change.

Benthic substrate ratio net change

All reeftypes (n = 18)

Independent variables	Slope	SE	Intercept	R^2	P-Value	AIC
log(exposure)	3.81	1.64	−0.76	0.20	0.03	50.8
log(exposure) × herb_size	0.62	0.27	1.33	0.23	0.05	44.6

Interstitial and spur-and-groove reefs (n = 12)

log(exposure)	5.46	2.71	−2.26	0.22	0.07	35.3
log(exposure) × urchin	0.84	0.28	0.74	0.42	0.01	31.7
log(exposure) × herb_size	0.87	0.36	0.85	0.38	0.04	27.4
log(exposure) × herb_size × urchin	0.19	0.06	1.58	0.57	0.01	24.2

Spur-and-groove reefs (n = 7)

log(exposure)	9.23	3.50	−6.06	0.50	0.05	21.8
log(exposure) × herb_size	1.13	0.35	0.40	0.64	0.03	17.6
log(exposure) × urchin	1.11	0.20	0.08	0.83	0.003	14.4

Coral assemblage net change

All reeftypes (n = 12)

herb_size	0.66	0.32	1.37	0.28	0.08	25.0
poll_proxy^{-1}	3.89	1.4	7.14	0.38	0.02	32.2

Spur-and-groove reefs (n = 7)

poll_proxy^{-1}	4.29	1.71	7.5	0.39	0.04	27.2
herb_size	0.93	0.41	0.64	0.44	0.08	18.9

Summary of forward, stepwise regression models that examined the drivers of net change in the benthic substrate ratio and coral ordination scores across the study period. Methods describe the suite of predictor variables examined and the basis for their selection. Significant independent variables presented below include wave exposure, mean herbivore/detritivore size, mean grazing urchin density, and the water quality proxy. The water quality proxy was inversely scaled (i.e., low-bad/high-good) for consistence with other localized stressors. AIC-scores were used to indicate the relative likelihood of models being able to predict outcomes, and are only comparable within each reeftype grouping.

Field observations and photographs provided anecdotal evidence of competition for refuge within the reef matrix, but clearly these relationships remain speculative. In the event that disturbances not only diminished coral cover but structural complexity as well, (i.e., fewer large *Acropora*, *Pocillopora*, *Stylophora* colonies), a decrease in urchin densities was the anticipated ecological response following disturbance [73]. Yet, urchin declines were concomitant with the onset of disturbance, and continued to decline throughout the study period. Given these trends, deeper investigations into the cause(s) of urchin declines seem warranted.

Last, the water quality proxy was an influential driver of net coral assemblage dynamics across the 12 year study period as well, despite its diminished presence across the suite of regression models (i.e., diminished presence in Tables 1 and 2). These findings support previous relationships between water quality proxies and coral species richness in CNMI [18,20], and reinforce that coral species composition may be a sensitive metric of water quality. It is beyond the purview of this study to formally discuss the linkages between diversity and ecosystem function, however, diversity is well known to facilitate functional redundancy in ecological systems, thereby providing for enhanced resistance and recovery to disturbance. Pollution contribution appeared to be influential to reefs where high human development existed,

however, extensive human development was not common within the majority of CNMI's coastal watersheds. In addition, anchor points in the water quality regression models were driven by the presence of interstitial reefs associated with higher groundwater connectivity (Figure 1, Table 2, and Table S3). We conclude that watershed restoration strategies aimed at improving reef condition might obviously focus upon the largest urban centers, but less obviously, focus upon karst watersheds adjacent to high-value reef assemblages. We note that a major watershed restoration project addressing Laolao Bay (eastern Saipan) remains ongoing, and if successful would serve to address one of the key anchor points.

Conclusions

Over the past 12 years in CNMI, a period of high COTS densities led to significant coral declines. Yet, the failure of some reefs to recovery was best attributed to localized stressors, which transformed the substrates opened up by coral loss into persistent stands of turf and macroalgae, less conducive for coral replenishment and recovery. Declining trends were strongest for reefs that have favorable geomorphology (i.e., a gently sloping reef foundation), which disproportionally occur on Saipan alongside lower wave exposure. These same reefs represent centers for reef-based tourism that constitutes a major part of CNMI's economy, highlighting a need to improve upon compromised fish assemblages, grazing urchin populations, and specific localities where water quality concerns exist.

Supporting Information

Table S1 Monitoring site frequencies. Monitoring frequency for each of the long-term sites incorporated into the present study (*see Fig. 1*). Lowercase letters indicate the type of survey conducted in each year: (b) benthic substrate, (i) macroinvertebrate, (c) coral, and (f) fish.

Table S2 Site-based coral coverage and *Acanthaster* density data. Coral coverage and *Acanthaster planci* density summary statistics for each of the long-term monitoring sites incorporated into the present study. Site-based data formed the basis for regression modeling (Table 1). Reeftypes follow: "sg" - optimal spur-and-groove structures, "int" - high-relief, interstitial framework, "rot" - low relief Holocene framework found on Rota only, and "pl" - incipient coral assemblages residing upon a Pleistocene basement (*see methods*).

Table S3 Site-based summary statistics for regression models. Summary statistics for each of the long-term monitoring sites incorporated into the present study. Site-based data formed the basis for regression modeling. Dependent variables included the net change in the benthic substrate ratio and coral assemblages, noted as the sum of the percent decline (−) and subsequent recovery (+) of these ecological metrics (*see methods*). Reeftypes follow: "sg" - optimal spur-and-groove structures, "int" - high-relief, interstitial framework, "rot" - low relief Holocene framework found on Rota only, and "pl" - incipient coral assemblages residing upon a Pleistocene basement (*see methods*). (DOC)

Acknowledgments

We are grateful for continuous discussions and support from several local and federal resource management agencies, including the United States Environmental Protection Agency Region IX, National Oceanic and Atmospheric Administration Coral Reef Conservation Program, and the Commonwealth of the Northern Mariana Islands (CNMI) Bureau of Environmental and Coastal Quality. We especially thank CNMI agency directors Frank Rabauliman and Frances Castro for logistical support throughout our study. Finally, several reviewers and the topic editor provided constructive criticism that greatly benefitted our manuscript.

Author Contributions

Conceived and designed the experiments: PH. Performed the experiments: PH DB JI SJ RO. Analyzed the data: PH DB JI SJ RO. Contributed reagents/materials/analysis tools: PH. Wrote the paper: PH DB JI SJ RO.

References

1. Hughes TP, Rodrigues MJ, Bellwood DR, Ceccarelli D, Hoegh-Guldberg O, et al. (2007) Phase shifts, herbivory, and the resilience of coral reefs to climate change. Curr Biol 17: 360–365.
2. Baker AC, Glynn PW, Riegl B (2008) Climate change and coral reef bleaching: An ecological assessment of long-term impacts, recovery trends and future outlook. Estuar Coastal Shelf S 80: 435–471.
3. Graham NAJ, Nash KL, Kool JT (2011) Coral reef recovery dynamics in a changing world. Coral Reefs 30: 283–294.
4. Hughes TP, Graham NAJ, Jackson JBC, Mumby PJ, Steneck RS (2010) Rising to the challenge of sustaining coral reef resilience. Trends Ecol Evol 25: 633–642.
5. Fabricius K, De'ath G, McCook L, Turak E, Williams DM (2005) Changes in algal, coral and fish assemblages along water quality gradients on the inshore Great Barrier Reef. Mar Pollut Bull 51: 384–398.
6. Mumby PJ, Dahlgren CP, Harborne AR, Kappel CV, Micheli F, et al. (2006) Fishing, trophic cascades, and the process of grazing on coral reefs. Science 311: 98–101.
7. Costa OS, Leao Z, Nimmo M, Attrill MJ (2000) Nutrification impacts on coral reefs from northern Bahia, Brazil. Hydrobiologia 440: 307–315.
8. Lapointe BE, Barile PJ, Matzie WR (2004) Anthropogenic nutrient enrichment of seagrass and coral reef communities in the Lower Florida Keys: discrimination of local versus regional nitrogen sources. J Exp Mar Biol Ecol 308: 23–58.
9. Houk P, Musburger C, Wiles P (2010) Water Quality and Herbivory Interactively Drive Coral-Reef Recovery Patterns in American Samoa. Plos One 5: e13913.
10. Mumby PJ, Harborne AR (2010) Marine Reserves Enhance the Recovery of Corals on Caribbean Reefs. Plos One 5: e8657.
11. Walsh SM (2010) Ecosystem-scale effects of nutrients and fishing on coral reefs. J Mar Biol doi:10.1155/2011/187248.
12. Dulvy NK, Freckleton RP, Polunin NV (2004) Coral reef cascades and the indirect effects of predator removal by exploitation. Ecol Lett 7: 410–416.
13. Aronson RB, Precht WF (2006) Conservation, precaution, and Caribbean reefs. Coral Reefs 25: 441–450.
14. De'ath G, Fabricius K (2010) Water quality as a regional driver of coral biodiversity and macroalgae on the Great Barrier Reef. Ecol App 20: 840–850.
15. Connell JH, Hughes TP, Wallace CC (1997) A 30-year study of coral abundance, recruitment, and disturbance at several scales in space and time. Ecol Monogr 67: 461–488.
16. Gardner TA, Côté IM, Gill JA, Grant A, Watkinson AR (2005) Hurricanes and Caribbean coral reefs: impacts, recovery patterns, and role in long-term decline. Ecology 86: 174–184.
17. Done TJ (1999) Coral community adaptability to environmental change at the scales of regions, reefs and reef zones. Am Zool 39: 66–79.
18. Houk P, van Woesik R (2010) Coral assemblages and reef growth in the Commonwealth of the Northern Mariana Islands (Western Pacific Ocean). Mar Ecol 31: 318–329.
19. Chollett I, Mumby PJ, Müller-Karger FE, Hu C (2012) Physical environments of the Caribbean Sea. Lim Oceanogr 57: 1233.
20. Cooper TF, Gilmour JP, Fabricius KE (2009) Bioindicators of changes in water quality on coral reefs: review and recommendations for monitoring programmes. Coral Reefs 28: 589–606.
21. Perry CT, Murphy GN, Kench PS, Smithers SG, Edinger EN, et al. (2013) Caribbean-wide decline in carbonate production threatens coral reef growth. Nat Commun 4: 1402.
22. McClanahan TR (2008) Response of the coral reef benthos and herbivory to fishery closure management and the 1998 ENSO disturbance. Oecologia 155: 169–177.
23. Bak RPM, Meesters EH (1998) Coral population structure: the hidden information of colony size-frequency distributions. Mar Ecol Prog Ser 162: 301–306.
24. Bak RPM, Meesters EH (1999) Population structure as a response of coral communities to global change. Am Zool 39: 56–65.
25. Loya Y, Sakai K, Yamazato K, Nakano Y, Sambali H, et al. (2001) Coral bleaching: the winners and the losers. Ecol Lett 4: 122–131.
26. Brown BE, Clarke KR, Warwick RM (2002) Serial patterns of biodiversity change in corals across shallow reef flats in Ko Phuket, Thailand, due to the effects of local (sedimentation) and regional (climatic) perturbations. Mar Biol 141: 21–29.
27. Pratchett MS, Trapon M, Berumen ML, Chong-Seng K (2011) Recent disturbances augment community shifts in coral assemblages in Moorea, French Polynesia. Coral Reefs 30: 183–193.
28. van Woesik R, Sakai K, Ganase A, Loya Y (2011) Revisiting the winners and the losers a decade after coral bleaching. Mar Ecol Prog Ser 434: 67–76.
29. Guest JR, Baird AH, Maynard JA, Muttaqin E, Edwards AJ, et al. (2012) Contrasting Patterns of Coral Bleaching Susceptibility in 2010 Suggest an Adaptive Response to Thermal Stress. Plos One 7: e33353.
30. Houk P, Benavente D, Fread V (2012a) Characterization and evaluation of coral reefs around Yap Proper, Federated States of Micronesia. Biodivers Conserv 21: 2045–2059.
31. Blackwood JC, Hastings A, Mumby PJ (2011) A model-based approach to determine the long-term effects of multiple interacting stressors on coral reefs. Ecol App 21: 2722–2733.
32. McClanahan TR, Maina J, Starger CJ, Herron-Perez P, Dusek E (2005) Detriments to post-bleaching recovery of corals. Coral Reefs 24: 230–246.
33. Burkepile DE, Hay ME (2006) Herbivore vs. nutrient control of marine primary producers: Context-dependent effects. Ecology 87: 3128–3139.
34. Banse K (2007) Do we live in a largely top-down regulated world? J Bioscience 32: 791–796.

35. Mork E, Sjoo GL, Kautsky N, McClanahan TR (2009) Top-down and bottom-up regulation of macroalgal community structure on a Kenyan reef. Estuar Coastal Shelf S 84: 331–336.

36. Sjoo GL, Mork E, Andersson S, Melander I (2011) Differences in top-down and bottom-up regulation of macroalgal communities between a reef crest and back reef habitat in Zanzibar. Estuar Coastal Shelf S 91: 511–518.

37. Wilson SK, Graham NAJ, Fisher R, Robinson J, Nash K, et al. (2012) Effect of Macroalgal Expansion and Marine Protected Areas on Coral Recovery Following a Climatic Disturbance. Conserv Biol 26: 995–1004.

38. Houk P, Van Woesik R (2006) Coral reef benthic video surveys facilitate long-term monitoring in the commonwealth of the Northern Mariana Islands: Toward an optimal sampling strategy. Pac Sci 60: 177–189.

39. Starmer J, Houk P (2008) Marine and Water Quality Monitoring Plan for the Commonwealth of the Northern Mariana Islands. CNMI Division of Environmental Quality, Saipan, MP.

40. Houk P, Bograd S, van Woesik R (2007) The transition zone chlorophyll front can trigger Acanthaster planci outbreaks in the Pacific Ocean: Historical confirmation. J Oceanogr 63: 149–154.

41. Keats DW, Chamberlain YM, Baba M (1997) Pneophyllum conicum (Dawson) comb nov (Rhodophyta, Corallinaceae), a widespread Indo-Pacific non-geniculate coralline alga that overgrows and kills live coral. Bot Mar 40: 263–279.

42. Antonius A (1999) Metapeyssonnelia corallepida, a new coral-killing red alga on Caribbean Reefs. Coral Reefs 18: 301–301.

43. Antonius A (2001) Pneophyllum conicum, a coralline red alga causing coral reef-death in Mauritius. Coral Reefs 19: 418–418.

44. O'Leary JK, Potts DC, Braga JC, McClanahan TR (2012) Indirect consequences of fishing: reduction of coralline algae suppresses juvenile coral abundance. Coral Reefs 31: 547–559.

45. Veron JEN (2000) Corals of the World. Stafford-Smith, Townsville.

46. Bohnsack JA, Bannerot SP (1986) A stationary visual census technique for quantatively assessing community structure of coral reef fishes. National Oceanic and Atmospheric Administration Technical Report NMFS 41.

47. Ekebom J, Laihonen P, Suominen T (2003) A GIS-based step-wise procedure for assessing physical exposure in fragmented archipelagos. Estuar Coastal Shelf S 57: 887–898.

48. Zar J (1999) Biostatistics. Prentice Hall, New Jersey.

49. Zuur AF, Ieno EN, Walker NJ, Saveliev AA, Smith GM (2009) Mixed effects models and extensions in ecology with R. Springer, New York.

50. Anderson M, Gorley R, Clarke K (2008) PERMANOVA+ for PRIMER: Guide to software and statistical methods. PRIMER-E, Plymouth, UK.

51. R Development Core Team RDC (2008) R: A language and environment for statistical computing. R foundation for statistical computing, Vienna, Austria. Available: http://wwwR-projectorg.

52. McCook LJ, Ayling T, Cappo M, Choat JH, Evans RD, et al. (2010) Adaptive management of the Great Barrier Reef: A globally significant demonstration of the benefits of networks of marine reserves. Proc Nat Acad Sci USA 107: 18278–18285.

53. Gilmour JP, Smith LD, Heyward AJ, Baird AH, Pratchett MS (2013) Recovery of an isolated coral reef system following severe disturbance. Science 340: 69–71.

54. Adams TC, Schmitt RJ, Holbrook SJ, Brooks AJ, Edmunds PJ, et al. (2011) Herbivory, connectivity, and ecosystem resilience: response of a coral reef to a large-scale perturbation. PLoS ONE 5: e23717.

55. Glynn PW, Enochs IC, Afflerbach JA, Brandtneris VW, Serafy JE (2014) Eastern Pacific reef fish responses to coral recovery following El Nino disturbances. Mar Ecol Prog Ser 495: 233–247.

56. Graham T (1994) Biological analysis of the nearshore reef fish fishery of Saipan and Tinian. CNMI Division of Fish and Wildlife Technical Report 94–02, Saipan, MP.

57. Trianni MS (1998) Summary and Further Analysis of the Nearshore Reef Fishery of the Northern Mariana Islands. Technical report submitted to the CNMI Division of Fish and Wildlife, Saipan, MP.

58. Houk P, Rhodes K, Cuetos-Bueno J, Lindfield S, Fread V, et al. (2012b) Commercial coral-reef fisheries across Micronesia: A need for improving management. Coral Reefs 31: 13–26.

59. Maynard J, McKagan S, Johnson S, Houk P, Ahmadia G, et al. (2012) Coral reef resilience to climate change in Saipan, CNMI; field-based assessments and implications for vulnerability and future management. Technical report submitted to the CNMI Division of Environmental Quality, Saipan, MP.

60. Halford A, Cheal AJ, Ryan D, Williams DM (2004) Resilience to large-scale disturbance in coral and fish assemblages on the Great Barrier Reef. 85: 1892–1905.

61. Brodie J, Fabricius K, De'ath G, Okaji K (2005) Are increased nutrient inputs responsible for more outbreaks of crown-of-thorns starfish? An appraisal of the evidence. Mar Pollut Bull 51: 266–278.

62. Fabricius K, Okaji K, De'ath G (2010) Three lines of evidence to link outbreaks of the crown-of-thorns seastar Acanthaster planci to the release of larval food limitation. Coral Reefs 29: 593–605.

63. Bearden C, Chambers D, Okano R (2012) Commonwealth of the Northern Mariana Islands Integrated 305(b) and 303(d) Water Quality Assessment Report. CNMI Division of Environmental Quality Saipan, MP.

64. Cloud PE (1959) Geology of Saipan, Mariana Islands, Part 4. Submarine topography and shoal-water ecology. Geological Survey Professional Paper 280-K, Washington, DC.

65. Crossman DJ, Choat JH, Clements KD, Hardy T, McConochie J (2001) Detritus as food for grazing fishes on coral reefs. Limnol Oceanogr 46: 1596–1605.

66. van Beukering P, Haider W, Wolfs E, Liu Y, van der Leeuw K, et al. (2006) The economic value of the coral reefs of Saipan, Commonwealth of the Northern Mariana Islands. Report prepared by Cesar Environmental Consulting for the National Oceanic and Atmospheric Administration, Silver Springs, MD.

67. Peters RH (1983) The ecological implications of body size. Cambridge University Press.

68. Lokrantz J, Nyström M, Thyresson M, Johansson C (2008) The non-linear relationship between body size and function in parrotfishes. Coral Reefs 27: 967–974.

69. Birkeland C, Dayton PK (2005) The importance in fishery management of leaving the big ones. Trends Ecol Evol 20: 356–358.

70. Houk P, Musburger C (2013) Trophic interactions and ecological stability across coral reefs in the Marshall Islands. Mar Ecol Prog Ser 488: 23–34.

71. Bellwood DR, Hoey AS, Hughes TP (2012) Human activity selectively impacts the ecosystem roles of parrotfishes on coral reefs. Proc R Soc B 279: 1621–1629.

72. Rasher DB, Hoey AS, Hay ME (2013) Consumer diversity interacts with prey defenses to drive ecosystem function. Ecology 94: 1347–1358.

73. Graham NAJ, Nash KL (2013) The importance of structural complexity in coral reef ecosystems. Coral Reefs 32: 315–326.

19

A Phylogenetic Re-Analysis of Groupers with Applications for Ciguatera Fish Poisoning

Charlotte Schoelinck[1,2]*, Damien D. Hinsinger[1], Agnès Dettaï[1], Corinne Cruaud[3], Jean-Lou Justine[1]

1 UMR 7138 "Systématique, Adaptation, Évolution", Muséum National d'Histoire Naturelle, Département Systématique et Évolution, Paris, France, 2 Fisheries and Oceans Canada, Molecular biology, Aquatic animal health, Moncton, Canada, 3 Génoscope, Centre National de Séquençage, Évry, France

Abstract

Background: Ciguatera fish poisoning (CFP) is a significant public health problem due to dinoflagellates. It is responsible for one of the highest reported incidence of seafood-borne illness and Groupers are commonly reported as a source of CFP due to their position in the food chain. With the role of recent climate change on harmful algal blooms, CFP cases might become more frequent and more geographically widespread. Since there is no appropriate treatment for CFP, the most efficient solution is to regulate fish consumption. Such a strategy can only work if the fish sold are correctly identified, and it has been repeatedly shown that misidentifications and species substitutions occur in fish markets.

Methods: We provide here both a DNA-barcoding reference for groupers, and a new phylogenetic reconstruction based on five genes and a comprehensive taxonomical sampling. We analyse the correlation between geographic range of species and their susceptibility to ciguatera accumulation, and the co-occurrence of ciguatoxins in closely related species, using both character mapping and statistical methods.

Results: Misidentifications were encountered in public databases, precluding accurate species identifications. Epinephelinae now includes only twelve genera (vs. 15 previously). Comparisons with the ciguatera incidences show that in some genera most species are ciguateric, but statistical tests display only a moderate correlation with the phylogeny. Atlantic species were rarely contaminated, with ciguatera occurrences being restricted to the South Pacific.

Conclusions: The recent changes in classification based on the reanalyses of the relationships within Epinephelidae have an impact on the interpretation of the ciguatera distribution in the genera. In this context and to improve the monitoring of fish trade and safety, we need to obtain extensive data on contamination at the species level. Accurate species identifications through DNA barcoding are thus an essential tool in controlling CFP since meal remnants in CFP cases can be easily identified with molecular tools.

Editor: James P. Meador, Northwest Fisheries Science Center, NOAA Fisheries, United States of America

Funding: This study was funded by MNHN ATM Barcode (2010, 2011), MNHN ATM Biodiversité Actuelle et Fossile (2010) and MNHN BQR (2011), awarded to CS and JLJ. This work was supported by the "Consortium National de Recherche en Génomique", and the "Service de Systématique Moléculaire" of the Muséum National d'Histoire Naturelle (CNRS UMS 2700). It is part of the agreement n°2005/67 between the Genoscope and the Muséum National d'Histoire Naturelle on the project "Macrophylogeny of life" directed by Guillaume Lecointre. This work is part of the project @ SPEED-ID "Accurate SPEciEs Delimitation and IDentification of eukaryotic biodiversity using DNA markers" proposed by F-BoL, the French Barcode of life initiative. The funders had no role in study design, data collection and analysis, decision to publish, or preparation of the manuscript.

Competing Interests: The authors have declared that no competing interests exist.

* Email: schoelinck@mnhn.fr

Introduction

Large carnivorous fishes associated with coral reefs are frequently contaminated by toxins responsible for ciguatera fish poisoning (CFP) in tropical and subtropical waters [1,2]. CFP is a food-borne disease contracted by the consumption of finfish that have accumulated lipid-soluble toxins produced by microalgae (dinoflagellates) of the genus *Gambierdiscus* in their flesh and viscera. Dinoflagellates produce gambiertoxins which are first accumulated in the viscera of herbivorous fish and are further accumulated and converted to ciguatoxins in the flesh of larger carnivorous species. For the purposes of this report, we define ciguateric as possessing the ability to accumulate ciguatoxins and cause ciguatera fish poisoning. At least three groups of ciguatoxins have been identified: Pacific (P-CTX), Indian Ocean (I-CTX) and

Caribbean (C-CTX) [3,4]. While the gambiertoxin precursors for P-CTX have been identified, the corresponding precursors for I-CTX and C-CTX have yet to be identified, let alone a thorough examination of which dinoflagellates produce them. This disease produces several gastrointestinal, neurological and cardiac symptoms a few minutes to a few hours after ingestion of contaminated seafood [5]. Although there are reports of symptom amelioration with some interventions (e.g. IV mannitol), no efficient treatment exists so far [6]. It is a significant public health problem, especially in the South Pacific but also in the United States, where it is responsible for one of the highest reported incidence of seafood-borne illness [1]. Although CFP was historically restricted to tropical and sub-tropical regions, case reports are increasingly seen in higher latitudes with escalating global trade and movement of seafood products [1,7]. The incidence of ciguatera, as well as the species of fish that are potentially poisonous, vary from region to region [2]. Precise information about the distribution of ciguatera-carrying species can be obtained from epidemiological data collected by research and health organisations in each country or region, but this depends heavily on correct species identification and is highly dependent on the intensity of data collection. An additional problem pointed out by several authors is the role of recent climate change on harmful algal blooms (HAB), including *Gambierdiscus* ssp. [1,8,9]. The abundance of *G.* spp. correlates positively with elevated sea surface temperature [8]. CFP cases might therefore become more frequent and more geographically widespread as an indirect consequence of climate change (review in [1]). Moreover, coral reefs perturbations, such as hurricanes or bleaching events, also free up space for microalgae to colonize. Even human activities altering the environment such as petroleum production platform building can contribute to the HAB [10]. Therefore, populations from developing countries, already facing these disturbances, appeared to be particularly exposed to the intensification of CFP.

Since there is no appropriate treatment for CFP (for a review, see [6]), the most efficient solution is to regulate fish consumption [11,12]. Lewis [13], and more recently Clua et al. [11] recommended banning some specific species and sizes from fish markets. However, such a strategy can only work if the fish sold are correctly identified and labelled, and it has been repeatedly shown that misidentifications and species substitutions commonly occur in fish markets [14–16].

Groupers (Epinephelidae: rockcods, coralgroupers, hinds, and lyretails) are one of the families most commonly reported as a source of ciguatera poisoning [11]. Some grouper species, like *Plectropomus laevis* and *Cephalopholis argus*, are known to be especially contaminated by ciguatera toxins [17–19]. Large individuals are generally more toxic than small ones since ciguatoxins accumulate in fish via the food chain [11,20]. For instance, specimens belonging to the potentially ciguatoxic fish species *Epinephelus fuscoguttatus* and *Variola louti* are considered dangerous only if they weigh more than 13 and 1.7 kg respectively [20]. Because they are widely distributed in warm and temperate shore waters, from surface to deep-sea, and adults of some species reach 3 m in length and 400 kg [18], groupers represent a considerable economic value in tropical and subtropical regions and most particularly in south-east Asia [21–24]. They are a major component of the artisanal fisheries resource especially in the south Pacific [18]. Global capture fisheries production has increased from approximately 214,000 tons in 1999 to more than 275,000 tons in 2009 [25]. Grouper aquaculture was first introduced in the early 1970s and is now widely practised throughout Southeast Asia [21]. Global grouper aquaculture production has increased tremendously due to increasing demand, from 60,000 tons in 1990 to 200,000 tons in 2007. The premium price of groupers can reach US$ 100/kg in the Chinese live fish markets [21].

Although groupers are large fish and supposedly easily identifiable, comprehensive and reliable species identification tools are rare and a good taxonomic framework is also necessary. Even when intact adult specimens are available (which is generally not the case for food-borne poisoning cases) the morphological characters used to discriminate species can be subtle, making identification difficult even for trained taxonomists. Moreover, accessing the historical literature and assessing the validity of species with a controversial taxonomic history are challenging tasks, even for experts [26].

Some rapid and reliable species identification tools such as DNA barcoding have been developed to facilitate species identification

[26–30]. Given the estimated $US200 billion annual value of fisheries worldwide, the Fish Barcode of Life campaign (FISH-BOL) initiative, as a part of the International Barcode of Life Project (iBOL; http://www.ibolproject.org), is addressing socially relevant questions concerning market substitution and quota management of commercial fisheries (http://www.fishbol.org), with a special focus on developing countries [31]. However, species identification tools require complete and reliable databases. Indeed, DNA databases play a key role for the species identification of groupers, and more generally, for seafood, as non-specialists use essentially those databases to identify species for which they often have access to tissue samples only. The Epinephelidae comprise about 163 species [32] among which 106 are recorded in BOLD (942 public sequences in February 2014). The incompleteness of the reference datasets is a well-identified problem for species identification [33], which can be slightly alleviated if the marker used for identification is also relevant for phylogeny. In such a case, and if the taxonomic framework is accurate, species not represented in the database might still be assigned to clades or higher rank groups, like genera. Completing the largest molecular identification dataset (the cytochrome oxidase 1 of the Barcode of Life project), combined with an accurate study of the relationships of groupers, will help the management of grouper diversity through easier and more accurate identifications. Much remains to be done on both of these aspects, as the phylogenetic framework of the group has undergone many changes recently, and is yet incomplete.

The relationships of the Epinephelidae, recently raised to family rank by Smith and Craig [34], are indeed not yet totally resolved. Epinephelidae were previously a subfamily (Epinephelinae) included with Serraninae and Anthiinae among Serranidae [18,35]. The relationships within the former Serranidae, as well as the composition of the family, have been the object of much discussion. Two molecular studies including the Serranidae showed the non-monophyly of the family [34,36]. Smith and Craig [34] grouped Serraninae and Anthiinae in the Serranidae and raised the subfamily Epinephelinae to the family rank Epinephelidae. On the other hand, Lautrédou et al. [36] showed the polyphyly of Serranidae (with the Serraninae – Anthiinae composition), while recovering an Anthiinae and Epinephelidae clade. Craig et al. [32] defined four subfamilies in Epinephelidae: Diploprioninae, Epinephelinae, Grammistinae and Liopropomi-nae, corresponding to the four previous tribes Diploprionini, Epinephelini, Grammistini, Liopropomini. In their molecular phylogeny, Craig and Hastings [37] attempted to resolve the phylogeny of the Epinephelidae using an almost complete species sampling within the genus *Epinephelus*, and several specimens of other subfamilies, using two mitochondrial and two nuclear markers. They proposed taxonomic changes for species of the subfamily Epinephelinae to reflect their phylogenetic position. For instance, they included *Cromileptes altivelis* and *Anyperodon leucogrammicus* in *Epinephelus* and they moved *Epinephelus septemfasciatus* and *E. ergastularius* to *Hyporthodus*. However, many nodes of their phylogeny lacked robustness. Because of the absence of morphological differences between the genera *Anyperodon*, *Cromileptes* and *Epinephelus*, Craig et al. [32] retained the monotypic genera *Anyperodon* and *Cromileptes*.

To further the study of the relationships between the genera, we sequenced five markers, two mitochondrial, Cytochrome Oxidase Subunit I (COI) and 16S ribosomal RNA (16S) and three nuclear, Rhodopsin (Rh), Titin-like protein (TMO-4C4) and Polycystic kidney disease 1 protein (Pkd1). We choose to include the reference barcoding marker COI to provide a simple and reliable tool for species identification to the non-specialist community

Table 1. List of fish species, specimen vouchers localities and sequence accession numbers.

Family	Group: subfamily /tribe	Species	Accession Number Mitochondrial COI	16S	Nuclear TMO-4C4	Rhodopsin	Pkd1	Specimen Voucher	Locality
Epinephelidae	Diploprioninae	Belonoperca chabanaudi	JQ431484$_v$*	JX094024$_v$*	JX093971$_v$*			MNHN 2008-1159	Moorea, French Polynesia
		Diploprion bifasciatum	KM077912$_v$*	KM077970$_v$*	KM078001$_v$*			MNHN-icti-2815	Queensland, Australia
	Liopropominae	Liopropoma fasciatum	JX093903*	JX093999*	JX093972*	JX093952*	JX093928*		Ecuador
		Liopropoma lunulatum	JQ431888$_v$	JX094023$_v$*	JX093974$_v$*	JX093953$_v$*	JX093929$_v$*	MNHN 2008-1023	Moorea, French Polynesia
		Liopropoma pallidum	JQ431890$_v$	JX094020$_v$*	JX093973$_v$*			MNHN 2008-0793	Moorea, French Polynesia
	Grammistinae	Aporops bilinearis	JQ431457$_v$	JX094016$_v$*	AY949271			MNHN 2008-0307	Moorea, French Polynesia
		Grammistes sexlineatus	JQ431776$_v$	AY539050	AY539458			MNHN 2008-1105	Moorea, French Polynesia
		Grammistops ocellatus	JQ431778$_v$	JX094021$_v$*	X093975$_v$*			USNM 391102	Moorea, French Polynesia
		Pogonoperca punctata	JX093904*	JX093998*	JX093976*	JX093951*	JX093927*		
		Pseudogramma gregoryi	GU225013	AY947571	AY949213				
		Pseudogramma polyacantha	JQ432063$_v$	JX094018$_v$*	JX093977$_v$*			MBIO509.4	Moorea, French Polynesia
		Rypticus saponaceus	JX093905$_v$*	JX094000$_v$*	JX093978$_v$*	JX093956$_v$*	JX093932$_v$*	MNHN2002-158	Ghana
		Suttonia lineata	JQ432178						
	Epinephelinae	Cephalopholis argus	JQ431565$_v$	JX094015$_v$*	JX093979$_v$*	JX093958$_v$*	JX093934$_v$*	MNHN 2008-0229	Moorea, French Polynesia
		Cephalopholis boenak	KM077907$_v$*	KM077965$_v$*	KM077996$_v$*	KM077936$_v$*	KM077879$_v$*	MNHN-icti-2875	New Caledonia
		Cephalopholis colonus	GU440449						
		Cephalopholis cruentata	GU225172	AF297323	AY949266				
		Cephalopholis cyanostigma	KM077908$_v$*	KM077966$_v$*	KM077997$_v$*			MNHN-icti-2821	Queensland, Australia
		Cephalopholis formosa	FJ583004	AY947603	EF517741				
		Cephalopholis fulva	FJ583007	AF297292	AY949282				
		Cephalopholis hemistiktos	HQ149822						
		Cephalopholis igarashiensis	EU871685	AY947599	AY949292				
		Cephalopholis leopardus	FJ583010	AY947560	AY949323				
		Cephalopholis microprion	FJ237608						
		Cephalopholis miniata	KM077909$_v$*	KM077967$_v$*	KM077998$_v$*	KM077937$_v$*	KM077880$_v$*	MNHN-icti-3008	New Caledonia
		Cephalopholis rogaa	JQ349677	EF503626	EF517737				
		Cephalopholis sexmaculata	JQ431572$_v$	JX094019$_v$*	JX093980$_v$*	JX093959$_v$*	JX093935$_v$*	MNHN 2008-0754	Moorea, French Polynesia
		Cephalopholis sonnerati	JX093918*	JX094007*	JX093981*	JX093960*	JX093936*		New Caledonia
		Cephalopholis spiloparaea	KM077910$_v$*	KM077968$_v$*	KM077999$_v$*	KM077938$_v$*	KM077881$_v$*	MNHN-icti-2961	New Caledonia
		Cephalopholis urodeta	KM077911$_v$*	KM077969$_v$*	KM078000$_v$*	KM077939$_v$*	KM077882$_v$*	MNHN-icti-2886	New Caledonia
		Dermatolepis dermatolepis	JX093917*	JX094006*	JX093982*	JX093955*	JX093931*		Ecuador
		Epinephelus adscensionis	FJ583396	AY539049	AY949284				

Table 1. Cont.

Family	Group: subfamily /tribe	Species	Accession Number					Specimen Voucher	Locality
			Mitochondrial		Nuclear				
			COI	16S	TMO-4C4	Rhodopsin	Pkd1		
		Epinephelus aeneus	KM077913$_v$*	KM077971$_v$*	KM078002$_v$*	KM077940$_v$*	KM077883$_v$*	MNHN-icti-2842	Senegal[1]
		Epinephelus akaara	GBGC7768	DQ154107	EF517707				Origin unknown, Aquarium
		Epinephelus altivelis	JX093906*	JX094001*	JX093983*	JX093957*	JX093933*		South China Sea[1]
		Epinephelus amblycephalus	JX093910*	JX094009*	JX093984*	JX093961*	JX093937*		
		Epinephelus analogus	JX093915*	JX094003*	AY949220				Ecuador
		Epinephelus areolatus	KM077914$_v$*	KM077972$_v$*	KM078003$_v$*	KM077941$_v$*	KM077884$_v$*	MNHN-icti-2996	New Caledonia
		Epinephelus awoara	JX093913*	JX094013*	JX093985*	JX093969*	JX093945*		South China Sea[1]
		Epinephelus bleekeri	JX093911*	JX094011*	JX093986*	JX093962*	JX093938*		South China Sea[1]
		Epinephelus bruneus	JX093909*	JX094010*	JX093992*	JX093968*	JX093944*		South China Sea[1]
		Epinephelus chlorostigma	JQ412501$_v$	KM077973$_v$*	KM078004$_v$*	KM077942$_v$*	KM077885$_v$*	MNHN-icti-2145	New Caledonia
		Epinephelus clippertonensis	JX093914*	JX094002*	JX093987*	JX093964*	JX093940*		Clipperton Island
		Epinephelus coeruleopunctatus	KM077915$_v$*	KM077974$_v$*	KM078005$_v$*	KM077943$_v$*	KM077886$_v$*	MNHN-2006-1706	New Caledonia
		Epinephelus coioides	KM077916$_v$*	KM077975$_v$*	KM078006$_v$*	KM077944$_v$*	KM077887$_v$*	MNHN-icti-2994	New Caledonia
		Epinephelus corallicola	JX093908*	JX094008*	JX093988*	JX093967*	JX093943*		South China Sea[1]
		Epinephelus cyanopodus	JQ412502$_v$	KM077976$_v$*	KM078007$_v$*	KM077945$_v$*	KM077888$_v$*	MNHN-icti-2148	New Caledonia
		Epinephelus diacanthus	EF609517	AY947619	AY949274				
		Epinephelus fasciatomaculosus	EF607565	AY947622	EF517717				
		Epinephelus fasciatus	JX093907*	JX094005*	JX093990*	JX093970*	JX093946*		Gulf of Aqaba
		Epinephelus flavocaeruleus	SAIAB330-06$_b$	AY947607	EF517731				
		Epinephelus fuscoguttatus	EU600139	JF750752	EF517713				
		Epinephelus hexagonatus	JQ431719$_v$	JX094017$_v$*	JX093991$_v$*	JX093966$_v$*	JX093942$_v$*	MNHN 2008-0338	Moorea, French Polynesia
		Epinephelus howlandi	KM077917$_v$*	KM077977$_v$*	KM078008$_v$*	KM077946$_v$*	KM077889$_v$*	MNHN-icti-2981	New Caledonia
		Epinephelus lanceolatus	NC_011715	AY947588	EF517736				
		Epinephelus latifasciatus	EU014219	DQ088044	EF517724				
		Epinephelus leucogrammicus	KM077918$_v$*	KM077978$_v$*	KM078009$_v$*	KM077947$_v$*	KM077890$_v$*	MNHN-icti-2993	New Caledonia
		Ep.nephelus longispinis	EF609522	EF213704	EF517706				
		Epinephelus macrospilos	SAIAB547-07$_b$	AY731072	AY949238				
		Epinephelus maculatus	KM077919$_v$*	KM077979$_v$*	KM078010$_v$*	KM077948$_v$*	KM077891$_v$*	MNHN-icti-2907	New Caledonia
		Epinephelus malabaricus	KM077920$_v$*	KM077980$_v$*	KM078011$_v$*	KM077949$_v$*	KM077892$_v$*	MNHN-icti-3018	New Caledonia
		Epinephelus melanostigma	JQ349966	EF503624	EF517730				
		Epinephelus merra	KM077921$_v$*	KM077981$_v$*	KM078012$_v$*	KM077950$_v$*	KM077893$_v$*	MNHN-icti-2963	New Caledonia
		Epinephelus multinotatus	SAIAB149-06$_b$						

Table 1. Cont.

Family	Group: subfamily /tribe	Species	Accession Number					Specimen Voucher	Locality
			Mitochondrial		Nuclear				
			COI	16S	TMO-4C4	Rhodopsin	Pkd1		
		Epinephelus ongus	KM077922$_v$*	KM077982$_v$*	KM078013$_v$*	KM077951$_v$*	KM077894$_v$*	MNHN-icti-3030	New Caledonia
		Epinephelus polyphekadion	KM077923$_v$*	KM077983$_v$*	KM078014$_v$*		KM077895$_v$*	MNHN-icti-2979	New Caledonia
		Epinephelus polyphekadion			KM077952$_v$*			MNHN-icti-3015	New Caledonia
		Epinephelus quoyanus	DQ107861	JX094014$_v$*	JX093993$_v$*			MNHN-icti-2819	Queensland, Australia
		Epinephelus retouti	KM077924$_v$*	KM077984$_v$*	KM078015$_v$*	KM077954$_v$*	KM077896$_v$*	MNHN-icti-2967	New Caledonia
		Epinephelus rivulatus	KM077925$_v$*	KM077985$_v$*	KM078016$_v$*	KM077955$_v$*	KM077897$_v$*	MNHN-icti-2890	New Caledonia
		Epinephelus sexfasciatus	EF607564	DQ067310	EF517716				
		Epinephelus spilotoceps	KM077926$_v$*	KM077986$_v$*	KM078017$_v$*	KM077956$_v$*	KM077898$_v$*	MNHN-icti-3050	Maldives[1]
		Epinephelus tauvina	JX093916*	JX094004*	JX093994*	JX093965*	JX093941*		Gulf of Aqaba
		Epinephelus undulosus	EF609352	DQ088041	EF517704				
		Hyporthodus haifensis	CSFOM034-10$_b$						
		Hyporthodus septemfasciatus	DQ107851	AY947559	AY949247				
		Mycteroperca bonaci	GU225646	DQ267150	AY949270				
		Mycteroperca canina	CSFOM032-10$_b$	AY947585	AY949294				
		Mycteroperca costae	KM077928$_v$*	KM077988$_v$*	KM078019$_v$*	KM077957$_v$*		MNHN-icti-2868	Tunisia[1]
		Mycteroperca costae					KM077899$_v$*	MNHN-icti-2859	Tunisia[1]
		Mycteroperca interstitialis	FJ583668	AY947632	AY949221				
		Mycteroperca jordani	GU440412	AF297329	AY949303				
		Mycteroperca marginata	KM077929$_v$*	KM077989$_v$*	KM078020$_v$*	KM077958$_v$*	KM077900$_v$*	MNHN-icti-2854	Senegal
		Mycteroperca microlepis	JN021310	AF297312	AY949253				
		Mycteroperca morrhua	KM077930$_v$*	KM077990$_v$*	KM078021$_v$*	KM077959$_v$*	KM077901$_v$*	MNHN-icti-2947	New Caledonia
		Mycteroperca poecilonota	SAIAB576-07$_b$						
		Mycteroperca rubra	CSFOM051-10$_b$	AY947587	AY949255				
		Mycteroperca xenarcha	GU440413						
		Plectropomus areolatus	JN242591	EF213706	EF517750				
		Plectropomus laevis	KM077932$_v$*	KM077992$_v$*	KM078023$_v$*	KM077961$_v$*	KM077903$_v$*	MNHN-icti-3012	New Caledonia
		Plectropomus leopardus	KM077933$_v$*	KM077993$_v$*	KM078024$_v$*	KM077962$_v$*	KM077904$_v$*	MNHN-icti-2883	New Caledonia
		Plectropomus maculatus	DQ107911	JF750755	EF517751				
		Saloptia powelli	JQ432090$_v$	JX094022$_v$*	JX093995$_v$*	JX093954$_v$	JX093930$_v$*	MNHN 2008-1014	Moorea, French Polynesia
		Triso dermopterus	DQ107934						
		Variola albimarginata	KM077934$_v$*	KM077994$_v$*	KM078025$_v$*	KM077963$_v$*	KM077905$_v$*	MNHN-icti-2999	New Caledonia
		Variola louti	KM077935$_v$*	KM077995v*	KM078026$_v$*	KM077964$_v$*	KM077906$_v$*	MNHN-icti-2964	New Caledonia

Table 1. Cont.

Family	Group: subfamily/tribe	Species	Accession Number					Specimen Voucher	Locality
			Mitochondrial		Nuclear				
			COI	16S	TMO-4C4	Rhodopsin	Pkd1		
Anthiinae		Pseudanthias hypselosoma	JX093919*	JX094027*	JX093996*	JX093950*	JX093926*		New Caledonia[1]
		Pseudanthias pleurotaenia	JX093920*	JX094026*	AY949308*	JX093949*	JX093925*		
		Pseudanthias tuka	JX093921*	JX094025*	JX093997*	JX093948*	JX093924*		
Cirrhitidae		Cirrhitus pinnulatus	JQ431641			KC222240_b	JX628396		
		Cirrhitus rivulatus		AY539059	AY539467				
Harpigiferidae		Harpagifer kerguelenensis	EATF605_b	AY539063	AY539471	EATF605_b	JQ688766		
Niphonidae		Niphon spinosus	EF143386	AY947575	AY949210				
Percidae		Perca fluviatilis	AP040-12_b	GU018097		AP040-12_b	JX628360		
		Perca flavescens			AY539463				
Scorpaenidae	Congiopodidae	Zanclorhynchus spinifer	AP139-12_b	AY538999	AY539413	EU638021	JX628429		
	Cyclopteridae	Cyclopterus lumpus	AP041-12_b	AY539043	AY539451	AP041-12_b	JX628359		
	Scorpaeninae	Pontinus longispinis		AY538982	AY539398				
		Pontinus macrocephalus	JX093922*			JX093947*	JX093923*		Philippines
		Scorpaenopsis macrochir		AY538987	AY539402				
		Scorpaenopsis possi	JQ432137			KC222244_b	JX628403		
Sebastinae		Helicolenus dactylopterus	AP121-12_b	AY538975	AY539391	AP121-12_b	JX628383		
		Trachyscorpia cristulata	AP111-12_b	AY538980	AY539396	AP111-12_b	JX628374		
Synanceinae		Synanceia verrucosa	JQ432179	AY538995	AY539410	KC222242_b	JX628405		
Serraninae		Centropistis striata	HQ024935	AY072667	AY949216				
		Serranus tigrinus	FJ584106	AY072688	AY949259				
Trachinidae		Trachinus draco	AP104-12_b	AY539068	AY539476	AP104-12_b	JX628367		

(¹) Fish specimen collected at Nouméa fish market or obtained through colleagues, (*) new sequences, (b) BOLD accession number, (v) sequence corresponds to voucher indicated in Table.

(fisheries, governmental organisations, etc.). As ciguatera occurrence has not yet been studied with regard to the evolutionary relationships of the ciguateric fish replaced in their evolutionary context, the second aim of this paper is to map the high risk species for ciguatera fish poisoning into the phylogeny using published information about ciguatera-prone species. We statistically test whether high risk species are closely related and could therefore have inherited their susceptibility to ciguatera from their common ancestor.

Methods

Taxonomic sampling

Sequencing fresh material collection. Fishes collected from different localities (Table 1) in 2009–2011 were dead at the time we acquired them for study, having been commercially caught, and available for purchase at the Nouméa fish market. Each individual was morphologically identified according to Heemstra & Randall [18], measured, weighed, and photographed, and a tissue sample was collected and preserved in absolute ethanol until DNA extraction. Several specimens per species were sequenced to evaluate intraspecific variation and to corroborate identification (data not shown). Additional fish tissues were obtained from colleagues (see acknowledgements) or bought at fish markets, and also preserved in absolute ethanol. For these tissues, no photograph was available; the identification of these tissues was checked by a BLAST search in BOLD [38] followed by a thorough evaluation of the results. Samples and results not corresponding to a higher sequence and identification quality standard were discarded.

Publicly available sequences. All available COI, 16S and TMO-4C4 sequences of Epinephelidae and some outgroup sequences were downloaded from public sequence databases (Barcode of life Database, GenBank Nucleotide). All sequences were controlled for contamination, indels, and stop-codons indicating possible pseudogenes [39]. We followed the classification of Craig and Hastings [37]. When necessary, we amended the species name to agree with the genus gender (*Cephalopholis rogaa*, *Mycteroperca canina* and *M. marginata*) as per the recommendations of the International Code of Zoological Nomenclature (Fourth Edition).

Problems were identified within both GenBank and BOLD sequences, and some sequences available in those databases were therefore not integrated in our dataset. Database sequences presented either: (i) taxonomic problems, such as high genetic divergence within species or erroneous species identification and (ii) nomenclatural problems like the use of invalid species names.

The specimen identification for some sequences was problematic. While there were often several specimens attributed to a species in the databases, high intraspecific variability for COI between specimens was observed for some of them. For instance, the species *E. macrospilos* and *E. tauvina* have a COI genetic divergence within species of 7.4 and 6.2% respectively, values very largely above what is known of fish intraspecific diversity for this marker [40]. These species, and some others, were represented by multiple, disjoint clusters. For example, for *Epinephelus tauvina* one cluster of database sequences included our seven *E. tauvina* sequences, while another cluster matched with our eleven sequences of *E. coioides*. The species *E. akaara*, *E. amblycephalus*, *E. diacanthus*, *E. longispinis*, *E. macrospilos*, *E. sexfasciatus* and *Variola louti* were especially subject to such high intraspecific divergence, a well-known indication of misidentification or unresolved taxonomic issues. These reliability problems are all the more important when the molecular identification has medical applications, such as determining whether a sample can come from a ciguateric species. In order to select the more reliably identified specimens, we checked the supplementary information (voucher, geographic information) to corroborate the identification of the specimens. Moreover, in case of relatively high COI intraspecific variability within species (>1%), we selected the specimen collected closest to the type-locality to minimize taxonomic misidentifications and errors linked to possible cryptic species.

Second, the databases use some invalid names. While spelling errors are relatively straightforward to identify, like the use of *Cephalopholis miniatus* instead of *C. miniata* or *Pseudogramma polyacanthum* instead of *P. polyacantha*, nomenclaturally invalid names were also present in the databases. For instance, *E. 'fario'* (Thunberg, 1793), represented by eight specimens in GenBank including two COI sequences, has been pointed out by Heemstra and Randall [18] as a synonym of *E. longispinis* (Kner, 1864) and as a *nomen dubium*. Randall and Heemstra [41] also regarded this species as unidentifiable. We did not include these sequences in our dataset.

Ethics statement

Fish were dead at the time we acquired them for study, having been commercially caught, and available for purchase at the Nouméa fish market; no permits were required for the described study, which complied with all relevant regulations.

Table 2. List of the primers used in this study; T° of hyb: temperature of hybridisation used to amplify the marker.

Gene	Fragment size	Name	Primers	T° of hyb	Sources
COI	≈650 bp	FishF1	5'-TCAACCAACCACAAAGACATTGGCAC-3'	48°C	[30]
		FishR1	5'-TAGACTTCTGGGTGGCCAAAGAATCA3'		[30]
16S	≈410 bp	16SarL	5'-CGCCTGTTTATCAAAAACAT-3'	54°C	[75]
		16SbrH	5'-CCGGTCTGAACTCAGATCACGT-3'		[75]
Rhodo	≈720 bp	Rh193	5'-CNTATGAATAYCCTCAGTACTACC-3'	52°C	[76]
		Rh1039r	5'-TGCTTGTTCATGCAGATGTAGA-3'		[76]
Pkd1	≈850 bp	Pkd1F62	5'-CATGAGYGTCTACAGCATCCT-3'	50°C	[77]
		Pkd1R952	5'-YCCTCTNCCAAAGTCCCACT-3'		[77]
TMO-4C4	≈540 bp	TMOF1	5'-CCTCCGGCCTTCCTAAAACCTCTC-3'	55°C	[78]
		TMOR1	5'-CATCGTGCTCCTGGGTGACAAAGT-3'		[78]

DNA extraction, amplification, and sequencing

DNA was extracted from tissue samples using NucleoSpin 96 tissue kit (Macherey-Nagel, Düren, Germany) and five genes were amplified (Table 2).

Each PCR reaction was performed in a 20 µl final volume, containing 2 ng of DNA, 1X reaction buffer, 0.26 mM dNTP, 0.8 µM of each primer, 5% DMSO and 1.5 units of Taq polymerase (Qiagen). Thermocycles consisted of an initial denaturation step at 94°C for 2′, followed by 37–55 cycles of denaturation at 94°C for 30″, annealing at 48–55°C for 40″ (Table 2) and extension at 72°C for 1′. The final extension was conducted at 72°C for 10′. Purification and cycle-sequencing reactions were performed at the Génoscope (Évry, France), using the BigDye Terminator version 3 sequencing kit, the GeneAmp PCR System 9700 and a capillary ABI3730 DNA Analyser, all from Applied Biosystems. Sequences were edited and assembled using Sequencher 4.9 (Gene Codes Corporation, Ann Arbor, MI, USA). Sequences for this study were deposited in GenBank (Table 1).

Phylogenetic analyses and mapping

All markers were sequenced for both directions to confirm accuracy for each individual specimen. However, a few sequences could not be obtained (Table 1). Sequences were aligned using ClustalW as implemented in BioEdit version 7.0.5.3 [42] or by eye. The accuracy of automatic alignments was assessed by eye. 16S sequences were aligned manually and two portions (between positions 240–300 and 347–450) were removed from the alignment due to hypervariable regions that could not be aligned reliably.

94 species of Epinephelidae were included in our analyses, 2 species for Diploprioninae, 81 species for Epinephelinae, 8 species for Grammistinae and 3 species for Liopropominae.

To study (i) the monophyly of the Epinephelidae (i.e according to traditional classification Epinephelinae minus Niphon) and (ii) the relationships within this family, outgroups were chosen in the sub-families of Serranidae, the Serraninae and the Anthiinae [34]. We also included multiple, non-monophyletic outgroups: specimens from Perciformes (Cirrhitidae, Harpagiferidae, Niphonidae, Percidae, Trachinidae) and from Scorpaeniformes (Congiopodidae, Cyclopteridae, Scorpaeninae, Sebastinae, Synanceinae) placed as close relatives of Epinephelidae in the study of Lautrédou et al. [36].

Phylogenetic analyses of the Epinephelidae

All markers were first analysed separately, tested for incongruence and, since none was detected, concatenated in two different datasets. The combined and separate analyses have three different aims (i) obtaining the most robust phylogenetic reconstruction while including a maximal species representation, i.e. including downloaded sequences (dataset 1), (ii) maximising the number of markers, even if it includes less species (dataset 2), and (iii) best representing the diversity of the Epinephelidae with the COI gene alone (dataset 3). Dataset 1 included the concatenated COI, 16S and TMO-4C4 sequences. Dataset 2 included all five concatenated markers, i.e. COI, 16S, TMO-4C4, Rhodopsin and Pkd1 sequences.

The best-fitting models of nucleotide evolution for each gene and for the concatenation were determined based on the Akaike Information Criterion (AIC) implemented in ModelTest 2.3 [43] in conjunction with PAUP 4.0b10 [44]. The GTR + I + Γ model was selected for each marker and for the two datasets.

Trees were inferred using two probabilistic approaches: maximum likelihood with a non-parametric bootstrap (BP) using RAxML 7.2.8 [45,46] and Bayesian Inference [47]. Maximum likelihood (ML) analyses were carried out online on the CIPRES Science Gateway (The CIPRES Portals. URL: http://www.phylo.org/sub_sections/portal) with RAxML-HPC BlackBox (7.2.7) [45]. Datasets were partitioned by codon position for each marker (except 16S) and by marker and by codon position for datasets 2 and 3. BI analyses were performed with MRBAYES version 3.2.1 [47] using 75,000,000, 75,000,000, and 30,000,000 generations for datasets 1, 2 and 3 respectively, with sampling every 1,000 generations and four Metropolis-coupled Markov chains Monte Carlo (MCMCMC). The parameter estimates and convergence were checked using Tracer version 1.4 [48]. The first 25% of sampled trees were considered burn-in trees, and discarded prior to constructing a 50% majority rule consensus trees. Posterior probabilities (PP) (Bayesian analysis) and Bootstrap probabilities (BP) were used as indicators of node credibility; P95% was considered significant [49]. Two independent analyses were conducted to check for convergence of the results.

Mapping of potential ciguatera poisonous species

Potentially ciguatera affected species were mapped onto the phylogenetic tree inferred from all the available reliable sequences for the COI fragment. Although several relationships were not resolved compared to the combined analyses, the COI topology is congruent with the combined analyses topologies, and moreover best represents the diversity of the Epinephelidae.

A large number of studies on ciguatera were reviewed to establish a list of ciguatera affected species [7,11,12,19,20,50–59]. However, there were very few publications where species names were precisely indicated. We ended up by using five [11,20,53,54,58], as well as the available ciguatera data in Fishbase database [17]. To detect a potential pattern in the evolution of the occurrence of ciguatera in Epinephelinae, we quantified the strength of phylogeny-trait association. The MrBayes phylogeny from the dataset 3 was used as an input in the BaTS software [60], with the occurrence of the ciguatera coded as absent/present. 10.000 trees from the posterior set of trees from MrBayes analysis were randomly selected after removing the first 7.5 million generations as a burnin according to Tracer [48], and re-rooted using the outgroups and a custom-made R script [61] (script available on request). BaTS estimates several statistics: the parsimony score (PS), the association index (AI) and the monophyletic clade (MC) and tests their significance against a null distribution (obtained by reshuffling 100 times the ciguatera states on the tips). A strong phylogeny-trait association is identified by low PS and AI scores and a high MC score.

Evolution of the ciguatera fish poisoning (CFP)

Ancestral character state reconstructions for ciguatera fish poisoning were conducted using the maximum-likelihood method implemented in Mesquite 2.75 [62]. Recognized species were assigned different states (following a bibliographic survey, see above): absence (0) and presence (1) of CFP. Ancestral states were reconstructed for all Bayesian trees retained from the analysis of the combined data set and their mean likelihood was then plotted on the maximum clade credibility tree.

Results

Phylogenetic relationships within Epinephelidae

Dataset length and number of variable sites are reported in Table 3. 47 and 58 sequences were obtained for the mitochondrial markers COI and 16S, respectively. 54 sequences were obtained for TMO-4C4, 50 for Rhodopsin and 50 for Pkd1.

Table 3. Dataset composition, marker size and information.

Dataset	Number of species	Markers	Length (bp)	Conserved sites	Variable sites
1	102	COI	651	372	279
		16S	410	254	156
		TMO-4C4	511	271	240
2	60	COI	651	382	269
		16S	410	278	132
		TMO-4C4	536	330	206
		Pkd1	861	422	439
		Rhodopsin	723	470	253
3	112	COI	651	371	280

The concatenated phylogeny (Fig. 1A) based on the three markers with the largest sampling (COI, 16S and TMO-4C4) recovered the monophyly of Epinephelidae comprising the four subfamilies, Diploprioninae, Epinephelinae, Grammistinae, and Liopropominae with a weak support. The Grammistinae was monophyletic. The three species of the genus *Liopropoma* of the Liopromninae were grouped. The Diploprioninae and the Epinephelinae were together not monophyletic because of the inclusion of *Belonoperca chabanaudi* within Epinephelinae. However, to the exception of the insertion of *B. chabanaudi*, all Epinephelinae specimens were grouped. In the Epinephelinae, the genera *Variola* and *Plectropomus* were together monophyletic with a strong support. *Saloptia* was sister-group to *Plectropomus*. *Cephalopholis* was monophyletic but divided in two robust groups with the inclusion of *C. colonus* and *C. rogaa* (earlier combinations: *Paranthias colonus* and *Aethaloperca rogaa*). *Epinephelus leucogrammicus* and *E. altivelis* (earlier combinations: *Anyperodon leucogrammicus* and *Cromileptes altivelis*, both monotypic genera) were included within the clade *Epinephelus*. *Mycteroperca* was sister-group to *Epinephelus* and monophyletic with a strong support. *D. dermatolepis* was sister-group to *Hyporthodus*, *Mycteroperca* and *Epinephelus*. The subfamily Grammistinae was sister-group to the Liopropominae and *Diploprion bifasciatum* from the Diploprioninae.

The concatenated phylogeny (Fig. 1B) based on all five markers (COI, 16S, TMO-4C4, Rhodopsin and Pkd1) confirmed the monophyly of Epinephelinae as well as the three subfamilies Epinephelinae, Grammistinae and Liopropominae. No specimen from the subfamily Diploprioninae was available for the Rhodopsin and Pkd1 markers; the monophyly of this subfamily cannot be evaluated with this second dataset.

Within Epinephelinae and as suggested by dataset 1, the monophyly of *Mycteroperca*, *Plectropomus* and *Variola* were well supported. *Cephalopholis* also constituted a robust monophylum with two clades, one containing *C. argus* and *C. boenak* and the other the rest of the species (as observed with the previous tree). *Epinephelus* was monophyletic with a weak support. *D. dermatolepis* was sister-group to a clade including *Epinephelus* and *Mycteroperca*.

Although several relationships were not resolved compared to the combined analyses (Fig. 1), the COI topology (Figure S1) is congruent with them, and has a much larger species sampling. The subfamilies Grammistinae and Liopropominae are monophyletic.

With the exception of the insertion of the species *Epinephelus poecilonotus* and *E. haifensis* included in the genera *Mycteroperca* and *Hyporthodus* respectively, all *Mycteroperca* and *Hyporthodus* species were grouped.

Like in the concatenated analyses (Figs. 1A-B), the monotypic genera *Anyperodon* and *Cromileptes* were together included in *Epinephelus*. The subfamily Diploprioninae including the three genera *Beloperca*, *Diploprion* and *Aulacocephalus* was not monophyletic, with the exclusion of *Beloperca chabanaudi*. *Diploprion bifasciatum* and *Aulacocephalus temminckii* constituted a robust clade.

Mapping of potentially ciguatera affected species on the COI tree

Twenty nine ciguateric species were found in a review of the literature. Most of the species affected by ciguatera belong to the subfamily Epinephelinae (Fig. 2). Only two species, outside the Epinephelinae, *Grammistes sexlineatus* and *Rypticus saponaceus* are ciguateric. Within Epinephelinae, three genera contain multiple ciguateric species. All species included in the genera *Plectropomus* (4) and *Variola* (2) are ciguateric. In the genus *Epinephelus*, four species out of seven are ciguateric in one clade (*E. fasciatus*, *E. hexagonatus*, *E. melanostigma*, *E. merra*, *E. retouti*, *E. spilotoceps* and *E. tauvina*) (Fig. 2). In another, three species out of four are ciguateric (*E. quoyanus*, *E. macrospilos*, *E. howlandi* and *E. rivulatus*). The other ciguateric species are dispersed in clades where most species are not known to be affected. Within the other genera of Epinephelinae, *Mycteroperca* and *Hyporthodus*, only one species is ciguateric, *M. bonaci*.

The analysis of the presently available data on presence or absence of ciguatera fish poisoning (CFP) for each species mapped the absence of CFP as the ancestral state of Epinephelidae. The ancestral state analysis underlines the multiple appearances of CFP in this group with at least 10 events in the subfamily Epinephelinae (Fig. 2).

The Bayesian analysis of the phylogenetic signal (Table 4) revealed the occurrences of ciguatera in groupers species have a moderate non-random association with phylogeny; only the MC statistic for the occurrence of ciguatera shown significant P-value (p = 0.04), whereas the PS statistic shown a marginally significant P-value (p = 0.080). Both the AI statistic and the MC value for the ancestral state were not significant (p = 0.24 and p = 0.55, respectively).

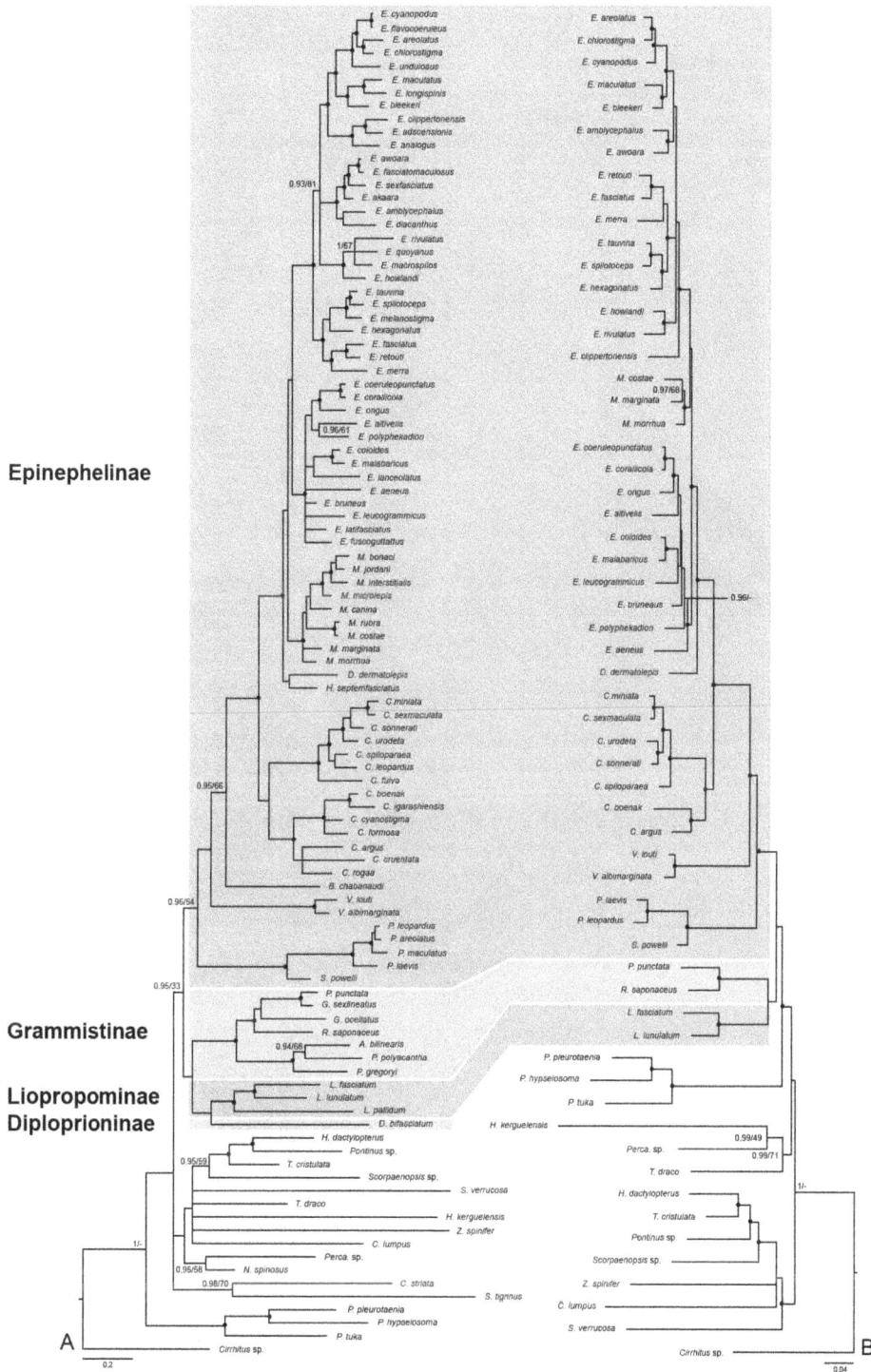

Figure 1. Phylogenetic relationships within the Epinephelidae. Bayesian inference phylogram obtained from phylogenetic analyses of the dataset 1 (A) based on the concatenation of three genes, COI, 16S and TMO-4C4, and the dataset 2 (B) obtained with the concatenation of five genes, COI, 16S, TMO-4C4, Rhodopsin and Pkd1, both under the GTR + I + Γ model. Epinephelidae are highlighted in colour (pink and blue). Each sub families are shown in alternate blue and pink colours. Values at nodes indicate Bayesian posterior probabilities (PP) and maximum likelihood bootstrap percentages (BP). Black circles indicate nodes supported by posterior probability ≥95% and ML bootstrap probability ≥75%.

Discussion

Systematics and taxonomy within Epinephelidae

As in Craig and Hastings [37] (but contradicted by Craig et al. [32], because of the absence of morphological synapomorphy), our result show the inclusion of *Epinephelus leucogrammicus* and *E. altivelis* (previous combinations *Anyperodon leucogrammicus* and *Cromileptes altivelis*) within *Epinephelus* but also the inclusion of *Cephalopholis rogaa* and *C. colonus* within *Cephalopholis* (previous combinations *Aethaloperca rogaa* and *Paranthias colonus*).

212

Marine and Aquatic Sciences

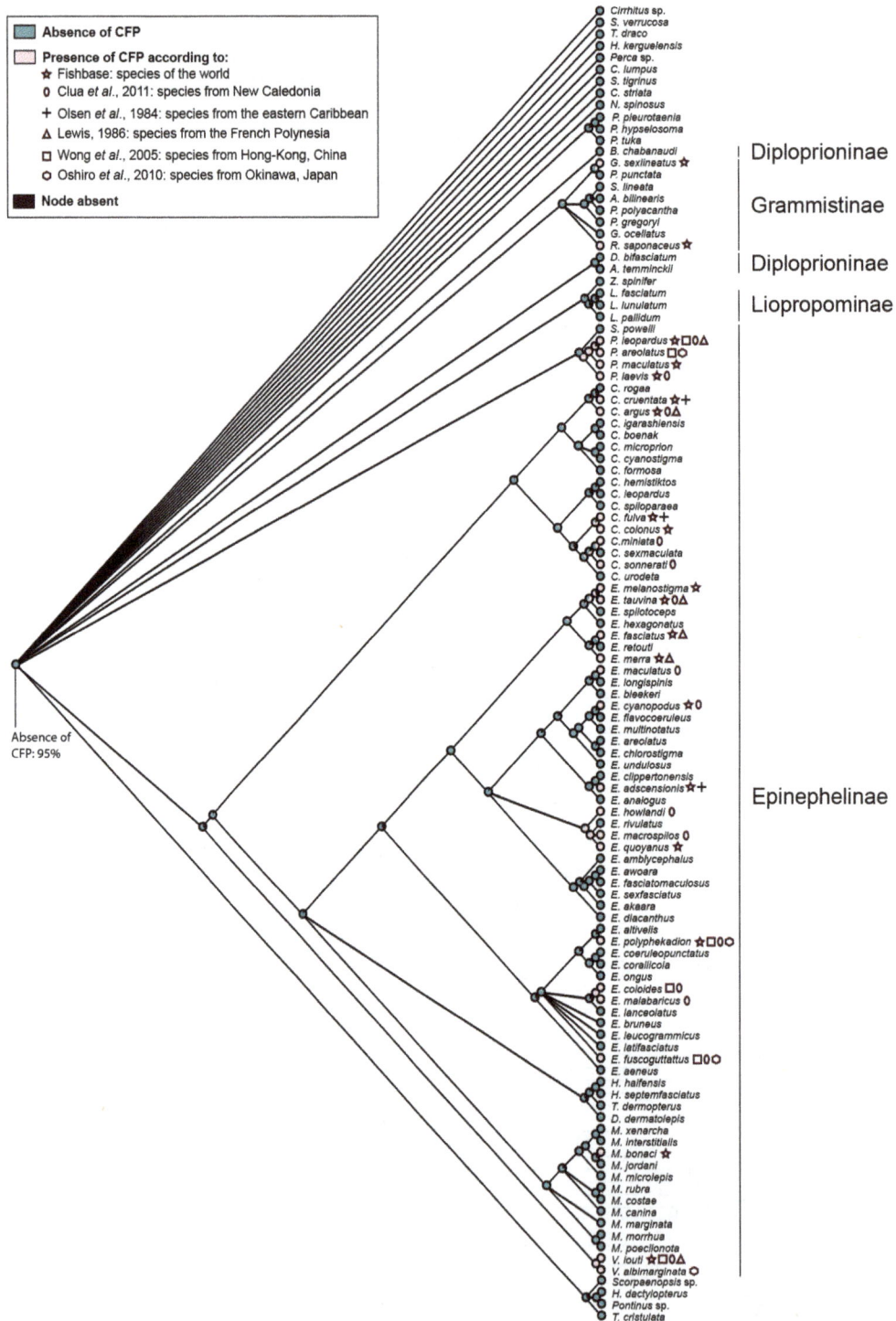

Figure 2. Ancestral ciguatera fish poisoning (CFP) reconstruction of Epinephelidae. Bayesian cladogram of the COI dataset with maximum likelihood estimates of ancestral CFP states. Pie charts correspond to average likelihoods for each state. Percentage values are given for nodes of interest.

Craig and Hastings [37] included in the genus *Mycteroperca* several species usually considered as members of *Epinephelus*. Based on our results, we confirm these new combinations. In addition, we proposed to transfer *E. poecilonotus* (Temminck and Schlegel, 1842) in *Mycteroperca* as *M. poecilonota*.

They also observed a monophyletic lineage distinct from the remaining species of *Epinephelus* and from *Mycteroperca*, for which they resurrected the oldest available generic name *Hyporthodus*. Our topologies corroborate this lineage, which includes at least two species, *H. septemfasciatus* and *H. haifensis* (previously *Epinephelus haifensis*).

Table 4. Results of the Bayesian phylogeny-trait association.

Statistic	Observed Distribution		Null Distribution		P-value
	observed mean	95% CI	null mean	95% CI	significance
AI	4.2	3.7–4.6	4.6	3.6–5.5	0.240
PS	23.7	23.0–24.0	26.3	23.3–28.7	0.080*
MC (absence of ciguatera)	6.6	6.0–9.0	6.1	4.1–7.9	0.550
MC (occurrence of ciguatera)	4.0	4.0–4.0	2.3	1.6–3.1	0.040**

Association index (AI), parsimony score (PS), and monophyletic clade (MC) and their significance. While AI and PS indices test for the overall phylogeny and all the characters at once, MC is drawn to specifically quantify the phylogenetic signal for each specific character (occurrence of ciguatera contamination). Asterisk indicates significant values (*: p≤0.1; **: p≤0.05).

The relationships between *Belonoperca chabanaudi*, *Diploprion bifasciatum* and *Aulacocephalus temminckii* in Figure S1 question the monophyly of the Diploprionini. In Figure 1A where *A. temminckii* was not represented, *B. chabanaudi* is included in the subfamily Epinephelinae (PP = 0.96, BP_{ML} = 54). *D. bifasciatum* constituted, with a weak support, the sister-group of the Grammistinae with the Liopropominae. Since the support for several deeper clades is low at least for maximum likelihood analyses, we prefer not to discuss further the phylogeny of these groups.

Within Epinephelidae, there is no modification of the subfamilies Grammistinae and Liopropominae. The Epinephelinae now include only twelve genera (minus *Anyperodon*, *Cromileptes* and *Paranthias* vs. 15 previously). In our results (Fig. 1), *Variola* and *Plectropomus-Saloptia* are in basal position relative to the other Epinephelinae genera.

Ciguatera fish poisoning

In the studies on ciguatera reviewed, most were fairly unhelpful to establish a list of ciguatera-affected species with scientific names. Most publications use vernacular names only, and many of these designate several fish species (see [2,6]). Vernacular names might be important for local communication and consumer warnings. However, scientific names should be systematically associated to them to enhance precision and communication between localities, as vernacular name use varies with geography. Feedback questionnaires [63,64] were provided to public health and fisheries department staff to collect ciguatera poisoning data and then to put more monitoring and research into place. Traceback investigations of fish associated with outbreaks provide valuable information regarding fishing areas associated with CFP. However, in a weekly report, the Centers for Disease Control and Prevention pointed to limitations in the personal feedbacks [65]; where physician reports were unavailable, the symptoms were based entirely on self-report or second hand reports from family members and may be wrong. Moreover, additional cases might have occurred but were unrecognized because involved physicians were not aware of the need to make an appropriate diagnosis and to report, especially in countries where the public health network is weakly organized.

With the limitations of our current knowledge in mind, 29 among 163 grouper species are considered ciguateric in the world. While there is no correlation between the phylogeny and the currently known ciguatera status of the species, some clades include a higher number of possibly ciguateric species, and might be interesting to investigate in order to determine the ciguateric status of species where it has not been described yet. The recent

changes in classification based on the reanalyses of the relationships within Epinephelidae ([37], this study) have an impact on the interpretation of the ciguatera distribution in the genera. Multiple species from the genus *Epinephelus* that were not reported to be ciguateric belong in fact to other genera that have very few ciguateric species (*Hyporthodus* and *Mycteroperca*). As for the genera *Plectropomus* and *Variola*, the species included in our datasets are all considered potentially ciguateric (4 and 2 species respectively). In the three other species contained in the genus *Plectropomus*, only one, *P. oligacanthus*, is ciguateric according to FishBase.

In our study, even if the COI gene gives little support for deeper nodes in phylogenetic analyses, it performs well in Epinephelidae for interspecific relationships, as already observed for other Teleosts and taxonomic groups [66,67]. Better phylogenetic performance means that even if an identical sequence is not available for identification in the database, the position of the unknown sequence in a phylogenetic analysis can provide information about the genus it belongs to, and which species it is most closely related to. Consequently, samples from species that cannot be readily identified, but fall within these groups in an analysis might best be considered potentially ciguateric when tested for ciguatera case suspicions. This, added to the large dataset already available, and the more stringent guidelines of the Barcode of Life project compared to other sequence databases, makes it a very good choice for identification in the group.

Many questions remain to be answered pertaining to the production and accumulation of ciguatoxins and the subsequent occurrence of ciguatera fish poisoning. For example, why does Ciguatera affect only some species in a given locality, why does Ciguatera not affect species over their whole range? Explanations might be found on the life history traits of groupers. The diet, the location on the reef and the size of grouper species could explain some of these ciguateric patterns. Yet no clear tendency based on these data appears. On the other hand, some species like *Epinephelus macrospilos* or *Cephalopholis miniata* are ciguateric, for example, in New Caledonia (Fig. 2) but not in their whole range. Richlen et al. [68] were some of the first authors to correlate the geographic patterns of fish toxicity with the preponderance of highly toxic strains of *Gambierdiscus* spp. They showed that *G. toxicus* was not a single cosmopolitan species, but instead was a species complex comprised of several distantly related groups co-occurring across geography. Thus the range of *G. toxicus* appears to be much smaller than the range of the fish species, but this also depends on the correctness of our knowledge about the systematics and distribution of the fish species. The presence and relative abundance of the members of this species complex among geographic regions may help explaining patterns of ciguatera toxicity, particularly if differences in physiology and/or toxin-producing capabilities also exist among these groups. In addition, a better understanding of the three groups of ciguatoxins produced by *Gambierdiscus* species will be very useful.

As suggested by several authors, due to recent climate change or human activities, harmful algal blooms (HAB), including *Gambierdiscus* spp. [1,8,9] might become more frequent and more geographically widespread. A consequence would be that ciguatera starts affecting fish species in other localities in their range where there was no previously recorded problem, affecting populations that will be not aware of the ciguateric risk of fish consumption. The first reports of consumer illness and detection of ciguatoxic fish from the Canary Islands [55] seems to be consistent with such a geographic expansion of ciguatera. Another recent expansion of dinoflagellates, and subsequent incidences of CFP illness, was recorded in the northern Gulf of Mexico, USA [10]. Of the 29 ciguateric grouper species, 16 have a very large distribution in the Indo-Pacific Ocean. The ranges extend from the East of Africa to the West of America, including the Red Sea. Current distribution areas might help us determine future changes in the prevalence area of ciguatera.

Barcoding, a useful tool for various applications

As all our knowledge on ciguateric species hinges on correct identification of the species involved in CFP, DNA barcoding is an essential tool in controlling CFP, but also investigating mislabelling of seafood or endangered species monitoring. Fishermen and restaurant owners do not hesitate to sell fish of lower quality under erroneous labelling [16], a problem compounded in groupers by overfishing. Because of the very high price of live groupers, the mislabelling of fishes and the dangers of ciguatera, accurate species identification is part of what makes the quality of the fish meat [16]. The misidentification of several species was largely discussed by Stewart et al. [12]. They encountered nine misidentifications in their molecular analyses, and remarked that eight of the nine misnamed fish contain ciguatera toxins. While the authors did not suggest deliberate substitutions, they wondered about this high rate. Some confusions were not surprising in regard of the similar morphology of the species (Spanish mackerel - *Scomberomorus commerson* and *Scomberomorus queenslandicus* for instance). However, they pointed out that selling the Spanish mackerel – a ciguatera-prone species banned in Platypus Bay, Queensland – as red snapper and swordfish, neither known to be problematic for ciguatera, represents a potential risk of bypassing ciguatera prevention strategies [12].

In such a case, DNA barcoding has proved to be a very useful tool to quickly and easily identify seafood species sold on fish markets or on restaurants, thus helping to avoid at least some of the risky species. The mislabelling studies have also had a positive effect on at least some parts of the retail sector [69]. Moreover, meal remnants in CFP cases often cannot be identified without molecular tools, where proper identification is critical for improving the ciguatera affected species list and warnings. DNA barcoding-based traceability procedures were implemented in several U.S. state and federal laboratories [65]. As the symptoms appear in minutes to hours after the ingestion of contaminated seafood [13], recovery of samples for sequencing is possible, at least in some cases. For most types of cooked seafood, the full barcode can be obtained. For severely degraded or heavily processed products (e.g. canned), the sequencing of shorter sequences (i.e. Mini barcodes, see [70]), also works for identification.

The key role of the public databases

For DNA barcoding to be a useful tool for identifying species, it needs complete (or at least large) and reliable reference sequences available in public databases. For an example of both the usefulness of barcoding methods and the limits of the available databases, a sample of fish tissue collected at the fish market in Nouméa (New Caledonia) showed 100% identity in the databases with those of two distinct species. Both are deep-sea species, as was the grouper from the market according to the fishmongers. *H. ergastularius* is not recorded from New Caledonia, but *H. octofasciatus* was recorded, sometimes under other names [71]. Unfortunately, identical sequences for *H. ergastularius* and *H. octofasciatus* in databases preclude an identification of this sample. Currently, available grouper sequences contain several cases of misidentification both in GenBank and in BOLD. The numerous grey-brown species with dots (*E. tauvina*, *E. akaara*, *E. diacanthus*, *E. amblycephalus*, *E. longispinis*, *E. sexfasciatus* and *E. macrospilos*) are particularly affected, with the additional problem that young specimens of some species resemble adults of smaller species. This is not an isolated case. According to Vilgalys [72] up to 20% of the named sequences in public databases may be misidentified. Hassanin et al. [73] suggested to annotate database sequences through an additional "external expertise" field, and there is indeed a possibility to add comments to data in BOLD. However, a lot remains to be done before all available identifications can be trusted, even with the additional geographical and voucher information required by BOLD. There is an ongoing effort by the BOLD crew to flag dubious sequences. They developed a system to grade the level of reliability of the identification in BOLD [74], but it has not been applied to all sequences yet. In the end, a morphological study of the voucher specimen remains necessary, but this is not practical when fast identification is needed such as for ciguateric sample identifications. While currently neither reliability nor comprehensivity are at hand for Epinephelinae sequences, our study has added 47 COI sequences for carefully identified and vouchered specimens.

Supporting Information

Figure S1 Phylogenetic relationships within Epinephelidae. Bayesian inference phylogram obtained from phylogenetic analyses of the COI under the GTR + I + Γ model. Values at nodes indicate Bayesian posterior probabilities (PP) and maximum likelihood bootstrap percentages (BP). Black circles indicate nodes supported by posterior probability ≥95% and ML bootstrap probability ≥75%.

Acknowledgments

Anne-Claire Lautrédou (MNHN, Paris), Paolo Galli (University of Milano, Italy: material from Red Sea), Philippe Béarez (MNHN, Paris: Ecuador, Philippines), Serges Planes (Moorea Biocode Project, French Polynesia), Monica Mwale (SAIAB, South Africa), Dominique Ponton (IRD, Nouméa, New Caledonia), Jean-Dominique Durand (IRD, Dakar, Senegal), Rishen Liang (Guangzhou University, China), Lassâd Neifar (Sfax University, Tunisia) provided tissues samples and/or sequences. The Australian Node of the CReefs Global Research Initiative, Thomas Cribb, Terrence Miller, and Holly Heiniger helped in obtaining tissue samples during a mission in Heron Island. Martine Desoutter (MNHN, Paris) and Philippe Bouchet (MNHN, Paris) discussed nomenclatural questions and Gaël Denys, Frédéric Busson and Gaël Lancelot contributed to the analyses, the data collection and text corrections respectively. Sincere thanks to the staff of the "Service de Systeématique Moleéculaire" (UMS2700 CNRS-MNHN). This work is part of the project @ SPEED-ID "Accurate SPEciEs Delimitation and IDentification of eukaryotic biodiversity using DNA markers" proposed by F-BoL, the French Barcode of life initiative, and the network "Bibliothèque du Vivant" funded by the CNRS, the Muséum National d'Histoire Naturelle, the INRA and the CEA (Genoscope). We thank the anonymous reviewers for their very helpful comments on the manuscript.

References

1. Dickey RW, Plakas SM (2010) Ciguatera: A public health perspective. Toxicon 56: 123–136.
2. Laurent D, Yeeting B, Labrosse P, Gaudechoux JP (2005) Ciguatera: a Field Reference Guide.(Ciguatera: un guide pratique). Secretariat of the Pacific Community. 91 p.
3. De Fouw JC, Van Egmond HP, Speijers GJA (2001) Ciguatera fish poisoning: a review. Available: http://wwwrivmnl/bibliotheek/rapporten/388802021pdf. Accessed 25 June 2014.
4. Lewis RJ, Jones A (1997) Characterization of ciguatoxins and ciguatoxin congeners present in ciguateric fish by gradient reverse-phase high-performance liquid chromatography mass spectrometry. Toxicon 35: 159–168.
5. Lewis RJ (2001) The changing face of ciguatera. Toxicon 39: 97–106.
6. Friedman MA, Fleming LE, Fernandez M, Bienfang P, Schrank K, et al. (2008) Ciguatera fish poisoning: Treatment, prevention and management. Marine Drugs 6: 456–479.
7. Kipping R, Eastcott H, Sarangi J (2006) Tropical fish poisoning in temperate climates: food poisoning from ciguatera toxin presenting in Avonmouth. Journal of Public Health 28: 343–346.
8. Chateau-Degat ML, Chinain M, Cerf N, Gingras S, Hubert B, et al. (2005) Seawater temperature, *Gambierdiscus* spp. variability and incidence of ciguatera poisoning in French Polynesia. Harmful Algae 4: 1053–1062.
9. Moore SK, Trainer VL, Mantua NJ, Parker MS, Laws EA, et al. (2008) Impacts of climate variability and future climate change on harmful algal blooms and human health. Environmental Health 7: (Suppl 2):S4.
10. Villareal TA, Hanson S, Qualia S, Jester ELE, Granade HR, et al. (2007) Petroleum production platforms as sites for the expansion of ciguatera in the northwestern Gulf of Mexico. Harmful Algae 6: 253–259.
11. Clua E, Brena PF, Lecasble C, Ghnassia R, Chauvet C (2011) Prevalence and proposal for cost-effective management of the ciguatera risk in the Noumea fish market, New Caledonia (South Pacific). Toxicon 58: 591–601.
12. Stewart I, Eaglesham GK, Poole S, Graham G, Paulo C, et al. (2010) Establishing a public health analytical service based on chemical methods for detecting and quantifying Pacific ciguatoxin in fish samples. Toxicon 56: 804–812.
13. Lewis RJ (2006) Ciguatera: Australian perspectives on a global problem. Toxicon 48: 799–809.
14. Carvalho DC, Neto DAP, Brasil BSAF, Oliveira DAA (2011) DNA barcoding unveils a high rate of mislabeling in a commercial freshwater catfish from Brazil. Mitochondrial DNA 22: 97–105.
15. Cline E (2012) Marketplace substitution of Atlantic salmon for Pacific salmon in Washington State detected by DNA barcoding. Food Research International 45: 388–393.
16. Wong EHK, Hanner RH (2008) DNA barcoding detects market substitution in North American seafood. Food Research International 41: 828–837.
17. Froese R, Pauly D (2012) FishBase. World Wide Web electronic publication. Available: www.fishbase.org.
18. Heemstra PC, Randall JE (1993) FAO Species Catalogue. Vol. 16. Groupers of the world (Family Serranidae, Subfamily Epinephelinae). An annotated and illustrated catalogue of the grouper, rockcod, hind, coral grouper and lyretail species known to date Rome: FAO. 382 p.
19. Laboute P, Grandperrin R (2000) Poissons de Nouvelle-Calédonie. Nouméa, New Caledonia: Éditions Catherine Ledru. 520 p.
20. Oshiro N, Yogi K, Asato S, Sasaki T, Tamanaha K, et al. (2010) Ciguatera incidence and fish toxicity in Okinawa, Japan. Toxicon 56: 656–661.
21. Harıkrıshnan R, Balasundaram C, Heo M-S (2010) Molecular studies, disease status and prophylactic measures in grouper aquaculture: Economic importance, diseases and immunology. Aquaculture 309: 1–14.
22. Ottolenghi F, Silvestri C, Giordano P, Lovatelli A, New MB (2004) Capture-based aquaculture. The fattening of eels, groupers, tunas and yellowtails. Rome: FAO. 308 p.
23. Pierre S, Gaillard S, Prevot-D'Alvise N, Aubert J, Rostaing-Capaillon O, et al. (2008) Grouper aquaculture: Asian success and Mediterranean trials. Aquatic Conservation-Marine and Freshwater Ecosystems 18: 297–308.
24. Sadovy de Mitcheson Y, Craig MT, Bertoncini AA, Carpenter KE, Cheung WWL, et al. (2013) Fishing groupers towards extinction: a global assessment of threats and extinction risks in a billion dollar fishery. Fish and Fisheries 14: 119–136.
25. FAO (2010) Capture Production 1950–2008. FAO Fisheries Department, Fishery Information, Data and Statistics Unit.
26. Ward RD, Hanner R, Hebert PDN (2009) The campaign to DNA barcode all fishes, FISH-BOL. Journal of Fish Biology 74: 329–356.
27. Collins RA, Armstrong KF, Meier R, Yi YG, Brown SDJ, et al. (2012) Barcoding and border biosecurity: Identifying cyprinid fishes in the aquarium trade. Plos One 7.
28. Hanner R, Becker S, Ivanova NV, Steinke D (2011) FISH-BOL and seafood identification: Geographically dispersed case studies reveal systemic market substitution across Canada. Mitochondrial DNA 22: 106–122.
29. Lakra WS, Verma MS, Goswami M, Lal KK, Mohindra V, et al. (2011) DNA barcoding Indian marine fishes. Molecular Ecology Resources 11: 60–71.
30. Ward RD, Zemlak TS, Innes BH, Last PR, Hebert PD (2005) DNA barcoding Australia's fish species. Philosophical transactions of the Royal Society of London Series B, Biological Sciences 360: 1847–1857.
31. Vernooy R, Haribabu E, Muller MR, Vogel JH, Hebert PDN, et al. (2010) Barcoding life to conserve biological diversity: beyond the taxonomic imperative. PLoS Biol 8.
32. Craig MT, Sadovy de Mitcheson YJ, Heemstra PC (2012) Groupers of the world: a field and market guide: NISC (Pty) Ltd, Grahamstown.
33. Ekrem T, Willassen E, Stur E (2007) A comprehensive DNA sequence library is essential for identification with DNA barcodes. Molecular Phylogenetics and Evolution 43: 530–542.
34. Smith WL, Craig MT (2007) Casting the percomorph net widely: The importance of broad taxonomic sampling in the search for the placement of serranid and percid fishes. Copeia: 35–55.
35. Nelson JS (2006) Fishes of the world: Fourth edition. 601 p.
36. Lautrédou AC, Motomura H, Gallut C, Ozouf-Costaz C, Cruaud C, et al. (2013) New nuclear markers and exploration of the relationships among Serraniformes (Acanthomorpha, Teleostei): the importance of working at multiple scales. Molecular Phylogenetics and Evolution 67: 140–155.
37. Craig MT, Hastings PA (2007) A molecular phylogeny of the groupers of the subfamily Epinephelinae (Serranidae) with a revised classification of the Epinephelini. Ichthyological Research 54: 1–17.
38. Ratnasingham S, Hebert PDN (2007) The Barcode of Life Data System (www.barcodinglife.org). Molecular Ecology Notes 7: 355–364.
39. Buhay JE (2009) "Coi-like" sequences are becoming problematic in molecular systematic and DNA barcoding studies. Journal of Crustacean Biology 29: 96–110.
40. Ward RD (2009) DNA barcode divergence among species and genera of birds and fishes. Molecular Ecology Resources 9: 1077–1085.
41. Randall JE, Heemstra PC (1991) Revision of Indo-Pacific groupers (Perciformes: Serranidae: Epinephelinae), with descriptions of five new species. Indo-Pacific Fishes 20: 1–332.
42. Hall TA (1999) BioEdit: a user-friendly biological sequence alignment editor and analysis program for Windows 95/98/NT. Nucl Acids Symp 41: 95–98.
43. Posada D, Crandall KA (2001) Evaluation of methods for detecting recombination from DNA sequences: Computer simulations. Proceedings of the National Academy of Sciences of the United States of America 98: 13757–13762.
44. Swofford DL (2002) PAUP*. Phylogenetic Analysis Using Parsimony* and other methods. Sinauer Associates.
45. Stamatakis A (2006) RAxML-VI-HPC: Maximum likelihood-based phylogenetic analyses with thousands of taxa and mixed models. Bioinformatics 22: 2688–2690.
46. Stamatakis A, Hoover P, Rougemont J (2008) A rapid bootstrap algorithm for the RAxML web servers. Systematic Biology 57: 758–771.
47. Ronquist F, Huelsenbeck JP (2003) MrBayes 3: Bayesian phylogenetic inference under mixed models. Bioinformatics 19: 1572–1574.
48. Rambaut A, Drummond A (2007) Tracer v1.4. Available: http://beast.bio.ed.ac.uk/Tracer.
49. Leache AD, Reeder TW (2002) Molecular systematics of the Eastern fence lizard (*Sceloporus undulatus*): A comparison of parsimony, likelihood, and Bayesian approaches. Systematic Biology 51: 44–68.
50. Baumann F, Bourrat M-B, Pauillac S (2010) Prevalence, symptoms and chronicity of ciguatera in New Caledonia: Results from an adult population survey conducted in Nouméa during 2005. Toxicon 56: 662–667.
51. Chateau-Degat M-L, Huin-Blondey M-O, Chinain M, Darius T, Legrand A-M, et al. (2007) Prevalence of chronic symptoms of Ciguatera disease in French Polynesian adults. American Journal of Tropical Medicine and Hygiene 77: 842–846.
52. Lehane L, Lewis RJ (2000) Ciguatera: recent advances but the risk remains. International Journal of Food Microbiology 61: 91–125.
53. Lewis ND (1986) Epidemiology and Impact of Ciguatera in the Pacific - a Review. Marine Fisheries Review 48: 6–13.
54. Olsen DA, Nellis DW, Wood RS (1984) Ciguatera in the eastern Caribbean. U S National Marine Fisheries Service Marine Fisheries Review 46: 13–18.
55. Perez-Arellano JL, Luzardo OP, Brito AP, Cabrera MH, Zumbado M, et al. (2005) Ciguatera fish poisoning, Canary Islands. Emerging Infectious Diseases 11: 1981–1982.
56. Quod JP, Turquet J (1996) Ciguatera in Réunion Island (SW Indian Ocean): Epidemiology and clinical patterns. Toxicon 34: 779–785.

Author Contributions

Conceived and designed the experiments: CS AD. Performed the experiments: CS DDH. Analyzed the data: CS DDH. Contributed reagents/materials/analysis tools: CC. Wrote the paper: CS DDH AD JLJ.

57. Schlaich C, Hagelstein J-G, Burchard G-D, Schmiedel S (2012) Outbreak of ciguatera fish poisoning on a cargo ship in the port of Hamburg. Journal of Travel Medicine 19: 238–242.

58. Wong CK, Hung P, Lee KLH, Kam KM (2005) Study of an outbreak of ciguatera fish poisoning in Hong Kong. Toxicon 46: 563–571.

59. Wong C-K, Hung P, Lee KLH, Mok T, Chung T, et al. (2008) Features of ciguatera fish poisoning cases in Hong Kong 2004–2007. Biomedical and Environmental Sciences 21: 521–527.

60. Parker J, Rambaut A, Pybus OG (2008) Correlating viral phenotypes with phylogeny: Accounting for phylogenetic uncertainty. Infection Genetics and Evolution 8: 239–246.

61. R Core Team (2013) R: A Language and Environment for Statistical Computing, R Foundation for Statistical Computing, Vienna, Austria. Available: http://www.R-project.org.

62. Maddison WP, Maddison DR (2011) Mesquite: a modular system for evolutionary analysis. Version 2.75. Available: http://mesquiteproject.org.

63. Skinner MP, Brewer TD, Johnstone R, Fleming LE, Lewis RJ (2011) Ciguatera fish poisoning in the Pacific islands (1998 to 2008). Plos Neglected Tropical Diseases 5.

64. Tester PA, Feldman RL, Nau AW, Faust MA, Litaker RW (2009) Ciguatera fish poisoning in the Caribbean. Smithsonian contributions to the Marine Sciences 38: 301–311.

65. Centers for disease control and prevention (2013) Ciguatera fish poisoning - New York city, 2010–2011. JAMA 309: 1102–1104.

66. Dettai A, Berkani M, Lautredou AC, Couloux A, Lecointre G, et al. (2012) Tracking the elusive monophyly of nototheniid fishes (Teleostei) with multiple mitochondrial and nuclear markers. Marine Genomics 8: 49–58.

67. Mueller RL (2006) Evolutionary rates, divergence dates, and the performance of mitochondrial genes in Bayesian phylogenetic analysis. Systematic Biology 55: 289–300.

68. Richlen ML, Morton SL, Barber PH, Lobel PS (2008) Phylogeography, morphological variation and taxonomy of the toxic dinoflagellate *Gambierdiscus toxicus* (Dinophyceae). Harmful Algae 7: 614–629.

69. Mariani S, Ellis J, O'Reilly A, Bréchon AL, Sacchi C, et al. (2014) Mass media influence and the regulation of illegal practices in the seafood market. Conservation Letters.

70. Meusnier I, Singer GAC, Landry JF, Hickey DA, Hebert PDN, et al. (2008) A universal DNA mini-barcode for biodiversity analysis. BMC Genomics 9.

71. Fricke R, Kulbicki M, Wantiez L (2011) Checklist of the fishes of New Caledonia, and their distribution in the Southwest Pacific Ocean (Pisces). Stuttgarter Beitrage zur Naturkunde A Neue Serie (Biol) 4: 341–463.

72. Vilgalys R (2003) Taxonomic misidentification in public DNA databases. New Phytologist 160: 4–5.

73. Hassanin A, Bonillo C, Bui XN, Cruaud C (2010) Comparisons between mitochondrial genomes of domestic goat (*Capra hircus*) reveal the presence of numts and multiple sequencing errors. Mitochondrial DNA 21: 68–76.

74. Steinke D, Hanner R (2011) The FISH-BOL Collaborators protocol. Mitochondrial DNA 22 (S1): 10–14.

75. Palumbi SR (1996) Nucleic acids 2: the polymerase chain reaction. Molecular systematics Second edition.pp. 205–247.

76. Chen W-J, Bonillo Cl, Lecointre G (2003) Repeatability of clades as a criterion of reliability: a case study for molecular phylogeny of Acanthomorpha (Teleostei) with larger number of taxa. Molecular Phylogenetics and Evolution 26: 262–288.

77. Lautrédou AC, Bonillo C, Denys G, Cruaud C, Ozouf-Costaz C, et al. (2010) Molecular taxonomy and identification within the Antarctic genus *Trematomus* (Notothenioidei, Teleostei): How valuable is barcoding with COI? Polar Science 4: 333–352.

78. Streelman JT, Karl SA (1997) Reconstructing labroid evolution with single-copy nuclear DNA. Proceedings of the Royal Society B-Biological Sciences 264: 1011–1020.

20

Spatial Structure and Distribution of Small Pelagic Fish in the Northwestern Mediterranean Sea

Claire Saraux[1]*, Jean-Marc Fromentin[1], Jean-Louis Bigot[1], Jean-Hervé Bourdeix[1], Marie Morfin[1], David Roos[1], Elisabeth Van Beveren[1], Nicolas Bez[2]

1 IFREMER (Institut Français de Recherche pour l'Exploitation de la MER), Research Unit EME (UMR 212), Sète, France, **2** IRD (Institut de Recherche pour le Développement), Research Unit EME (UMR 212), Sète, France

Abstract

Understanding the ecological and anthropogenic drivers of population dynamics requires detailed studies on habitat selection and spatial distribution. Although small pelagic fish aggregate in large shoals and usually exhibit important spatial structure, their dynamics in time and space remain unpredictable and challenging. In the Gulf of Lions (north-western Mediterranean), sardine and anchovy biomasses have declined over the past 5 years causing an important fishery crisis while sprat abundance rose. Applying geostatistical tools on scientific acoustic surveys conducted in the Gulf of Lions, we investigated anchovy, sardine and sprat spatial distributions and structures over 10 years. Our results show that sardines and sprats were more coastal than anchovies. The spatial structure of the three species was fairly stable over time according to variogram outputs, while year-to-year variations in kriged maps highlighted substantial changes in their location. Support for the McCall's basin hypothesis (covariation of both population density and presence area with biomass) was found only in sprats, the most variable of the three species. An innovative method to investigate species collocation at different scales revealed that globally the three species strongly overlap. Although species often co-occurred in terms of presence/absence, their biomass density differed at local scale, suggesting potential interspecific avoidance or different sensitivity to local environmental characteristics. Persistent favourable areas were finally detected, but their environmental characteristics remain to be determined.

Editor: Konstantinos I. Stergiou, Aristotle University of Thessaloniki, Greece

Funding: Surveys were cofunded by IFREMER and the European Union through the Data Collection Framework (DCF). This study is part of the program EcoPelGol (Study of the Pelagic ecosystem in the Gulf of Lions), funded by France Filière Pêche (FFP). The funders had no role in study design, data collection and analysis, decision to publish, or preparation of the manuscript.

Competing Interests: The present study is part of a more general program investigating changes in small pelagic populations in the Gulf of Lions, named EcoPelGol. This program is partly funded by the interprofessional association France Filière Pêche. The authors are free to fully develop and publish their research without consulting France Filiè`re Pe^che, as per previous signed agreement with the association. France Filiè`re Pe^che has not yet seen these results, which will be communicated to them once published only.

* Email: claire.saraux@ifremer.fr

Introduction

Because animal spatial distribution is often strongly associated with population dynamics, spatial indices may provide valuable tools for assessing the status of these populations, and in particular the status of endangered or exploited species [1–4]. At the population scale, spatial distribution can be seen, in the absence of substantial anthropogenic impacts, as the emergent property of habitat selection. In addition to social motivation (presence *vs.* absence of conspecifics; [5]), spatio-temporal aggregation patterns in animal populations may be explained by individuals sharing similar needs and dealing with similar biotic and abiotic pressures, such as prey abundance *vs.* predation risk [6], propitious *vs.* detrimental environmental conditions [7]. Nonetheless, as population density increases, so does intra-specific competition [8], and individuals are expected to spread towards less suitable habitats when a certain threshold is reached ('Basin hypothesis'; [9]). Both population density and its occupation area should then vary with its abundance [10–12]. If their density-dependent and density-independent drivers are difficult to disentangle (but see [13]),

spatial distributions in themselves offer valuable information for both ecological understanding and management.

Indeed, besides an obvious fundamental interest in terms of population dynamics and marine ecosystem functioning [14], information on fish biomass location is also of crucial importance for stock management. For instance, the effective implementation of Marine Protected Areas (MPAs) requires detailed knowledge on species spatio-temporal dynamics, *i.e.* on temporal variability in fish spatial distributions [15]. In addition, knowledge on interspecific interactions, such as the co-occurrence or repulsion of species based on their spatial distributions, may greatly help scientists to understand competition or predator-prey processes but also policy makers to implement ecosystemic rather than single species management measures, as presently done.

In this study, we investigated the spatio-temporal distribution of small pelagic fish in the Gulf of Lions using a unique dataset of 10 years of acoustic surveys. Because small pelagic fish are key species of the pelagic ecosystem due to their central place in the food web, transferring energy from the lowest trophic levels (plankton) towards top-predators [16], and because their life-history traits (short lifespan, large fecundity) make them strongly dependent on the abiotic environment [17], changes in their distribution and abundance have been a major source of concern for scientists and managers worldwide ([18]; Global Ocean Ecosystem Dynamics program - GLOBEC;).

Over the past years, important changes in the two main target species (the European sardine, *Sardina pilchardus* and the European anchovy, *Engraulis encrasicolus*) biomass [19], along with a shift in the size-distribution of these species towards smaller individuals [20] have been observed in the Northwestern Mediterranean Sea (Gulf of Lions), resulting in important economic losses for fisheries. Both stocks are now considered to have a low biomass and a low fishing mortality by General Fisheries Commission for the Mediterranean [19]. In parallel, a third small pelagic species, the sprat (*Sprattus sprattus*), which is not commercially exploited in the Western Mediterranean has appeared in the system with a steadily increasing biomass since 2007 [19]. This unexpected situation offered us the opportunity to describe and compare simultaneous spatio-temporal distributions of three species sharing the same trophic level, with similar feeding behaviours [21–22], but with completely different trends in biomass, through the analysis of 10 years of scientific acoustic surveys. First, we studied the aggregation patterns and spatial structures of these species separately. Second, we investigated species spatial dynamics by considering spatio-temporal changes in biomass, and defined optimal recurrent areas based on a combination of biomass levels and their variability. Finally, we focused on interspecific relationships by studying the overlap between species at different spatio-temporal scales.

Methods

Ethic statement

The study was conducted in the Gulf of Lions (Longitude in [3.05°; 5.20°] and Latitude in [42.44°; 43.44°]), a public sea area. No sampling was operated from private land and field studies did not involve endangered or protected species. All data used in this study came from PELMED acoustic surveys, which comply with the MEDIAS (Mediterranean Acoustic Survey) protocol. The sampling has been performed under repeated international standardized surveys where the research vessel had full permission to sample from all relevant national public authorities (governments). Acoustic data were collected at a distance, which does not require any particular ethic approval. Further data used in this study came from scientific trawls conducted during these same PELMED surveys. Again, no approval by an ethic committee was required as the targeted species are exploited species and trawling methods done according to international standard trawl surveying.

Data collection and survey design

Every July since 2003, the French Research Institute IFRE-MER has been carrying out acoustic surveys of the pelagic resources present in the Gulf of Lions, Mediterranean Sea. The summer period of the survey corresponds to contrasted biological periods for our 3 species. Indeed, this is the peak of reproduction for anchovies, while sprats and sardines reproduce in winter [23]. Due to the biological cycle of these species, it is important not to extrapolate our results outside the summer period. Sampling was performed along 9 parallel transects, regularly spaced by 12 nautical miles (nm) (see Figure S1 in File S1). Acoustic data were recorded every 1 nm using multi-frequency echosounders (Simrad EK500 and ER60), while travelling at a constant speed of 8 nm.h^{-1}. All 4 frequencies were visualized during sampling to help deciding when to trawl for species identification. However, only energies from the 38 kHz (typical frequency used for fish)

channel were used to estimate fish density. Acoustic data analyses, such as bottom correction, were later performed using *Movies+* [24] and *FishView* IFREMER softwares. Species discrimination and echo-partitioning were performed by the combination of echo trace classification and trawl outputs [25]. Species biomass and abundance were finally estimated from species energy using specific target strength (TS = 20 log(L) - 71.2, where L is the length of the fish for all 3 species, see [26] for more details on acoustic surveys and analyses). Main survey features are summarised in Table 1.

Analyses

We defined and used different spatial indicators and geostatistical methods described in details below. An overview of these indicators, including their calculation formula, representative scale and biological meaning are summarised in Table 2. A key concept in statistics in general and in spatial statistics in particular is the support of the information, i.e. the geographical area over which measures are recorded [27–28]. In this study, the support size was the size of the Elementary Sampling Distance Unit, that is 1 nm. All statistics derived from these data are thus associated to the 1 nm sampling support and, for all of them, their values would have been different if computed at another support. Even though some metrics are labelled like spatial statistics (e.g. space selectivity index, local index of collocation), they are not sensitive to the location of the points in space (i.e. any exchange between two data does not change the result). They are sensitive to the spatial patterns that exist at scales smaller than the support size and that are integrated in the observations. For some of them (e.g. Empirical Orthogonal Functions), it remains possible to represent them as geographical distributions making the confusion even worth.

Aggregation patterns and spatial structure. Presence Area (PA) was calculated as the percentage of sampled points at which the species was found, independently of its abundance.

A space selectivity index was defined on the basis of annual concentration curves [29–30]. Like Lorenz curves (as used in [31]), concentration curves represent the maximum proportion of biomass as a function of the proportion of samples [30]. For instance, a proportion of biomass of 0.6 for a proportion of sampled area of 0.2 would indicate that 60% of population total biomass was found in only 20% of the total area sampled. For a homogeneously distributed population, each proportion of biomass should be found in the same proportion of total area (i.e. 10% of biomass in 10% of area, etc.), so that the annual concentration curve of a homogeneously distributed population equals the first diagonal. It follows that the more concentrated a population, the further the curve falls from the first diagonal, *i.e.* a higher proportion of biomass is situated in a given proportion of the total area. Consequently, the space selectivity index is defined as twice the area between the concentration curve and the first diagonal [29–30], so that the higher the space selectivity index, the more concentrated the spatial distribution of the fish.

Spatial structure was defined as the spatial autocorrelation between values at different locations. In other words, spatial structure considers whether the correlation between biomass in two locations depends on the distance separating them. It was investigated using variograms [27] that calculate the value of the spatial autocorrelation at different lags (distance intervals). As is often the case, the frequency distributions of species biomass were highly skewed with a large proportion of zeros or small values, and few extremely large values contributing importantly to total biomass (coefficients of variation CV were 2.6, 3.8 and 4.2 for anchovies, sprats and sardines, respectively). Besides, the variance

Table 1. Main PELMED survey features from 2003 to 2012.

Year	Sampling dates	Nb of nm	Nb of EDSU	Nb of trawls	Anchovy biomass	Sardine biomass	Sprat biomass
2003	07/07/03-06/08/03	1274	284	27	27 860	126 120	685
2004	05/07/04-04/08/04	1174	285	29	25 953	215 560	786
2005	08/07/05-07/08/05	1462	294	33	15 962	264 024	1 955
2006	09/07/06-08/08/06	1473	288	39	25 658	102 276	772
2007	11/07/07-10/08/07	1500	290	43	13 654	88 297	15
2008	19/06/08-30/07/08	2151	273	60	23 395	91 546	5 002
2009	24/06/09-29/07/09	2173	284	43	30 424	52 977	7 845
2010	24/06/10-29/07/10	2000	276	39	23 514	51 819	15 760
2011	27/06/11-31/07/11	1704	282	42	25 906	44 926	26 638
2012	27/06/12-31/07/12	1172	279	37	39 061	80 537	70 263

Biomasses are indicated in tons, nm stands for nautical mile.

in biomass increased proportionally with mean biomass in our data. This led us to log-transform the studied variables (the optimal Box-Cox transformation was $Y = ln\ (B+c)$, with c a positive constant added to insure positive values). This constant was set to the smallest biomass observed in our data, $i.e.$ 0.05 tons. While distributions of the log-biomass were still skewed, the proportionality effect disappeared. The log-transformed data was used to calculate empirical variograms for all three species for each year of survey. To account for the anisotropy of the sampling (only along parallel transects) and for a potential anisotropic structure due to a bathymetry gradient, bidirectional variograms (along transects and perpendicularly to them) were calculated and modelled by automatic fitting using a least square method [32]. The temporal variability of the spatial structure was assessed on annually standardised data by comparing annual variograms to the mean variogram (average across all years of survey, pairs being retained only if the two sampling locations belong to the same year) and its 95%-confidence interval (adaptation from [12] for bidirectional variograms). The confidence interval was obtained by simulating 1000 random fields according to the mean modelled variogram using the turning-bands method [33] and estimating the 1000 variograms associated with the sampling points extracted from these random fields.

Species spatial distributions. To capture the spatial patterns of the 3 populations as simply as possible and to investigate year-to-year variations, we calculated the centres of gravity of their biomass along with their associated inertia [34–35]. The centre of gravity represents the mean location of the population, while inertia describes the dispersion of the population around its centre of gravity. Because fish populations are often aggregated in a given area and therefore spatial distributions heterogeneous, we also investigated the presence of spatial patches of high biomass densities, by adapting the recursive algorithm developed by [35]. This was adapted to take into account the anisotropy of our sampling by setting two different threshold distances (one parallel to the transect and another one perpendicular). The sensitivity to the threshold distances was tested using different values. Very small distances did not enable us to identify patches with more than 10% of biomass, while increasing the distance reduced the number of patches to one which comprised the whole presence area. The final distances (i.e. 6 nm in the direction of the transect and 24 perpendicular to it) were set in the interval where the resulting number of patches was stable and were kept constant both across species and years. The number of patches, centres of gravity and inertias of all patches were then used to study year-to-year and interspecific variations.

Kriging maps were produced for each species on a 1 nm*1 nm-grid over the study area, using the modelled mean or annual variograms depending on the results of interannual variability tests. In order to account for anisotropic sampling, we used an anisotropic search of neighbours when conducting kriging, $i.e.$ the distance at which neighbours were looked for was smaller in the transect direction than perpendicular to it to make sure that each point comprised neighbours in two different transects.

Two different methods were used to quantify the temporal stability/variability of spatial distributions at 2 different scales (Table 2). First, to assess the stability/variability of the entire distribution, we calculated the Empirical Orthogonal Functions (EOF) on the raw acoustic data for each species. EOF analysis is a decomposition of a spatio-temporal dataset in terms of orthogonal basis functions. It is thus very similar to a Principal Component Analysis, except that it is applied to spatio-temporal data, using time as a descriptor and space (i.e. sampling locations) as objects [36]. EOF was performed on standardized data to compare

Table 2. Spatial indicators used in the study.

Indicator name	Formula	Spatial scale	Time scale	Biological meaning
Presence area				
Presence area	$PA = \frac{\sum 1_i}{n} * 100$	Global	Annual	Area of presence(in %) of the species over the studied area
Aggregation and spatial structure				
Space selectivity index		Global	Annual	Spatial compactness
Annual variograms			Annual	Spatial autocorrelation
Mean variogram			Decadal	Mean Spatial autocorrelation
Spatial distribution				
Centre of gravity	$CG = \frac{\int x\,z(x)\,dx}{\int z(x)\,dx}$	Global	Annual & decadal	Mean location of the species
Inertia	$I = \frac{\int (x - CG)^2\,z(x)\,dx}{\int z(x)\,dx}$	Global	Annual & decadal	Spatial dispersion around CG
Patches		Intermediate	Annual	Concentration of biomass
EOF		Global	decadal	Spatio-temporal variability
CV	$CV_x = \frac{\sigma_{z_a}(x)}{Mean(z_{a\,(x)})}$	Local	decadal	Temporal variability
Interspecific relationships/co-occurence				
Overlap	$O_{i,j} = \frac{A_{E_i} \cap E_j}{A_{E_i} \cup E_j}$	Global	Annual decadal	Overlap between the spatial distribution envelops of two species
Overlap of patches	$O_{i,j} = \frac{\sum_{P_i}\sum_{P_j} A_{E_{P_i}} \cap E_{P_j}}{A\bigcup_{P_i,P_j} E_{P_i} E_{P_j}}$	Intermediate	Annual decadal	Proportion of patches where both species present
Co-occurrence index	$Co_{i,j} = \frac{\sum 1_{ij}}{\sum 1_i + \sum 1_j}$	Local	Annual decadal	Level of local co-occurrence between 2 species based on presence/absence
Local index of collocation	$LIC_{ij} = \frac{\int z_i(x)\,z_j(x)\,dx}{\sqrt{\int z_i^2(x)\,dx}\sqrt{\int z_j^2(x)\,dx}}$	Local	Annual decadal	Level of local cooccurence between species based on densities
Dominance index	$Dom_{i,j} = \frac{z_i(x) - z_j(x)}{z_i(x) + z_j(x)}$	Local	Annual decadal	Relative biomass in each point
Mapping				
Kriging	Anisotropic neighbourhood	Global	Annual	Distribution maps
Area classification				
	Variability map * Average map	Global	Decadal	Identification of recurrent, occasional & unfavourable areas

n: Number of sampling points, z(x): biomass density in x, z_a (x): annual biomass density in x.

annual spatial distributions independently of annual total levels of biomass. The first axis (eigen vector) of the EOF is the linear combination of years which maximizes the percentage of the interannual variance of spatial distributions. When all yearly contributions to this first axis share the same sign, the greater the percentage of variance explained by the first EOF, the more persistent the spatial distributions [37;12]. In order to have a local estimation of temporal variability/stability and see whether some areas were more stable than others, we also calculated the Coefficient of Variation (CV) of the 10 annual values in each sampling point. As biomass varied substantially between years, we used relative biomass (i.e absolute biomass divided by the total

annual biomass) instead of absolute biomass, so that the observed CV corresponded to a change in geographical location rather than a change in overall biomass.

Finally, we defined recurrent, occasional and unfavourable areas by combining average and variability maps [38]. To produce average and variability maps, we calculated the mean and the standard deviation of the 10 annual kriged maps node by node (1 nm*1 nm). Then, each 1 nm*1 nm pixel was assigned one of the 3 categories (recurrent, occasional and unfavourable) depending on its value in the average and variability maps. Pixels with both low mean and variability (i.e. inferior to the median) were considered unfavourable; those with high mean and low variability recurrent, while pixels with high variability were defined as occasional.

Interspecific indices: collocation index and overlap between species. We quantified the collocation at three different spatial scales in a coherent manner: from a global scale comparing the distribution over the whole study area, to a local scale comparing species at each sampling point, passing by an intermediate scale comparing previously defined patches.

Global overlap was defined based on centres of gravity and associated inertia ellipses (see above), as follows:

$$O_{i,j} = \frac{A_{E_i \cap E_j}}{A_{E_i \cup E_j}}$$

where E_i and E_j are the inertia ellipses of species i and j respectively and A is the area. The index varies from 0 when the two ellipses are totally separated to 1 when the two ellipses are identical.

At the intermediate scale, we used the same index but calculated from the inertia ellipses associated with the patches of the monospecific distributions:

$$O_{i,j} = \frac{\sum_{p_i} \sum_{p_j} A_{E_{p_i} \cap E_{p_j}}}{A_{\bigcup_{p_i,p_j} E_{p_i} E_{p_j}}}$$

where E_{pi} and E_{pj} are the ellipses associated with the patches of species i and j respectively.

To characterize the co-occurence of two species at the local scale, we used two different indices: (*i*) an index of co-occurrence defined as a proportion of sampling points where both species co-occur (taking into account only the sampling points where at least one of the two species is present), and (*ii*) a local index of collocation (LIC; [39]) defined as:

$$LIC_{ij} = \frac{\int z_i(x) \, z_j(x) \, dx}{\sqrt{\int z_i^2(x) \, dx} \sqrt{\int z_j^2(x) \, dx}}$$

where $z_i(x)$ and $z_j(x)$ are the biomass density of the species i and j in the location x. Both indices vary between 0 (no location where the 2 species are found simultaneously) and 1 (the two species are situated in exactly the same locations). Yet, they differ in the fact that the former is based only on presence/absence while the second is based on biomass density.

To evaluate the deviation of these indices from random expectation, we used a randomization test by non-parametric bootstrap of 1000 values. For species of strong local overlap (i.e. anchovies and sardines), we also calculated a dominance index based on the relative difference in biomass of the two species in each point (Table 2). This index varies between -1, where only the species j is present and 1 where only the presence i is present, 0 corresponding to a situation where species i and j are equally abundant [40]. This index enabled us to better represent which species dominate where.

Softwares. Statistics were conducted using R v. 2.15.0 [41]. Geostatistical analyses were performed with the package RGeoS [42]. Spatial data are given in WGS84 coordinate system. Analyses were conducted on both raw and kriged data. As both methods yielded very similar results, we only presented results on raw data for clarity purposes. Kriging was produced only for mapping purposes.

Results

Aggregation patterns and spatial structure

Species were absent from several sampled locations, regardless of whether anchovies (absent from 9.3% of sampled locations), sardines (12.9%) or sprats (51.0%) were considered. However, these absences were not spatially consistent across years. When pooling all data, anchovies and sardines could be observed at least once in all sampled locations; sprats in contrast were never found after the 200 m isobath (Fig. S1 in File S1). During each of the 10 study-years, sardines and anchovies occupied most of the Gulf of Lions, and presence area only varied slightly for these two species (PA: 85.0 to 94.7% and 67.2 to 99.6% for anchovies and sardines, respectively). Variability was far greater for sprats, which were almost completely absent in 2007 (PA = 0.3%), but covered most of the Gulf of Lions in 2012 (PA = 91.8%). Presence area was positively correlated with total log-biomass index for sprats (LM: $R^2_{adj} = 0.87$, p<0.001), but not for anchovies (p = 0.68) or sardines (p = 0.10) (Figure 1), *i.e.* the area occupied by the population expanded with total logbiomass in sprats but not in the other two species. Log-biomass and presence area were far less variable for sardines and anchovies than for sprats (CVs being at least 4 to 7 fold smaller for PA and biomass, respectively). The mean density in presence area also increased with log-biomass for anchovies ($R^2_{adj} = 0.71$, P = 0.001) and sprats ($R^2_{adj} = 0.69$, P = 0.002). The same trend was observed in sardines, but the probability was lower ($R^2_{adj} = 0.27$, P = 0.07).

The degree of aggregation appeared to be different in anchovies compared to sprats or sardines (Figure 1; Figure S2 in File S1). Anchovies were, on average, less concentrated than the two other species. This was confirmed by the space selectivity index, which was lower in anchovies (0.722) than in sardines and sprats (0.850 and 0.867 respectively). Over the 10 study-years, yearly space selectivity indices decreased significantly with increasing log-biomass in anchovies (LM: $R^2_{adj} = 0.43$, P = 0.02,) and in sprats ($R^2_{adj} = 0.94$, P<0.001), but not in sardines (P = 0.41, Figure 1).

Mean variograms of log biomass exhibited spatial structure for the three species in both directions (i.e, increasing variance when the distance between sampling points increased before levelling off; Figure 2), meaning that the spatial distribution was not random and that two close points had a higher probability to have similar values than distant points. While sardine and anchovy variograms shared similar main characteristics, sprat mean variogram differed slightly. In particular, it differed in the range at which variance stabilized (Figure 2). Variance stabilized rapidly in sprats (range = 4 and 12 nm respectively for the first and second spherical component), while the first structure appeared around 15 nm for sardines and anchovies and the second was not apparent at the scale of our study, *i.e.* variance did not stabilize completely within the 50 nm limit that we fixed. Further, the mean structure seemed similar between the two perpendicular directions in sprats, while it differed slightly in anchovies and sardines, though the mean variogram of each direction was mostly included in the confidence interval around the mean variogram of the other direction.

Annual variograms were computed for each of the three species (Figure 2). All exhibited clear spatial structure and were included in the 95% confidence interval around the mean variogram for anchovies and sardines (see Figure 2). This revealed the absence of a year effect on the spatial structure of these 2 species. In sprats however, some variograms (especially along transect) were not included in the confidence interval, suggesting interannual variability in spatial structure.

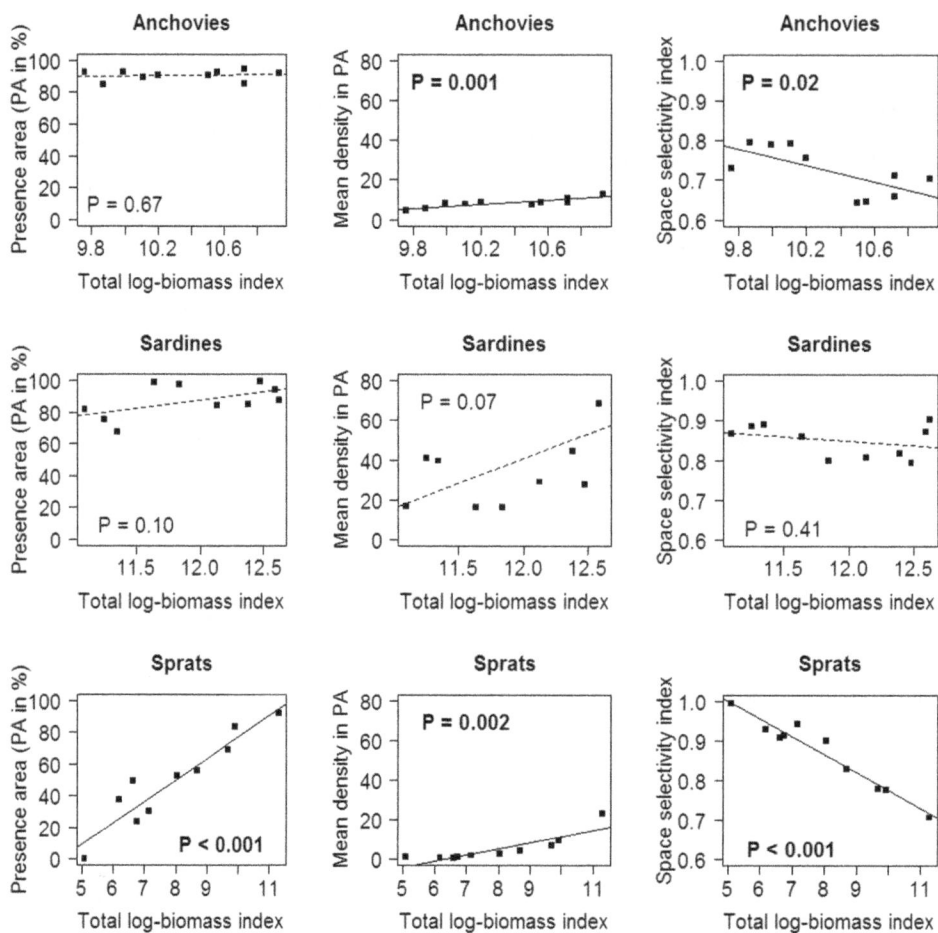

Figure 1. Presence areas, biomass densities and space selectivity indices relatively to total log-biomass indices in anchovies, sardines and sprats. Lines represent the linear regressions. Significant linear regressions are represented by plain lines and Pvalues are indicated in bold. For non-significant relationships, the trend is shown by a dotted line.

Species spatial distribution

Log-biomass distributions. The annual spatial distributions exhibited similarities between species in some years (*e.g.* in 2011 a large area in the middle of the shelf was unoccupied both for anchovies and sardines), but not in others (Figures S3, S4, S5 in File S1). Some consistent differences between species also appeared. For instance, anchovies occupied the centre of the continental shelf, while sardines were more coastal. For the three species, kriged annual distributions also revealed the important interannual variability of these populations both in terms of biomass levels and their spatial repartitions.

Centres of gravity of spatial distributions did not vary much according to species or year and were situated close to the geometrical centre of gravity of the sampled area (8.2±0.8 nm). Interestingly, the associated inertia was high for anchovies and sardines (inertia = 1152±70 nm^2), meaning that the mean distance between a presence point and the centre of gravity of its population was 34 nm. This distance was remarkably close to that between a sampling point and the geometrical centre of gravity of the total area (37 nm). As a consequence, the ellipse associated with this inertia covered most of the sampled area for these two species (Figure S6 in File S1), *i.e.* the populations were dispersed in the entire Gulf of Lions. In sprats, while inertia was quite low in the first 6 years (447±101 nm^2; Figure S6 in File S1)

due to their low presence, it reached values almost as high as the ones in sardines and anchovies, in the last 4 years (991±75 nm^2).

The number of patches present per year and per species varied between 1 and 4 (Figure 3). As inferred by the square root of the inertia associated with the patch, the mean distance between a presence point in a patch and its centre of gravity ranged from 6 nm to 17 nm (excluding 2007 in sprats for which the species had only been observed in 1 location). The centres of gravity of anchovy patches were slightly more offshore than those of sardines (14.9±0.8 vs. 11.1±1.0 nm, Wilcoxon test: W = 662, P = 0.001), confirming the tendency of sardines to be more coastal than anchovies.

Instability of the spatial distribution. EOF of the 10 annual maps of each species confirmed the visual inspection highlighting high temporal variations in the spatial distributions. The first component of the EOF explained no more than 20% of the variance (14%, 20% and 17% for anchovies, sardines and sprats, respectively). Further, no trend could be detected from the contributions of each year to the first 2 axes of the EOF. The sign of the correlations between annual maps and the two first axes varied depending on the year for the three species (Figure S7 in File S1).

Coefficients of variation (calculated in each sampling point) were higher in sprats than in sardines and anchovies. Most coefficients of variation were higher than 1 (100%, 92.3% and

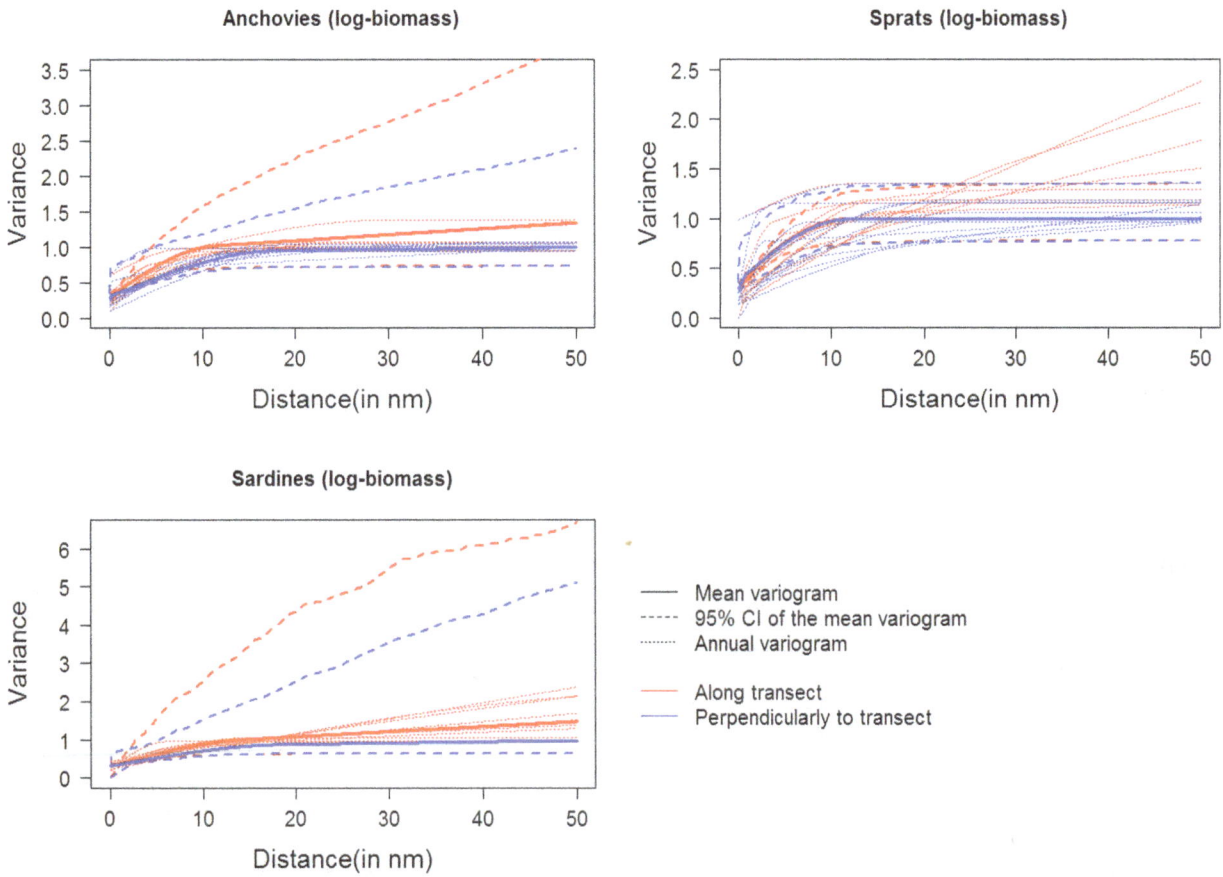

Figure 2. Annual (black) and mean (red) modelled variograms of anchovies, sardines and sprats. The red dotted lines correspond to the 95% confidence interval of the mean variogram deduced from 500 simulations.

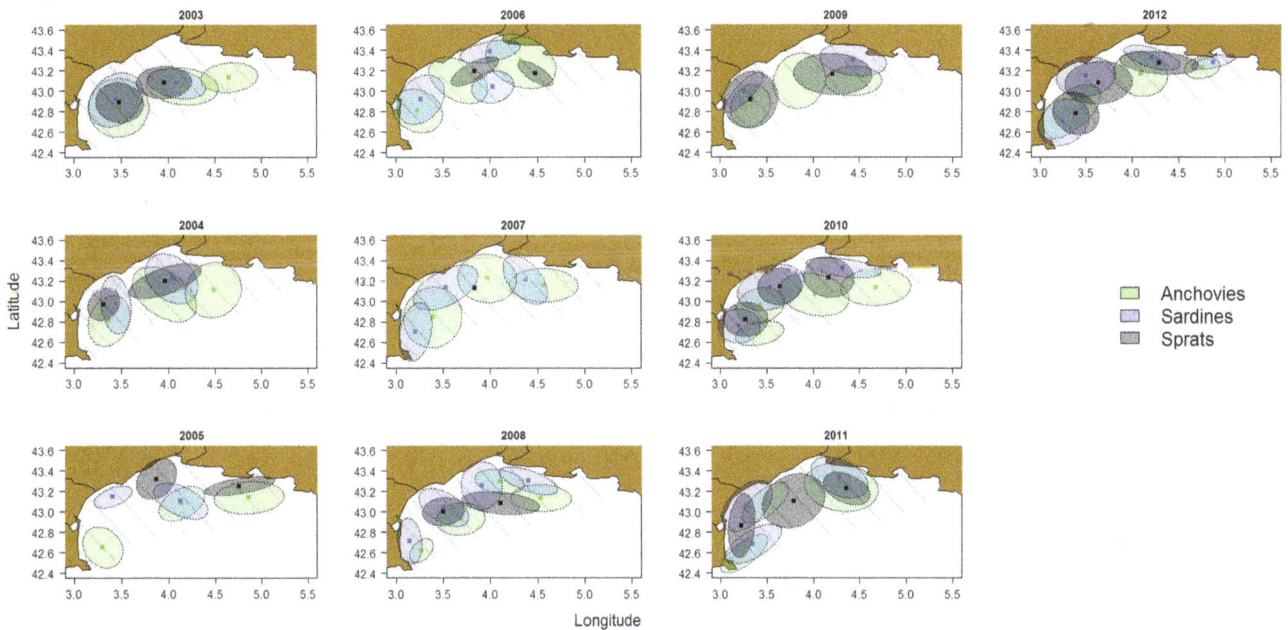

Figure 3. Annual maps of centres of gravity and inertia of patches for anchovies (in green), sardines (in blue) and sprats (in black).

81.6% of CV values were higher than 1 in sprats, sardines and anchovies, respectively), meaning that the distributions were overdispersed and again suggesting high temporal variability of the spatial distributions for all three species.

Recurrent, occasional and unfavourable areas. Average maps confirmed the tendency for sardines to be quite coastal, while anchovies occupied most of the shelf (Figure 4). Sprats seemed in an intermediate position where high biomass areas were situated in the centre of the Gulf of Lions, neither too coastal nor too offshore, but, as noted above, sprat biomass was also more variable. The combination of average and variability maps enabled us to detect a recurrent (persistent) area for anchovies in the centre of the Gulf, slightly West of the Rhone estuary, and, for sardines, in western coastal areas. Recurrent areas were minimal for sprats as the period encompassed some years with a quasi-absence of this species. Finally, deep waters represented unfavourable areas both for sardines and sprats.

Interspecific indices: collocation index and overlap between species

Regardless of the scale at which we investigated it, collocation indices were higher between sardines and anchovies than between any of these 2 species and sprats (Table S1 in File S1), meaning that sardines and anchovies co-occurred more often than they did with sprats. Additionally, collocation indices were higher, on average, at the global scale than at the intermediate scale (*e.g.* between anchovies and sardines 0.53 ± 0.04 *vs.* 0.30 ± 0.04 for the global and intermediate scales, respectively), suggesting that species globally lived in the same areas, but were not always found together in each given sampling location. Collocation indices, whether global, local or intermediate, varied substantially from year-to-year (Table S1 in File S1). Collocation indices at the three different scales were not correlated with each other, except for the global and intermediate indices between sprats and anchovies. For instance in 2004, the index of collocation between sprats and sardines was fairly high at the global scale (2nd highest observed), while the associated LIC was low and not significantly different from random expectation. The two local collocation indices were not correlated for anchovy and sardine association or anchovy and sprat association ($\rho = -0.61$, $P = 0.07$; $\rho = 0.44$, $P = 0.20$ respectively), suggesting that the co-occurrence of species (i.e. species found in the same location), did not necessarily concur with a co-occurrence of their hotspots or peaks of biomass. They were however slightly correlated when looking at the association between sardines and sprats ($\rho = 0.72$, $P = 0.02$).

Finally, although the co-occurrence between sardines and anchovies was high (Table S1 in File S1), in most sampled points one of the two species clearly dominated the other (Figure 5). The

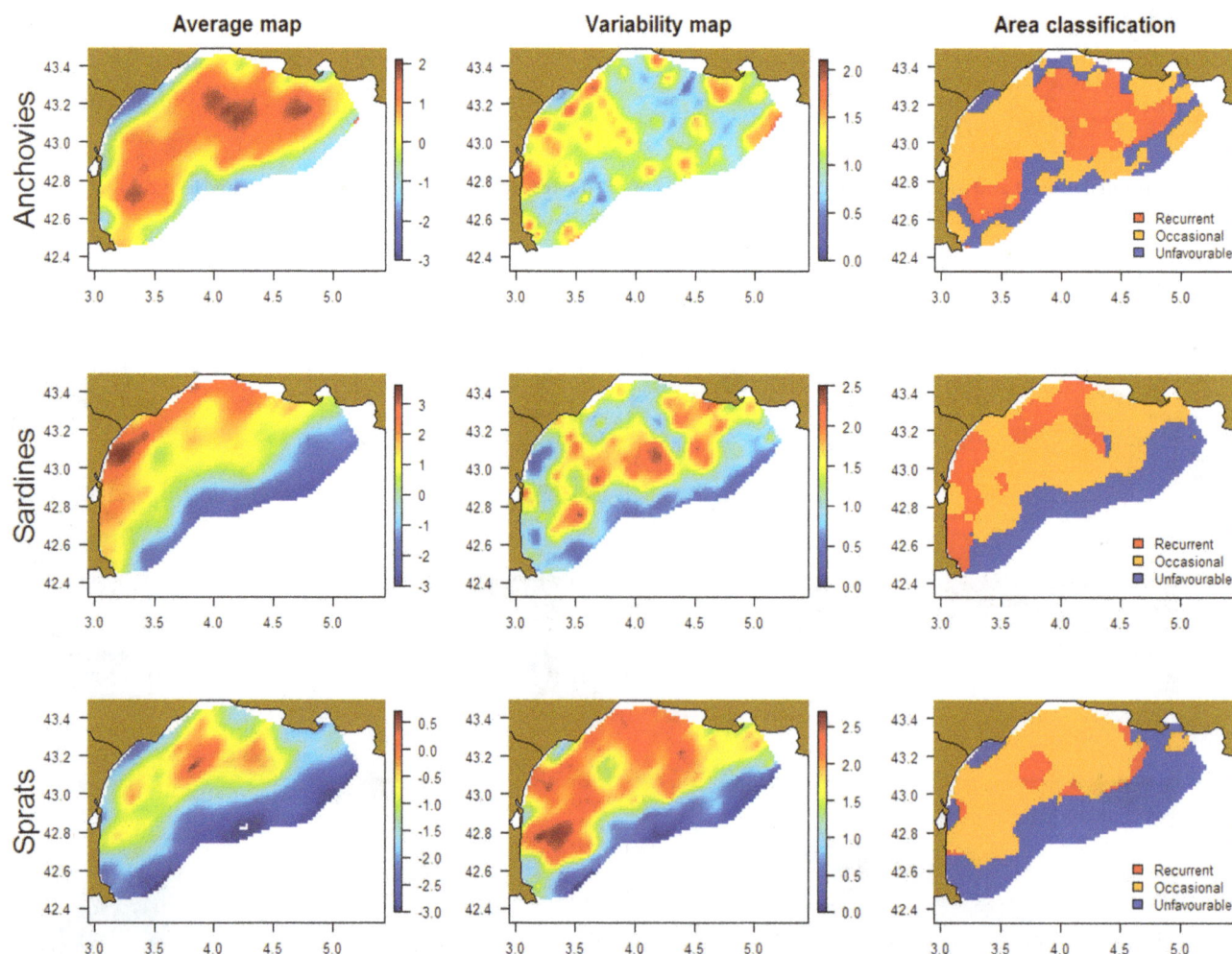

Figure 4. Average and variability maps and area classification for anchovies, sardines and sprats.

dominance index increased with depth, *i.e.* anchovies dominated sardines in deeper offshore waters, while sardines were dominant in shallow waters close to the coast (see also statistics in Figure S8 in File S1). Besides, the dominance index also increased with longitude, anchovies dominating towards the East (see statistics in Figure S8 in File S1).

Discussion

The strong increase in sprat biomass (from almost 0 to 70,000 tons from 2007 to 2012) that has paralleled a strong decline in anchovy and sardine biomass over the past few years [19], highlighted the urgent need to better understand the cross and mutual dynamics of those three species. This study documents the spatio-temporal patterns observed in these three species in the Gulf of Lions. However, it should be noted that surveys were conducted only in summer, corresponding to peak reproduction of anchovies but resting period for sardines and sprats. The results presented here thus correspond to different biological phases for the three species and should not be extrapolated to other non-monitored seasons.

Our main objective was to investigate the temporal variability of biomass and spatial distribution in the three species. Biomass estimates obtained by direct acoustic stock assessment substantially varied during the 10 years of survey. However, sprat was the only species in which a clear trend in biomass could be identified over the study period. For anchovies and sardines, biomass variability corresponded, respectively, either to fluctuations around a low value or a succession of an increasing and decreasing trends. The increase in sprat biomass resulted in both an increase of fish local densities and an expansion of fish towards new areas, in agreement with the McCall basin hypothesis [9]. On the contrary, the Gulf of

Lions being a favourable area for small pelagics [43], sardine and anchovy presence area was large and stable over the entire study period and fluctuations in biomass only resulted in fluctuations of local densities. Changes in local densities could be a buffer for fluctuating biomasses of already well implanted species. However, if a species continuously increase or decrease its biomass, this might well not be enough anymore. For instance, sprats which were almost absent from the area at the start of the study had to expand their presence area to cope with their increasing biomass. Inversely, a declining population would see local areas disappear from its presence area as these species are gregarious and have to stay in shoals. Our results may thus support the hypothesis that interannual variability may not be enough to observe contraction/expansion, and that a directional trend would be needed to investigate the link between abundance and distribution [44–45]. Still, using space selectivity indices to investigate aggregation patterns in more details, we found that sprats and anchovies became less selective with increasing biomass, suggesting that they selected less suitable areas rather than increased their local density. This is in contrast with sardines for which the space selectivity index did not vary according to total biomass and densities 5 fold higher could be reached. This could suggest different social constraints/benefits, sardines being able to aggregate in much denser shoals than the other 2 species.

Similarly to what was found in demersal species [12], we highlighted the stability of the spatial structure (i.e. spatial autocorrelation) of small pelagic fish in the Gulf of Lions over time. Only for sprats did some interannual variability appear, but this is probably due to a reduced confidence interval around the mean variogram due to low and distance-independent variance between sampling points in the first half of the study, when sprats were quasi-absent. Spatial structure is defined as the spatial

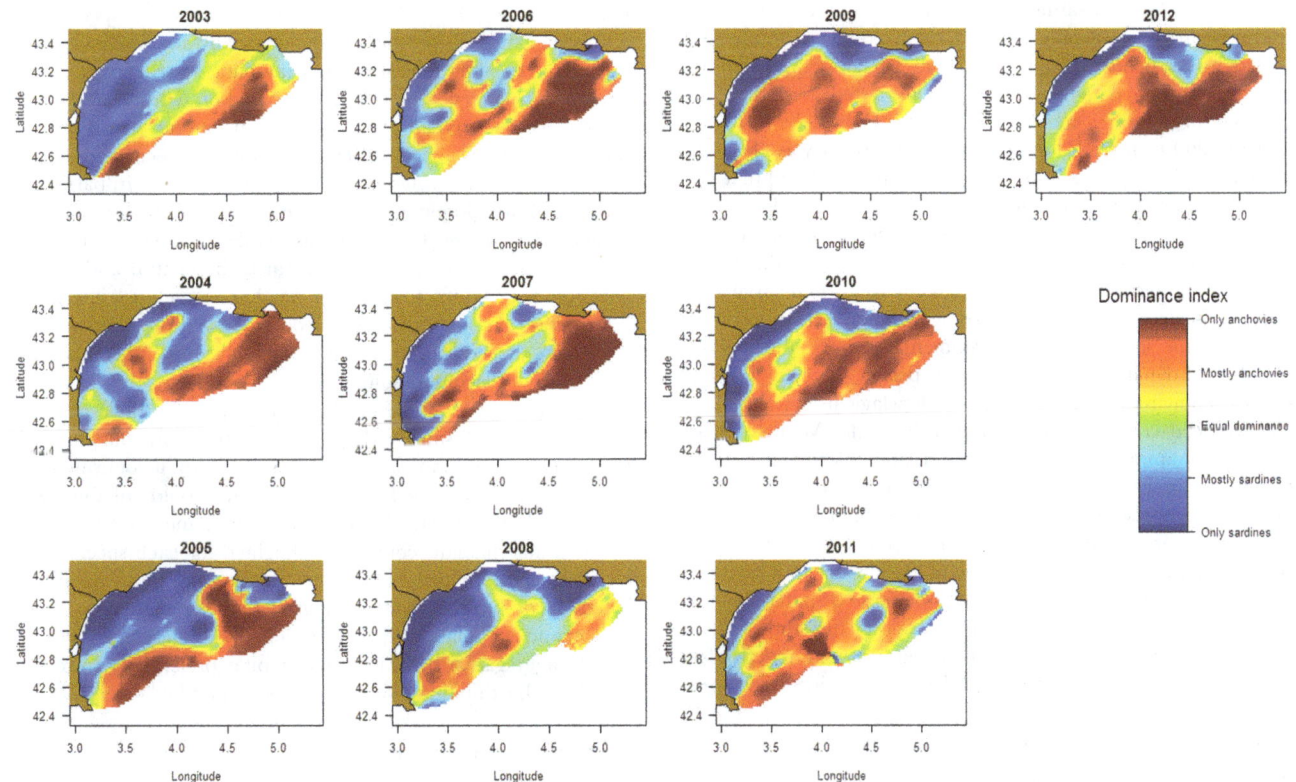

Figure 5. Annual maps of relative biomass between anchovies and sardines.

autocorrelation between values at different locations, i.e. in our case whether biomasses taken at 2 close points were more correlated than biomasses of points situated far apart. From an ecological point of view, spatial structure could thus indicate the occurrence of concentration in certain areas due to similar needs (e.g. habitat preferences) or aggregation patterns due to species social constraints (formation of shoals for instance here). Such stability of the spatial structure could result from important aggregative behaviours of small pelagic species, giving them intrinsic aggregation properties, independently from their abiotic environment.

Though necessary, the study of the spatial structure offers no insight on the locations where these aggregations took place or where the population was absent. In a second step, we thus investigated temporal changes in the geographic locations of biomass. In contrast to the stable spatial structure, the geographical distributions of small pelagics were highly variable from year-to-year. Although such variability is clearly visible when comparing the maps, it is more difficult to quantify it. Therefore, we resorted to spatial indicators (CVs and empirical orthogonal functions) for summarizing annual information and exploring possible spatial shifts in population distribution [35;46]. These indicators confirmed the visual inspection of the annual maps (high interannual variability) and thus the expected mobility of small pelagic fish, which contrasted to the stable and area-specific spatial distributions of demersal species of the same area [12]. Such result was to be expected, as demersal species are restricted to the bottom and very dependent on substrates or bathymetry. By contrast, small pelagic fish live in the water column, are much more mobile and should be affected mostly by dynamic environmental factors, which change from year to year. Previous studies on small pelagic fish habitats conducted in the Mediterranean [43] have identified the Gulf of Lions as a potential favourable habitat for sardines and anchovies. In this smaller-scale study, we tried to identify whether there could be within Gulf variations in habitat quality for small pelagic fish. From the presence/absence data pooled over the entire study period, we saw that the three species could use the entire Gulf of Lions (except past the 200 m isobath for sprat). Nonetheless, it was clear that spatial distributions were not uniform over the Gulf and that they exhibited more than one patch of high biomass. The location of patches varied across years and between species, suggesting that they reacted to some external variable factors. In the future, linking spatio-temporal distributions to environmental variables at a minute scale while accounting for aggregation structures should help in understanding the environmental drivers affecting the population dynamics of those three pelagic species. Previous studies on environmental drivers of small pelagic fish habitats have already been conducted at a large scale in the Mediterranean, showing the importance of temperature and chlorophyll concentration as forcing factors of the spatial distribution (e.g. [43]). However, processes that drive the small pelagic populations may be different at smaller scales (i.e. within the Gulf of Lions). In particular, the relative importance of density-dependent *versus* density-independent (environmental) processes on spatial distributions is likely to be dependent on the scale of the study [47]. A study accounting for both density-dependent and independent variables is thus of necessity to understand the identified areas [13].

Despite the important variability in spatial distributions, we defined unfavourable areas close to the coast for anchovies and on the 150–200 m-deep stratum for sprats and sardines. In contrast, we could assess recurrent areas (high average biomass and low variability) for sardines and anchovies in the Gulf of Lions. For sardines, recurrent areas were situated near the coast and in the Western part of the Gulf of Lions. For anchovies, two recurrent areas were identified further from the coast (isobaths of ~70 to 100 m). One was situated on the west part while the other one could be associated with the Rhone river and its plume on the east. As the surveys occurred during peak reproduction for anchovies [23], it is likely these areas corresponded to the spawning grounds of anchovies in this region. Interestingly, these 2 areas are very similar to the spawning areas detected from egg surveys in the 60 s [48]. This gives us important insights on their spawning grounds, exhibiting their stability despite several changes in external variables between these 2 periods (e.g. fishing effort and captures, environment). Also, such information is important for the elaboration of spatially explicit management plan (e.g. MPA, etc.). Indeed, it shows that despite the high mobility of small pelagic fish and high interannual variability in peak biomass locations, some areas consistently offer favourable spawning grounds for this species, making the task easier if one wanted to protect them.

Finally, we investigated interspecific relationships between 3 species sharing similar trophic level (i.e. zooplankton and especially copepods constitute the bulk of their diet at that period, though phytoplankton is also consumed in particular by sardines [21–22]). Different results were obtained according to the scale at which it was studied. Indeed, the index of overlap that we proposed decreased from the global (the population envelop) to the intermediate scale, indicating that if the three species co-habited in the Gulf of Lions, their hotspots were not always situated in the same area. This suggests that global environmental factors may have driven the three species to inhabit the Gulf of Lions, but that local environmental factors and/or inter-specific competition may have resulted in segregation within the Gulf. This was confirmed by the difference obtained between the two local collocation indices used. The collocation index translating species co-occurrence was surprisingly high (especially for anchovies and sardines), revealing that at a 2-nm sampling scale, species almost always co-occurred. The LIC index (based on density rather than presence/absence) had lower values, suggesting that for most sampled locations a given species predominantly occurred in each location (as also indicated by the dominance index). In particular, sardines clearly dominated the first depth strata (0–50 m) and the western part of the Gulf of Lions, while anchovies were more abundant in deeper waters and towards the central and eastern part of the Gulf. Collocation analyses give us interesting information on potential competition of these three planktivorous species, as their co-occurrence is a potential source of competition if/when food becomes limiting.

In summary, this study exhibited the relatively stable aggregation structure of small pelagic species, probably due to their inherent social structure (shoals, etc.). In contrast, it confirmed that the three species were highly mobile and could inhabit every stratum of the Gulf of Lions, though some preferred and unfavourable habitats could be highlighted for each species. The environmental drivers of both these habitats and the variability in locations of peak biomass remain to be investigated to better understand population dynamics. Finally, differences between species aggregative behaviour and habitat preferences were also highlighted, exhibiting potential competition avoidance.

Supporting Information

File S1 Supplementary material. Figure S1. Presence of anchovies, sardines and sprats in the Gulf of Lions. **Figure S2**. Annual aggregation curves for each species. **Figure S3**. Annual maps of log-biomass for anchovies. **Figure S4**. Annual maps of log-biomass for sardines. **Figure S5**. Annual maps of log-biomass for sprats. **Figure S6**. Centres of gravity and inertia. **Figure S7**. Empirical Orthogonal Function analysis. **Figure S8**. Dominance index depending on depth or longitude strata. **Table S1**. Yearly and global collocation indices at three different time scales.

Acknowledgments

We are very grateful to the crew of the Ifremer RV *L'Europe* and everyone who participated to the PELMED surveys between 2003 and 2012. PELMED surveys are cofunded by IFREMER and the European Union through the Data Collection Framework (DCF). This study is part of the program EcoPelGol (Study of the Pelagic ecosystem in the Gulf of Lions), funded by France Filière Pêche (FFP). We also wish to thank Vincent A. Viblanc for his helpful suggestions and his English-proofing of the manuscript. Finally, we thank 2 anonymous reviewers for their helpful comments.

Author Contributions

Conceived and designed the experiments: CS JLB DR. Performed the experiments: JLB JHB DR EVB CS. Analyzed the data: CS NB JMF. Contributed reagents/materials/analysis tools: CS MM. Wrote the paper: CS JMF NB. Made useful suggestions on the manuscript: MM EVB.

References

1. Woillez M, Petitgas P, Rivoirard J, Fernandes P, Hoftstede Rter, et al. (2006) Relationships between population spatial occupation and population dynamics. ICES CM 2006/O:05
2. Agostini VN, Hendrix AN, Hollowed AB, Wilson CD, Pierce SD, et al. (2008) Climate-ocean variability and Pacific hake: a geostatistical modeling approach. Journal of marine systems 71: 237–248.
3. Bacheler NM, Bailey KM, Ciannelli L, Bartolino V, Chan KS (2009) Density-dependent, landscape and climate effects on spawning distribution of walleye Pollock Theragra chalcogramma. Marine Ecology Progress Series 391: 1–12.
4. Cardinale M, Hagberg J, Svedang H, Bartolino V, Gedamke T, et al. (2010) Fishing through time: population dynamics of plaice (Pleuronectes platessa) in the Kattegat-Skagerrak over a century. Population Ecology 52: 251–261.
5. Boulinier T, Danchin E (1997) The use of conspecific reproductive success for breeding patch selection in terrestrial migratory species. Evolutionary Ecology 11: 505–517.
6. Hixon MA, Anderson TW, Buch KL, Johnson DW, McLeod JB, et al. (2012) Density dependence and population regulation in marine fish: a large-scale, long-term field manipulation. Ecological Monographs 82: 467–489.
7. Petitgas P, Alheit J, Peck MA, Raab K, Irigoien X, et al. (2012) Anchovy population expansion in the North Sea. Marine Ecology Progress Series 444: 1–13.
8. Lewis S, Sheratt TN, Hamer KC, Wanless S (2001) Evidence of intra-specific competition for food in pelagic seabird. Nature 412: 816–819.
9. MacCall AD (1990) Dynamic Geography of Marine Fish Populations. Seattle: University of Washington Pr. 153 p.
10. Paloheimo JE, Dickie LM (1964) Abundance and fishing success. Rapport et Proces Verbaux des Reunions du Conseil International pour l'Exploration de la Mer 155: 152–163.
11. Bertrand A, Segura M, Gutiérrez M, Vásquez L (2004) From small-scale habitat loopholes to decadal cycles: a habitat-based hypothesis explaining fluctuation in pelagic fish populations off Peru. Fish and Fisheries 5: 296–316.
12. Morfin M, Fromentin JM, Jadaud A, Bez N (2012) Spatio-temporal Patterns of Key Exploited Marine Species in the Northwestern Mediterranean Sea. PlosOne 7 (5): e37907.
13. Ciannelli L, Bartolino V, Chan KS (2012) Non-additive and non-stationary properties in the spatial distribution of a large marine fish population. Proceedings of the Royal Society B 279: 3635–3642.
14. Ciannelli L, Fisher JAD, Skern-Mauritzen M, Hunsicker ME, Hidalgo M, et al. (2013) Theory, consequences and evidence of eroding population spatial structure in harvested marine fishes: a review. Marine Ecology Progress Series 480: 227–243.
15. Apostolaki P, Milner-Gulland EJ, McAllister MK, Kirkwood GP (2002) Modelling the effects of establishing a marine reserve for mobile fish species. Canadian Journal of Fisheries and Aquatic Sciences 59: 405–415.
16. Cury P, Bakun A, Crawford RJM, Jarre-Teichmann A, Quinones R, et al. (2000) Small pelagics in upwelling systems: patterns of interaction and structural changes in "wasp-waist" ecosystems. ICES J. Mar. Sci. 57: 603–618.
17. Bakun A (1996) Patterns in the ocean: ocean processes and marine population dynamics. University of California Sea Grant, San Diego, California, USA in cooperation with Centro de Investigaciones Biologicas de Noroeste, La Paz, Baja California Sur, Mexico. 323 pp.
18. Checkley DM, Ayon P, Baumgartner TR, Bernal M, Coetzee JC, et al. (2009) Habitats. In: Climate Change and Small Pelagic Fish. A. Checkley, J. Alheit, Y. Oozeki & C. Roy (eds) New York: Cambridge University Press
19. General Fisheries Commission for the Mediterranean (2012) Report of the working group on stock assessment of small pelagic species. Split, Coratia.
20. Van Beveren E, Bonhommeau S, Fromentin JM, Bigot JL, Bourdeix JH, et al. (2014) Rapid changes in growth, condition, size and age of small pelagic fish in the Mediterranean. Marine Biology 161: 1809–1822.
21. Costalago D, Navarro J, Álvarez-Calleja I, Palomera I (2012) Ontogenetic and seasonal changes in the feeding habits and trophic levels of two small pelagic fish species. Marine Ecology Progress Series 460: 169–181
22. Nikolioudakis N, Isari S, Somarakis S (2014) Trophodynamics of anchovy in a non-upwelling system: direct comparison with sardine. Marine Ecology Progress Series 500: 215–229.
23. Palomera I, Olivar MP, Salat J, Sabates A, Coll M, et al. (2007) Small pelagic fish in the NW Mediterranean Sea: An ecological review. Progress in Oceanography 74: 377–396.
24. Weill A, Scalabrin C, Diner N (1993) MOVIES-B: An acoustic detection description software. Application to shoal species classification. Aquatic Living Resources, 6: 255–267
25. Simmonds J, MacLennan D (2005) Fisheries Acoustics, Theory and Practice, 2nd edn. Oxford: Blackwell Publishing, 437 pp.
26. Doray M, Masse J, Petitgas P (2010) Pelagic fish stock assessment by acoustic methods at Ifremer, Internal report, pp.18, *epic.awi.de/33739/1/ifremer-acoustic-methods.pdf*.
27. Matheron G (1963) Principles of geostatistics. Economic Geology 58: 1246–1266.
28. Dungan JL, Perry JN, Dale MRT, Lengendre P, Citron-Pousty S, et al. (2002) A balanced view of scale in spatial statistical analysis. Ecography 25: 626–640.
29. Gini C (1921) Measurement of inequality of income. Economic Journal 31: 22–43.
30. Petitgas P (1998) Biomass-dependent dynamics of fish spatial distributions characterized by geostatistical aggregation curves. ICES Journal of Marine Science 55: 443–453.
31. Myers RA, Cadigan NG (1995) Was an increase in natural mortality responsible for the collapse of northern cod? Canadian Journal of Fish and Aquatic Science 52: 1274–1285.
32. Rivoirard J, Simmonds J, Foote KG, Fernandes P, Bez N (2000) Gesotatistics for estimating fish abundance. Blackwell Publishing, Oxford.
33. Chilès J-P, Delfiner P (1999) Geostatistics: Modeling Spatial Uncertainty. 1st ed. New York: Wiley-Interscience. 720 p.
34. Bez N, Rivoirard J (2001) Transitive geostatistics to characterise spatial aggregations with diffuse limits: an application on mackerel ichtyoplankton. Fisheries Research 50: 41–58.
35. Woillez M, Poulard J-C, Rivoirard J, Petitgas P, Bez N (2007) Indices for capturing spatial patterns and their evolution in time, with application to European hake (Merluccius merluccius) in the Bay of Biscay. ICES Journal of Marine Science 64: 537–550.
36. Lorenz EN (1956) Empirical orthogonal functions and statistical weather prediction. Department of Meteorology, MIT, Scientific Report n°1.
37. Korres G, Pinardi N, Lascaratos A (2000) The Ocean Response to Low-Frequency Interannual Atmospheric Variability in the Mediterranean Sea. Part II: Empirical Orthogonal Functions Analysis. Journal of Climate 13: 732–745.
38. Bellier E, Planque B, Petitgas P (2007) Historical fluctuations in spawning location of anchovy (Engraulis encrasicolus) and sardine (Sardina pilchardus) in the Bay of Biscay during 1967–73 and 2000–2004. Fisheries Oceanography 16: 1–15.
39. Bez N, Rivoirard J (2000) Indices of collocation between populations. In: Checkley DM, Hunter JR, Motos L, von der Lingen CD (editors).Workshop on the Use of Continuous Underway Fish Egg Sampler (CUFES) for mapping spawning habitat of pelagic fish. GLOBEC Rep., pp. 48–52.
40. Barange M, Coetzee JC, Twatwa NM (2005) Strategies of space occupation in anchovy and sardine in the southern Benguela: the role of stock size and intra-species competition. ICES Journal of Marine Science 62: 645e654.
41. R Development Core Team (2012) R: A language and environment for statistical computing. R Foundation for Statistical Computing, Vienna, Austria.
42. Renard D, Bez N, Desassis N, Beucher H, Ors F (2013) RGeoS: Geostatistical Package. R package version 9.1.6. (Ecoles des Mines website, available at http://cg.ensmp.fr/rgeos, accessed 2014 October 10th).
43. Giannoulaki M, Iglesias M, Tugores MP, Bonanno A, Patti B, et al. (2013) Characterizing the potential habitat of European anchovy Engraulis encrasicolus in the Mediterranean Sea, at different life-stages. Fisheries Oceanography 22: 69–89.

44. Gaston KJ, Blackburn TM, Gregory RD (1999) Intraspecific abundance–occupancy relationships: case studies of six bird species in Britain. Div. Distr. 5: 197–212. (doi:10.1046/j.1472-4642.1999.00054.x)

45. Fisher JAD, Frank KT (2004) Abundance–distribution relationships and conservation of exploited marine fishes. Mar. Ecol. Prog. Ser. 279: 201–213. (doi:10.3354/meps279201)

46. Atkinson DB, Rose GA, Murphy EF, Bishop CA (1997) Distribution changes and abundance of northern cod (*Gadus morhua*), 1981–1993. Canadian Journal of Fisheries and Aquatic Sciences 54: 132–138.

47. Shepherd T, Litvak M (2004) Density-dependent habitat selection and the ideal free distribution in marine fish spatial dynamics: considerations and cautions. Fish and Fisheries 5: 141–152.

48. Aldebert Y, Tournier H (1971) La reproduction de la sardine et de l'anchois dans le Golfe du Lion. Revue de Travail de l'Institut des Pêches maritimes 35: 57–75.

Permissions

List of Contributors

Guoli Zhu1, Wenqiao Tang, Dong Liu and Jinquan Yang
College of Fisheries and Life Science, Shanghai Ocean University, Shanghai, China

Liangjiang Wang
Department of Genetics and Biochemistry, Clemson University, Clemson, South Carolina, United States of America

Ian R. Bradbury
Department of Fisheries and Oceans, St. John's, Newfoundland, Canada
Ocean Sciences Center, Memorial University of Newfoundland, St. John's, Newfoundland, Canada

Jeffrey A. Hutchings, Jackie Lighten, Daniel E. Ruzzante and Paul Bentzen
Marine Gene Probe Laboratory, Department of Biology, Dalhousie University, Halifax, Nova Scotia

Sharen Bowman and Tudor Borza
The Atlantic Genome Centre, Halifax, Nova Scotia, Canada

Paul V. R. Snelgrove
Ocean Sciences Center, Memorial University of Newfoundland, St. John's, Newfoundland, Canada

Paul R. Berg
Centre for Ecological and Evolutionary Synthesis, Department of Biology, University of Oslo, Oslo, Norway

Naiara Rodríguez-Ezpeleta
AZTI-Tecnalia, Marine Research Division, Txatxarramendi ugartea z/g, Sukarrieta, Spain

Christopher Taggart
Department of Oceanography, Dalhousie University, Halifax, Nova Scotia, Canada

Annette Taugbøl, Tina Arntsen and Leif Asbjørn Vøllestad
Centre for Ecological and Evolutionary Synthesis (CEES), Department of Biosciences, University of Oslo, Blindern, Norway

Kjartan Østbye
Centre for Ecological and Evolutionary Synthesis (CEES), Department of Biosciences, University of Oslo, Blindern, Norway

Hedmark University College, Department of Forestry and Wildlife Management, Campus Evenstad, Elverum, Norway

Ross M. Thompson
Institute for Applied Ecology, University of Canberra, Bruce, Australian Capital Territory, Australia
School of Biological Sciences, Monash University, Clayton, Victoria, Australia

Alissa Monk
Institute for Applied Ecology, University of Canberra, Bruce, Australian Capital Territory, Australia
School of Biological Sciences, Monash University, Clayton, Victoria, Australia
Dolphin Research Institute, Hastings, Victoria, Australia

Kate Charlton-Robb
School of Biological Sciences, Monash University, Clayton, Victoria, Australia
Australian Marine Mammal Conservation Foundation, Hampton East, Victoria, Australia

Saman Buddhadasa
National Measurement Institute, Commonwealth Government, Port Melbourne, Victoria, Australia

Michael J. Emslie, Alistair J. Cheal and Kerryn A. Johns
Australian Institute of Marine Science, Townsville, Queensland, Australia

Yanhong Zhang, Huixian Zhang and Qiang Lin
Key Laboratory of Tropical Marine Bio-resources and Ecology, South China Sea Institute of Oceanology, Chinese Academy of Sciences, Guangzhou, China

Nancy Kim Pham and Junda Lin
Vero Beach Marine Laboratory, Florida Institute of Technology, Vero Beach, Florida, United States of America

Yifei Dong, Hua Tian, Wei Wang, Xiaona Zhang, Jinxiang Liu and Shaoguo Ru
Marine Life Science College, Ocean University of China, Qingdao, Shandong Province, The People's Republic of China

Marlee A. Tucker and Tracey L. Rogers
Evolution and Ecology Research Centre, School of Biological, Earth and Environmental Sciences, The University of New South Wales, Sydney, New South Wales, Australia

Maria João Lança
Escola de Ciências e Tecnologia, Departamento de Zootecnia, Universidade de Évora, Évora, Portugal
Instituto de Ciências Agrárias e Ambientais Mediterrânicas, Universidade deÉvora, Évora, Portugal

Maria Machado
Instituto de Ciências Agrárias e Ambientais Mediterrânicas, Universidade deÉvora, Évora, Portugal

Marta Lourenço and Ana F. Ferreira
Centro de Oceanografia, Faculdade de Ciências, Universidade de Lisboa, Lisboa, Portugal

Catarina S. Mateus
Centro de Oceanografia, Faculdade de Ciências, Universidade de Lisboa, Lisboa, Portugal
Museu Nacional de História Natural e da Ciência & Centro de Biologia Ambiental, Universidade de Lisboa, Lisboa, Portugal

Bernardo R. Quintella
Centro de Oceanografia, Faculdade de Ciências, Universidade de Lisboa, Lisboa, Portugal
Departamento de Biologia Animal, Faculdade de Ciências, Universidade de Lisboa, Lisboa, Portugal

Pedro R. Almeida
Museu Nacional de História Natural e da Ciência & Centro de Biologia Ambiental, Universidade de Lisboa, Lisboa, Portugal
Escola de Ciências e Tecnologia, Departamento de Biologia, Universidade de Évora, Évora, Portugal

Richard D. Pillans and Russel C. Babcock
CSIRO Marine and Atmospheric Research, Brisbane, Queensland, Australia

Ryan Downie and Toby A. Patterson
CSIRO Marine and Atmospheric Research, Hobart, Tasmania, Australia

Douglas Bearham and Damian P. Thomson
CSIRO Marine and Atmospheric Research, Floreat, Perth, Western Australia, Australia

Andrew Boomer
SIMS/IMOS Animal Tagging and Monitoring, Mosman, New South Wales, Australia

Stephen D. Simpson
Biosciences, College of Life and Environmental Sciences, University of Exeter, Exeter, United Kingdom

Hugo B. Harrison
Australian Research Council Centre of Excellence for Coral Reef Studies, James Cook University, Townsville, Queensland, Australia

Michel R. Claereboudt
Department of Marine Science and Fisheries, Sultan Qaboos University, Al-Khod, Oman

Serge Planes
Le Centre de Biologie et d'Ecologie Tropicale et Méditerranéenne, l'Université de Perpignan, Perpignan, Pyrénées-Orientales, France
Laboratoire d'Excellence "CORAIL", Centre de Recherches Insulaires et Observatoire de l'Environnement, Moorea, French Polynesia

Rosalind A. Leggatt, Tanya Hollo, Wendy E. Vandersteen, Kassandra McFarlane, Benjamin Goh, Joelle Prevost and Robert H. Devlin
Fisheries and Oceans Canada, West Vancouver Laboratories, West Vancouver, BC, Canada

Monica Landi and Filipe O. Costa
Centre of Molecular and Environmental Biology (CBMA), Department of Biology, University of Minho, Braga, Portugal

Mark Dimech
Malta Centre for Fisheries Science (MCFS), Fort San Lucjan Marsaxlokk, Malta

Marco Arculeo, Girolama Biondo and Sabrina Lo Brutto
Dipartimento di Scienze e Tecnologie Biologiche, Chimiche e Farmaceutiche (STEBICEF), University of Palermo, Palermo, Italy

Rogelia Martins and Miguel Carneiro
Modelling and Management Fishery Resources Division (DIV-RP), Instituto Português do Mar e da Atmosfera, Lisboa, Portugal

Gary Robert Carvalho
Molecular Ecology and Fisheries Genetics Laboratory, School of Biological Sciences, Bangor University, Bangor, United Kingdom

Anna de Polo, Anne E. Lockyer and Mark D. Scrimshaw
Institute for the Environment, Brunel University, London, United Kingdom

Luigi Margiotta-Casaluci
Institute for the Environment, Brunel University, London, United Kingdom
AstraZeneca, Global Health, Safety and Environment, Freshwater Quarry, Brixham, United Kingdom

Andrea Anton
Curriculum for the Environment and Ecology, University of North Carolina, Chapel Hill, North Carolina, United States of America

Michael S. Simpson and Ivana Vu
Department of Biology, University of North Carolina, Chapel Hill, North Carolina, United States of America

Nathan M. Bacheler
National Marine Fisheries Service, Southeast Fisheries Science Center, Beaufort, North Carolina, United States of America

Lewis G. Coggins Jr.
National Marine Fisheries Service, Southeast Fisheries Science Center, Beaufort, North Carolina, United States of America
United States Fish and Wildlife Service, Yukon Delta National Wildlife Refuge, Bethel, Alaska, United States of America

Daniel C. Gwinn
Biometric Research, and Program for Fisheries and Aquatic Sciences, School of Forest Resources and Conservation, University of Florida, Gainesville, Florida, United States of America

Fabien Pointin and Mark R. Payne
Centre for Ocean Life, National Institute of Aquatic Resources (DTU-Aqua), Technical University of Denmark, Charlottenlund, Denmark

Peter Houk
University of Guam Marine Laboratory, UOG Station, Mangilao, Guam
Pacific Marine Resources Institute, Saipan, Northern Mariana Islands

David Benavente, John Iguel, Steven Johnson and Ryan Okano
CNMI Bureau of Environmental and Coastal Quality, Saipan, Northern Mariana Islands

Damien D. Hinsinger, Agnès Dettaï and Jean-Lou Justine
UMR 7138 "Systématique, Adaptation, E'volution", Muséum National d'Histoire Naturelle, Département Systématique et Évolution, Paris, France

Charlotte Schoelinck
UMR 7138 "Systématique, Adaptation, E'volution", Muséum National d'Histoire Naturelle, Département Systématique et Évolution, Paris, France
Fisheries and Oceans Canada, Molecular biology, Aquatic animal health, Moncton, Canada

Corinne Cruaud
Génoscope, Centre National de Séquençage, Évry, France

Claire Saraux, Jean-Marc Fromentin, Jean-Louis Bigot, Jean-Hervé Bourdeix, Marie Morfin, David Roos and Elisabeth Van Beveren
IFREMER (Institut Français de Recherche pour l'Exploitation de la MER), Research Unit EME (UMR 212), Séte, France

Nicolas Bez
IRD (Institut de Recherche pour le Développement), Research Unit EME (UMR 212), Séte, France

Index

www.ingramcontent.com/pod-product-compliance
Lightning Source LLC
Chambersburg PA
CBHW080251230326
41458CB00097B/4269